RESEARCH HANDBOOK ON INTERNATIONAL MIGRATION AND DIGITAL TECHNOLOGY

Research Handbook on International Migration and Digital Technology

Edited by

Marie McAuliffe

Head, Migration Research Division, International Organization for Migration, Switzerland and Sir Roland Wilson Fellow, School of Demography, The Australian National University, Australia

ELGAR HANDBOOKS IN MIGRATION

Edward Elgar
PUBLISHING

Cheltenham, UK • Northampton, MA, USA

Published by
Edward Elgar Publishing Limited
The Lypiatts
15 Lansdown Road
Cheltenham
Glos GL50 2JA
UK

Edward Elgar Publishing, Inc.
William Pratt House
9 Dewey Court
Northampton
Massachusetts 01060
USA

A catalogue record for this book
is available from the British Library

Library of Congress Control Number: 2021947938

This book is available electronically in the **Elgar**online
Sociology, Social Policy and Education subject collection
http://dx.doi.org/10.4337/9781839100611

Printed on elemental chlorine free (ECF)
recycled paper containing 30% Post-Consumer Waste

ISBN 978 1 83910 060 4 (cased)
ISBN 978 1 83910 061 1 (eBook)
Printed and bound in the USA

Contents

Figures

Tables

Boxes

Contributors

Guy Abel is a professor at the Asian Demographic Research Institute and School of Sociology and Political Sciences at Shanghai University. He is also a researcher at the International Institute for Applied Systems Analysis. His research focuses on techniques for estimating migration patterns and applying statistical methods to better forecast components of population change.

William Allen is a British Academy Postdoctoral Research Fellow at the Department of Politics and International Relations (DPIR) of the University of Oxford. His research, which has won recognition from the UK's Political Studies Association and the American Political Science Association, examines political communication, public attitudes, and policymaking in migration and integration domains. He also serves as Associate Editor for the journal *Evidence & Policy*.

Georgios A. Antonopoulos obtained his PhD from the University of Durham. He is currently Professor of Criminology at Teesside University. His research interests include ethnicity, crime and justice, and 'organized crime'. He is board member of the Cross-Border Crime Colloquium, and Editor-in-Chief of *Trends in Organized Crime*.

Farah Azhar graduated with a PhD in Media and Communication Studies from the University of Oregon's School of Journalism and Communication. She holds a Master's in Economics from Lahore University of Management Sciences, Pakistan and a Master's in Social Policy from University of Pennsylvania. Her research is at the intersection of Development Communication, Gender and Development, and Health Communication. She has over five years of international experience of working with central bank and non-profit organizations.

Céline Bauloz (PhD) is head of the Research Unit in the Migration Research Division in the International Organization for Migration (IOM) and editor of IOM's Migration Research Series, and a senior fellow at the Global Migration Centre, Graduate Institute of International and Development Studies. She previously worked as lecturer/researcher in migration in academic institutions, including the University of Bocconi, the University of London, the University of Fribourg, and Harvard Law School, and as consultant for international organizations.

Laurie Berg is Associate Professor in the Law Faculty at the University of Technology, Sydney and Co-Director of the Migrant Worker Justice Initiative. She has led major studies on low-waged migrant workers in Australia, including wage theft of international students and backpackers, as well as global studies on the use of technology to advance access to justice for migrant workers. She is the author of *Migrant Rights at Work: Law's Precariousness at the Intersection of Immigration and Labour* (Routledge, 2016).

Jacqueline Bhabha is the Director of Research at the FXB Center for Health & Human Rights at Harvard University, Professor of the Practice of Health and Human Rights at the Harvard T.H. Chan School of Public Health, the Jeremiah Smith Jr. Lecturer in Law at Harvard Law

School, and an Adjunct Lecturer in Public Policy at the Harvard Kennedy School. She has published extensively on issues of transnational child migration, refugee protection, children's rights, and citizenship.

Abhishek Bhatia is the IDHN Data Science Fellow at Harvard University, where his interests lie at the intersection of information technology, data science, and disaster response. His research focuses on assessing the analytical and translational readiness of data for equitable and crisis planning and response, and leveraging novel data streams from mobility data, remote sensing data, and other digital phenotypes to quantify the public health burden of war and disasters.

Jenna Blower is a PhD candidate at York University in the Department of Social Anthropology and Research Consultant in the Migration Research Division, International Organization for Migration (IOM). Jenna has an MA in Immigration and Settlement Studies from Ryerson University and previously worked at Cities of Migration where she contributed to the My City of Migration Diagnostic (MyCOM) Tool and led the Immigrant Futures project.

Cecilia Cannon is a researcher at the Graduate Institute of International and Development Studies, Geneva. She directs the Graduate Institute's Executive Master in International Negotiation and Policy-making, and lectures in the Master of International Development and International Affairs. She served as Academic Adviser to the United Nations for its 75th anniversary in 2020. Her research focuses on the design, reform, and effectiveness of international organizations; the United Nations; non-state actors in international policy processes; and migration policy, including immigration detention.

Eileen Culloty is an assistant professor in the School of Communications at Dublin City University. Eileen's research has been published in the *European Journal of Communication, Environmental Communication, Digital Journalism*, and *Critical Studies on Terrorism*. A co-authored book with Jane Suiter, *Disinformation and Manipulation in Digital Media*, is published by Routledge.

Valentin Danchev is Lecturer in Computational Social Science in the Department of Sociology at the University of Essex. He received his DPhil (PhD) from the University of Oxford and held postdoctoral positions at the University of Chicago and at Stanford University. His research interests include computational social science, network analysis, human mobility, global migration, and open and reproducible research.

Parisa Diba is a PhD researcher and a part-time project manager and research associate in Public Health at Teesside University. Parisa has a strong background and a reputable track record in research. Parisa possesses close to seven years of experience as an inter-disciplinary social researcher in academia and the third sector. Parisa has worked as lead researcher on a diverse range of internally and externally funded research projects across the disciplines of Anthropology, Criminology, Public Health, Sociology, and Social Policy.

Huub Dijstelbloem is Professor of Philosophy of Science, Technology and Politics and Director of the Institute for Advanced Study of the University of Amsterdam. He is co-founder of the UvA's Platform for the Ethics and Politics of Technology and one of the initiators of the movement Science in Transition. His current research concerns the politics of border control and long-term climate policy. His work has been published in *Nature, Security Dialogue,*

Geopolitics, the *Journal of Borderlands Studies*, *International Political Sociology*, *Sociology of Health and Illness* and the *Journal of Environmental Policy & Planning*. His most recent book is *Borders as Infrastructure: The Technopolitics of Border Control* (MIT Press, 2021).

Bassina Farbenblum is Associate Professor in the Faculty of Law & Justice, UNSW Sydney and Co-Director of the Migrant Worker Justice Initiative. Bassina has led research teams on migrant workers across Asia, the Middle East, the US, and Australia, and frequently advises governments, UN agencies, and NGOs globally. She practised law at the New York Bar and as a solicitor in NSW, and previously worked for the American Civil Liberties Union, the Public Interest Advocacy Centre, and the Australian Human Rights Commission.

Ricardo Gomez is Associate Professor at the University of Washington Information School, and faculty affiliate with the Latin American & Caribbean Studies Program, the Harry Bridges Center for Labor Studies, and the UW Center for Human Rights. His research interests focus on the uses of information and communication technologies in the context of migration, human rights, and social justice. He specializes in social dimensions of the use (or non-use) of communication technologies, and how they contribute (or not) to well-being and social justice.

Roland Happ is substitute Professor of Business and Economics Education at the University of Leipzig. His research interests lie in the modelling of economic and financial literacy of young adults. He has published numerous publications on the relationship between the migration background of young adults and business and economic knowledge.

G. Harindranath is Professor of Information Systems at Royal Holloway, University of London. Hari holds a PhD from the London School of Economics and his research interests centre on the social and organisational implications of digital technologies, including ICT4D. His recent work examines digital technologies' role in relation to migration and inequalities in the Global South, with MIDEQ, a UK Government-funded project that brings together 40+ partners from 12 countries to investigate South–South migration, inequality, and development.

Shaminda Kanapathi is a human rights activist and Sri Lankan refugee who was held in Australia's offshore processing facilities in PNG between 2013 and 2020. He has published numerous newspaper articles and opinion pieces detailing his experience in immigration detention and offshore processing, including in *The Guardian* and *The Saturday Paper*.

Camille Kasavan is Project Manager at Samuel Hall, where she supports multi-country coordination of migration-related research projects. She holds an MA in Human Rights and Humanitarian Action with a concentration in Research Methods from Sciences Po – Paris. Camille's research focuses on protection, reintegration, migration and human rights, and migrant aspirations. Her work has covered a diverse range of countries including Tunisia, Ethiopia, Sudan, Kenya, Somalia, and Afghanistan.

Binod Khadria is a former Professor of Economics of Education and Chairperson of Zakir Husain Centre for Educational Studies, School of Social Sciences, Jawaharlal Nehru University (JNU). In 2017, he held the inaugural Indian Council for Cultural Relations (ICCR) Chair at Rutgers University, and was the Thematic Expert at the Second UN Debate on Global Compact for Migration (GCM) 2018. Currently, he is the President of the Global Research Forum on Diaspora and Transnationalism (GRFDT), and a co-convener of Metropolis Asia-Pacific, a regional branch of Metropolis International.

Angela Kintominas is a Scientia PhD scholar and Teaching Fellow at the Faculty of Law & Justice, UNSW Sydney. Her doctoral project explores the futures and histories of gigs in migrant domestic work in Australia and her research interests include gender, migration, and reproductive labour. Angela is Research Associate with the Migrant Worker Justice Initiative, contributing on projects relating to technology and migration, and the experiences of low-waged migrant workers in Australia.

Adrian Kitimbo is a Research Officer in the Migration Research Division, International Organization for Migration (IOM). He is also a Research Associate at the Gordon Institute of Business Science, University of Pretoria, where he previously held a full-time research position. Adrian has also worked at the Brenthurst Foundation in Johannesburg, where he was the Machel-Mandela Fellow.

Emre Eren Korkmaz is a political scientist, and he has been working as an academic at the University of Oxford since October 2016. He is a Departmental Lecturer in Migration and Development and teaching the MSc in Migration Studies. He has been granted Junior Research Fellowship at St Edmund Hall and, before the lecturer position, he was a British Academy Newton International Fellow in the first two years at the same department.

Rey Koslowski is Professor of Political Science at the University at Albany. His books and edited volumes include *Migrants and Citizens: Demographic Change in the European States System* (2000) and *Global Mobility Regimes* (2011). His work on border control and international travel security includes *Real Challenges for Virtual Borders: The Implementation of US-VISIT* (2005), *The Evolution of Border Controls as a Mechanism to Prevent Illegal Immigration* (2011), and "International Travel Security and the Global Compacts on Refugees and Migration" (2019).

Koen Leurs is Assistant Professor in Gender, Media and Migration Studies at the Graduate Gender Program, Department of Media and Culture, Utrecht University, the Netherlands. Leurs' research and teaching interests include migration, identity, infrastructures, personal archives as well as research ethics, creative, participatory and digital methods. His publications include the books *Digital Passages. Migrant Youth 2.0* (Amsterdam University Press, 2015), and *Digital Migration Studies* (forthcoming with Sage 2022). He also co-edited the *Sage Handbook of Media and Migration* (Sage, 2020) and special issues 'Forced migration and digital connectivity' for *Social Media + Society* and 'Connected migrants' for *Popular Communication*.

Sun Sun Lim is Professor of Communication and Technology and Head of Humanities, Arts and Social Sciences at the Singapore University of Technology and Design. She has extensively researched the social impact of technology, studying technology domestication, digital disruptions, and smart city technologies. She recently published *Transcendent Parenting: Raising Children in the Digital Age* (Oxford University Press, 2020). She serves on eleven journal editorial boards and frequently offers her expert commentary in diverse outlets including *Nature* and *Scientific American*.

Nassim Majidi is the co-founder and director of Samuel Hall, a social enterprise that conducts research in countries affected by issues of migration and displacement. She leads empirical research, evidence-based programming, and policy development. Covering three continents (Africa, Asia, Europe), her research has documented post-return outcomes and informed reintegration programmes based on empirical evidence and interviews with refugees, migrants, and returnees, across countries of origin and transit in Africa and Asia. Nassim holds a PhD from Sciences Po Paris.

Marie McAuliffe (PhD) is head of IOM's Migration Research Division and Editor of IOM's flagship World Migration Report. She is a senior fellow/associate at the Geneva Graduate Institute, the Australian National University and the Center for Strategic and International Studies (CSIS), and co-chairs the World Bank's KNOMAD thematic working group on migration data and demography. She is also a member of MIT's Global Technology Review Panel, IUSSP's panel on international migration and curates the World Economic Forum's Migration Transformation Map. Marie serves on the editorial board of *International Migration* and *Migration Studies*, and is an associate editor of the *Harvard Data Science Review*. In 2018, Marie was awarded the Charles Price Prize in demography for outstanding doctoral research in migration.

Ratnam Mishra is an Assistant Professor of Economics in the University School of Management and Entrepreneurship (USME), Delhi Technological University (DTU). She has a PhD degree from Jawaharlal Nehru University (JNU), New Delhi, with specialization in Economics of Education, and a Masters in Economics and Rural Development from Dr. Ram Manohar Lohia Awadh University, Ayodhya. Her areas of interest are economics of migration, entrepreneurship, vocational education and skill development, and issues in the Indian Economy.

Petra Molnar is a lawyer, researcher, and the Associate Director of the Refugee Law Laboratory, York University. She is also co-creating the Migration and Technology Monitor, an interdisciplinary initiative to investigate the use of surveillance and automated technologies on people crossing borders.

Sarah Nell-Müller is a research associate at the Chair of Business and Economics Education of Prof. Olga Zlatkin-Troitschanskaia at the Johannes Gutenberg-University in Mainz (Germany). She has been working on the SUCCESS research project since 2017. She focuses on qualitative studies and analyses of the educational pathways of refugees with a study interest.

Bryce Clayton Newell, PhD, JD, is an assistant professor in the School of Journalism and Communication at the University of Oregon (USA). He has studied the information practices of undocumented immigrants and migrant-aid workers along the US–Mexico border since 2014. From 2010 to 2014, he produced and directed a documentary film (*The Tinaja Trail*) about humanitarian and artistic response to migrant deaths in southern Arizona and California. Beyond migration, his research focuses on information privacy, surveillance, and the regulation of police work.

Henrietta Nyamnjoh is a researcher with South–South Migration, Inequality and Development Hub at the University of Cape Town. Her research interests include migration and mobility, transnational studies, and migration and health. Additionally, she is also interested in understanding religion in the context of migration. Henrietta has researched and published widely on religious healing among migrants in South Africa, migrants' appropriation of Information and Communication Technologies, Hometown Associations, and migrants' economy and everyday lives.

Markus Ojala (DSocSc) is a political communication researcher with a specific interest in the implications of new communication technologies on the public sphere. His research has been published in *New Media and Society*, *Journalism*, *European Journal of Cultural Studies*, and *Media, Culture & Society*, among other journals.

Georgios Papanicolaou is Associate Professor at Northumbria Law School, Northumbria University, UK. He studied Law and Penal Sciences at the University of Athens, and Criminology & Criminal Justice at the University of Edinburgh. His research investigates the political economy of policing and the policing of illicit markets. Georgios is the author of *Policing Sex Trafficking in Southeast Europe* (Palgrave, 2011) and co-author of *Organised Crime: A Very Short Introduction* (OUP, 2018).

Sam Peisch is Project Manager at the FXB Center for Health and Human Rights at Harvard University who supports the Center's leadership in a research, administrative, and financial support capacity. Sam is currently completing his MPH degree at the Harvard T.H. Chan School of Public Health concentrating on Public Health Leadership (PHL). He also is the President of Zamfund, a non-profit 501c3 organization dedicated to supporting women's education in Livingstone, Zambia. His interests are in equity and social justice.

Mason A. Porter is a professor in the Department of Mathematics at UCLA and an external professor at the Santa Fe Institute. His research interests lie in networks, complex systems, nonlinear systems, and their applications. Mason did his undergraduate education at Caltech and his graduate education at Cornell. He joined the faculty of University of Oxford in 2007 and moved to UCLA in 2016. Mason was born in Los Angeles and is excited to be living in his hometown.

Franziska Reinhardt is a research associate at the Department of Economics Education at the Johannes Gutenberg University in Mainz, Germany. Her research focuses on higher education and online education for disadvantaged learner groups. She has published numerous papers on the integration of refugees into higher education.

Ibrahim L. Saïd is an adjunct professor at the International University in Geneva and a research associate at the Centre on Conflict Development and Peacebuilding (CCDP) in Geneva. He has a PhD in Anthropology and Sociology of Development from the Graduate Institute of International and Development Studies and a Masters degree from the Department of Social Policy and Interventions at the University of Oxford. His research interests span across the fields of legal and political anthropology, gender, human rights, sociology of translation, settler-colonial and postcolonial studies.

Albert Ali Salah is Professor of Social and Affective Computing in the Department of Information and Computing Sciences, Utrecht University, and Adjunct Professor in the Department of Computer Engineering, Boğaziçi University. He has co-authored over 200 publications on computer analysis of human behaviour. Albert chaired the Data for Refugees Challenge, and leads a workpackage of the HumMingBird EU Horizon2020 project. He is a senior member of IEEE and ACM, and a research affiliate of DataPop Alliance.

Adam Sawyer has an MS in Geospatial Information Science from The University of Texas at Dallas and an MA in International Relations from Syracuse University. Adam is also a certified public educator with several years of instructional experience in bilingual classrooms.

Jane Suiter is a professor at Dublin City University with a focus on scaling up deliberation and disinformation. She is the Senior Research Fellow on the Irish Citizens' Assembly, a member of the OECD's FutureDemocracy network, and she represents the Royal Irish Academy on ALLEA's 'Fact or Fake' project on scientific disinformation. She was the winner of the 2019 Brown Democracy Medal and the Irish Research Council's Researcher of the Year 2020. A co-authored book with Eileen Culloty, *Disinformation and Manipulation in Digital Media*, is published by Routledge.

Sara Vannini is Lecturer (Assistant Professor) in Information Management and Information Systems at the University of Sheffield Information School. Her research interests are at the intersection of critical studies of technology and society and social change. Her main focus is social appropriation and embodied experiences of technologies by different social groups, information privacy in the context of migration, the role of public access to information in mis/disinformation, and participatory and visual methodologies of inquiry.

Yang Wang is a research fellow at the Lee Kuan Yew Centre for Innovative Cities, Singapore University of Technology and Design. She received her PhD in Communications and New Media from the National University of Singapore. Her research interests include ICT domestication, transnational communication, and digital transformation in the workplace. She has published in leading international journals including the *Journal of Computer-Mediated Communication*, *New Media and Society*, and *Journal of Children and Media*.

Arkadiusz Wiśniowski is Senior Lecturer in Social Statistics at the Department of Social Statistics, and co-leads the Statistical Modelling Research Group at the Cathie Marsh Institute, University of Manchester. His main research interest is in developing statistical methods for combining traditional and new forms of data to model and forecast complex social processes, with a particular focus on international migration. He is also involved in interdisciplinary projects on transportation, child labour, and impacts of the COVID-19 pandemic.

Saskia Witteborn is Associate Professor in the School of Journalism and Communication at CUHK. She specializes in transnational migration and technologies and has worked with migrants in the United States, in Europe, and in East Asia. She has written on the political economy of mobility, technology and space, data privacy and migration, and AI and ethics. Her research has appeared in leading journals and in edited collections. She is co-editor of *The SAGE Handbook of Media and Migration* (2020).

Dilek Yildiz is a research scholar at the International Institute for Applied Systems Analysis and a researcher at Vienna Institute of Demography. She holds a PhD in Social Statistics and Demography from the University of Southampton. Her research interests are statistical demography with a focus on Bayesian projections/reconstruction of multistate populations, population count and migration estimates, and investigating the use of big data sources.

Olga Zlatkin-Troitschanskaia is Professor of Business and Economics Education at Johannes Gutenberg University in Mainz, Germany. Her research focuses on modelling and measuring domain-specific and generic competencies, e.g. in the domains of digital education, and teacher education in national and international comparative studies.

1. International migration and digital technology: an overview

Marie McAuliffe

INTRODUCTION

In an age of information overload and online connectivity, there is a strong sense that we are more closely linked transnationally than ever before in human history. Routine activities in day-to-day lives around the world speak volumes of the recent emergence of modern-day digital transformations. Sending e-mails, transferring money to family back home, searching online for information and advice, engaging in online shopping, posting comments on social media and listening to music are all fairly mundane and routine, so much so that such activities have become taken for granted in many societies around the world as access to digital technology has expanded at an unprecedented rate. Terms such as e-business, e-commerce, e-banking, e-trade, e-shopping, e-crime, e-book, e-reader, e-learning, e-form, e-dating, e-health, e-visa, e-passport and e-mail have become increasingly common in our lexicon. Over time, our social and economic interactions are increasingly mediated through digital systems and processes, placing much greater emphasis on digital technologies and reducing the need for personal interaction (Hirsh-Pasek et al., 2018).

The modern-day international migration narrative is deeply intertwined with the notion of betterment, whether that relates to individual attainment, household income or community resilience and coping strategies (Boyd, 1989; Castles, 2010; Massey, 1987; Stark and Bloom, 1985; Todaro, 1989). People migrate for better lives. This has long been the cornerstone of international migration research, analysis and policy (Massey et al., 1998, 2):

> Like many birds, but unlike most other animals, humans are a migratory species. Indeed, migration is as old as humanity itself …. A careful examination of virtually any historical era reveals a consistent propensity towards geographic mobility among men and women, who are driven to wander by diverse motives, but nearly always with some idea of material improvement.

It can also be said that, in most recent and current discussions on migration, the starting point is usually numbers. Understanding changes in scale, emerging trends and shifting demographics related to global social and economic transformations, such as migration, helps us make sense of the changing world we live in and plan for the future. The current global estimate is that there were around 272 million international migrants in the world in 2019, which equates to 3.5 per cent of the global population. This is a very small minority of the world's population, meaning that staying within one's country of birth overwhelmingly remains the norm. The great majority of people do not migrate across borders, and it is thought that much larger numbers of people migrate within countries, although reliable statistics are not available on global estimates of internal migration.

The increase in international migrants has been evident over time – both numerically and proportionally – and at a slightly faster rate than previously anticipated, including because of the prevailing forces of globalisation (Czaika and de Haas, 2015; IOM, 2019b; McAuliffe and Goossens, 2018; Triandafyllidou, 2018). Figure 1.1 shows the increase in international migrants over time since 1970. While the number of international migrants has increased dramatically, in the context of the world's growing population, the proportional increases remain relatively modest.

This *Research Handbook* focuses on the dynamics that link international migration and digitalisation from economic, social, political and normative perspectives with reference to temporal and spatial dimensions. Temporality is particularly relevant in this domain as it is widely acknowledged that the pace of technological change through increasing digitalisation is intensifying at an unprecedented rate, creating uncertainty and disrupting societal norms (Muggah and Goldin, 2019; Skog et al., 2018). Long-held assumptions about societal values and their manifestations in economic, social and political systems and structures are being questioned, if not undermined, via recent technological advances in telecommunications technology, such as personal electronic devices and social media platforms, as the business sector scrambles to adapt to increasingly pervasive digitalisation (Skog et al., 2018). The digitalisation of communications that are reshaping the societies in which we live and the nature of our daily lives (and deaths) are causing us to re-think even the most fundamental concepts underpinning humanity, including its interaction with the natural world.

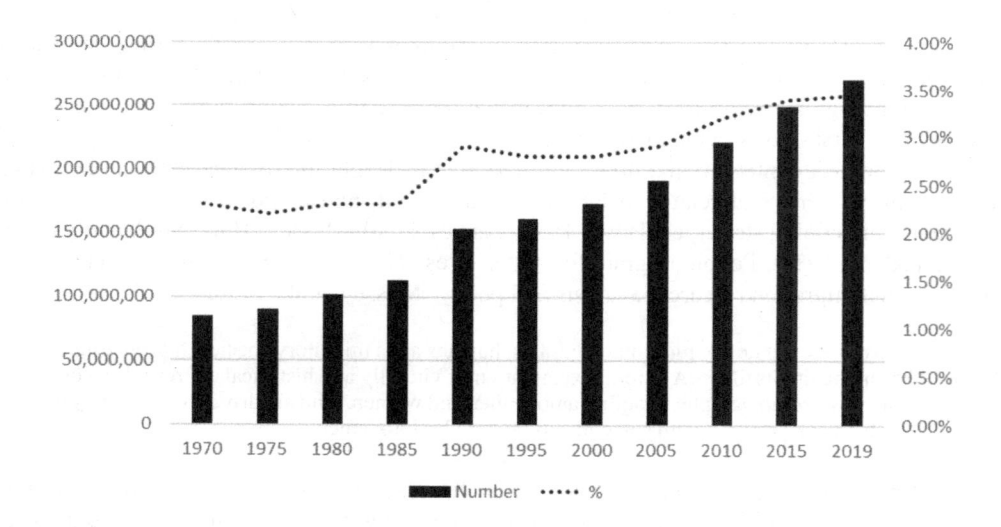

Source: UNDESA, 2019.

Figure 1.1   International migrants, 1970–2019

AN AGE-OLD PHENOMENON IN A RAPIDLY EVOLVING DIGITAL LANDSCAPE

The notion of migration is as old as humanity itself, and its complexities and multiple manifestations have been the subject of philosophical and scholarly enquiry, as well as law, for centuries. In the United Kingdom, for example, difficulties in determining parish responsibility as well as concerns over migration between parishes in search of more generous social support resulted in the Settlement Act of 1662, which sought to limit the ability of people to move between parishes (Bloy, 2002). More than a century later, and while the restrictions on internal migration were being softened, the first legislation to exclude foreign immigrants from entering the United Kingdom during peace time was enacted. The Aliens Act of 1793 allowed for the executive to exclude foreigners and was introduced in response to an increase in French nationals moving to Britain following the beginning of the French Revolution; the British government suspecting that some of the French immigrants would potentially bring 'anarchist' ideas with them (Aliverti, 2013). This was decades before the invention of the telegram and at a time during which news, information and ideas still had to be physically carried, rather than transmitted electronically. Even after the invention of the telegram in the mid-1800s, which greatly enabled international communications (although at great expense), news could be slow to travel as infrastructure was limited and, at times, diverted to national and international efforts. The experience of Paolina Roccanello, whose father had emigrated to Australia from Italy ahead of his family in order to set up the new home before his wife and children arrived—a common migration story—is briefly outlined in the text box below, highlighting just how much has changed in communications over the last 70 years.

BOX 1.1   WAR-TIME SEPARATION: MIGRANTS' COMMUNICATIONS IN THE 1940s

Eleven-year-old Paolina Roccanello arrived in Melbourne in April 1947 from Italy with her mother Elisabeta and younger brother Giuseppe on the SS Misr in the shadow of WWII (Huxley, 2007). The SS Misr carried over 600 passengers from 26 countries across Europe, the Middle East, East Africa and Southern Africa, including returning Australians, family migrants, displaced persons, Jewish refugees and South African footballers. The trip was long and arduous taking months and stopping in at many ports along the way (Boyatzis, 2010). Paolina was lucky to be reunited with her father who had emigrated to Australia eight years before, expecting his family to follow soon after. For all their war-time separation they received only one of his letters, which had taken five years to reach them (Huxley, 2007). Back then, there was no Internet, there were no mobiles or fax machines, and postal services were slow and often disrupted. Telegram and telephone communication was limited and costly, and had mostly been diverted by States to support fighting the war. After the war refugee movements beyond war-torn Europe were also heavily regulated by States (including under the United Nations). The UN coordinated repatriation, returns and resettlement of refugees to third countries. In today's terms, movements were extremely slow, highly regulated and very selective. Information on migration options for those who had been displaced was largely the monopoly of States, and opportunities for migrating to other regions were limited to formal channels (McAuliffe and Goossens, 2018).

What is Digital Technology?

The concept of digital technology has historically been closely related to computer technology, and in particular the significant advances in computerisation and increasing reliance on computers in human and machine labour from the mid-twentieth century onwards (Rabinovitz and Geil, 2004). However, digital technology is broader, encompassing computer technology and non-computer technological innovations, such as those associated with smart phones and other devices (Ensmenger, 2012). And while many associate digital technologies with quantitative data, important nuances are central to understanding digital technology and its uptake (Ensmenger, 2012, 769):

> It is important to note that this process of digitization is not the same as quantification .... The defining motivation of quantification is measurement; the principal goal of digitization, however, is manipulation .... Although the digital data in an MP3 file is numeric data, these numbers are not so much a measurement of sound as a model of sound. The value of that model is not so much that it is accurate as it is manipulable; MP3 data is valuable because it is easy to capture, store, communicate, analyze, and transform. It is only within a digital ecosystem of networks and devices that digital data becomes truly significant; but the rapidly increasing scale and scope of this ecosystem makes the imperative to digitize almost irresistible.

Digitalisation involves the creation of digital records or versions of all sorts of materials, such as paper/electronic documents and other information, photos, videos, music and other audio, biometric scans/imagery. Since 2005, we have witnessed an intensification of automatic digital capture throughout business and other sectors, resulting in massive increases in the number of digital interactions globally (Degryse, 2016). For example, it is estimated that in 2019 almost 300 billion emails were sent across the world, equivalent to around 805 million emails per day (Desjardins, 2019). The text box below places the recent estimates in a historical context that signposts major developments in the digitalisation of telecommunications over the last century.

BOX 1.2   DIGITALISATION OF TELECOMMUNICATIONS OVER THE LAST CENTURY

1930s – Invention of the modern computer.	In 2019 digitalisation was estimated to have resulted in:
1960s – E-mail entering use.	● 500 million tweets
1980s – Cell phones introduced to the public.	● 294 billion emails
1983 – GPS made available for public use.	● 4 petabytes of data created on Facebook
1989 – Invention of the World Wide Web.	● 4 terabytes of data created by connected cars
1993 – Internet is shared in the public domain.	● 65 billion messages on WhatsApp
1994 – Start of blogs and social networks.	● 5 billion searches made.
2001 – Start of 3G smart phones.	
2004 – Facebook is launched.	
2006 – Twitter is launched.	
2007 – Cloud computing; first iPhone launched.	
2009 – Start of 4G smart phones.	
2010 – First iPad; rapid growth of tablet usage.	
2018 – Start of 5G technology.	

Sources: McAuliffe et al., 2017; Desjardins, 2019.

Key data on digital connectivity indicate that these rapid transformations are resulting in massive uptake globally, with 93 per cent of the world's population now having access to a mobile broadband network (ITU, 2020). In fact, as shown in Figure 1.2 there are now more mobile phone subscriptions globally than there are people.

Notwithstanding this massive growth, we are also witnessing the further entrenchment of a number of 'digital divides' along economic, geographic, demographic and gender lines. Digital technology usage is greater in developed than developing countries globally, while we also see higher uptake in urban settings over rural areas globally, by males compared to females and by youth over aged persons (ITU, 2020). Beyond telecommunications, the race for digitalisation is impacting industry (formal and informal) globally as well as the government sector as commercial and geopolitical competition intensifies (OECD, 2020; Skog et al., 2018; Wramsmyr, 2018).

Migration and Digitalisation

Alongside other key domains, international migration as a growing phenomenon in recent years and decades is increasingly affected by digitalisation processes and related technological advances. Migration scholarship has resulted in a rich body of knowledge on the impacts of technology through history, including during the industrial revolution, which spawned Ravenstein's 'laws of migration' in the late 19th century (Ravenstein, 1885, 1889). Much more recently a key focus of 'digitalisation' has been in the realm of information and communication, and how migrants, potential migrants, and their families and networks engage with migration through ICTs (Andersson, 2019; Metykova, 2010; Nedelcu, 2009). This was heightened further during the 2015–16 migration 'crisis' in Europe when the 'appification' of migration became highly visual and contested during public debates (McAuliffe, 2016; Sanchez, 2018; Zijlstra and van Liempt, 2017). Many of the chapters in this *Research Handbook* explore

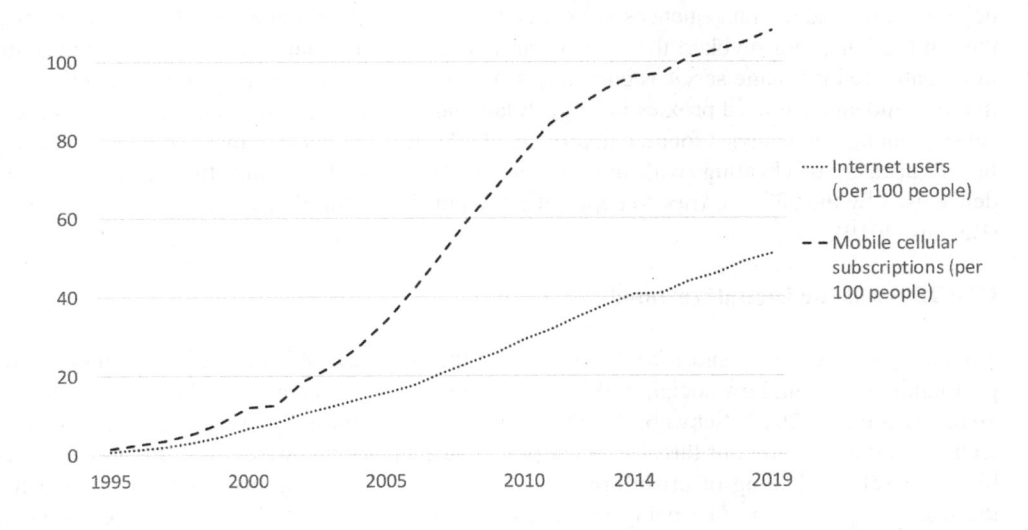

Source: ITU, 2020.

Figure 1.2   Global internet and mobile telephone access, 1995–2019

the latest research findings on ICTs throughout the migration cycle, and for specific migrant groups (see discussion below).

In migration policy and practice, on the other hand, there has been significant investment by States in digitalisation and automation over recent years (and in some cases, decades), including so as to realise efficiencies and manage significant increases in volume. Australia, for example, embarked on its 'global working' program in the mid-1990s in order to move from a paper-based application processing system to a digital platform supported by service delivery partners in locations with limited/no online accessibility (Rizvi, 2004). One of the early online visa application systems resulted in between 15 and 20 basic application checks being automated, thereby significantly reducing processing time and staffing costs (ANAO, 2008; PwC, 2011). Initially, human visa officers were required to make final decisions on applications. However, the online platforms were further developed over time so that they provided automated decisions of low-risk applications, including through the use of profiling techniques, that did not require a human visa officer to be involved (Rizvi, 2004).

The consequences of rapid and ongoing digitalisation in the migration sphere can be profound, with significant impacts on people's lives, including as they relate to human rights. Digitalisation can result in privacy breaches of applicants, for example, such as has been experienced multiple times by immigration authorities in recent years on a mass scale (Farrell and Laughland, 2014; Karp, 2020). In addition to privacy and confidentiality issues, digitalisation can exacerbate existing inequalities of access to regular pathways by limiting visa/migration services to those who have online access and capabilities; the 'digital divide' occurs in many settings globally, including as it relates to migration (Cherewka, 2020; ITU, 2020). The gender dimensions are also highly relevant, with digital access being markedly different in gender terms in many locations around the world, most notably in more traditional cultures with patriarchal systems of power (Brinkerhoff, 2012; Said, this volume).

Further, digitilisation combined with machine learning automation can have extremely negative unintended consequences for officials and individuals alike. In 2016, for example, the United Kingdom revoked the visas of around 36,000 international students on the basis of a contracted language services company's AI human voice recognition analysis indicating that the students had used proxies in English language tests needed to secure visas. However, subsequent human analysis found that around 7,000 (or 20 per cent) of these students had been falsely accused of cheating, with the UK immigration appeals tribunal finding that the evidence used by the Home Office to deport the students had multiple frailties and shortcomings (Baynes, 2019).

COVID-19 as the Digital 'turbo-charger'

Technological advances since 2005 resulting in the so-called '4th industrial revolution' are profoundly changing how social, political and economic systems operate globally (Friedman, 2016; McKenzie, 2017; Schwab, 2016). We have been witnessing the rising power of 'big tech' (and the concomitant threat to State power structures), the increasing production capability for self-publishing of misinformation, the race by businesses to 'digitalise or perish', the massive increase in data being produced (mainly through user interactions) resulting in increasing 'datafication' of human interactions, and the rapid development and roll-out of AI capabilities within business and governments sectors (Beduschi and McAuliffe, 2021). Profound technological change was deepening before COVID-19, but has significantly intensified during the pandemic as States, industry and communities have quickly needed to adapt to physical isolation and immobility, which has presented challenges but also demonstrable opportunities and efficiencies. Deep digitalisation of an already digitalising world will most likely be one of the most significant long-term effects of COVID-19. What this will mean for migration and for migrants is of intense interest, and early reflections are offered in this volume (see McAuliffe and Blower), noting that the contents of this *Research Handbook* were compiled largely before or in the early stages of the pandemic.

CONTENTS OF THIS *HANDBOOK*

This *Research Handbook* is organised into six parts corresponding to the migration cycle (moving, borders, integration/reintegration, trends) as well as key domains of thematic examination and enquiry (research and analysis, public debates, migration futures). Central to the Handbook is recognition of the need to extend analysis beyond the more dominant realms in migration research, such as the heavy focus on (geo)political and related policy responses, while bringing together currently fragmented research and analysis covering perhaps the most significant technological change of the modern era, that of digital transformations.

The first part, entitled 'Understanding Migration Patterns and Processes: Digital Technology and Migration Research and Analysis', looks at the use of digital technology in migration research and analysis. This includes how digital technology and data-driven knowledge production may pose opportunities as well as risks for migrants and researchers, notwithstanding the increasing prevalence of digital technologies in the interdisciplinary research area of migration (Leurs and Witteborn in this volume). Two complementary chapters in this part also explore the significant challenges in quantifying migration patterns and processes through the analysis of migration data, which is moving from traditional sources into non-traditional digitally derived sources that increasingly encompass more diverse forms of mobility (Yildiz and Abel; McAuliffe and Sawyer). The application of network analysis, for example, to global migration patterns provides for new methods to understand macroscale patterning, however, it also presents challenges and limitations, including ethical ones, methodological ones, socio-technological ones and research reproducibility (Danchev and Porter). Likewise, the fast-moving and increasingly influential area of data visualisation is outlined in this part, with a particular focus on how migration data visualisations are connected to the long-standing 'politics of statistics' and their representations (Allen in this volume). As digitalisation deepens and sectors adapt, the production of data visualisation becomes increasingly important, most especially in the education sector as the next generations move away from static textbooks to interactive platforms able to present analytical representations of statistical and other forms of data in real time. Overall, the implications of digital technologies for fieldwork, analysis and related enquiry into migration all around the world are profound. The ongoing 'datafication' of policy, practice and research may risk moving even scholarly endeavours that draw upon rigorous methods into the realm of social disconnectedness and into a sphere dominated by 'data points'.

The second part of this *Research Handbook*, entitled 'Digital Technology and the Act of Moving: (Im)Mobility, Barriers and Borders', looks into the first stage of the migration cycle and how digital technologies are increasingly facilitating, but also disrupting and preventing, movements. The use of ICTs by migrants in navigating migration journeys in perilous circumstances is examined, with reference to the extent of social networking and reducing potential vulnerability through the migration process (Azhar, Vannini, Newell and Gomez). On the one hand ICTs enable pre-departure considerations of migration and the gathering of information and advice right the way through to navigating issues in destination countries. On the other hand, ICTs also have the potential to mask or obscure the very high risks and dangers involved in some migration journeys but considerably reducing social proximity and increasing the perception of feasibility—a serious double-edged sword for migrants, authorities and the civil society groups involved in supporting migrants' human rights. In exploring a subset of this broader topic, the role of ICTs in migrant smuggling processes is critically examined with reference to the latest research and analysis available (Papanicolaou, Diba and Antonopoulos). The 2015 and 2016 mass movements into and through Europe raised the issue of migrant smuggling to new heights in political, policy and operational spheres as hundreds of thousands of migrants (including refugees) made their way to specific destination countries in Europe from adjacent areas, principally Turkey. The research community was similarly affected and the long-standing scholarship on migrant smuggling dynamics saw heightened interest and output in the wake of Syrian refugees' (and others) use of smart phones during their journeys and to integrate into their new communities—the so-called 'appification' of migration (McAuliffe, 2016). The growing virtual mountain of user-generated data is being increasingly utilised to critically analyse migration and mobility patterns and processes in a range of geographies, including the movements of forced migrants (such as refugees). A key source is call data record analysis, which is explored in a case study chapter on how CDR has been utilised to support the Data for refugees Challenge in Turkey, including the data gaps, biases, ethical and privacy issues that need to be addressed when drawing upon such sensitive data (Salah, this volume).

Digital technologies are increasingly being deployed by migration authorities in a variety of settings, most notably at borders and prior to entry, through automated visa processing. However, while efficiencies can be realised and increases in volume better managed, the lack of transparency and oversight raises concerns about development and deployment in light of the profound human rights ramifications and real impacts on human lives (Molnar). These issues are also arising given the increasing use of drones in the management and control of land and sea borders, with drones being used not only by government authorities but also by criminal groups in their illicit operations. There has been a proliferation of drone technology as the costs have reduced dramatically, opening up opportunities for others to access and use the technology in migration settings, including greater numbers of non-state actors (Koslowski). This presents major political and ethical challenges, and with drone technology only becoming more sophisticated by the day, the implications for drone technology use in a wider variety of settings are profound.

Part III of this *Research Handbook*, entitled 'Integration, Reintegration and Migrants' (Digital) (Virtual) (Transnational) Identities', examines the impacts of digital technology during migrants' stay and integration in destination countries as well as after their return home to origin. The part explores how migrants themselves utilise digital technology as they adapt to new communities and seek to forge safe and meaningful lives in unfamiliar locations while remaining connected to home.

In most settings, digital technology has become the 'lifeline' for migrants, and in some cases can mean a matter of survival in highly dangerous settings—a matter of life or death (Gallagher and McAuliffe, 2016; Sanchez, 2018). For those wanting to settle into a new society and feel a sense of inclusion, ICTs can present a short-cut to belonging through facilitating migrants' agency during integration journeys. Digital tools and expanding uptake represent opportunities for inclusion in destination countries while also supporting connectedness and knowledge transfer to origin communities (Bauloz). The particularities of refugees' inclusion through the provision of virtual higher education extends this discussion to the examination of a particularly vulnerable and significant group (Reinhardt, Zlatkin-Troitschanskaia, Happ and Nell-Müller). On the other hand, ICTs and social media platforms can also be used by migrants to be 'here and not here', including in the context of religious transnationalism and the realisation of digitally based faith in migration settings (Nyamnjoh). While in other settings, such as those involving international students and parents in transnational households, connectivity can raise rather than reduce parental burdens in sociocultural contexts that have not yet fully adjusted to new familial roles and practices (Wang and Lim).

The role of digital technology is markedly under-researched in the study of return migration and reintegration dynamics, and yet return migration processes are increasingly reliant on digital technology to take effect (especially on the part of State actors). This under-explored area is examined in depth with reference to existing evidence on both the benefits and risks offered by ICTs in return and reintegration contexts, while also highlighting knowledge gaps and potential policy responses (Majidi, Kasavan and Harindranath).

The next part in the *Research Handbook*—Part IV on 'Connectivity and Migration: Trends and Impacts'—takes a big picture look at how digital technology is impacting particular migrant groups and migration corridors, including migrant workers, high-skilled tech professionals and also along gender lines. Digital technology holds the promise to support the engagement of migrant workers more rapidly, cheaply and at greater scale, however, despite being often seen as a panacea, digital solutions rarely improve migrant workers' precarity (Kintominas, Berg and Farbenblum). A key aspect of labour migration that has risen in prominence in recent years is the analysis of international remittances of migrant workers, and most notably the development impacts over the longer term. The growth in remittances has come about during an era of rapid digitalisation, and yet the economic and technological barriers that have resulted in uneven remittance flows and wide variance in transaction costs have been of modest interest in migration scholarship. The links between remittance flows and development can be traced to mobile money and financial–digital inclusion (including along gender lines), and there is much room for improvement before many communities globally, but most especially in sub-Saharan Africa, are able to realise the maximum benefits of international remittances from diaspora worldwide (Kitimbo). On the other hand, and in other geographic contexts, long-term policy planning by India has enabled and facilitated the emigration of Indian tech professionals on a major scale, requiring realignments and adjustments to education and skills development streams both inside and outside India over time (Khadria and Mishra).

The feminisation of migration has long featured in scholarship as well as more recently in policy and programmatic responses to aspects of international migration (IOM, 2019a; Piper, 2008). Less prominent are examinations of the intersections of digital technology and gender in the context of migration and displacement, however, with the increasing awareness of the centrality of gender in the development, production and deployment of tech globally, deeper examination of this topic in the field of migration studies is much needed (Saïd). Closely related, but conceptually distinct, the issue of digital technology and family life in migrant settings highlights the complex array of opportunities and threats to transnational families posed by connectivity and the regulation and implementation thereof. Adaptable and rights-based regulation and cooperation frameworks and regimes are central if countries and regions are to realise the full potential of technology to support the increasing number of transnational families globally (Bhabha, Bhatia and Peisch).

Increasingly important in all realms of modern societies, digital technology has come to dominate the *means* and *manner* of political, social and economic discourse, and increasingly the *content* of such discourses as unfiltered self-publishing of (ill-informed) opinion becomes the norm. With this in mind, the next part focuses on technology and public debates on migration. The exploitation of online platforms to spread disinformation challenges accurate and balanced public debates to manipulate segments of the public in order to further political ambitions of power, particularly those of the far-right (Culloty and Suiter). The logical extension of such tactics of political actors, combined with the organic engagement of 'networked' publics of all types and from all walks of life, is ultimately resulting in polarisation between anti-immigrant conservatives and pro-immigrant liberal publics that undermines democratic processes (Ojala). In contrast, platforms are also means of human rights advocacy by migrants themselves, including those caught up in highly (geo)politicised contexts such as Nauru and Papua New Guinea. Asylum seekers and refugees detained as part of Australia's offshore processing policy recount experiences using new media platforms to improve specific situations and encourage the formation of social movements to champion human rights in real time (Cannon and Kanapathi).

In the final part, entitled 'Digital Migration Futures', a range of strategic issues underpinning the digitalisation of migration and mobility are contemplated. With digital technology increasingly pervasive in nearly all public domains, migration cannot be immune from deeper transformations of societies currently underway. Technologies appear throughout migration processes, ultimately transforming the categories of place, time and action, and giving rise to a multiplicity of actors and institutions related to migration issues (Dijstelbloem). To better understand the impacts of the digital transformations of migration, forecasting methodologies are embracing technological innovation, especially in the information and communication domain, but also by providing new ways of measuring migration. Technical and conceptual limitations seek to combine traditional and new forms of migration data; however, forecasting remains indicative rather than quantitative in light of the inherent complexity and uncertainty of migration processes (Wiśniowski). Notwithstanding the limitations of forecasting migration flows and trends, the technological power of the State greatly outweighs the tech capacity of migrants (including refugees) through all aspects of the migration cycle. The recent experiences of migrants vis-à-vis large tech providers provide worrying signs of surveillance and control, rendering individuals and vulnerable groups open to exploitation and abuse during migration (Korkmaz). In the final chapter of the *Research Handbook*, an early glimpse of how digital technology during COVID-19 is reshaping migration governance and practice is outlined. Written in the midst of the pandemic, and while national and international dynamics were evolving rapidly, the early signs indicate that securitisation rhetoric was bolstered during the initial weeks and months, thereby enabling the uptake of digital technology for both pandemic and non-pandemic purposes, such as broader population surveillance. Technological innovations can undoubtedly help mitigate the impacts of COVID-19; however, considerable concerns are being raised by scholars and civil society actors on the privacy and human rights implications of data collection and usage. The full implications for migration and mobility, privacy and security are as yet unknown; that said, alarming developments regarding the extent to which human rights principles have been so rapidly set aside during the pandemic are cause for major concern (McAuliffe and Blower).

ACKNOWLEDGEMENTS

Thanks are due to Edward Elgar Publishing for its interest in this research topic and for commissioning the *Research Handbook*, and with special thanks to Daniel Mather for his advice and encouragement throughout the process. Thanks also to IOM leadership for its support of migration research in a wide variety of settings, including academic and applied settings globally, and for support to enable this volume in particular (especially to Jill Helke, Gervais Appave and Wen Li). Thanks also to colleagues at the School of Demography at the Australian National University (Edith Gray and James Raymer) and at the Graduate Institute in Geneva (Cecilia Cannon and Vincent Chetail). Enormous thanks to all contributing authors, who persevered with this volume to see it through to the end notwithstanding the pandemic disruptions and delays (which for some people included coronavirus illness), and to Jenna Blower for helping to get the final manuscript ready for submission.

NOTE

The opinions, comments and analyses expressed in this chapter are those of the author and do not necessarily represent the views of any of the organisations or institutions with which the editor/author is affiliated.

REFERENCES

Aliverti, A. (2013) *Crimes of mobility: Criminal Law and the regulation of immigration*. Hoboken, NJ: Taylor & Francis.
Andersson, K. (2019) Digital diasporas: an overview of the research areas of migration and new media through a narrative literature review, *Human Technology*, 15(2): 142–180.
Australian National Audit Office (ANAO) (2008) *DIAC's Management of the Introduction of Biometric Technologies*. Department of Immigration and Citizenship, Australia.
Baynes, C. (2019) Government 'deported 7,000 foreign students after falsely accusing them of cheating in English language tests', *The Independent*, 14 June.
Beduschi, A. and M. McAuliffe (2021) AI, migration and mobility: implications for policy and practice, M. McAuliffe and A. Triandafyllidou (eds) *World Migration Report 2022*, IOM: Geneva. *Forthcoming*.
Bloy, M. (2002) The 1662 Settlement Act, *The Victorian Web: Literature, history and culture in the age of Victoria*, at http://www.victorianweb.org/history/poorlaw/settle.html (accessed on 20 December 2020).
Boyatzis, P. (2010) A new life, *The Empire Patrol Disaster website*, at http://www.empirepatrol.com/index.htm (accessed on 20 December 2020).
Boyd, M. (1989) Family and personal networks in international migration: recent developments and new agendas, *International Migration Review*, 23(3): 638–670.
Brinkerhoff, J.M. (2012) Digital diasporas' challenge to traditional power: the case of TibetBoard, *Review of International Studies*, 38(1): 77–95.
Castles, S. (2010) Understanding global migration: a social transformation perspective, *Journal of Ethnic and Migration Studies*, 36(19): 1565–1586.

Cherewka, A. (2020) *The Digital Divide Hits U.S. Immigrant Households Disproportionately during the COVID-19 Pandemic*, Migration Policy Institute: Washington DC, at https://www.migrationpolicy .org/article/digital-divide-hits-us-immigrant-households-during-covid-19 (accessed on 20 December 2020).

Czaika, M. and H. de Haas (2015) The globalization of migration: has the world become more migratory?, *International Migration Review* 48(2): 283–323.

Degryse, C. (2016) *Digitalisation of the Economy and its Impact on Labour Markets*, ETUI Working Paper 2016.02, European Trade Union Institute, at https://www.etui.org/Publications2/Working -Papers/Digitalisation-of-the-economy-and-its-impact-on-labour-markets (accessed on 20 December 2020).

Desjardins, J. (2019) *How Much Data is Generated Each Day?* eNewsWithoutBorders, 23 April, at https://enewswithoutborders.com/2019/04/23/how-much-data-is-generated-each-day/ (accessed on 20 December 2020).

Ensmenger, N. (2012) The digital construction of technology: rethinking the history of computers in society, *Technology and Culture* 53(4): 753–776.

Farrell, P. and O. Laughland (2014) Review blames Immigration for data breach exposing 10,000 detainees, *The Guardian*, 12 June, at https://www.theguardian.com/world/2014/jun/12/review-blames -immigration-data-breach-detainees (accessed on 20 December 2020).

Friedman, T. (2016) *Thank you for Being Late: an Optimist's Guide to Thriving in the Age of Accelerations*. New York, NY: Farrar, Straus and Giroux.

Gallagher, A. and M. McAuliffe (2016) South-East Asia and Australia, in *Migrant Smuggling Data and Research: A Global Review of the Emerging Evidence Base* (eds M. McAuliffe and F. Laczko), IOM, Geneva, at https://publications.iom.int/system/files/smuggling_report.pdf (accessed on 20 December 2020).

Hirsh-Pasek, K.M. Schlesinger, R. Michnick Golinkoff and E. Care (2018) *The New Humanism: Technology should enhance, not replace, human interactions*, The Brookings Institution, 11 June, at https://www.brookings.edu/blog/education-plus-development/2018/06/11/the-new-humanism -technology-should-enhance-not-replace-human-interactions/ (accessed on 20 December 2020).

Huxley, J. (2007) *When a Boat Came In: The First Wave: Beyond a White Australia*, at www.smh.com .au/multimedia/misr/story.html (accessed on 20 December 2020).

International Organization for Migration (2019a) *Supporting Brighter Futures: Young Women and Girls and Labour Migration in South-East Asia and the Pacific*. IOM: Geneva.

International Organization for Migration (2019b) *World Migration Report 2020*. IOM: Geneva.

International Telecommunication Union (ITU) (2020) *Measuring Digital Development: Facts and Figures 2020*. Geneva: ITU.

Karp, P. (2020) Government investigates data breach revealing details of 774,000 migrants, *The Guardian*, 4 May, at https://www.theguardian.com/australia-news/2020/may/05/government-investigates-data -breach-revealing-details-of-774000-migrants (accessed on 20 December 2020).

Massey, D.S. (1987) Understanding Mexican migration to the United States, *American Journal of Sociology*, 92: 1372–1403.

Massey, D.S., J. Arango, G. Hugo, A. Kouaouci and A. Pellegrino (1998) *Worlds in Motion: Understanding International Migration at the End of the Millennium*. Oxford: Oxford University Press.

McAuliffe, M. (2018) The link between migration and technology is not what you think, *Agenda*, 14 December, World Economic Forum, Geneva, at www.weforum.org/agenda/2018/12/socialmedia-is -casting-a-dark-shadow-over-migration/ (accessed on 20 December 2020).

McAuliffe, M. (2016) The appification of migration. A million migrants? There are apps for that. *APPS Policy Forum*, Asia and the Pacific Policy Society, 20 January, at www.policyforum.net/the -appification-of-migration (accessed on 20 December 2020).

McAuliffe, M. and A.M. Goossens (2018) Regulating international migration in an era of increasing interconnectedness, in *Handbook of Migration and Globalisation* (ed. A. Triandafyllidou). Cheltenham, UK and Northampton, MA, USA: Edward Elgar Publishing, pp. 86–104.

McAuliffe, M., A.M. Goossens and A. Sengupta (2017) Mobility, migration and transnational connectivity, in *World Migration Report 2018* (eds M. McAuliffe and M. Ruhs). Geneva: IOM, pp. 149–169, at www.iom.int/wmr/chapter-6 (accessed on 20 December 2020).

McKenzie, F. (2017) *The Fourth Industrial Revolution and International Migration*, Sydney: Lowy Institute for International Policy.

Metykova, M. (2010) Only a mouse click away from home: transnational practices of Eastern European migrants in the United Kingdom, *Journal for the Study of Race, Nation and Culture*, 16(3): 325–338, at https://doi.org/10.1080/13504630.2010.482418

Muggah, R. and I. Goldin (2019) How to survive and thrive in our age of uncertainty, *Agenda*, World Economic Forum, Geneva, 7 January, at www.weforum.org/agenda/2019/01/how-to-survive-ourage -of-uncertainty-muggah-goldin/ (accessed on 20 December 2020).

Nedelcu, M. (2009) Migrants' new transnational habitus: rethinking migration through a cosmopolitan lens in the digital age, *Journal of Ethnic and Migration Studies*, 12(9): 1339–1356, at https://doi.org/10.1080/1369183X.2012.698203

OECD (2020) *Digital Government Index (DGI): 2019*, OECD, Paris, at http://www.oecd.org/gov/digital -government/oecd-digital-government-index-2019.htm (accessed on 20 December 2020).

Piper, N. (2008) Feminisation of migration and the social dimensions of development: the Asian case, *Third World Quarterly*, 29(7): 1287–1303.

Pricewaterhouse Coopers (PwC) (2011) *Policy study on an EU Electronic System for travel Authorization (EU ESTA)*, PwC, Brussels, at https://ec.europa.eu/home-affairs/sites/homeaffairs/files/e-library/docs/pdf/esta_annexes_en.pdf (accessed on 20 December 2020).

Rabinovitz, L. and A. Geil (2004) *Memory bites: history, technology and digital culture*. Durham, NC, and London: Duke University Press.

Ravenstein, E.G. (1885) The laws of migration, *Journal of the Statistical Society of London*, 48(2): 167–235.

Ravenstein, E. (1889) The laws of migration, *Journal of the Royal Statistical Society*, 52: 241–305.

Rizvi, A. (2004) Designing and delivering visas. *People and Place*, 12(2): 45–52.

Sanchez, G. (2018) Critical perspectives on clandestine migration facilitation: an overview of migrant smuggling research, *Journal on Migration and Human Security*, 5(1): 9–27.

Schwab, K. (2016) The Fourth Industrial Revolution: what it means, how to respond, *Agenda*, 14 January, World Economic Forum, Geneva.

Skog, D.A., H. Wimelius and J. Sandberg (2018) Digital Disruption, *Business & Information Systems Engineering*, 60(5): 431–437, at https://doi.org/10.1007/s12599-018-0550-4.

Stark, O. and D. Bloom (1985) The new economics of labor migration, *American Economic Review*, 75: 173–178.

Todaro, M. (1989) *Economic Development in the Third World*. Fourth edition. New York, NY: Longman.

Triandafyllidou, A. (ed.) (2018) *Handbook of Migration and Globalisation*. Cheltenham, UK and Northampton, MA, USA: Edward Elgar Publishing.

United Nations Department of Economic and Social Affairs (UNDESA) (2019) *International Migrant Stock 2019*, at www.un.org/en/development/desa/population/migration/data/estimates2/estimates19 .asp (accessed on 20 December 2020).

Wramsmyr, A. (2018) Competition in the digitalization race, *Adnavem*, 13 March, at https://www .adnavem.com/blog/i-welcome-competition-in-the-digitalisation-race (accessed on 20 December 2020).

Zijlstra, J. and van Liempt, I. (2017) Smart(phone) travelling: understanding the use and impact of mobile technology on irregular migration journeys, *International Journal for Migration and Border Studies*, 3(2–3): 174–191.

PART I

UNDERSTANDING MIGRATION PATTERNS AND PROCESSES: DIGITAL TECHNOLOGY AND MIGRATION RESEARCH AND ANALYSIS

2. Digital migration studies

Koen Leurs and Saskia Witteborn

INTRODUCTION

Digital migration studies is here taken to refer to an emerging interdisciplinary research area. It is concerned with ontological questions linked to the growing digitization and datafication of human mobilities within and across spatial scales. It is also concerned with methodological and ethical questions emerging from digital ways of studying these mobilities. Ontology, epistemology, and ethics are inherently related as "migrants" come into being as a category through spatial and legal moves, such as crossing borders, work visas, or refugee status determination. Migrants also come into being as datafied subjects. The digitalization of migrants and migration processes through top-down datafication and bottom-up digital connectivity results in a proliferation of datasets that provides new research and analysis of opportunities as well as challenges. While migration is inherently related to the concept of the border – national, legal, ideological, urban-rural, or digital – and with it to evaluative notions of culturalizing and politicizing the subject crossing those borders, mobility remains a rather apolitical concept with positive connotations (Lehnert & Lemberger, 2014). Even more, migration refers mostly to people and their politically choreographed mobilities (Lehnert & Lemberger, 2014), and mobility to all types of movements of people, objects, and ideas (Larsen, Urry, & Axhausen, 2006). This distinction between mobility and migration is already a theological one, which is compounded by but can also be resolved with empirical moves.

In this chapter, we focus on methodology and the various practices of *doing* migration studies digitally. We will overview how scholars are adopting digital technologies and methods to study migration processes and practices. Our aim is to raise awareness for how digital technology and data-driven knowledge production may pose opportunities and risks for mobile subjects. Information scientists, governance actors, and the humanitarian community have embraced the opportunities digital technologies afford. Those include Geographic Information Systems (GIS) technologies to map displacement and camps (Tomaszewski, 2018; Macias, 2020), mobile and real-time data gathering in the field using tablets, smartphones, and online surveys, and empowering migrant respondents during research through new forms of participation (Kaufmann, 2020; Maitland, 2018). Technology changes have also created new opportunities for researchers to examine social media use by migrants, including ways to better understand risky migrations and protection issues (e.g., Zijlstra & Van Liempt, 2017). While new technology developments might help us grasp migration dynamics differently and ensure the acknowledgement of migrants' perspectives on the ground, those developments also come with challenges. For example, the new opportunities for big data research to forecast mass movements or the biometric identification of asylum seekers (Kingston, 2018) raise protection, privacy, and ethical considerations, in addition to questions about data verification, ownership, and access.

In taking stock of digital methodological approaches in researching mobility and migration, our focus is broad, covering research on irregular migrants, refugees, expatriates, internally displaced populations, and international students. The chapter is structured as follows. First, we discuss various epistemological approaches and their strengths and weaknesses. Second, we focus on selective examples of methodologies, tools, and techniques adopted in digital migration research. Third, we reflect on the ethics of digitization, datafication, and data use. We argue that migration and mobilities need to be linked further to critically examine the conditions – political, socio-economic, cultural, technological, and ideological – through which the migrant is produced.

EPISTEMOLOGIES

In this section, we focus on epistemology to map how the digitization of migration has led to a proliferation of views on what counts as proper knowledge about migration and the digital. A myriad of actors have adopted digital technologies and methods to achieve a greater understanding of migration practices and processes. The "digital is a game changer for migration" (EU EPSC, 2017, p. 12). Digital technologies and analysis methods are seen as disruptive tools, offering innovative solutions for policy makers, humanitarian organizations, the migration industry, and border control agencies.

The epistemologies of actors producing knowledge about migration range from positivist objectivity to social constructionist subjectivity. Causal explanation, aggregation, and quantification to interpretation, meaning-making, and discourse reflect the epistemological breadth of migration studies (e.g., Iosifides, 2018). The digitization of migration processes and practices has resulted in an array of "migration traceability" data (Diminescu, 2020, p. 77), which have been used for quantitative and big data studies as well as interpretatively oriented small data studies. On the one hand, we see how projects adopt "digital methods," which are "techniques for the study of societal change and cultural condition with online data," emerging from the computational turn in the humanities and social sciences (Rogers, 2019, p. 3). In order to aggregate digital data points to understand larger human processes, researchers need to develop tools and programs to clean, filter, sort, analyze, and visualize data. On the other hand, we see researchers maintaining a holistic commitment to address everyday experiences of mobile populations across offline and online settings. Small data scholars build on the adagium, "If you want to get at the internet, don't start from there," put forward by Miller and Slater in their work on Trinidadians' transnational communicative practices (2000, p. x). Such small data approaches, now typically understood as non-digital-media-centric, try to avoid the assumption of "communications technologies themselves [being] the main agents of social change" (Morley, 2017, p. 196).

Recent examples illustrate how migration is studied digitally from various epistemological perspectives. Big data epistemologies highlight felt urgencies of tapping into new digital sources for the purpose of obtaining large-scale, statistical data. These urgencies are based on the push for evidence-driven policies, as exemplified by the *Compact for Safe, Orderly and Regular Migration*, adopted by UN member states in 2018. The push for numerical understanding of human mobility is also reflected in the "big data for migration" (BD4M) approach launched by the *European Commission's Knowledge Centre on Migration and Demography* (KCMD) and the International Organization for Migration (IOM)'s *Global Migration Data Analysis Centre* (GMDAC) (Rango & Vespe, 2017). Another example for the push for numbers is the recent IOM report *More Than Numbers: How Migration Data Can Deliver Real-Life Benefits for Migrants and Governments* (IOM & McKinsey, 2018).

Initial academic big data studies seem to confirm the potential of social media data for being proxies for human cross-border mobility. Spyratos and co-authors (2019) analysed Facebook network data (Facebook, Instagram, Messenger, and the Audience Network) by looking at the number of people categorized as living in another country than they had lived previously. The authors maintained that such commercial "lived in country X" data can be repurposed as a proxy for quantifying mobility practices for forecasting physical movement. Similarly, computer scientists suggested researching Facebook-targeted advertising audiences to describe and predict Venezuelans' "real-time" cross-border mobility. Using Facebook user classifications, they proposed to "estimate and validate national and sub-national numbers of refugees and migrants and break-down their socio-economic profiles to further understand the complexity of the phenomenon" (Palotti et al., 2020, p. 1). Such analyses of large-scale migration movements draw on epistemologies of big data knowledge production, based on pattern detection and self-explanatory numerical aggregation. The empirical goal is facticity, objectivity, and neutrality.

Simultaneously, we have seen an embrace of interpretative and critical approaches among scholars, taking digital methods and technologies as a means to conduct ethnographic, creative, and/or participatory research on migration processes. For example, smartphone practices are taken as an entry-point to understand mobile subjects' everyday life experiences. Greene, in her study with irregular migrant women living in a refugee camp in mainland Greece, captures how mobile phones' video and photography features are taken up by the women in creative and personal ways to emotionally cope with displacement (2019). Interpretivist and critical epistemological stances prioritize in-depth practices of meaning-making. Those allow for an elaboration of ambiguities, attuned to the rich textures of lived experiences and the role of affect when engaging with digital technologies. In such studies, digital practices are commonly understood as part of the wider context of embodied, situated, everyday life.

In sum, received debates surrounding post-positivistic, interpretivist, and critical epistemologies reappear in migration research with digital methods and digital datasets. But scholars have begun to bridge paradigms. For example, Alinejad et al. (2019) show how their quantitative data gathered through digital methods cannot speak for themselves. They bring their findings on Turkish–Dutch migration into dialogue with interpretative methods and critical theory. Theory, conceptualization, and critical reflection are necessary to tease out the conceptual hierarchies and biased assumptions that shape how questions are asked and how data are extracted, processed, and interpreted.

METHODOLOGIES, METHODS, AND TECHNIQUES: BIG/SMALL DATA RESEARCH

Although big data remain incomplete, noisy, messy, and systematically biased, migrant regulatory bodies and scholars alike are embracing big data for renewed purposes of modelling, experimentation, and conceptualization (Tazzioli, 2020). Martin and Singh, for example, draw on local news sources and Twitter feeds to algorithmically "identify determinants and triggers of mass displacement" (2018, p. 193). Datafication and automation are the basis for new developments that seek to improve, augment, or substitute human decision-making through overlapping developments, including artificial intelligence (AI), machine learning, and predictive analytics. Within big/small data research, we have identified biometrics and geospatial mapping as well as social media and mobile phone data analytics. We highlight that those approaches are not exhaustive but a snapshot of current big and small data research.

Biometrics and Geospatial Mapping

Biometrics refers to the "measurement of life," combining the Greek words "bios" (life) and "metron" (measurement). In the area of digital migration, biometric technologies refers to the statistical processing of biological and biographic data extracted from the body. Biometrical identification can benefit mobile subjects, especially those on the privileged side. Digital borders augmented with biometric eye, fingerprint, and facial scanning technologies ensure frictionless cross-border movement. Moreover, the process of digitization has also resulted in new economic incentivized plans, such as Estonian e-Residency. Based on the distribution of E-residency cards for biometric authentication, Estonia's is the world's first transnational digital identity scheme providing "virtual residency" to non-residents, irrespective of their place of citizenship and residence (Tammpuu & Masso, 2018, p. 543).

Biometric data can be especially powerful during disaster and crisis situations, facilitating the rapid distribution of humanitarian aid and assisting in Refugee Status Determination. Personal data of refugees can be collected, recorded, organized, stored, retrieved, and transferred in digital, oral, and written form. In line with growing governmental, as well as supranational organizations' and donors' calls for evidence-based policy-making and accountability metrics, biometric registration of refugees is growing rapidly. This registration combines scans of fingerprints, faces, and irises or researching bone marrow, DNA, voices, and keystrokes. The result has been described as the datafication of lives (Kingston, 2018; Maitland, 2018; Tazzioli, 2020). The acceleration of big data employment in contexts of humanitarian crisis has unintended negative implications. Behnam and Crabtree state: "donors' thirst for data is increasingly undermining security and confidentiality, putting both survivors of violence and staff at risk" (2019, p. 4). Non-state actors and humanitarian organizations alike promote biometric technologies as solutions for refugees' needs and security threats under the headings of validity, neutrality, and exactness. However, these assemblages are ambiguous and uncertain, as they are the result of an obscured myriad of human and non-human decisions and processes. These processes are further obfuscated by "substantial subjective interpretation, translation and brokering" by migrants, border guards, humanitarian agents, and corporate actors (Olwig et al., 2020, pp. 1–2). Information and security scholars have addressed the risks of the many noted data leaks, hacks, and breaches. Audits also showed discrepancies between overarching policies of data encryption, storage, and sharing and increasingly outsourced on-the-ground ad-hoc operations (Stavinoha & Fotiadis, 2020). The latter function under constraints of scarcity of personnel, time, and resources and are thus prone to mistakes (Leurs, 2020; Macias, 2020).

The spatial aspect of human mobility "uniquely positions Geographic Information Systems (GIS) for documenting, representing, analyzing, curating, and understanding displacement situations" (Tomaszweski, 2018, p. 165). In refugee and post-disaster contexts, GIS is readily incorporated into larger information systems of service providers. The purposes are rapid-response, longitudinal camp management, planning, decision-making, and field operations. ICT4D and digital humanitarianism advocates champion smartphone tools and applications for data collection. Examples are the smartphone mapping tool *Collector for ArcGIS*, the Harvard University *Kobo Toolbox*, or the *Open Data Kit (ODK) Collect* tool for camp monitoring. Notwithstanding their potential, the stakes are high when using data as statistical indicators, as they are increasingly employed for remote decisions and become tools of political power as well as surveillance (Macias, 2020; Tomaszweski, 2018). GIS technology holds grassroots potential for "maptivism." Specialist NGOs and projects like *MapAction* and *RefuGIS* demonstrate how refugees and non-UN actors can be heard and consulted in shared decision-making processes. However, class, gender, or ability-based digital divides, among others, preclude less privileged mobile subjects to have control over which data are collected by whom and for which purposes. "GIS use by refugees is far less prevalent than use by staff," states Tomaszweski (2018, p. 167). GIS data is increasingly shared with specialists and general publics through dashboards and interactive visualizations, both by activists and NGOs. An example is the *Migrant Files* (https://www.themigrantsfiles.com/), which main-tained and visualized a crowd-sourced open access database of migrant and refugee deaths en route to Europe. With the *Missing Migrants* dashboard (https://missingmigrants.iom.int/), the International Organization of Migration has scaled up this initiative, listing and visualizing recorded incidents and deaths of international migrants across continents. It is the praise of digital advancements, like biometrics and GIS, that makes some scholars question their implications. The hope projected on technological advancement risks overlooking underlying systemic conditions, which allow mobility for some and halt it for others: "digital humanitari-anism represents just the latest example of 'humanitarian neophilia'" (Roth & Luczak-Roesch, 2020, p. 558). In other words, technological innovation risks strengthening existing power relations without questioning how aid is distributed and who benefits.

Social Media and Mobile Phone Data Analytics

Across disciplines, from informatics to geography and media and communication research, we can observe the growing leveraging of mobile phone and social media data to study mobility. National governments are experimenting with these datasets as part of their border controls. There was a public outcry over the U.S. State Department's decision to require work or study visa applicants to the US to disclose their social media handles over the past 5 years. According to the American Civil Liberties Union – a civil rights group – there is "no evidence that such social media monitoring is effective or fair," expecting it to lead to self-censorship (BBC, 2019). However, it is important to note that governments are already using social media data for refugee status determination. In the case of Norway, government agents locate public profiles as well as construct fictional profiles to access Facebook data. In their analysis, Brekke and Balke Staver (2019) highlight that visual, textual, and geolocation information on social media can be decisive for asylum case decisions.

In addition, collecting geo-tagged data has become a desirable method to trace global mobilities. Messias et al. (2016), for example, explored the value of "places lived" data gathered from the now defunct Google+ platform to chart cross-border mobilities between the clusters of United Arab Emirates, India, and Singapore; Brazil, Mexico, and the US; and Spain, France, and Italy. More recently, Sánchez-Querubín and Rogers put forward an approach to migration studies using digital devices (2018). They discuss the example of taking TripAdvisor, a traveler's review platform, to explore evolving geographic sentiments about migration. Subsequently, media and communication scholars have shown the key roles social media platforms play in the framing of migration. Siapera et al. (2018) studied the use of refugee-related hashtags on Twitter during the so-called European refugee crisis to locate important actors, voices, and the workings of networked framing. Pöyhtäri et al. (2019) adopted a cross-platform big data approach to study the representation of refugees in the Finnish "hybrid media environment," including news outlets, social media platforms as well as blogs. They found an overemphasis on threat and crime, and noted the role of social media in shaping the polarized debate through framing and the sharing of hyperlinks.

Various methodological experiments and innovations on cellular digital trace data have emerged from the release of the Call Detail Record (CDR) of one million Turk Telekom customers in Turkey, including their phone calls and SMS messages as part of the Data 4 Refugees (D4R) challenge. Marquez et al. (2019), for example, leveraged this data, alongside Twitter data, to study the development of communication and segregation between Turkish citizens and Syrian refugees. The authors note the variation across time and neighborhoods, which would have been difficult to capture through traditional survey-based methods. The combination of spatial mining, census, and phone provider data thus show promise in linking transnational migration processes to internal mobilities.

Limitations of Big Data Research

The problem with big data research is the black-boxed workings of social media platforms, algorithms, and the inherent limitations of datasets. As Spyratos et al. (2019, p. 1) maintained, "the exact methods the FN uses for classifying its users are not known, and might change over time." Amoore (2013) has warned that the predictive modeling of mobility and security can lead to mobile norms that fluently adjust scenarios and feed probable irregularities back into existing datasets. The result is the creation of hypothetical categories or types of people moving across space, thus constantly feeding loops of risk in which every person can become a potential threat over time. Therefore, Messias et al. (2016, p. 8) argue for a combination of various data sources, including web-derived as well as traditional demography/census datasets.

The question becomes how data narratives are constituted and linked to existing data ideologies as providing ontological truths about people. Related, data systems established during a period of crisis remain in periods of calm and can be used with intended or unintended consequences. The World Food Programme, for example, aggregates data to check whether refugees have a "balanced diet." In the case of Rohingya refugees residing in Bangladesh, the Bangladeshi government has stated it will use biometric registrations to ensure all refugees return to Myanmar (Leurs, 2020). These are a few examples of possible "function creep" – the gradual widening of the use of a system beyond its original purpose, "often with unintended consequences" (Kingston, 2018, p. 42). Drawing on Rango and Vespe (2017, p. 6), we sum up with a concise comparison of big data challenges and opportunities (see Table 2.1).

ETHNOGRAPHIC METHODS

Apart from research with and on big and small data, research on mobile methods has accelerated as has digital ethnography. We discuss those two methods in one section as they share a common ground: Ethnography. Digital ethnography and mobile methods build strongly on received ways of collecting data in the ethnographic spirit, including interviewing, participant

Table 2.1   Big data for migration

Challenges	Opportunities
• Data access limitations (proprietary data, technological sustainability, costs)	• High spatial resolution
• Noisy data, black-boxed algorithms and protocols, tools to process and filter need to be adaptable	• High frequency of update
	• Timeliness, virtually real-time
• Big data studies on migration are pilots and case studies, systematic approaches lacking	• Near global coverage, even in areas with limited/no migration statistics
• Confidentiality, security, privacy, and ethical concerns	• Not limited to statistical definitions: potential to better understand temporary and irregularized forms of migration and mobility
• Sampling bias given the persisting digital divides	
• Difficulty in applying official UN definitions of international migrants.	• Larger sample sizes compared to traditional migration survey research.

observation, or creating proxy accounts for research collaboration. Many principles applying to ethnographic research are important in mobile methods and digital ethnography as well, including reflexivity on the part of the researcher about how one's presence can change social interaction patterns or providing collaborators with opportunities to use the research for advancing collective purposes (Iosifides, 2018). Likewise, there are challenges, including a techno-centric or methodological nationalism. Another shared feature between mobile methods and digital ethnography is the focus on agency for migrants in co-producing, checking, or meta-communicating about their data and collaborating on research projects involving digital methodologies.

Digital Ethnography

Digital ethnography refers to ethnographic studies of experiences, practices, relationships, things, localities, social worlds, and events that increasingly take place in digitized environments (Pink et al., 2016). Not only do these environments have different meanings for different communities and individuals, but these environments are also inherently dynamic. They change as a result of technological innovations, evolving user preferences, community norms, and social relations. Therefore, Pink et al. maintain that digital ethnography demands scholars to work with the following five principles: "multiplicity, non-digital-centric-ness, openness, reflexivity and [being] unorthodox" (2016, p. 8).

Digital ethnography encompasses a variety of methods for data gathering and analysis. Since the turn of the millennium, research has explored how migrants and diasporas constitute and present themselves in digital spaces. Initially, the majority of scholars engaged with a discursive reading of verbal and visual messages produced on discussion forums, mailing lists, and in chatrooms. Studying Indians in the diaspora, Mitra (2001) concluded already two decades ago that "the internet can be used to voice the unspeakable stories and eventually construct powerful connections that can be labeled 'cyber communities'" (p. 30), a notion that resonated with other scholars and their work on understanding how diasporas use the digital as resistant space. Ade-Odutola argues that the online presence of Nigerians in the diaspora allows them to "cyber-frame" Nigerian nationhood differently vis-à-vis the West (2012). In methodological terms, the Internet, in such studies, becomes another type of "field" to do research through received discourse and thematic analysis of mostly written text (Candidatu et al., 2019; Witteborn, 2019a).

In recent digital ethnographies, researchers usually start with interviewing face to face for people to share and reflect upon their digital media practices. Twigt (2018), for example, began her research with Iraqi refugees in Jordan with interviews and participant observations in people's homes. Researchers also contact people directly through their social media accounts and ask them to participate in research on a particular subject. Gajjala (2019) located diasporic research participants online. She interviewed them using teleconferencing software, observing their digital practices remotely. Storyboards and having individuals and/or transnational families write about their migration experiences are another form of ethnographic data collection (Marlowe, 2019; Winarnita, 2019). Triangulation of methods is especially productive, such as doing interviews on Skype or WhatsApp, co-producing Facebook data with collaborators, writing and reflecting on digitally produced narratives, and studying visual archives (Alinejad et al., 2019; Greene, 2019).

Many studies in this line of research conceptualize diasporas in ethno-nationalistic terms, which might be a consequence of the ease of pin-pointing the groups in discussion fora, on community organization websites, and in public Facebook groups. As a result, scholarship shows how diaspora and migrant groups reproduce themselves in digital space. Interactions among diasporas and between diasporas and societies remain understudied.

Mobile Methods

Much of the literature since the refugee movement into Europe in 2015 has focused on mobile methods. The goals are to capture how migrants, people seeking asylum in particular, use technology for particular purposes before, during, or after the journey. In their "trajectory ethnography" of Iranian and Afghan migrants' journeys towards Europe, Zijlstra and van Liempt followed research participants and conducted impromptu interviews, both face to face and through digital channels. This combination, they conclude, worked as "a useful tool that allows us to understand how mobile technology shapes and facilitates parts of the journey" (2017, p. 174). The research employed traditional ethnographic field research like interviewing and participant observation. But the scholars shifted from a physical place-based to a movement-based paradigm, complementing it with digital interviewing and text analysis to create a robust dataset on how people use technology while being physically mobile.

Smartphones have also been adopted to innovate mobile methods in migration research. Kaufmann (2020) recognizes two methodological approaches in her research with migrants and their smartphone communication: self-reported data (p. 169) and data-elicitation in interviews or focus groups (p. 173). Elicitation can be based on automatic log-data (such as connectivity, geo-tags, or duration of app-use) or data produced by informants (p. 173). In a methodologically related study, Cabalquinto explored mobile photography to understand how overseas Filipino workers in Melbourne, Australia, maintain family life in the Philippines (2019). Leurs, in his study with young refugees and expatriates, approached smartphones as meaningful digital pocket archives, inviting informants to share and discuss material from their phones to reconstruct experiences and identification processes (2017).

In another line of research, scholars have focused on the ways material objects, including phones and SIM cards, are acquired and distributed. In her article on the migration industry of connectivity services, Gordano Peile (2014) illustrated through ethnographic fieldwork that a socio-economic ecology has developed, based on the needs of poor, forced migrants in Europe. Unlike the majority of research celebrating how the migrant became modern through technology, this research started to show how technology becomes a stress factor in the lives of those having crossed borders for political or economic reasons, something that Awad and Tossell (2019) have emphasized in their more recent work.

Other scholars have used culturally grounded approaches to capture the logics of technology use and content production in contexts of displacement. Witteborn (2019b), for example, illustrated the importance of digital gifting by applying principles of the Ethnography of Communication, an approach concerned with situated communicative practice and the meanings it has for the people engaging with it (Philipsen & Coutu, 2005). By analyzing the practice of gifting, the researcher explored situated cultural logics without imposing epistemological lenses. Related, a Speech Codes Theory approach (Philipsen, 1997; Philipsen, Coutu & Covarrubias, 2005) proved productive for identifying culturally grounded notions of data privacy and safety (Witteborn, 2020). Through the triangulation of interviewing, participant observations of mobile phone use while moving through urban space, textual analysis of Facebook posts and geolocation, the research highlights that the displaced framed data privacy as data *safety*. The findings were put to practice through the creation of a data safety workshop for the concerned population.

In sum, collecting data on how people move through space through received ethnographic research in combination with textual and visual analysis of social media content is effective for linking migration to mobility research and politicizing mobility processes.

ETHICS

"Classic ethical concepts" demand reconsideration in a "socio-technical context that is persistent, replicable, scalable, and searchable" (Tiidenberg, 2018, p. 467). Tiidenberg (2018) pointed out the "grey areas" within ethical principles developed in the pre-internet era and transposed to the digital domain. Those principles are still worked out. And they require ongoing debates, including who and what counts as a human subject, the entities, and places able to gather, store, and process data, the sharing of sensitive information, data privacy, and informed consent. Reflexivity in decision-making when researching the digital domain is a pivotal process, as Leurs (2017) maintained, especially in research with already marginalized populations, such as refugees, sans papier, the homeless, trafficked, and poor labor migrants. We focus briefly on three areas, which we regard as particularly important when it comes to ethics: the digitization of mobile subjects, data privacy, and the development of categories.

The ethical implications of the increased digitization of mobile subjects are outlined by scholars like Sandvik (2020). She discusses technological experimentation in the humanitarian sector by the example of "humanitarian wearables" (2020, p. 87). These include devices inserted in or worn on beneficiary bodies that track, measure, and represent these bodies digitally through various metrics. In the "transactional data collection process" of humanitarian aid, data volunteered by beneficiaries is the product, not the wearable itself. For Sandvik, this new development demands critical ethical scrutiny based on a "techno-legal consciousness" (2020, p. 88). The three approaches she proposes are valuable for digital migration studies: (1) a renewed engagement with top-down humanitarian frameworks; (2) a reconsideration of data from a bottom-up, rights-based perspective; and (3) a data justice approach attuned to relevant stakeholders (2020).

Data privacy is another area demanding continued ethical attention. Humanitarian and social media organizations do not have full control over data. There is the risk of data not being able to be traced back to their origins and shared datasets not being updated (Privacy International, 2011). An example is a cloud service used by the UN being hacked and refugee data being stolen (Parker in Kaurin, 2019). Overall, there is little transparency about asylum seekers on data collection, processing, access, or deletion of data (Latonero et al., 2019 on refugees in Italy). While IOM's (2010) ethical guidelines for giving consent are important, the implementation is problematic at times (Latonero et al., 2019). A previously discussed problem is mobile phone metadata. The records bear the risk of violations of data privacy on the one hand, and creating surveillance databases with the help of researchers that individuals are not aware of, on the other (Salah et al., 2019, p. 10).

Finally, we arrive at the question of categorization. Using categories such as *migrant*, *refugee*, and *asylum seeker* unreflexively in data-driven digital studies of migration risks taking them as "naturally given." There have been calls for the "de-migranticization" of a "nation-state- and ethnicity-centred epistemology that informs a large share of migration and integration research" (Dahinden, 2016, p. 2208). Pre-assumed categories like *nationality*, *ethnicity*, or migrant, are in danger of normalizing inequalities by depoliticizing mobilities. This view had already been expressed in 2014 by critical migration researchers Lehnert and Lemberger who argued for dissecting categories of migrancy and to link them to the political and economic conditions engineering mobilities. Moreover, accelerating forms of technological categorization are harmful for some mobile subjects. Technologies render categories invisible and neutral. As Sim and Cheesman note: "de-politicised approaches to humanitarian digitalisation stabilise categorisation practices as objective, eclipsing intersectional issues about category-making, discrimination and mobility control" (2020, n.p.). The problem is compounded by the fact that in the majority of categorization practices, mobile subjects have no say over their categorization (Sim & Cheesman, 2020).

CONCLUSION

In this chapter, we have discussed selective research methods and how those methods constitute and are reflective of different epistemologies and approaches to "doing digital migration research." Received methodologies – from post-positivistic survey research to participatory action research – are reflected in studies on migration with digital objects, digitally mined data, big data visualization, and digital ethnographies. Given the current interest in AI, big data, and visualization, we predict that big data research on migration will gain further grounds in the future. On a global scale, digital migration studies may further entrench long-standing biases between researchers with social and financial capital dominating the field on the basis of their access to datasets and people whose bodies and minds serve as mining fields for data excavation. This trend is likely to gain traction in countries where people are not aware of their data rights and/or where political and legal systems do not protect personal and collective information. On a disciplinary level, prioritizing migrants as data points and patterns risks sidelining non-digital research on mobilities and borders. The contextualized individual migrant story might be pushed from the social sciences towards the humanities, with literature, film studies, and the arts being the fields left where storytelling "counts" as evidence of a lived reality.

Big data research supports and fuels the current trends in predictive mobilities (Amoore, 2013) and the felt need by political and economic actors to predict labor demands and population growth. With economies and states competing for skilled talent and balancing skilled and humanitarian migration, big and small data research in migration will become indispensable to predict and channel migratory movement. On the policy level, we expect that biometric evidence – alongside social media evidence – will be increasingly incorporated into frameworks governing and managing migration, impacting not only refugee status determination but also visa and work-permit applications of other mobile subjects, including expatriates. This trend raises the question of how migrant categories come into being and on which political, legal, and ethical basis they are maintained. Like other scholars before us, we argued in the chapter for politicizing mobilities processes to understand the conditions under which *the migrant* is produced.

Related, the chapter raised ethical issues in terms of collecting data on migrants and for which purposes. In situ case studies and ethnographic and participatory action approaches have to be included to gather contextualized evidence and to avoid a domination of a big data narrative that silences personal and situated narratives by those crossing borders. Overall, our discussion has made an argument for keeping mixed-methods approaches. It has also made an argument for using logged and self-reported data materials shared by participants to create granular as well as generalizable perspectives on the intersections between migration, mobilities, and technologies. With a growing awareness of surveillance – particularly among migration populations who increasingly realize data traces are further jeopardizing them – digital and mobile ethnography research will become more challenging and will have to constantly reflect its ethics. Scholars and practitioners are in the privileged position to join the debate around ethics, data privacy, and anonymity, a debate that is linked to accountability and which requires the input from many disciplines (Daly et al., 2019). The recent *Report of the Special Rapporteur on contemporary forms of racism, racial discrimination, xenophobia and related intolerance* is noteworthy here, as it offers an overview of how digital technologies used in border enforcement can enhance racial discrimination and undermine human rights of migrants (OHCR, 2021).

REFERENCES

Alinejad, D., L. Candidatu, M. Mevsimler, et al. (2019), "Diaspora and mapping methodologies: Tracing transnational digital connections with 'mattering maps'", *Global Networks*, **19**(1), 21–43.

Amoore, L. (2013), *The Politics of Possibility: Risk and Security beyond Probability*. Durham, NC: Duke University Press.

Awad, I. & J. Tossell (2019), "Is the smartphone always a smart choice? Against the utilitarian view of the 'connected migrant'", *Information, Communication & Society*, online first: DOI: 10.1080/1369118X.2019.1668456.

Behnam, N. & K. Crabtree (2019), "Big data, little ethics: Confidentiality and consent", *Forced Migration Review*, **61** (June), 4–6.

Brekke, J.P. & A. Balke Staver (2019), "Social media screening: Norway's asylum system", *Forced Migration Review*, **61** (June), 9–11.

Cabalquinto, E.C. (2019), "'They could picture me, or I could picture them': 'Displaying' family life beyond borders through mobile photography", *Information, Communication & Society*, online first: 10.1080/1369118X.2019.1602663.

Candidatu, L., K. Leurs, & S. Ponzanesi (2019), "Digital diasporas: beyond the buzzword", in J. Retis & R. Tsagarousianou (eds), *The Handbook of Diasporas, Media and Culture*, Hoboken, NJ: Wiley Blackwell, pp. 31–48.

Dahinden, J. (2016), "A plea for the 'de-migranticization' of research on migration and integration", *Ethnic and Racial Studies*, **39** (13), 2207–2225.

Daly, A., T. Hagendorff, H. Li, et al. (2019), *Artificial Intelligence, Governance and Ethics: Global Perspectives*. The Chinese University of Hong Kong Faculty of Law Research Paper No. 2019-15, accessed 20 July 2020 at https://ssrn.com/abstract=3414805 or http://dx.doi.org/10.2139/ssrn.3414805.

Diminescu, D. (2020), "Researching the connected migrant", in K. Smets, K. Leurs, M. Georgiou, S. Witteborn, & R. Gajjala (eds), *Sage Handbook of Media and Migration*, London, UK: Sage, pp. 74–78.

Dourish, P. & E.G. Cruz (2018), "Datafication and data fiction: Narrating data and narrating with data", *Big Data & Society*, Jul–Dec, 1–10.

EU EPSC (2017), "10 trends shaping migration", *European Commission – European Political Strategy Centre*, accessed 27 January 2020 at https://ec.europa.eu/epsc/publications/other-publications/10-trends-shaping-migration_en.

Gajjala, R. (2019), *Digital Diasporas: Labor and Affect in Gendered Indian Digital Publics*. London: Rowman & Littlefield.

Gordano Peile, C. (2014), "The migration industry of connectivity services: a critical discourse approach to the Spanish case in a European perspective", *Crossings: Journal of Migration and Culture*, **5** (1), 57–71.

Greene, A. (2019), "Mobiles and 'making do': exploring the affective, digital practices of refugee women waiting in Greece", *European Journal of Cultural Studies*, online first: https://doi.org/10.1177/1367549419869346.

IOM and McKinsey & Company (2018), "More than numbers: how migration data can deliver real-life benefits for migrants and governments", IOM Global Migration Data Analysis Centre and McKinsey & Company.

Iosifides, T. (2018), "Epistemological issues in qualitative migration research", in R. Zapata-Barrero & E. Yalaz (eds), *Qualitative Research in European Migration Studies*. IMISCOE Research Series. Cham: Springer, pp. 93–112.

Kaufmann, K. (2020), "Mobile methods: doing migration research with the help of smartphones", in K. Smets, K. Leurs, M. Georgiou, S. Witteborn & R. Gajjala (eds), *Sage Handbook of Media and Migration*, London, UK: Sage, pp. 167–179.

Kaurin, D. (2019), "Data protection and digital agency for refugees", *World Refugee Council Research Paper No. 12*, 12 May, accessed 20 July 2020 at: https://www.cigionline.org/sites/default/files/documents/WRC%20Research%20Paper%20no.12.pdf.

Kingston, L.N. (2018), "Biometric identification, displacement and protection gaps", in C. Maitland (ed.), *Digital Lifeline? ICTs for Refugees and Displaced Persons*. Cambridge, MA: MIT Press, pp. 35–54.

Larsen, J., J. Urry, & K. Axhausen (2006), *Mobilities, Networks, Geographies: Transport and Society*. Aldershot: Ashgate.

Latonero, M.. L. Hiatt, A. Napolitano, G. Clericetti, & M. Penagos (2019), "Digital identity in the migration and refugee context", *Data & Society*, 4 April, accessed 29 January 2020 at https://datasociety.net/output/digital-identity-in-the-migration-refugee-context/.

Lehnert, K. & B. Lemberger (2014), "Mit Mobilitaet aus der Sackgasse de Migrationsforschung?" in Labor Migration (ed), *Vom Rand ins Zentrum*. Berlin: Panama Verlag, pp. 45–61.

Leurs, K. (2020), "Migration infrastructures", in K. Smets, K. Leurs, M. Georgiou, S. Witteborn, & R. Gajjala (eds), *Sage Handbook of Media and Migration*, London, UK: Sage, pp. 91–102.

Leurs, K. (2017), "Communication rights from the margins: politicising young refugees' smartphone pocket archives", *International Communication Gazette*, **79** (6–7), 674–698.

Macias, L. (2020), "Digital humanitarianism in a refugee camp", in K. Smets, K. Leurs, M. Georgiou, S. Witteborn, & R. Gajjala (eds), *Sage Handbook of Media and Migration*, London, UK: Sage, pp. 334–345.

Maitland, C. (2018), "The ICTs and displacement research agenda and practical matters", in C. Maitland (ed.), *Digital Lifeline? ICTs for Refugees and Displaced Persons*, Cambridge: MIT Press, pp. 239–258.

Marlowe, J. (2019), "Social media and forced migration: the subversion and subjugation of political life", *Media and Communication*, **7** (2), 73–183.

Marquez, N., K. Garimella, O. Toomet, et al. (2019), "Segregation and sentiment: estimating refugee segregation and its effects using digital trace data", in A.A. Salah, A. Pentland, B. Lepri, & E. Letouzeé (eds), *Guide to Mobile Data Analytics in Refugee Scenarios*, Cham: Springer, pp. 3–28.

Martin, S.F. & L. Singh (2018), "Data analytics and displacement: using big data to forecast mass movements of people", in C.F. Maitland (ed.), *Digital Lifeline? ICTs for Refugees and Displaced Persons*, Cambridge, MA: MIT Press, pp. 185–206.

Messias, J., F. Benevenuto, I. Weber, & E. Zagheni (2016), "Clusters: The value of Google+ data for migration studies", *ASONAM '16*, accessed 20 July 2020 at https://arxiv.org/pdf/1607.00421v1.pdf.

Miller, D. & D. Slater (2000), *The Internet: An Ethnographic Approach*, Oxford: Berg.

Mitra, A. (2001), "Marginal voices in cyberspace", *New Media & Society*, **3** (1): 29–48.

Morley, D. (2017), *Communications and Mobility*, Oxford: Wiley.

Odutola, K. (2012), *Diaspora and Imagined Community*. Durham, NC: Carolina University Press.

OHCR (2021), Contemporary forms of racism, racial discrimination, xenophobia and related intolerance. United Nations Human Rights Office of the High Commissioner, accessed 30 September 2021 at https://undocs.org/A/75/590

Olwig, K.F., K. Grünenberg, P. Møhl, et al. (2020), *The Biometric Border World*, London, UK: Routledge.

Palotti, J., N. Adler, A. Morales-Guzman, J. Villaveces, V. Sekara, M. Garcia Herranz, et al. (2020), "Monitoring of the Venezuelan exodus through Facebook's advertising platform", *PLoS ONE* **15**(2): e0229175. https://doi.org/10.1371/journal.pone.0229175

Philipsen, G. (1997), "A theory of speech codes", in G. Philipsen and T. Albrecht (eds), *Developing Communication Theory*, Albany: State University of New York Press, pp. 119–156.

Philipsen, G. & L. Coutu (2005), "The ethnography of speaking", in K.L. Fitch & R.E. Sanders (eds), *Handbook of Language and Social Interaction*. Mahwah, NJ: LEA, pp. 355–379.

Philipsen, G., L. Coutu and P. Covarrubias (2005), "Speech codes theory: restatement, revisions, and a response to criticisms", in W.B. Gudykunst (ed.), *Theorizing about Intercultural Communication*, Thousand Oaks, CA: Sage, pp. 55–68.

Pink, S., H. Horst, J. Postill, et al. (2016). *Digital Ethnography*. London: Sage.

Pöyhtäri, R., M. Nelimarkka, K. Nikunen, M. Ojala, M. Pantti, & J. Pääkkönen (2021), Refugee debate and networked framing in the hybrid media environment, *International Communication Gazette* **83**(1): 81–102.

Privacy International (2011), "Why we work on refugee privacy", accessed 27 March 2020 at https://privacyinternational.org/news-analysis/1322/why-we-work-refugee-privacy.

Rango, M. and M. Vespe (2017), "Big Data and alternative data sources on migration: from case-studies to policy support", European Commission – Joint Research Centre (JRC). accessed 20 July 2020 at: https://bluehub.jrc.ec.europa.eu/bigdata4migration/uploads/attachments/cjdelbdgo00hnqa zv3u7xi6pd-big-data-workshop-draft-summary-report.pdf.

Rogers, R. (2019). *Doing Digital Methods*. London: Sage.

Roth, S. & M. Luczak-Roesch (2020), "Deconstructing the data life-cycle in digital humanitarianism", *Information, Communication & Society*, **23**(4), 555–571.

Salah, A.A., A. Pentland, B. Lepri, et al. (2019), "Introduction to the data for refugees challenge on mobility of Syrian refugees in Turkey", in A.A. Salah, A. Pentland, B. Lepri, & E. Letouzeé (eds), *Guide to Mobile Data Analytics in Refugee Scenarios*. Cham: Springer, pp. 3–28.

Sánchez-Querubín, N. & R. Rogers (2018), "Connected routes: Migration Studies with digital devices and platforms", *Social Media + Society*, Jan–Mar, 1–13.

Sandvik, K.B. (2020), "Humanitarian wearables: digital bodies, experimentation and ethics", in D. Messelken & D. Winkler (eds), *Ethics of Medical Innovation, Experimentation, and Enhancement in Military and Humanitarian Contexts*. Cham: Springer, pp. 87–104.

Siapera, E., M. Boudourides, S. Lenis, & J. Suiter (2018), "Refugees and network publics on Twitter: networked framing, affect, and capture", *Social Media + Society*, Jan–Mar: 1–21.

Sim, K. and M. Cheesman (2020), "What's the harm in categorisation? Reflections on the categorisation work of Tech 4 Good", *Big Data & Society* blog, accessed 20 July 2020 at http://bigdatasoc.blogspot.com/2020/03/whats-harm-in-categorisation.html.

Spyratos, S., M. Vespe, F. Natale, et al. (2019), "Quantifying international human mobility patterns using Facebook Network data", *PLoS ONE* **14** (10): e0224134. *Qualitative Data Collection*. London: Sage, pp. 466–841.

Stavinoha, L. and A. Fotiadis (2020), "Asylum outsourced: McKinsey's secret role in Europe's refugee crisis", *BalkanInsight*, accessed 20 July 2020 at https://balkaninsight.com/2020/06/22/asylum-outsourced-mckinseys-secret-role-in-europes-refugee-crisis/.

Tazzioli, M. (2020), *The Making of Migration: The Biopolitics of Mobility at Europe's Borders*. London: Sage.

Tiidenberg, K. (2018), "Ethics in digital research", in U. Flick (ed.), *The Sage Handbook of Qualitative Data Collection*. London: Sage, pp. 466–481.

Tomaszewski, B. (2018), "Geographic Information Systems (GIS) and displacement", in C. Maitland (ed.), *Digital Lifeline? ICTs for Refugees and Displaced Persons*. Cambridge, MA: MIT Press, pp. 165–184.

Twigt, M.A. (2018), "The mediation of hope: digital technologies and affective affordances within Iraqi refugee households in Jordan", *Social Media + Society*, Jan–Mar: 1–14.

Winarnita, M. (2019), "Digital family ethnography: lessons from fieldwork amongst Indonesians in Australia", *Migration, Mobility & Displacement; Victoria*, **4** (1): 105–117.

Witteborn, S. (2019a), "Digital diaspora: social alliances beyond the ethno-national bond", in J. Retis & R. Tsagarousianou (eds), *The Handbook of Diasporas, Media and Culture*. Malden, MA: Wiley-Blackwell, pp. 179–192.

Witteborn, S. (2019b), "The digital gift and aspirational mobility", *International Journal of Cultural Studies*, **22** (6): 754–769.

Witteborn, S. (2020), "Data privacy and displacement: a cultural approach", *Journal of Refugee Studies*. Online first: https://doi.org/10.1093/jrs/feaa004

Zijlstra, J. and van Liempt, I. (2017), "Smart(phone) travelling: understanding the use and impact of mobile technology on irregular migration journeys", *International Journal of Migration and Border Studies*, **3** (2/3), 174–191.

3. Migration stocks and flows: data concepts, availability and comparability

Dilek Yildiz and Guy Abel

INTRODUCTION

Migration is increasingly becoming an important driver of population and socio-economic change and of relevance to policymakers. Consequently, good data on migration are increasingly required to monitor the impacts of migration and inform related policies. However, good migration data are rarely available, especially outside of rich, Western countries. Where data do exist, inconsistent definitions and collection methods lead to inconsistent numbers of the migrants, prohibiting cross-national comparisons. In this chapter we attempt to map out the range of data availability to monitor migration within or between multiple countries. In particular we focus on discussing data on migration flows and stocks. These two measures allow users to study the contribution of migration to demographic changes of the entire population of a country or sub-national regions and its impact on the size and structure of migrant populations.

Challenges in collecting migration data have always existed. As we discuss in this chapter, there are a range of data sources, none of which provides a perfect measurement of the number of migration events or migrant population sizes. In recent decades, a number of new data sources have emerged to measure migration flows and stocks, many of which are based on technologies that had not previously existed. These new data sources have the potential to provide some improvements in the measurements of migration over traditional data sources, but often with important caveats that need to be considered if they are to be adopted, such as restrictions on the availability of data to potential users or biases in their population coverage. In the next section we first provide an overview of migration measurement and data collection methods. These two areas are commonly linked, where migration data are often collected as part of a wider data collection exercise rather than specifically for the sole study of migration, and hence the resulting measurement of migration does not always conform to a standard definition. In the following section we outline available migration stock and flow data from traditional data sources. This includes a comparative analysis of international migration flow statistics in Europe, which due to their political importance and clear inconsistencies has often been the focus of efforts to harmonize via regulations and statistical models. In the penultimate section we outline recent research that has utilized emerging data sources to monitor migration flows and stocks. In the final section we conclude with a review of the potential future of migration data spanning both traditional and new sources.

MIGRATION MEASUREMENT AND DATA COLLECTION METHODS

Unlike other demographic processes, migration can occur multiple times during a lifetime and can be defined in a number of different ways. How both the temporal (what qualifies as a time period) and spatial (what qualifies as a new or different place) aspects of the migration definition are specified is not standardised, with different data producers using different definitions for each. Typically, internal migration involves changes of places of residence that involve moving to a different administrative area such as a new province, state or county, where the administrative geography of a given country is developed for purposes not related to migration. International migration is more straightforward, where migration involves a change of countries of residence.

The United Nations developed recommendations regarding the temporal aspect of a migration definition. It suggests a 12-month period to determine the place of usual residence in line with either of the criteria: (a) the place at which the person has lived continuously for most of the last 12 months (i.e., for at least 6 months and 1 day), not including temporary absences for holidays or work assignments, or intends to live for at least 6 months; or (b) the place at which the person has lived continuously for at least the last 12 months, not including temporary absences for holidays or work assignments, or intends to live for at least 12 months (United Nations Department of Economic and Social Affairs Population Division 1998). As we discuss in further depth in this chapter, the adoption of this recommendation is far from universal, prohibiting effective measures and comparisons of migration across multiple countries.

Censuses, population registers, administrative data, sample surveys and border crossings are the principal sources of data on migration and migrants. There also exist a wide range of data sources, which we do not cover in this chapter, related to more specific migration events and migrant groups such as internally displaced populations, migrant deaths and trafficking or media representations of migrants, integration metrics and remittances, to name but a few. Censuses typically involve one or more direct questions on the respondent's place of birth; place of last residence; duration of residence in the place of enumeration; place of residence on a specific date before the census. Based on answers to these questions, detailed data on retrospective migration can be created for the population.

National population registers that cover the whole (or a large majority of the) population provide another data source on migration that is timelier and more frequent than censuses. However, population registers are not always specifically designed for migration data collection. Consequently, the migration data they provide may meet only their specific administrative purposes. Further, migrants leaving an area or country may not have to or may not wish to inform the register of their outward move subduing out-migration statistics.

Administrative data sets can also provide data on migration if details on the place of residence are updated when members of the population change addresses. As with population registers, migration statistics tend to be a byproduct from administrative records that are not the primary design for collecting migration data. Surveys provide a fourth major source of migration data. Nationally representative surveys typically collected by statistical offices may include questions on migration periodically or in one-off supplements. Surveys tend to allow for more detailed questions on migrants' experiences such as their previous migration history, reasons for moves or intentions for future migration. If migration data are collected periodically, surveys can also provide a more regular time series than censuses. However, migration tends to be a rare event and sample sizes of nationally representative surveys that collect migration data are usually too small to provide reliable detailed data on migration that are representative of the population as a whole, especially in smaller geographic units. General-purpose surveys also miss emigrants and hence are biased towards migrants arriving into the population being surveyed.

Border agencies are able to collect data on arrival and departure information, usually related to international migration. Arrival data from potential immigrants are commonly based on the declarations of intention to migrate (a change in their place of residence to their new country) beyond a specified time period (such as 12 months). Arrival data are thus subject to errors when migrants leave a country permanently before their intended length of stay is complete. Data on migrants leaving countries are comparatively less common than data on migrants arriving into countries.

A final, new, data source for migration has begun to emerge in recent years. Geo-located data from mobile phones and internet-based activities have the potential to provide timely migration data based on their customer or user populations. These new data sources are typically controlled by their private operating companies and in most cases are not freely available to researchers but offer enormous potential to monitor migration.

MIGRATION DATA FROM TRADITIONAL SOURCES

The availability of migration data varies over time, countries and migration measures. In recent decades, different organizations have begun to collect and organize data on the two most commonly available migration measures: stocks and flows. Migrant stocks are counts of persons at a given time point who reside in a usual place of residence that is different to their place of birth. Migration flows are counts of population over a given time period that have changed their places of usual residence.

Migrant Stock Data

Migration stock data, also known as lifetime migration data, are considerably easier to collect than migration flow data. Migrant stock data are a static figure at a fixed point of time, where estimates can be gathered via records or responses to questions on place of birth. When questions on place of birth do not exist, often censuses have questions on citizenship as a means to measure migrant stocks. In the 2010 round of censuses a question on nationality or citizenship was included in 112 national census questionnaires (Juran and Snow 2018) that enabled the enumeration of international migrant stock data. More general questions on place of birth, which can be used to gather migrant stock statistics for internal migration, are also frequently included in censuses. During the 2000 round of censuses, 122 countries produced internal lifetime migration data based on a question on place of birth (Bell and Charles-Edwards 2013).

The abundance of international migration stock data has motivated efforts to collate published figures from all countries and impute counts of estimated foreign-born populations in those where there was no census or other data sources were unavailable. Researchers at the University of Sussex were the first to provide detailed bilateral migrant stock data on the counts of number of foreign-born residents in each country broken down by their place of birth (Migration Globalisation and Poverty DRC 2007). Their bilateral migration matrix has since been adopted, improved and expanded over a number of years and additional dimensions such as sex (e.g., Özden et al. 2011) and skill level (e.g., Artuç et al. 2015) by both the World Bank and the United Nations Population Division.

The United Nations Population Division has regularly produced estimates for foreign-born populations for all countries through a *Trends in International Migrant Stock* series (United Nations Department of Economic and Social Affairs Population Division 2019, 2015, 2013, 2012, 2009, 2006, 2005, 2004). Earlier versions of the Population Division data provided estimates of the total foreign-born populations for each country back to 1960. Totals from these series show a jump after 1985 predominantly from the disintegration of the former Union of Soviet Socialist Republics (USSR), which led to the creation of 15 independent states and resulted in the reclassification of millions of persons, who had been internal migrants within the USSR, into international migrants because their place of birth became part of a different country. Later versions of the United Nations Population Division data begin later, in 1990, but with more detail including statistics on the age and sex structure of the foreign-born population in each country and counts of number of foreign-born residents broken down by their place of birth.

Data developed by the World Bank have been released on a less regular basis than those from the United Nations Population Division: the majority of the datasets is based on a single year, with the exception of the series from Özden et al. (2011), who provide bilateral migrant stock data from 1960 to 2000 for all countries. They treat migrants born in one Soviet Republic, but resident in another, as foreign born prior to the breakup of the USSR.

There are no global sets of migrant stock data related to internal migration outside of displacement. However, the IMAGE project based at the University of Queensland identified 122 counties that produce lifetime migration data from censuses and 34 from surveys (Bell, Charles-Edwards, Kupiszewska, et al. 2015). Despite the common occurrence of a place of birth question in census and other data collection systems, the availability of internal migrant stock data is not as prominent as it might be.

Migration Flow Data

Migration flow data provide a measure of the number of people who change their usual place of residence over a specific period rather than an individual's lifetime. In this sub-section, we discuss available migration flow data, regulations and recommendations to help harmonize reported flow data, the current comparability of international migration statistics, internal migration flow data, and the efforts to estimate both internal and international migration flow data.

A number of international organizations provide collection of international migration flows obtained from national statistics offices. The United Nations Population Division has regularly published international migration annual time series flows data to and from a selected number of countries. Their most recent version in 2015 covered 45 countries starting from 1980 (United Nations Department of Economic and Social Affairs, Population Division 2015). A large majority of the countries are European, where in recent decades countries have built capacity to collect and publish international migration data. Eurostat, the statistical office of the European Union, provides a number of measures of annual international migration flows from its member states. These include data on immigration by age group, sex and country of previous residence and emigration by age group, sex and country of next usual residence starting from 1998 (Eurostat 2021a, 2021b). The Organisation for Economic Co-operation and Development (OECD) also regularly provides annual international migration flow data from its members including total inflows and outflows of foreign population by nationality but not by place of previous residence (typically used to study migration patterns) starting from 2000 (OECD 2020).

In addition to international organizations, researchers in the DEMIG project based at the University of Oxford collected data from national statistics offices to create a longitudinal bilateral migration flow database (DEMIG C2C) for over 34 countries and annual total immigration and emigration flows from 161 countries (Vezzoli, Villares-Varela and de Haas 2014). In the DEMIG C2C dataset, migration flow data stretched back to 1946 for some countries, whilst in the totals database the earliest statistics come from 1815, though for many countries, in both data sets, the majority of data is for more recent years.

As mentioned in Section 1, another source of migration flow data is cross-border movements data, which includes counts of border crossings. However, only a handful of countries collects such information on both entries and departures (International Organization for Migration 2019).

The most notable feature in each of the aforementioned migration flow datasets (from international organization and research groups alike) is that no two countries are using the same definition of migration because of different policies and data needs. This feature likely hinders any cross-national comparisons of the data as both the volume and spatial distribution of migration flows can be exaggerated or subdued by the way it is measured.

Beyond the timing criteria in defining a migration flow event, further difficulties in comparing migration flow are created from a number of different issues including the use of actual or intended duration of stay; the definition of place of origin; the use of different data collection methods. The use of intended or actual duration of stay is not specified in the UN Recommendation. Grundström (1993, cited in Lemaitre 2005) showed intention of stay and actual stay change not only by intended period of stay but also by country of destination. Countries typically collect international migration flow data based on a change of country of residence. However, a number of countries prefer to collect and publish flow data disaggregated by country of citizenship or country of birth that can distort the migration patterns in comparison to a country of last residence, as discussed.

In countries where cross-border arrival and departure data, population registers or residence permits are not available to collect migration statistics, such as Cyprus and the United Kingdom, alternative sources such as passenger surveys are used. Elsewhere, household surveys are used. For example, Ireland uses a quarterly national household survey as the basis for its international migration flow statistics. In comparison to population registers, these data sources might incur problems of sampling bias and inadequate population coverage leading to issues in their reliability. However, as discussed in International Organization for Migration (2019) population registers provide proxies of migration flows, and they are neither free from coverage problems nor fully up to date. These problems are more prominent in developing countries.

Regulation 862/2007

In 2007, Regulation 862/2007 passed by the European Parliament provided a legal framework for the improved statistics on migration by EU Member States (European Commission 2007), including requirements to report international migration flows data that comply with a harmonized 12 month definition.

In order to investigate the impact of the regulation on the comparability of European migration flow statistics in Figure 3.1, we compare the differences in migration statistics as reported by both sending and receiving countries for the same migration corridor (origin–destination pair). In a perfect setting, the immigration figures reported by the receiving country should match with emigration figures reported by the sending country. However, as Kupiszewska and Nowok (2008) showed, receiving countries often report higher figures than the receiving countries for the same migration flow. This is due to the relative ease in collecting data on migrants arriving in countries due to the lack of incentives for individuals to notify authorities when emigrating.

The panel on the left shows the relationships between the reported migration flows by receiving and sending countries prior and after Regulation 862/2007. If all data were harmonized, they would lie on a single diagonal line. The panel on the right shows the same combination of sending and receiving data but after Regulation 862/2007 was passed. It is noticeable that there exists a lower degree of variation between the points in the more recent data. Having considered this visual, we used two approaches to statistically investigate a change in relationship over time. First, we used a Regression Discontinuity Analysis, to test whether the regulation has an effect. We did not find any discontinuity in cut-off points formed by a discrete parameter to indicate whether observations were made before or after the regulation. Second, we used a Fisher's F-test for a change in variance between the two datasets. The test indicates there is evidence to suggest a decline in the variance in the relationships between the sending and receiving of data before and after the regulation.

Internal Migration Flow Data

Internal migration flow data are more plentiful in comparison to international migration flows. Although measurement issues do exist, data harmonization is a less prominent issue as typically single data sources are used to collect migration data for all regions (as opposed to international migration where each country is or is not collecting data following national policies). The IMAGE project provides a large collection of internal migration flow data close to the 2000 round of censuses. Migration flows are defined differently across the set of countries: 29 countries collected data based on a one-year migration interval, 52 based on a five-year interval and 32 based on another fixed-period interval (Bell, Charles-Edwards, Kupiszewska, et al. 2015). In addition, 55 countries collected data on the time since a respondent's last move and 71 on the duration of time at current residence. Comparisons of internal migration across countries is not straightforward. Not only are countries using different migration intervals to measure migration, they also examine the different administrative geographies with a wide range in the number, size and densities of sub-national units (the Modifiable Aerial Unit Problem). Bell, Charles-Edwards, Ueffing, et al. (2015) outlined a method and number of indicators that can overcome some of these issues to enable comparisons of internal migration intensities across countries.

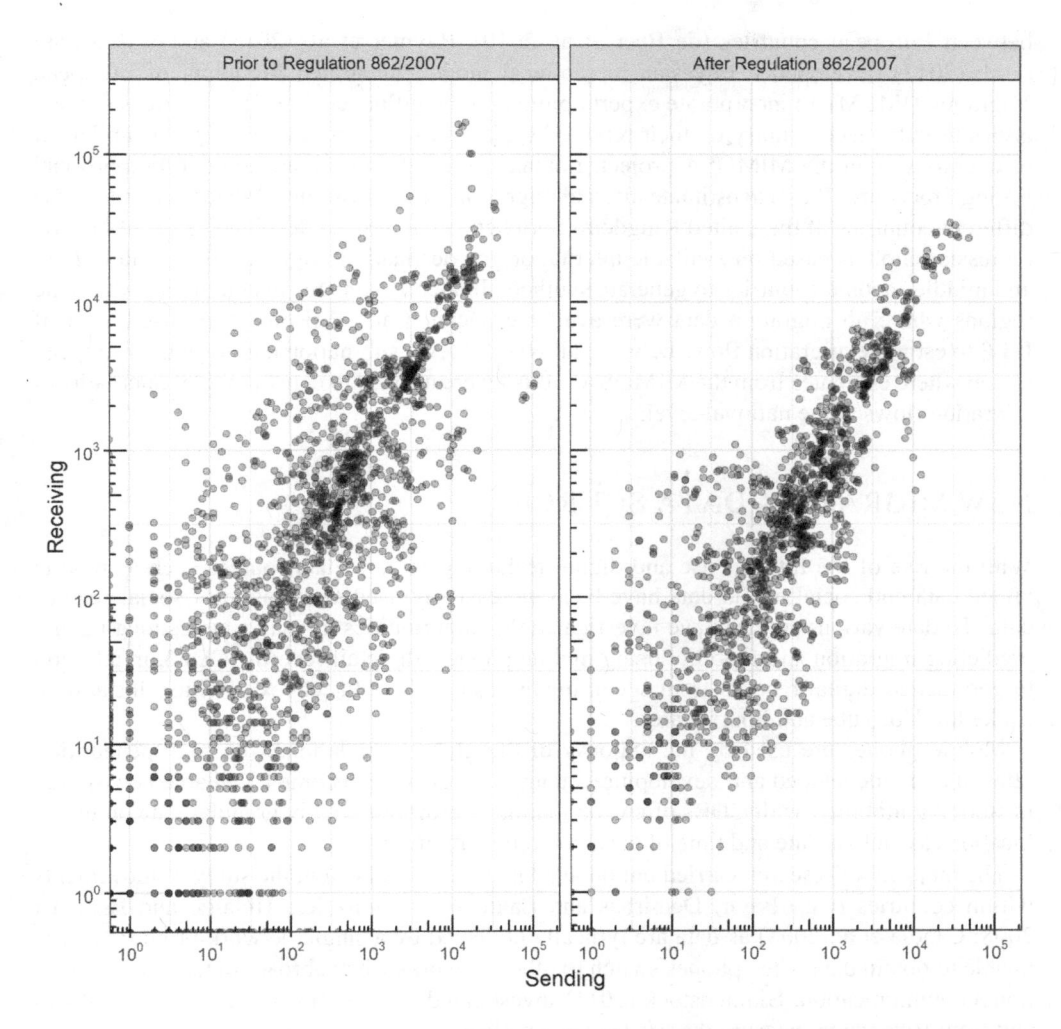

Figure 3.1   *International migration flows from receiving and sending countries prior and after Regulation 862/2007 (plotted on logarithmic axis)*

Synthetic Migration Flow Data

In recent years, there have been a number of efforts to develop synthetic migration flow data sets to overcome the problems of missing and unharmonized data. Methods to estimate global migration flows between all countries have been developed by several researchers based on changes in the global bilateral migration stock discussed in the previous section. Abel and Cohen (2019) compare six of these methods and propose a validation strategy based on the unharmonized reported migration flows where available. Estimation methods have also been developed to estimate European migration flows based on the unharmonized data collected by Eurostat. Researchers in the Migration Modelling for Statistical Analyses (MIMOSA) project used optimization procedure to estimate annual migration flows with a one-year definition between European countries (de Beer et al. 2010). Raymer et al. (2013) and Wiśniowski et al. (2013) developed a Bayesian hierarchical model, Integrated Modeling of European Migration (IMEM), to incorporate expert opinion on the influence of a range of measurement issues used by each country on their reported statistics to generate estimates for a similar set of countries as in the MIMOSA project. Lomax et al. (2013) used an Iterative Proportional Fitting Procedure (IPFP) to estimate internal migration flows between sub-national units in the different countries of the United Kingdom. Sorichetta et al. (2016) developed a set of logistic regression models based on available internal origin–destination migration flow data in low- and middle-income countries to generate synthetic bilateral estimates in countries in the same regions where no migration data were available. Dennett and Wilson (2013) used a set of IPFP to estimate migration flows between all NUTS2 level sub-national units in the European Union where estimates from the MIMOSA project were utilized to constrain total international migration flows at the national level.

NEW MIGRATION DATA SOURCES

With the rise of the Internet use and digital technologies, new data sources such as mobile phone data and social media data have been increasingly used as a source of social science data. To date various studies have investigated the opportunities in analysing, estimating and predicting migration and mobility using mobile phone call detail records (CDR), in addition to geo-tagged digital records from social media data such as Facebook, Twitter, Instagram, LinkedIn, YouTube and Foursquare.

Mobile phones are used by the majority of the global population, with high penetration rates in many developed and developing countries. They have become a valuable new source in studying mobility, with CDRs often containing near-instant, highly granular data on users' locations as well as date and time of each communication.

The majority of research carried out on available CDR has been in the study of movements within countries (e.g., Bayir, Demirbas and Eagle 2009; González, Hidalgo and Barabási 2008; Candia et al. 2008) as data are typically collected by a single network provider that is unable to obtain data when phones switch to other networks when abroad. In terms of international communication, Blumenstock (2011) investigated the patterns of international calls to and from Rwanda to measure the ties between nations.

Whilst providing great opportunities to better understand real-time population movements, CDR are typically expensive and not publicly available or provided for free as a one-off sample. Therefore, most of the previous studies have explored a range of topics including short term mobility, seasonal patterns and population displacement using CDR data for rather short periods of time, less than or equal to one year (e.g., Wesolowski et al. 2012; Wesolowski et al. 2015; Ruktanonchai et al. 2016). Few studies that employed longer period data demonstrated that CDR data can be used to produce estimates comparable to traditional sources. For example, Lai et al. (2019) used multiannual data from Namibia to estimate and predict migration with one year definition. They showed that CDR derived migration estimates are as accurate as census data derived estimates. Although CDR data hold a lot of possibilities, often they do not include details on the phone users, such as age, sex or education.

Despite holding numerous benefits, CDR data also have some limitations when they are used for migration research. They can be divided into three broad groups: (1) the privacy and ethical issues arising from using individual data collected without consent, (2) the unrepresentativeness and selection bias due to having access to data from only mobile phone owners with a certain provider, and (3) the short-term data availability.

Digital data from internet-based activities provide another source of data to monitor migration. Zagheni and Weber (2012) and State et al. (2013) used geo-location data from IP addresses from Yahoo Mail users' log-in records to estimate the trends in international migration. Google Location History data provide high spatial graduality equivalent to GPS tracker data (Ruktanonchai et al. 2018).

Social network services that allow sharing geo-locations with recommendations or photos such as CouchSurfing and Flickr, have been used to track mobility patterns of tourists (Pultar and Raubal 2009; De Choudhury et al. 2010). Social media platforms such as Facebook and LinkedIn provide advertising platforms that have enabled researchers to analyse migration patterns based on freely available audience data (values published by the companies on the potential reach of advertisements aimed at a specific subset of their user bases). Characteristics such as hometown information and current location from the Facebook advertisement platform have been used to measure international migrant stocks (Zagheni, Weber, and Gummadi 2017; Gendronneau et al. 2019).

More computationally intensive methods have been used to gather histories of geo-located Tweets using the Application Programming Interface (API) of Twitter, a micro-blogging platform. Aggregated data from resulting scrapes of geo-located Tweets have then been used to study mobility patterns (Zagheni et al. 2014; Mazzoli et al. 2020) and to analyse the role of different interval durations when defining migration events on the overall level of migration (Fiorio et al. 2020, 2017).

Migration data from internet-based activities have a number of potential biases when used to analyse migration patterns and trends. Traditional data sources usually either aim to cover the entire population, a population of interest or an event, and produce representative estimates, sometimes with degrees of uncertainty (e.g., household surveys report estimates with their statistical confidence intervals). However, internet access is not universal, especially in less developed countries. Even though there has been substantial leapfrogging, gender and class inequalities still exist, and the coverage of each user platform can also vary between regions, over time and between different subsets of the populations (McAuliffe, Goossens and Sengupta 2017). Access to data via APIs or advertising platforms, and the black-box methods to generate audience estimates, can be altered or removed without notice.

THE FUTURE OF MIGRATION DATA

Migration is likely to continue to be an important issue for the demographic, social and economic wellbeing of future nations. As a result, a number of opportunities exist for the improvement of the collection of migration data from both traditional and new data sources.

Tomas et al. (2009) proposed five steps towards better (international) migration data: ask basic census questions and make data publicly available; compile and release existing administrative data; centralize Labor Force Surveys; provide access to microdata, not just tabulations; include migration modules on more existing household surveys. As the authors note, each of the steps is politically and technically practical and would allow countries to greatly improve their migration data at low cost, and with existing mechanisms. As noted in a report by the

International Organization for Migration (2016), one means to also speed the improvement of migration data is through illustrating the potential financial value of good migration data by enabling policymakers to better monitor as well as direct funds and legislation to (for example) fill known labour market shortages or increase remittances.

The emergence of new data sources is likely to continue as new technologies develop and more users gain access to existing social media platforms and mobile phones, increasing the digital traces that potentially allow for migration estimates to be derived. Access to data collected in the private sector will continue to be dependent on the user's technical skills and capacity to access free data, monetary resources to purchase data being sold or developing partnerships to obtain data not publicly available. The increase in the number of data sources comes with a potentially large number of datasets that provide complementary information across countries, over time and at different levels of granularity, that will vary in size and populations. As discussed by Hughes et al. (2016) the development of methods to combine data sources will be key in producing an infrastructure that is robust to unanticipated changes in the use of technology. Doing so will potentially allow for estimates of migration data that can utilize the broad coverage and robustness of traditional data sources and combine them with the temporal and spatial detail from data derived from new technologies.

BIBLIOGRAPHY

Abel, Guy J., and Joel E. Cohen (2019), 'Bilateral International Migration Flow Estimates for 200 Countries', *Scientific Data* 6 (1): 82. https://doi.org/10.1038/s41597-019-0089-3.

Artuç, Erhan, Frédéric Docquier, Çağlar Özden and Christopher R. Parsons (2015), 'A Global Assessment of Human Capital Mobility: The Role of Non-OECD Destinations', 6863, *World Development*, Vol. 65, Policy Research Working Paper. https://doi.org/10.1016/j.worlddev.2014.04.004.

Bayir, Murat Ali, Murat Demirbas and Nathan Eagle (2009), 'Discovering Spatiotemporal Mobility Profiles of Cellphone Users', in *2009* IEEE International Symposium on a World of Wireless, Mobile and Multimedia Networks & Workshops, 1–9. Kos, Greece: IEEE. https://doi.org/10.1109/WOWMOM.2009.5282489.

Beer, Joop de, James Raymer, Rob van der Erf and Leo van Wissen (2010), 'Overcoming the Problems of Inconsistent International Migration Data: A New Method Applied to Flows in Europe', *European Journal of Population / Revue Européenne de Démographie* 26 (4): 459–81. https://doi.org/10.1007/s10680-010-9220-z.

Bell, Martin, and Elin Charles-Edwards (2013), 'Cross-National Comparisons of Internal Migration: An Update on Global Patterns and Trends'. Population Division Technical Paper No. 2013/1. New York, NY.

Bell, Martin, Elin Charles-Edwards, Dorota Kupiszewska, Marek Kupiszewski, John Stillwell and Yu Zhu (2015), 'Internal Migration Data around the World: Assessing Contemporary Practice', *Population, Space and Place* 21 (1): 1–17. https://doi.org/10.1002/psp.1848.

Bell, Martin, Elin Charles-Edwards, Philipp Ueffing, John Stillwell, Marek Kupiszewski and Dorota Kupiszewska (2015), 'Internal Migration and Development: Comparing Migration Intensities around the World', *Population and Development Review* 41 (1): 33–58. https://doi.org/10.1111/j.1728-4457.2015.00025.x.

Blumenstock, Joshua E. (2011), 'Using Mobile Phone Data to Measure the Ties between Nations' in iConference '11: Proceedings of the 2011 iConference, 195–202. ACM Press. https://doi.org/10.1145/1940761.1940788.

Candia, Julián, Marta C. González, Pu Wang, Timothy Schoenharl, Greg Madey and Albert-László Barabási (2008), 'Uncovering Individual and Collective Human Dynamics from Mobile Phone Records', *Journal of Physics A: Mathematical and Theoretical* 41 (22): 224015. https://doi.org/10.1088/1751-8113/41/22/224015.

De Choudhury, Munmun, Moran Feldman, Sihem Amer-Yahia, Nadav Golbandi, Ronny Lempel and Cong Yu (2010), 'Automatic Construction of Travel Itineraries Using Social Breadcrumbs', in HT '10: Proceedings of the 21st ACM conference on Hypertext and hypermedia, 35. ACM Press. https://doi.org/10.1145/1810617.1810626.

Dennett, Adam, and Alan G. Wilson (2013), 'A Multilevel Spatial Interaction Modelling Framework for Estimating Interregional Migration in Europe', *Environment and Planning A* **45** (6): 1491–1507. https://doi.org/10.1068/a45398.

European Commission (2007), 'Regulation (EC) No 862/2007 of the European Parliament and of the Council of 11 July 2007 on Community Statistics on Migration and International Protection and Repealing Council Regulation (EEC) No 311/76 on the Compilation of Statistics on Foreign Workers', *Official Journal of the European Union*, no. L 199/23. Available at https://eur-lex.europa.eu/legal-content/EN/ALL/?uri=CELEX%3A32007R0862 (accessed 9 August 2021).

Eurostat (2021a), 'Emigration by Age Group, Sex and Country of next Usual Residence[Migr_emi3nxt]', Data set. Available at http://appsso.eurostat.ec.europa.eu/nui/show.do?dataset=migr_emi3nxt&lang=en (accessed 9 August 2021).

Eurostat (2021b), 'Immigration by Age Group, Sex and Country of Previous Residence[Migr_imm5prv]'. Data set. Available at https://appsso.eurostat.ec.europa.eu/nui/show.do?dataset=migr_imm5prv&lang=en (accessed 9 August 2021).

Fiorio, Lee, Guy Abel, Jixuan Cai, Emilio Zagheni, Ingmar Weber and Guillermo Vinué (2017), 'Using Twitter Data to Estimate the Relationship between Short-Term Mobility and Long-Term Migration', in WebSci '17: Proceedings of the 2017 ACM on Web Science Conference. https://doi.org/10.1145/3091478.3091496.

Fiorio, Lee, Emilio Zagheni, G. J. Abel, J. Hill, G. Pestre, E. Letouze and J. Cai (2020), 'Analyzing the Effect of Time in Migration Measurement Using Geo-Referenced Digital Trace Data', MPIDR Working Paper WP-2020-024.

Gendronneau, Cloe, Arkadiusz Wisniowski, Dilek Yildiz, Emilio Zagheni, Lee Fiorio, Yuan Hsiao, Martin Stepanek, Ingmar Weber, Guy Abel and Stijn Hoorens (2019), 'Measuring Labour Mobility and Migration Using Big Data: Development of a Method to Measure Mobility Flows and Stocks within the EU Based on Data from Social Media', European Commission Directorate – General for Employment, Social Affairs and Inclusion.

González, Marta C., César A. Hidalgo and Albert-László Barabási (2008), 'Understanding Individual Human Mobility Patterns', *Nature* **453** (7196): 779–82. https://doi.org/10.1038/nature06958.

Grundström, C. (1993), 'Report on Nordic Immigrants and Migration', Copenhagen: Nordic Statistical Secretariat.

Hughes, Christina, Emilio Zagheni, Guy J. Abel, Arkadiusz Wiśniowski, Alessandro Sorichetta, Ingmar Weber and Andrew J. Tatem (2016), 'Inferring Migrations: Traditional Methods and New Approaches Based on Mobile Phone, Social Media, and Other Big Data', European Commission project #VT/2014/093, February.

International Organization for Migration (2016), *More than Numbers: How Migration Data Can Deliver Real-Life Benefits for Migrants and Governments*, report International Organization for Migration (IOM) and McKinsey & Company.

International Organization for Migration (2019), *World Migration Report 2020*. IOM World Migration Report. International Organization for Migration. Available at https://www.un-ilibrary.org/content/books/9789290687894c002 (accessed 9 August 2021).

Juran, Sabrina, and Rachel C. Snow (2018), 'The Potential of Population and Housing Censuses for International Migrant Analysis', *Statistical Journal of the IAOS* **34** (2): 203–13. https://doi.org/10.3233/SJI-170359.

Kupiszewska, Dorota, and Beata Nowok (2008), 'Comparability of Statistics on International Migration Flows in the European Union', *International Migration in Europe: Data, Models and Estimates*, edited by F. Willekens and J. Raymer, pp. 41–73. London, England: Wiley.

Lai, S., E. Erbach-Schoenberg, C. Pezzulo, N. W. Ruktanonchai, Alessandro Sorichetta, J. Steele, T. Li, C. A. Dooley and A. J. Tatem (2019), 'Exploring the Use of Mobile Phone Data for National Migration Statistics', *Palgrave Communications* **5** (1): 34. https://doi.org/10.1057/s41599-019-0242-9.

Lemaitre, G (2005), 'The Comparability of International Migration Statistics Problems and Prospects', 9, Statistics Brief. OECD. Available at https://www.oecd.org/migration/49215740.pdf (accessed 9 August 2021).

Lomax, Nik, Paul Norman, Philip Rees and John Stillwell (2013), 'Subnational Migration in the United Kingdom: Producing a Consistent Time Series Using a Combination of Available Data and Estimates', *Journal of Population Research* **30** (3): 265–88. https://doi.org/10.1007/s12546-013-9115-z.

Mazzoli, Mattia, Boris Diechtiareff, Antònia Tugores, Willian Wives, Natalia Adler, Pere Colet and José J. Ramasco (2020), 'Migrant Mobility Flows Characterized with Digital Data', edited by Jordi Paniagua. *PLOS ONE* **15** (3): e0230264. https://doi.org/10.1371/journal.pone.0230264.

McAuliffe, M., A.M. Goossens and A. Sengupta (2017), 'Mobility, Migration and Transnational Connectivity', in *World Migration Report 2018*. Geneva: IOM.

Migration Globalisation and Poverty DRC (2007), *Global Migrant Origin Database, University of Sussex*.

OECD (2020), 'OECD International Migration Database and Labour Market Outcomes of Immigrants'. Available at https://www.oecd.org/els/mig/keystat.htm (accessed 9 August 2021).

Özden, Çağlar, Christopher R. Parsons, Maurice Schiff and Terrie L Walmsley (2011), 'Where on Earth Is Everybody? The Evolution of Global Bilateral Migration 1960–2000', *World Bank Economic Review* **25** (1): 12–56. https://doi.org/10.1093/wber/lhr024.

Pultar, Edward, and Martin Raubal (2009), 'A Case for Space: Physical and Virtual Location Requirements in the CouchSurfing Social Network', in LBSN '09: Proceedings of the 2009 International Workshop on Location Based Social Networks, 88. New York, NY: ACM Press. https://doi.org/10.1145/1629890.1629909.

Raymer, James, Arkadiusz Wiśniowski, Jonathan J. Forster, Peter W.F. Smith and Jakub Bijak (2013), 'Integrated Modeling of European Migration', *Journal of the American Statistical Association* **108** (503): 801–19. https://doi.org/10.1080/01621459.2013.789435.

Ruktanonchai, N. W., C. W. Ruktanonchai, J. R. Floyd and A. J. Tatem (2018), 'Using Google Location History Data to Quantify Fine-Scale Human Mobility', *International Journal of Health Geographics* **17** (1): 28. https://doi.org/10.1186/s12942-018-0150-z.

Ruktanonchai, Nick W., Patrick DeLeenheer, Andrew J. Tatem, Victor A. Alegana, T. Trevor Caughlin, Elisabeth zu Erbach-Schoenberg, Christopher Lourenço, Corrine W. Ruktanonchai and David L. Smith (2016), 'Identifying Malaria Transmission Foci for Elimination Using Human Mobility Data', edited by Katia Koelle. *PLOS Computational Biology* **12** (4): e1004846. https://doi.org/10.1371/journal.pcbi.1004846.

Sorichetta, Alessandro, Tom J. Bird, Nick W. Ruktanonchai, Elisabeth zu Erbach-Schoenberg, Carla Pezzulo, Natalia Tejedor, Ian C. Waldock, et al. (2016), 'Mapping Internal Connectivity through Human Migration in Malaria Endemic Countries', *Scientific Data* **3** (August): 160066. https://doi.org/10.1038/sdata.2016.66.

State, Bogdan, Ingmar Weber and Emilio Zagheni (2013), 'Studying Inter-National Mobility through IP Geolocation', in *Proceedings of the Sixth ACM International Conference on Web Search and Data Mining*, 265–274. WSDM '13. New York, NY: ACM. https://doi.org/10.1145/2433396.2433432.

Tomas, Patricia A. Santo, Lawrence H. Summers, Michael Clemens, and Commission on International Migration Data for Development Research and Policy (2009), 'Migrants Count: Five Steps toward Better Migration Data: Report of the Commission on International Migration Data for Development Research and Policy', Washington D.C., USA.

United Nations Department of Economic and Social Affairs Population Division (1998), 'Recommendations on Statistics of International Migration. Revision 1', *Statistical Papers Series M*. New York, NY: United Nations Department of Economic and Social Affairs/Population Division.

United Nations Department of Economic and Social Affairs Population Division (2004), 'Trends in Total Migrant Stock: The 2003 Revision (United Nations Database, POP/DB/MIG/Stock/Rev.2003)', New York, NY: United Nations Department of Economic and Social Affairs/Population Division.

United Nations Department of Economic and Social Affairs Population Division (2005), 'Trends in Total Migrant Stock: The 2005 Revision (United Nations Database, POP/DB/MIG/Stock/Rev.2005)', New York, NY: United Nations Department of Economic and Social Affairs/Population Division.

United Nations Department of Economic and Social Affairs Population Division (2006), 'World Population Prospects: The 2004 Revision. Volume III Analytical Report'. ST/ESA/SER.A/246. Vol. III. New York, NY: United Nations Department of Economic and Social Affairs/Population Division.

United Nations Department of Economic and Social Affairs Population Division (2009), 'Trends in International Migrant Stock: The 2008 Revision (United Nations Database, POP/DB/MIG/Stock/Rev.2008)', New York, NY: United Nations Department of Economic and Social Affairs/Population Division.

United Nations Department of Economic and Social Affairs Population Division (2012), 'Trends in International Migrant Stock: Migrants by Destination and Origin (United Nations Database, POP/DB/MIG/Stock/Rev.2012)', New York, NY: United Nations Department of Economic and Social Affairs/Population Division.

United Nations Department of Economic and Social Affairs Population Division (2013), 'Trends in International Migrant Stock: Migrants by Destination and Origin (United Nations Database, POP/DB/MIG/Stock/Rev.2013)', New York, NY: United Nations Department of Economic and Social Affairs/Population Division.

United Nations, Department of Economic and Social Affairs Population Division (2015), 'International Migration Flows to and from Selected Countries: The 2015 Revision', New York, NY. Available at https://www.un.org/en/development/desa/population/migration/data/empirical2/docs/migflows2015documentation.pdf (accessed 9 August 2021).

United Nations Department of Economic and Social Affairs Population Division (2015), 'Trends in International Migrant Stock: The 2015 Revision (United Nations Database, POP/DB/MIG/Stock/Rev.2015)', New York, NY: United Nations Department of Economic and Social Affairs/Population Division.

United Nations Department of Economic and Social Affairs Population Division (2019), 'International Migrant Stock 2019 (United Nations Database, POP/DB/MIG/Stock/Rev.2019)', New York, NY: United Nations Department of Economic and Social Affairs/Population Division.

Vezzoli, Simona, Maria Villares-Varela and Hein de Haas (2014), 'Uncovering International Migration Flow Data: Insights from the DEMIG Databases', *International Migration Institute (IMI), Oxford Department of International Development (QEH), University of Oxford Working Papers*, no. 88.

Wesolowski, A., N. Eagle, A. J. Tatem, D. L. Smith, A. M. Noor, R. W. Snow and C. O. Buckee (2012), 'Quantifying the Impact of Human Mobility on Malaria', *Science* 338 (6104): 267–70. https://doi.org/10.1126/science.1223467.

Wesolowski, A., T. Qureshi, M. F. Boni, P. R. Sundsøy, M. A. Johansson, Syed Basit Rasheed, Kenth Engø-Monsen and Caroline O. Buckee (2015), 'Impact of Human Mobility on the Emergence of Dengue Epidemics in Pakistan', *Proceedings of the National Academy of Sciences* 112 (38): 11887–92. https://doi.org/10.1073/pnas.1504964112.

Wiśniowski, Arkadiusz, Jakub Bijak, Solveig Christiansen, Jonathan J. Forster, Nico Keilman, James Raymer and Peter W.F. Smith (2013), 'Utilising Expert Opinion to Improve the Measurement of International Migration in Europe', *Journal of Official Statistics* 29 (4): 583–607. https://doi.org/10.2478/jos-2013-0041.

Zagheni, Emilio, Venkata Rama Kiran Garimella, Ingmar Weber and Bogdan State (2014), 'Inferring International and Internal Migration Patterns from Twitter Data', in *Proceedings of the 23rd International Conference on World Wide Web*, 439–44. WWW '14 Companion. New York, NY, USA: ACM. https://doi.org/10.1145/2567948.2576930.

Zagheni, Emilio, and Ingmar Weber (2012), 'You Are Where You E-Mail: Using e-Mail Data to Estimate International Migration Rates', in *Proceedings of the 4th Annual ACM Web Science Conference*, 348–51. ACM. https://doi.org/10.1145/2380718.2380764.

Zagheni, Emilio, Ingmar Weber, and Krishna Gummadi (2017), 'Leveraging Facebook's Advertising Platform to Monitor Stocks of Migrants', *Population and Development Review* 43 (4): 721–34. https://doi.org/10.1111/padr.12102.

4. The roles and limitations of data science in understanding international migration flows and human mobility

Marie McAuliffe and Adam Sawyer[1]

INTRODUCTION

The starting point when it comes to discussions and critical examinations of international migration has always been numbers. Right from the first "Laws of Migration"—as articulated by Ravenstein in 1885 during the upheaval of industrialization in England that resulted in (then) unprecedented internal migration—regulators, analysts and academics have sought to understand scale at the outset. Scale is important, both in absolute numbers as well as longer-term trends, given that the size of a migration-related phenomenon will, to a large degree, influence impacts, policy responses as well as the likelihood of indirect effects.

We know, for example, that there were almost 272 million international migrants living outside their country of birth in 2019 (equivalent to 3.5% of the global population), up from 153 million in 1990 (or 2.9%) (UNDESA, 2019), representing a gradual modest increase in international migration over the last three decades. The vast majority of people globally remain within their country of birth. On the other hand, we have witnessed an international travel increase over the same period, but at a much greater rate. For example, there were an estimated 435 million tourist arrivals in 1990, which has more than tripled to reach 1,322 million international tourist arrivals in 2017 (UNWTO, 2018).[2] Tourism has emerged as one of the largest industries in the world, accounting for one in eleven jobs worldwide and 7 per cent of world exports (UNWTO, 2016). The tourism sector has become increasingly important for many countries, especially small island states that have an abundance of natural beauty but few other resources (Skeldon, 2018). Massive changes in transportation and communications technology in recent decades has opened up and fueled movements globally as part of broader globalization processes that produce increasing interconnections across time and space (Triandafyllidou, 2018).

As part of these broader globalization processes, we have also seen momentous changes in ICT as increasingly powerful distance-shrinking technologies elevate individualization of information and publishing *en masse*. Mobile phone penetration has increased quickly, with some developing countries experiencing great leaps technologically, forgoing older forms of ICT, such as landlines and personal computers, to take up smartphones (McAuliffe, Goossens and Sengupta, 2017). As the volume of data generated globally emanating from the expansion of new forms of ICTs skyrockets, the pressure to quantify every aspect of people's day-to-day lives intensifies. More data were created in a two-year period than in the entire previous history of the human race (Marr, 2015). We have well and truly entered a data-driven age. We are also increasingly witnessing the move from evidence-based policy to data-driven policy (van Veenstra and Kotterink, 2017); however, in moving in this direction we argue that we are becoming more and more distant from the underlying meaning of data variables, thereby increasing the risk of ineffective data-driven policy that misunderstands and misinterprets social phenomena such as migration. We step through key issues of migration data highly relevant to policy processes, with particular reference to international migration flows.

The purpose of this chapter is to examine how "new data" and data scientists can contribute to quantifying migration flows globally as well as the limitations involved. First, a brief discussion on context and concepts is provided. The chapter then examines the realities and challenges of quantifying migration flows over time, including with new technology. The chapter then articulates the main issues and dilemmas being faced by analysts in the pursuit of new tech solutions to understanding migration flows. It also briefly outlines areas most in need of further enquiry and examination.

DATA CONTEXT AND CONCEPTS

On the broad topic of "data", discussions often proceed with crossed purposes. The definition of what constitutes data is not always agreed upon, and the conceptualization of data utility is frequently contested. The definition of "data", as defined in the *Cambridge English Dictionary*, is "information, especially facts or numbers, collected to be examined and considered and used to help decision-making, or information in an electronic form that can be stored and used by a computer" (CED, 2020). Some argue that the term "data" has evolved to the point that it has expanded beyond previous definitions that were based on statistics, quantitative (and qualitative) research, and computer science to the point that it is has become potentially problematic at times or even dangerous (Ramanathan, 2016).

A significant part of this evolution in the term "data" is due to the substantial changes in the very nature of data, and the emergence of new technologies that produce them. In recent years we have heard the head-spinning statistics about how much data are produced globally every year (see text box 4.1). There has also been the realization that, because of the massive amount of data that are being generated, data scientists are needed in order to make sense of it, with some arguing that "data" is increasingly becoming a buzzword designed to attract talented scientists from various disciplines to work for business (Liffreing, 2018; Press, 2013).

BOX 4.1   ESTIMATED DAILY DATA GENERATION GLOBALLY, 2019

- 500 million tweets sent
- 294 billion emails sent
- 4 petabytes of data created on Facebook
- 4 terabytes of data created from each connected car
- 65 billion messages sent on WhatsApp
- 5 billion searches made

By 2025, total daily data generated: 463 exabytes of data globally (equivalent of 212,765,957 DVDs per day).

Source: Desjardins, 2019.

New technology that has emerged in the last two decades has changed the way we talk about, as well as use, data, including to better understand migration. As in other fields, there are increasing concerns that the prerequisites of statistical knowledge, computing skills and substantive expertise (see Figure 4.1) are giving way to a much greater focus on ICT skills without depth of knowledge in statistics or the topic being analysed (Liffreing, 2018; Ramanathan, 2016).

This is particularly noticeable in migration. The application of so-called "new data science" in the study of migration, which is at its most fundamental, is a highly social phenomenon that is rooted in normative regulatory systems (Castles, 2010; McAuliffe and Goossens, 2018), often fails to take into account the most basic understanding of the substantive topic. Without relevant content knowledge of migration, errors can be made at one or more of the basic steps in data science methods, leading to misspecification and misinterpretation. As noted by Kalev Leetaru, "for a field populated by statisticians, it is extraordinary that somehow we have accepted the idea of analyzing data [of which] we have no understanding" (Leetaru, 2019).

In the context of a rapidly evolving global data landscape, the very notion of "migration data" has changed (see Table 4.1). What was once focused more on traditional statistical data collection, reporting and analysis has increasingly expanded to also encompass user-generated data relevant to migration (and mobility). Examples include user-generated as part of closed systems, such as digital border processing, but also other user-generated data collection not initially designed for mobility purposes, such as social media platforms.

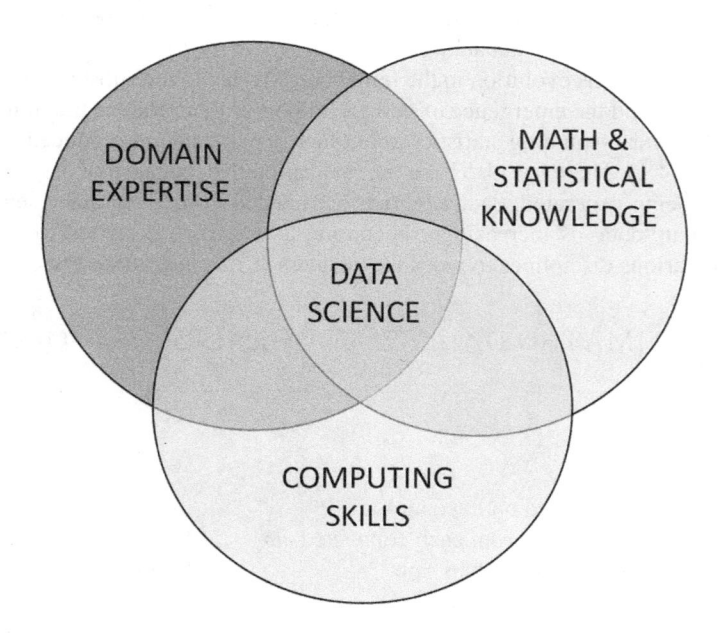

Source: Ramanathan, 2016.

Figure 4.1   The Data Science Venn Diagram

Table 4.1   *Migration data—the "old" and the "new"*

Term	Characteristics	Examples
Statistical data	• Frequently curated, managed and in a technical format • Presented with assumptions of understanding of quantitative vocabulary	• Population surveys and census • Administrative statistics (housing, employment, etc.) • Academic research surveys • Datasets released by agencies of the United Nations
User/event-generated data	• Little curation in the collection phase • Interaction of someone or something with a system, in an online or offline environment	• Smart Gate border processing • Biometrics capture • Online visa applications • Online hotel bookings • GPS tracking • Social media usage • ISP data

As shown in Table 4.1, the types of migration data that fall under the category of "statistics" are frequently curated and stored in databases with varied degrees of accessibility. By contrast, advancements in communication technology have allowed for exponential growth in user/event-generated data, resulting in massive databases with complex structures that update rapidly (see Davenport, 2014).[3] Notwithstanding the organizations and individuals who have worked to keep data freely available and transparent, much of what is constituted in user and event-generated data is of limited accessibility for commercial, proprietary or security reasons (Davies et al., 2019).[4]

Migration Context and Concepts

While it goes without saying that migration-related concepts are central to the formulation of accurate estimates and models produced by data scientists in the area of migration flows, the complexity of definitions is often glossed over, or not adequately investigated and articulated. Terms that often are based on legal definitions tend to be used interchangeably and without precision, causing confusion and eroding credibility of estimates. At worst, misleading estimates can be produced that can be fuel for policy considerations and/or promote inaccuracies in heavily contested public and political discourses. A recent example of this is Pew Research's 2019 estimates of irregular migrants in Europe (Connor and Passel, 2019), in which there was a disregard of basic normative settings, resulting in highly inaccurate estimates. In this instance the researchers included asylum seekers whose applications were being processed, despite the fact that this cohort had the right to remain while applications are being processed and so were not irregular migrants (Suro, 2019). The result was an inaccurate and inflated estimate. This is not an isolated case unfortunately, with recent estimates of irregular migrants globally also inflated on spurious grounds (see Box 4.2 below).

BOX 4.2   WHAT NOT TO DO: THE ESTIMATED GLOBAL POPULATION OF IRREGULAR MIGRANTS

In an August 2019 report on irregular migration, the authors come up with a global estimate of the number of irregular migrants that is based on a lack of understanding of migration and displacement policy, practice and normative settings (CSIS, 2019). In arriving at an erroneous figure of 106.9 million people, the authors included categories of people who were not irregular given their legal situations, such as internally displaced persons, stateless persons and Venezuelan migrants, including refugees and asylum-seekers. The important lessons in this example include:

- categories of migrants (even while overlapping at times) and limitations on definitions must be well understood before analysis commences;
- qualified and experienced analysts with an understanding of the topic must be engaged to lead such work;
- the advice and feedback of knowledge specialists in the field must be sought prior to publication (commonly referred to as "peer review").

There is also a tendency to conflate "migration" with "migrants", which are linked but conceptually distinct (IOM, 2019). In a general sense, migration is the process of moving from one place to another in order to reside there. To migrate internationally is to move from one country to a new country: It involves action. In contrast, a migrant is a person described as such for one or more reasons, depending on the context. While in many cases, migrants do undertake some form of migration, this is not always the case. In some situations, people who have never undertaken migration may be referred to as migrants—children of people born overseas, for example, are commonly called second- or third-generation migrants (Neto, 1995; Fertig and Schmidt, 2001). *Migration* flows, therefore, can be seen to be quite distinct from *migrant* stocks (or the number of migrants in specific populations). Please refer to Yildiz and Abel in this volume for discussion of stocks and flows.

Perhaps even more important than the distinction between "migration" and "migrant" are the differences between "migration" and "mobility". The main difference in a general sense is that migration involves a longer-term move (e.g., 12 months or more, as per the UN definition [UNDESA, 1998]) as it involves a change in usual residence whereas mobility involves moving regardless of length of time; international migration also comes with the extra requirement that it involves migration between two countries, rather than within a country. It can be seen, then, that migration is a subset of mobility, and while migration has always been difficult to define and capture, including because of administrative systems used to collect statistics but also because of people's agency (McAuliffe and Goossens, 2018; McAuliffe, 2017), the growing intersection with mobility has further highlighted conceptual issues, as argued by Skeldon (2018, 6):

> While migration is seen as a change in the usual place of residence of an individual, that is rarely a single, simple movement. People move on and back; they move over the short-term as well as for longer-term sojourns … one other form of mobility, mainly international but also internal, needs to be introduced into the discussion: the movement of tourists …. Yet, the emergence of the "gap year" and programmes for working holidaymakers has extended this category into a grey area that begins to overlap with other circular forms of mobility.

While Skeldon's analysis is at the international level, the COVID-19 pandemic has brought internal mobility into sharp focus. With unprecedented limitations on international and internal mobility arising from the pandemic, including quarantine and other severe restrictions on freedom of movement (McAuliffe, 2020), the (im)mobility of people is being monitored closely through new technologies (such as tracking and tracing apps). Data capture is occurring at international, national, local and even micro levels as isolation/quarantine measures are adopted and monitored.

A CLOSER LOOK AT MIGRATION FLOW STATISTICS

Even when considering the advancements made in data collection as described above, capturing international migration flow statistics has historically been very difficult (IOM, 2017). Exploration of international migration flows involves the analysis of migration "events", not the size and composition of migrant stocks or sub-populations. Nor do migration flow statistics involve all types of cross-border movements, such as that which relates to tourism, or cross-border workers who might live in one country but work in another.

While some primary sources of flow statistics are collected by a few States in the form of cross-border events—Australia, USA and Canada—very few countries capture *and* report cross-border data (Koser, 2010). Some countries collect entry data but not departures, and some countries employ "proxy" methods such as aggregating administrative data such as residency changes to infer migration flows (Koser, 2010; Skeldon, 2018). Furthermore, migratory movements are often very hard to separate from non-migratory travel, such as for tourism, business or cross-border work. Tracking migratory movements also requires considerable resources, infrastructure and IT/knowledge systems, posing sometimes insurmountable challenges for many countries.

Statistics related to a few specific migration corridors and certain types of flows of irregular migration (such as via the Mediterranean Sea) are among the subsets of international migration flow data that are available (McAuliffe and Mence, 2017); however, none is global in nature (IOM, 2017). Some academic literature also tracks specific migration flows, but access to meta-level data is limited. In general, collecting this kind of data is difficult for many countries and regions, as it requires significant technical resources and capacity, and can go against informal migration practices already in place (Gallagher and McAuliffe, 2016).

There are currently two main sources of international migration flow statistics at the global level. First, UNDESA publishes migration flow statistics by compiling the inflows and outflows reported by States. As of 2015, the latest report at the time of writing, only 45 countries reported these statistics (UNDESA, 2015). This compares with 232 entities (countries, territories and administrative units) for which statistics on the number of international migrants were available for the 2019 edition of UNDESA's *Revision of International Migrant Stocks* (UNDESA, 2019). It is also important to note that there has been a huge increase over time in the coverage of the migrant stock estimate, increasing from 135 entities in 1970 to 232 entities in 2019, whereas progress for migration flow statistics has been limited, from 29 entities in 2008 to 45 in 2019 (IOM, 2019). To overcome the dearth in migration flow statistics globally, some scholars have derived a global estimate of migration flows by conducting a change-analysis over the "snapshots" of total migrants in each State, as documented in international migration stock data (Abel and Cohen, 2018). However, since UNDESA's international migration stock data are themselves based on a carefully calibrated set of estimates based on demographic trends, global migration flow estimates based on stock data are built on inferred net movements, not aggregated events.

Second, the OECD's International Migration Database includes flow data with a particular focus on inflows into destination countries. OECD has updated this international migration dataset annually since 2000 by compiling the registries, surveys and estimates that are reported by States. These figures pass through a standardization protocol to be able to compare across varying migrant definitions and technical capacities for record-keeping from country to country (OECD, 2019). The sophistication in data compilation testifies to the difficulty of measuring the movement of people across borders at the international level of analysis.

With such limited migration flow statistics available at the global level, it is understandable that two phenomena have been observed: (1) the migration "hotspots" involving irregular migration flows have become a key focus of attention of States (especially destination countries); (2) the increasing opportunities posed by user-generated data in capturing and analysing movements appears extremely attractive to states, business and researchers. We now address these two aspects in turn.

Irregular Migration Flows

Capturing and analysing international migratory movements of people who are using regular visa and migration pathways is inherently difficult, as relatively few countries systematically collect and report such information, as highlighted above. The difficulties are pronounced when it comes to capturing data on irregular migration flows, although the consequences of not monitoring and analysing irregular flows in particular are perceived to be very significant for destination countries, many of whom have invested heavily in data collection, surveillance and border management as well as bilateral and multilateral cooperation on combatting irregular migration (Andersson, 2016; Koser and McAuliffe, 2013; Triandafyllidou and McAuliffe, 2018). In this sense, and despite the inherent challenge in data collection on clandestine and (at times) illegal processes, there has been significant investment in detecting and monitoring irregular migration flows, although there remains a distinct unevenness globally with some geographic areas heavily monitored and others neglected.

Many countries around the world have invested in border management capabilities and technologies allowing for the capture of data related to irregular migration flows. Some commentators question the utility of attempting to quantify irregular migration, citing the practical difficulties as well as the underlying rationale for collecting and citing such statistics, which can amount to alarmism (Castles, 2002; Clarke, 2000). There are, however, benefits in attempting to quantify irregular movements from the perspectives of national governments, regional and local governments, humanitarian service providers and others. A better understanding of the nature and extent of irregular migration flows facilitates the development of more effective responses and mitigation strategies, particularly for populations at risk of displacement or irregular migration (McAuliffe, 2017). Refugees and asylum seekers, for example, may make up a substantial proportion of people moving irregularly along particular migration corridors due to life-threatening events, such as conflict and violence. Irregular migration can exacerbate threat by placing migrants in dangerous and life-threatening situations during their migration journeys, as tragically demonstrated by the estimated number of migrant fatalities and disappearances globally recorded by IOM's Missing Migrants Project (IOM, 2020). The need to capture such statistics, including as a means to highlight and advocate for humanitarian responses, relates to the need to find more effective solutions underpinned by human rights while supporting migration governance.

The Allure of the Tech Solutions

With traditional statistical collection hamstrung by feasibility, technical capacity and resource constraints, the possibilities of new data sources and analysis appears particularly attractive to some. Given the explosion in user-generated data (as discussed earlier)—most especially that which is generated via mobile phones—businesses, governments and data scientists have sought to exploit the growing (virtual) mountain of mobile phone data in order to understand where, when and how often people move (see Salah in this volume). However, this more often than not relates to mobility—movements that do not necessarily involve migration to a new country of residence but do involve travelling locally (e.g., within a city), internally (i.e., within a country) or internationally (i.e., between countries). The collection of this granular "mobility" data is distinct in several key ways: it is collected/available in real time, it can produce data on movements involving short and long distances, it is collected at the individual level (and may or may not be confidentialized), it is collected predominantly by the business sector (although this differs significantly depending on the much broader political system of a country), and it is often connected to other non-mobility data (such as call, browse, purchase, social media post data). The genesis of this user-generated data is of course commercial, with tech companies such as Apple seeking to gain and expand market share globally by developing self-curated digital experiences via portable devices. Tracking individuals' mobility was likely not part of the initial design intention, but was a by-product that resulted from uptake on a mass scale. In a public policy context, these data are extremely valuable, informing a range of policy areas, such as urban planning, infrastructure, housing and services, and even internal migration, but there are serious limitations on international migration analysis, including because of the national limits of mobile service operators (see Salah in this volume for discussion). Such data are also potentially highly problematic from a human rights perspective (see Molnar in this volume).

In this context, geospatial information science has emerged as both a subdiscipline and a toolset of methods for data scientists interested in the study of human mobility. GIS focuses on the "where" in data. That is, GIS puts emphasis on the patterns of location, distance and concentrations using methods that identify patterns that can be visually rendered on maps. In the context of migration, GIS has helped the spatial dynamics of flows to become visualized and comprehensible. For example, many of the myriad visual stories developed during the outbreak of the COVID-19 pandemic utilized mapping or other figures to illustrate the spatial components of mobility and viral transmission (Wu et al., 2020; Freedman et al., 2020).

Data collection for GIS analysis of mobility can come from the same sources that make up a demographer's toolkit but must necessarily include some kind of geo-referenced attributes, such as longitude and latitude. Examples of GIS techniques to represent migration flows include micro-level studies on localized labor mobility or, conversely, migration flows can be analysed, visualized and interpreted at a continental or global scale. Additionally, GIS allows analysts to subset migration flow data over a specific region or a specific type of migration (see, e.g., Ballas and Clarke, 2000; Lai et al., 2014; IOM, 2020). Advanced modeling that includes GIS methods has begun attempting predictive forecasts of migration flows, but there remains significant uncertainty involved in these analyses (Sorichetta, 2016; Barbosa-Filho et al., 2018).

BOX 4.3   MACHINE LEARNING IN ANALYSIS OF MIGRATION

The term "machine learning" is used with great frequency among AI users and increasingly among immigration law enforcement officials. Yet its meaning can seem quite obscure to the audience whose data is applied most often: the general public. Centrally, the premise of machine learning rests on an interesting observation: that much of human thinking is based upon making instantaneous probability calculations, rather than applying abstracted theories (Meserole, 2018).

Machine learning can be summarized in two steps. First, an algorithm identifies a pattern. Second, the algorithm applies the discovered pattern over more data. Usually the computer "learns" by utilizing a set of statistical probabilities to pull out patterns in step one before applying these patterns to produce extensive amounts of data in step two (Hao, 2018).

Researchers have already started applying machine learning techniques to identify trends in mobility. For officials in immigration law enforcement, pattern-testing algorithms are already in use as part of facial-recognition technology at many customs and border checkpoints and is part of a broader movement toward the automated collection of people's data during migration events (Matyus, 2020).

ISSUES AND DILEMMAS

There has been little progress over recent years in both the collection of international migration flow data, such as reported by OECD and UNDESA, and approaches to modeling international migration flows, which are still largely based on migrant stock data. There has, however, been significant expansion in the collection of user-generated data that are increasingly analysed from a mobility perspective. These technological advances raise a number of issues and dilemmas for data scientists as well as the recipients of mobility analysis (such as States, businesses, humanitarian agencies) and the subjects of the analysis (be they commuters, tourists, migrants, asylum seekers, refugees, cross-border workers, etc.).

Source: Mason and Wiggins, 2010.

Figure 4.2   *The Data Science Process (OSEMI Framework)*

This section will explore the implications of these advancements by beginning at what is typically the last step of the data science workflow (see Figure 4.2)—the interpretation of analytical findings—before proceeding "backwards" by outlining some of the issues in the analysis stages and then concluding with concerns about the collection of mobility data itself.

Interpreting Data Outputs in an Age of Misinformation

Many analyses, including some in migration studies, are guided by algorithmic programs that are poorly understood, even by researchers. In the worst cases, improper representations of data and statistics lead to confusion or the perpetuation of misinformation. The outputs from faulty analysis of surveilled or surveyed migration data can have direct impact on major life events, such as migration adjudication decisions. For example, mapped visualizations of migration flows can be potentially misleading without recognizing certain guidelines, such as visually rendering the margin-of-error (Lunetta et al., 1991). Across all methods in data science, making an effort to communicate the level of certainty is an essential task to prevent misinterpretation.

For researchers applying data science techniques, a lack of familiarity with either the conceptual literature on migration or the underlying processes of data science can lead to studies of migration flows that can confuse the public more than clarify. If a figure or statistic lacks clarity, the audience can be quite creative at coming up with ways to misinterpret or disinterpret the meaning of the analysis. When discussing the situation of migrants, especially those in vulnerable situations, unclear charts or figures can lead to media representations that are passively or actively racist, xenophobic, or otherwise biased. The concept of "motivated reasoning"—that individuals engage in new ideas with an intent to confirm deeper held beliefs—extends to how people think about migration and migrants (Banulescu-Bogdan, 2018).

Scrubbing, Exploring and Modeling: Mobility and Migration Flows

The question of how to analyse mobility data in a transparent, publicly anonymous and understandable format remains an open one. Legislation that authorizes automated immigration procedures frequently invokes national security and counterterrorism as the fundamental reasons for employing algorithms that sort individuals into categories based on similar attributes (see textbox on Machine Learning above). This kind of data grouping is ripe for discrimination. Numerous studies have shown that algorithms can have human biases baked into the hardwiring of AI processes to such a degree that it is "mathematically inevitable" for human bias to infiltrate algorithms, at least in the case of calculating an individual's criminal risk (Angwin and Larson, 2016).

The importance of adhering to sound data science ethics is even higher in the field of geospatial intelligence, where GIS capabilities are applied to issues of national security, counterterrorism, and sometimes, immigration enforcement. States have long recognized the analytical power of GIS, but with the integration of machine learning and GIS, monitoring individual decisions has become both automated and mappable. In many cases, immigration law enforcement agencies employ various GIS technologies and data collection systems to monitor mobility patterns (Słomczyńska and Frankowski, 2016; Martin, 2018).

BOX 4.4   COVID-19 MOBILITY DATA

The unprecedented global pandemic caused a "scramble" for data by epidemiologists and public health officials needing to understand coronavirus transmission patterns and determinants. Global tech firms, such as Google and Facebook, began releasing COVID-19 "mobility" data based on users' movements early in the pandemic with the stated aim of helping public policy responses. However, and as recognized/stated by both Google and Facebook, user data that are based on "opt in" functions is of limited public value as it cannot reflect underlying populations (nor does it appear to have been re-weighted to help overcome this issue). Further, the limited information on methods provided by the tech firms highlights that there is a lack of clarity/understanding regarding the data collected and reported. For example, discussion of variables reported does not reflect basic underlying terms related to measures implemented by governments around the world. Further to the highly concerning basic technical aspects, privacy and human rights concerns have been raised by many commentators (see McAuliffe and Blower in this volume).

Even prior to the exploratory or modeling steps, mistakes in the scrubbing of data—the process of modifying and removing incomplete entries—can set up an erroneous model before the analysis even begins. If entries undergo unwarranted manipulation, or when an algorithm is even slightly mis-specified, biased or prejudiced against a particular group in a systematic way, it operates as "an instrument of power". If the criteria of a visa-admission rest upon an automated computer procedure that takes place on a server hundreds of miles away from a formal migration adjudication procedure, how does a migrant provide evidence of discrimination? Questions of due process—a legal right expressed in constitutional frameworks around the world—beg to be asked (Broeders and Dijstelbloem, 2016). So called "smart borders" are not necessarily just borders (see Molnar in this volume for discussion).

Obtaining Mobility Data: Privacy and Ownership

With the rise of automated surveillance in data science, advocates increasingly push for a change to the current regime governing the rights to data. Such advocates cite the profits garnered by multinational corporations based on the individual characteristics, attributes and behaviors that are tied to an individual user's identity. A proposed revision would include a detailed provision of information for each individual about how their data is being collected, what is being collected, and how that collection and use can be modified or stopped (Chen and Zhao, 2012).[5] In data science, the storage and accessibility of data can be just as important as the analysis. Monitoring who can access what aspects of a person's data under what circumstances can have real consequences for people, including international migrants.

Under the current scheme, users of social media platforms, search engines and other applications give permission for a company to use their data in exchange for the use of a product or for a more convenient interface system. A noted problem with this system is the information asymmetry inherent in many technological applications today: users have far less conception of the algorithms or electronic mechanisms underlying everyday internet tools while companies and government agencies have in-depth knowledge of the data that is collected (Broeders and Dijstelbloem, 2016; Beduschi, 2020).

Issues of data privacy are especially relevant for people engaging in international migration, who not only have an interest in the commercial collections of data, but must also interface with some apparatus of the State to first gain admission to the State and then to obtain necessities such as health care, education and other government services (Haan and McDonald, 2018). Recently, there has been evidence that only the most porous data "firewalls" isolate personal data registered with service agencies apart from law enforcement in the United States. A batch search query by field office supervisors can result in access to a documentation status database that is meant to be restricted from border management officials (Lind, 2020).

To monitor one's personal data "hygiene", the international migrant must become familiar with the data policy context of the destination country, which might differ significantly from that of the origin country. Location data or purchase data that may be commercial in one country may be accessible by law enforcement or government officials in another context. Finally, if data privacy regulations are violated, or even if a government algorithm has a biased or prejudicial output that is unfavorable to a migrant, there is little opportunity to seek compensation or remedy (Greenfield, 2020).

As humanitarian organizations have begun integrating data science into policy and decision-making, some have called for additional safeguards in these activities, as "humanitarian information activities" carry inherent risks from both technical and human standpoints (Greenwood et al., 2017). Vannini, Gomez and Newell argue that it is unreasonable to expect vulnerable populations, including migrants, to consistently engage in effective personal data management practices across so many systems, especially as the guidelines for humanitarian organizations in handling personal data of vulnerable populations remains informal instead of codified in specific international law. Recommendations include limiting the collection of personal information to only what is necessary and monitoring security and data protection mechanisms from technological and human risks (Vannini, Gomez and Newell, 2019). The issue is potentially even more pressing for those in the most vulnerable situations, such as victims of human trafficking, internally displaced persons and refugees, who have a very limited ability to avoid or limit data collection of personal information when receiving assistance and integrating into new communities (see Korkmaz in this volume and Bauloz in this volume).

CONCLUSION

Notwithstanding new technologies in recent years resulting in massive increases in data globally, the ongoing challenge to quantify international migration flows globally remains stubbornly elusive. While our understanding of migration flow patterns has not been greatly enhanced by recent technological advances, the explosion in user-generated data from mobile phones provides enormous potential for better understanding mobility patterns across a range of geographies. This technology may not have been originally intended as a means of analysing mobility, however, it has become clear over time and especially in the COVID-19 context that authorities (national, sub-national and local) are increasingly drawing upon these data to inform policy and programmatic responses.

Given the nature of user-generated data and some of the privacy issues it raises, collecting and analysing such data for mobility purposes raises a number of issues from a human rights perspective. Some countries are utilizing high-tech monitoring to engage in large-scale population surveillance in addition to, or instead of, monitoring mobility for public health and other purposes. In the United States, for example, immigration authorities have been using facial recognition technology to detect irregular migrants in traffic (Matyus, 2020). The coronavirus pandemic has also brought to light, through the track and tracing measures used, many of the countries with sophisticated surveillance systems in place that use mobile phone and ISP data to monitor people within their territory (Sonn, 2020).

There remain key challenges for data science and data scientists in analysing migration flows and human mobility. Ensuring that legal-policy definitions and normative frameworks underpinning concepts related to migration and mobility are not set aside is critical. Understanding what is being measured, and accurately reflecting these constructs, is particularly important at a time in which those very concepts are being refined and reshaped by events such as COVID-19. For data scientists and other researchers whose work focuses on sensitive issues such as irregular migration, smuggling, trafficking and displacement, the additional need to safeguard the privacy of those being studied as well as taking into account the ethical implications of the research is particularly important. Data scientists may be somewhat removed from frontline settings, but their research and analysis can have a profound impact on the lives of many. The datafication of policy further de-emphasizes the complex human and social aspects of migration and mobility at a time in which much greater emphasis on safeguarding fundamental human rights in migration and mobility is required.

NOTES

1. The opinions, comments and analyses expressed in this chapter are those of the authors and do not necessarily represent the views of the organizations or institutions with which the authors are affiliated.
2. These are not measures of individual tourists but of arrivals who spent at least a night in the destination country.
3. Some have labeled this revolution with the name "Big Data". Others have eschewed monikers and commented that these are technological progressions in data analytics or data science.
4. See the Open Data Impact Map from OpenData4Development at https://www.opendataimpactmap.org/ (accessed on 25 May 2020).
5. The ID2020 Alliance—a partnership of private companies, foundations and civil-society organizations working toward an ethical approach to digital identification—collaborated with UNHCR to produce a manifesto that sets forth a set of 10 principles to guide their mission. Point five states, "We believe that individuals must have control over their own digital identities, including how personal data is collected, used and shared."

REFERENCES

Abel, G. and Cohen, J.E. (2018) Bilateral international migration flow estimates for 200 countries. *Scientific Data*, 6(1): 1–13.

Andersson, R. (2016) Europe's failed "fight" against irregular migration: ethnographic notes on a counterproductive industry. *Journal of Ethnic and Migration Studies*, 42(7): 1055–1075.

Angwin J., and Larson, J. (2016) Bias in criminal risk scores is mathematically inevitable, researchers say. *ProPublica*, 30 December. Available at https://www.propublica.org/article/bias-in-criminal-risk-scores-is-mathematically-inevitable-researchers-say (accessed on 25 May 2020).

Ballas, D. and Clarke, G. (2000) GIS and microsimulation for local labour market analysis. *Computers, Environment and Urban Systems*, 24(4): 305–330. Available at https://doi.org/10.1016/S0198-9715(99)00051-4.

Banulescu-Bogdan, N. (2018) *When Facts Don't Matter: How to Communicate More Effectively about Immigration's Costs and Benefits*. Washington, D.C.: Migration Policy Institute.

Beduschi, A. (2020) International migration management in the age of artificial intelligence. *Migration Studies*. Available at https://doi.org/130.1093/migration/mnaa003.

Broeders, D. and Dijstelbloem, H. (2016) The datafication of mobility and migration management: the mediating state and its consequences, in *Digitizing Identities: Doing Identity in a Networked World*, I. Van der Ploeg and J. Pridmore (eds), London: Routledge, pp. 242–260.

Cambridge University (2020) *Cambridge English Dictionary*, Cambridge University Press. Available at https://dictionary.cambridge.org/dictionary/english/data (accessed on 25 May 2020).

Castles, S. (2002) Migration and community formation under conditions of globalization. *International Migration Review*, 36: 1143–1168.

Castles, S. (2010) Understanding global migration: a social transformation perspective. *Journal of Ethnic and Migration Studies*, 36(19): 1565–1586.

Center for Strategic and International Studies (CSIS) (2019) *Out of the Shadows: Shining a Light on Irregular Migration*, Washington, D.C. and Lanham, MT: CSIS and Rowman & Littlefield. Available at https://csis-prod.s3.amazonaws.com/s3fs-public/publication/190826_RundeYaybokeGallego_IrregularMigrations.pdf (accessed on 25 May 2020).

Chen, D. and Zhao, H. (2012) Data security and privacy protection issues in cloud computing, ICCSEE, March, 1: 647–651. doi: 10.1109/ICCSEE.2012.193.

Clarke, J. (2000) The problems of evaluating numbers of illegal migrants in the European Union, in *Regularisations of Illegal Immigrants in the European Union*, P. de Bruycker (ed.). Brussels: Bruylant, pp. 13–23.

Connor, P. and Passel, J.S. (2019) *Europe's Unauthorized Immigrant Population Peaks in 2016, Then Levels Off*, Pew Research, Washington D.C.

Davenport, T. (2014) Stop using the term "big data". *Deloitte Insights*, 6 November. Available at https://www2.deloitte.com/us/en/insights/topics/analytics/big-data-buzzword.html (accessed on 25 May 2020).

Davies, T., Walker, S.B., Rubinstein, M. and Perini, F. (2019) *State of Open Data*, Cape Town, African Minds. Available at http://www.africanminds.co.za/state-of-open-data/ (accessed on 25 May 2020).

Desjardins, J. (2019) *How Much Data is Generated Each Day?* eNewsWithoutBorders, 23 April. Available at https://enewswithoutborders.com/2019/04/23/how-much-data-is-generated-each-day/ (accessed on 25 May 2020).

Fertig, M. and Schmidt, C. (2001) First- and second-generation migrants in Germany: What do we know and what do people think? IZA Discussion Papers, 286: 1–48.

Freedman, A., Muyskens, J., Alcantara, C. and Ulmanu, M. (2020) How coronavirus grounded the airline industry. *The Washington Post*, 1 April. Available at https://www.washingtonpost.com/graphics/2020/business/coronavirus-airline-industry-collapse/ (accessed on 25 May 2020).

Gallagher, A. and McAuliffe, M. (2016) South-East Asia and Australia, in *Migrant Smuggling Data and Research: A Global Review of the Emerging Evidence Base*, McAuliffe, M. and Laczko, F. (eds), Geneva: IOM, pp. 211–241.

Greenfield, C. (2020) As Governments Build Advanced Surveillance Systems to Push Borders Out, Will Travel and Migration Become Unequal for Some Groups? Migration Policy Institute. Available at https://www.migrationpolicy.org/article/governments-build-advanced-surveillance-systems (accessed on 25 May 2020).

Greenwood, F., Howarth, C., Escudero Poole, D., Raymond, N.A. and Scarnecchia, D.P. (2017) The Signal Code: A human rights approach to information during crisis. Available at https://hhi.harvard.edu/sites/default/files/publications/signalcode_final.pdf (accessed on 25 May 2020).

Haan, M. and McDonald, J.T. (2018) Migration and immigration: recent advances using linked administrative data. *Journal of Popular Research*, 35: 319–324. Available at https://doi.org/10.1007/s12546-018-9214-y (accessed on 25 May 2020).

ID2020 Alliance (2018) The Alliance Manifesto. Available at https://id2020.org/manifesto.

International Organization for Migration (IOM) (2017) *World Migration Report 2018*, M. McAuliffe and M. Ruhs (eds). IOM, Geneva.

IOM (2019) *World Migration Report 2020*, McAuliffe, M. and Khadria, B. (eds). IOM, Geneva.

IOM (2020) Migration Data Portal, accessed on 23 April 2020. Available at https://migrationdataportal .org/ ?i=stock_abs_&t=2019 (accessed on 25 May 2020).

Koser, K. (2010) Dimensions and dynamics of irregular migration. *Population, Space and Place*, 16(3): 181–193.

Koser, K and McAuliffe, M. (2013) *Establishing an Evidence-Base for Future Policy Development on Irregular Migration to Australia*, Irregular Migration Research Programme Occasional paper series. Department of Immigration and Border Protection, Canberra.

Lai, S., zu Erbach-Schoenberg, E., Pezzulo, C., Ruktanonchai, N.W., Sorichetta, A., Steele, J., Li, T., Doole, C.A. and Tatum, A.J. (2014) Exploring the use of mobile phone data for national migration statistics. *Palgrave Communications*, 5(34): 1–10. Available at https://doi.org/10.1057/s41599-019 -0242-9.

Leetaru, K. (2019) How data scientists turned against statistics. *Forbes*, 7 March. Available at https://www.forbes.com/sites/kalevleetaru/2019/03/07/how-data-scientists-turned-against-statistics/ #471afdd257c8 (accessed on 25 May 2020).

Liffreing, I. (2018) Confessions of a data scientist: "Marketers don't know what they're asking for", *Digiday*, 19 November 2018. Available at https://digiday.com/marketing/confessions-data-scientist -marketers-dont-know-theyre-asking/ (accessed on 25 May 2020).

Lind, D. (2020) ICE has access to DACA recipients' personal information despite promises suggest- ing otherwise, internal emails show, 21 April. *Propublica*. Available at https://www.propublica .org/article/ice-has-access-to-daca-recipients-personal-information-despite-promises-suggesting -otherwise-internal-emails-show (accessed on 25 May 2020).

Lunetta, R.S., Congalton, R.G., Fenstermaker, L.K., Jensen, J.R., McGwire, K.C. and Tinney, L.R. (1991) Remote sensing and geographic information system data integration: error sources and research issues. *Photogrammetric Engineering & Remote Sensing*, 57(6): 677–687.

Marr, B. (2015) Big data: 20 mind-boggling facts everyone must read, *Forbes*, 30 September. Available at https://www.forbes.com/sites/bernardmarr/2015/09/30/big-data-20-mind-boggling-facts-everyone -must-read/#6719e28917b1 (accessed on 25 May 2020).

Martin, H. (2018) The effects of geospatial–intelligence on United States–Mexico border security. PhD Dissertation. University of Southern Mississippi, Hattiesburg.

Mason, H. and Wiggins, C. 2010. A taxonomy of data science. Dataists.com. Available at http://www .dataists.com/2010/09/a-taxonomy-of-data-science/ (accessed on 25 May 2020).

Matyus, A. (2020) ICE weaponizes state-issued licenses against Maryland's undocumented immigrants, 27 February, *Digital Trends*. Available at https://www.digitaltrends.com/news/ice-weaponizes-state -licences-against-undocumented-immigrants/ (accessed on 25 May 2020).

McAuliffe, M. (2020) *Immobility as the Ultimate "Migration Disrupter": An Initial Analysis of COVID-19 Impacts through the Prism of Securitization*. Migration Research Series, No 63, IOM, Geneva.

McAuliffe, M. (2017) *Self-agency and Asylum: A Critical Analysis of the Migration Patterns and Processes of Hazara Irregular Maritime Asylum Seekers to Australia*, Australian National University PhD thesis. Available at https://openresearch-repository.anu.edu.au/handle/1885/164239 (accessed on 25 May 2020).

McAuliffe, M. and Goossens, A.M. (2018) Regulating international migration in an era of increas- ing interconnectedness, in *Handbook of Migration and Globalisation*, A. Triandafyllidou (ed.). Cheltenham, UK and Northampton, MA, USA: Edward Elgar Publishing, pp. 86–104.

McAuliffe, M., Goossens A.M. and Sengupta, A. (2017) Mobility, migration and transnational connec- tivity, in *World Migration Report 2018*, M. McAuliffe and M. Ruhs (eds). Geneva: IOM, pp. 148–169.

McAuliffe, M. and Mence, V. (2017) Irregular maritime migration as a global phenomenon, in *A Long Way to Go: Irregular Migration Patterns, Processes, Drivers and Decision Making*, McAuliffe, M. and Koser, K. (eds). Canberra: ANU Press, pp. 11–47.

Meserole, C. (2018) What is machine learning? *A Blueprint for the Future of AI*. Brookings Institution, 4 October. Available at https://www.brookings.edu/research/what-is-machine-learning/ (accessed on 25 May 2020).

Neto, F. (1995) Predictors of satisfaction with life among second generation migrants. *Social Indicators Research*, 35(1): 93–116.

OpenData4Development (2020) *Open Data Impact Map*. Available at https://opendataimpactmap.org/ (accessed on 25 May 2020).

Organisation for Economic Co-operation and Development (OECD) (2019) *International Migration Outlook 2019*. Forty-third edition, OECD Publishing, Paris. Available at https://www.oecd-ilibrary .org/social-issues-migration-health/international-migration-outlook-2019_c3e35eec-en (accessed on 25 May 2020).

Press, G. (2013) Data science: what's the half-life of a buzzword. *Forbes*, 19 August. Available at http://www.forbes.com/sites/gilpress/2013/08/19/data-science-whats-the-half-life-of-a-buzzword/ (accessed on 25 May 2020).

Ramanathan, A. (2016) The Data Science Delusion, *Medium*, 18 November. Available at https://medium .com/@anandr42/the-data-science-delusion-7759f4eaac8e (accessed on 25 May 2020).

Skeldon, R. (2018) *International Migration, Internal Migration, Mobility and Urbanization: Towards More Integrated Approaches*. Migration Research Series, No. 53. IOM, Geneva. Available at https:// publications.iom.int/books/mrs-no-53-international-migration-internal-migration-mobility (accessed on 25 May 2020).

Słomczyńska, I. and Frankowski, P. (2016) Patrolling power Europe: the role of satellite observation in EU border management. In *EU Borders and Shifting Internal Security*. Springer, Cham, pp. 65–80.

Sonn, J.W. (2020) Coronavirus: South Korea's success in controlling disease is due to its acceptance of surveillance, *The Conversation*, 19 March. Available at https://theconversation.com/coronavirus -south-koreas-success-in-controlling-disease-is-due-to-its-acceptance-of-surveillance-134068 (accessed on 25 May 2020).

Suro, R. (2019) The Trouble with Pew's estimates of the "unauthorized" migrant population in Europe, *Medium*, 14 November. Available at https://medium.com/@suro_26975/the-trouble-with -pews-estimates-of-the-unauthorized-migrant-population-in-europe-8df6be3972e9 (accessed on 25 May 2020).

Triandafyllidou, A. (ed.) (2018) *Handbook of Migration and Globalisation*. Cheltenham, UK and Northampton, MA, USA: Edward Elgar Publishing.

Triandafyllidou, A. and M. McAuliffe (eds) (2018) *Migrant Smuggling Data and Research: A Global Review of the Emerging Evidence Base (volume 2)*. IOM, Geneva. Available at https://publications .iom.int/system/files/pdf/migrant_smuggling_data_vol2_0.pdf (accessed on 25 May 2020).

UNDESA (2019) International Migrant Stock 2019. United Nations, New York. Available at https:// www.un.org/en/development/desa/population/migration/data/estimates2/estimates19.asp (accessed on 25 May 2020).

UNDESA (1998) *Recommendations on Statistics of International Migration: Revision 1*. United Nations, New York.

UNDESA (2015) International migration flows to and from selected countries. United Nations, New York. Available at https://www.un.org/en/development/desa/population/migration/data/empirical2/ migrationflows.asp (accessed on 25 May 2020).

UNWTO (2016) UNWTO Tourism Highlights, 2016 edition. Available at https://www.e-unwto.org/doi/ pdf/10.18111/9789284418145 (accessed on 25 May 2020).

UNWTO (2018) *2017 International Tourism Results: the highest in seven years*. Press release, UNWTO, Madrid.

van Veenstra, A.F. and Kotterink, B. (2017) Data-driven policy making: the policy lab approach. In Electronic Participation. ePart 2017. *Lecture Notes in Computer Science*, vol. 10429 (Parycek P. et al., eds). Springer, Cham. Available at https://link.springer.com/chapter/10.1007/978-3-319-64322 -9_9 (accessed on 25 May 2020).

Vannini S., Gomez, R. and Newell, B.C. (2019) "Mind the five": guidelines for data privacy and security in humanitarian work with undocumented migrants and other vulnerable populations. *Journal of the Association for Information Science and Technology*. Available at doi: 10.1002/asi.24317.

Wu, J., Cai, W., Watkins D. and Glanz, J. (2020) How the Virus Got Out. *The New York Times*, 22 March. Available at https://www.nytimes.com/interactive/2020/03/22/world/coronavirus-spread.html (accessed on 25 May 2020).

5. The practice and politics of migration data visualization

William Allen

INTRODUCTION

Recent computational advances have provided social scientists with new tools and techniques. One of those tools is data visualization, comprising a wide range of visual outputs that aim to enhance both understanding and communication of key patterns and insights emerging from primarily, though not exclusively, quantitative sources of information. Although the practice of visualization may not be new, the forms it now takes in the age of Big Data, unprecedented computational capabilities, and technologies that bring visualization within reach to greater numbers of people have certainly changed.

Some argue that we visualize data to save time in making sense of huge volumes of varied information (Chen, Floridi, and Borgo 2014). In some ways, this would be correct: visualization's efficiency over conventional tabular presentations of data is undeniable. But such a simplistic and exclusive view potentially obscures deeper questions not only about *why* data visualization is perceived to be so powerful, but also *how, for what purposes*, and *in which circumstances* this power is expressed.

My central argument in this chapter is that the visualization of data—particularly in the domain of migration and mobility—is inextricably linked with questions of data politics. By bringing together examples from my own work as well as collaborative projects with colleagues, I aim to illustrate the value of, and need for further discussion about, the practice and politics of visualization. I make my case in three parts. First, I outline what I mean by data visualization, and introduce a typology of visualization from the world of design practice that illustrates how there is no single "right" way to visualize, but rather a range of options that emerge from decisions about the audience and purpose of a visualization. Second, I present some exemplars to illustrate contemporary approaches to visualization involving migration data, linking these outputs to the broader typology introduced earlier. Third, I consider how these visualizations both represent and generate new kinds of politics—what some call "data politics" (Bigo, Isin, and Ruppert 2019)—that are worth investigating further.

To be clear, this is not a methodological chapter that covers how to make visualizations. For readers interested in this aspect of data visualization, there are excellent resources from social science (Healy 2018) and design perspectives (Kirk 2019) that offer guidance in a step-by-step fashion, as well as overviews of available software and techniques with discussion of their advantages and limitations (Gatto 2015; Healy and Moody 2014; Kennedy and Allen 2017). Instead, my goal is to illustrate how data visualization as a set of practices, actors, and technologies raises political questions with implications for how diverse groups make sense of migration and mobility.

WHAT IS DATA VISUALIZATION?

From Historical Roots to Contemporary Uses

Visuals have been central to science communication throughout history (Bucchi and Saracino 2016). In this chapter, I am focusing on one particular type of visual object: the data visualization. Defined by Andy Kirk, a leading visualization designer, as "the visual representation and presentation of data to facilitate understanding" (Kirk 2019, 15), data visualization is often associated with seemingly straightforward—if numerous—design choices such as selecting colors, arranging shapes, and distinguishing among typefaces. Of course, these are important aspects of visualization practice that undoubtedly contribute to the ways and likelihood that people engage with a chart or graph (Kennedy, Hill, Allen, et al. 2016). But visualization is more than just about producing attractive and well-designed visual outputs. Rather, it involves multiple processes and contributors that exist within particular social, cultural, technological, and political contexts. This is even more the case for *data* visualizations that—unlike infographics, flowcharts, or other more illustrative outputs—have some basis in forms of quantitative information that are themselves generated in particular circumstances.

Edward Tufte (1983, 2006) is probably one of the best-known authors on visualization. His work and legacy have heavily contributed to received wisdom about what makes "effective" data visualizations, much of which still remains relevant and continues to feature in current guidelines (Kelleher and Wagener 2011). Two characteristics in particular, clarity and simplicity, stand out as being especially important for visualization: the data and their associated meanings should be given the greatest priority, not so-called "chartjunk" that potentially distracts and confuses readers (Wainer 1984). Although this is probably still good advice in general, it does suggest that what is clear or simple for one type of reader or context may not be suitable or appropriate in other settings. For example, the editorial team at the scientific journal *Nature* recognizes that both specialists and non-specialists may engage with its visual outputs, and therefore its visualization practice must align with an article's intended audience (Krause 2017). What is more, new technologies—including digital platforms and data collection methods—add greater complexity to the construction and communication of visualization. This is a point I will return to in the second part of the chapter.

As a result, current scholarship increasingly acknowledges how data visualizations, just like the various forms of data on which they are based (boyd and Crawford 2012), are "multifaceted and 'multitruthed'" (Welles and Meirelles 2015, 37). They are not neutral windows onto self-evident patterns within data. Rather, through the actions of anyone handling data as well as the affordances of technologies themselves, visualizations can transform, frame, or shape understanding of the concepts or populations represented by and in datasets (Hullman and Diakopoulos 2011). Intentionally or not, these transformations and framings can prioritize certain values and ideologies over others not only through explicit features such as titles but also through unspoken features such as design conventions (Kennedy, Hill, Aiello, et al. 2016) and software settings that may determine what data are sampled or included in the visualization in the first place (Kitchin and Lauriault 2015).

Distinguishing among Visualization Types and Purposes: Explaining versus Exploring, Reading versus Feeling

If data visualization has such potential power, how can we make sense of for whom and in which circumstances this might happen? Kirk (2019) provides a helpful analytical tool for understanding the variety of options that designers can use when visualizing data, and what these imply for intended audiences and communication purposes. Summarized in Figure 5.1, he argues visualizations have two main dimensions: on the one hand, outputs that aim to either *explain* key patterns in the data or enable readers to *explore* the data following their own curiosities or interests; and on the other hand, outputs that either facilitate *reading* data for particular values or *feeling* overall patterns in an impressionistic way.[1] Of course, visualizations might occupy several places along these axes simultaneously, especially if they are interactive objects with multiple layers of visualizations. Yet this simplified grid is a useful way of generating some archetypical visualizations to draw out their implications for audiences and usages.

For example, some charts might make specific claims from a dataset by highlighting individual values distinguished by time period, population type, or geographic region. These kinds of visualizations, demonstrating characteristics that are closer to the "explain" and "read" categories, might include official censuses or government statistical offices. This is in contrast to other visualizations that may use strong visual metaphors while allowing viewers to independently dig into the chart and its data—approaches that would be closer to the "explore" and "feel" categories.

From a design perspective, where visualizations lie on these axes depends on their purposes and audiences. Explanatory visualizations would suit situations where a clear message is involved, whereas exploratory visualizations would be better in settings where the aim is to demonstrate multiple possible conclusions rather than advancing particular claims, especially on partisan issues. Meanwhile, the ability to read data from visualizations may be useful for subject specialists or in situations where the message involves comparing discrete values. However, when audiences might benefit from having an overall impression of a dataset's key

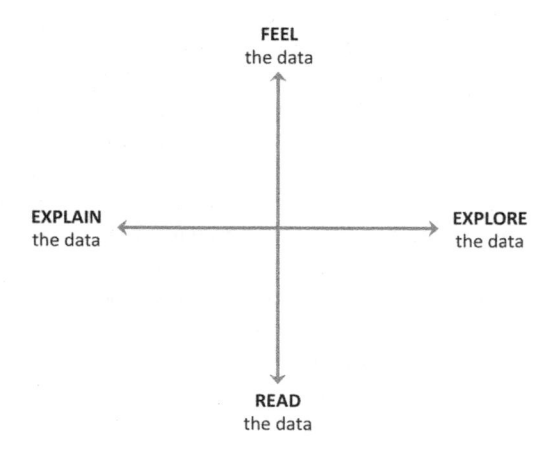

Figure 5.1   A typology of data visualization based on Kirk (2019)

trends without a need for picking out individual datapoints, the ability to feel the data could be much more valuable.

It is important to emphasize this typology is an idealized way of organizing the variety of possible options for visualizing data. Since it is meant to serve as a guide, it does not make normative claims about whether some ways of visualizing are preferable over others. Of course, there are some features that are universally viewed as improving visualization practice, notably advances in accessibility through the use of colorblind-safe palettes. On a conceptual level, visualizations are more successful when there are clear links between the chart types used and the "story" being told through the data at hand, whether that is some kind of change over time, differences among groups, or deviations from a norm. To make these links clearer, the visualizer Alan Smith (2019) created a "visual vocabulary," which is a useful guide that shows how some chart types are better at conveying certain kinds of data, or patterns within the data to be more precise. In the next section, I provide examples of visualization involving human migration and mobility that illustrate these approaches.

DATA VISUALIZATION IN MIGRATION AND MOBILITY DOMAINS

A recent review observed how new advances in digital technologies have afforded researchers working in migration studies the abilities to "interrogate and visualise the spatial, temporal and interaction components of migration data in ways which were hitherto impossible" (Dennett 2015, 143). It goes on to argue these visualizations have obvious appeal across a variety of audiences including "a policymaker with a government to inform, an academic with research to conduct, or a journalist with an article to write" (Dennett 2015, 152). But how and why? Using the previous typology of visualization practice, I disentangle these different usages to draw attention to the ways that visualization can support—and complicate—each set of efforts. This is not meant to be a comprehensive survey of all migration visualizations. Rather, I have chosen examples that demonstrate how migration data can be visualized for different purposes and with different audiences in mind.

Making Sense of Migration Data and Trends

The first set of tasks is mostly diagnostic and analytical, where the main questions of interest are likely to be descriptive ones. How many people are moving? To and from where? Over what time periods—and are these patterns changing? These kinds of questions likely demand the ability to read specific values, as well as explore a dataset without predetermined specifications.

An example of this kind of visualization is the "Local Data Guide" produced by The Migration Observatory at the University of Oxford and available at https://dataguide. migrationobservatory.ox.ac.uk. Based on local-level data produced by the UK's Office for National Statistics and National Records Scotland, this set of interactive visualizations is comprehensive and detailed, providing a great deal of information in a non-directed way. As demonstrated in the screenshot displayed in Figure 5.2, users can search the dataset for particular local authorities, choose to focus on particular subgroups, and download the customized visualization for their own use.

Figure 1: Percent of the population who are foreign born, by local authority in Great Britain (2018)

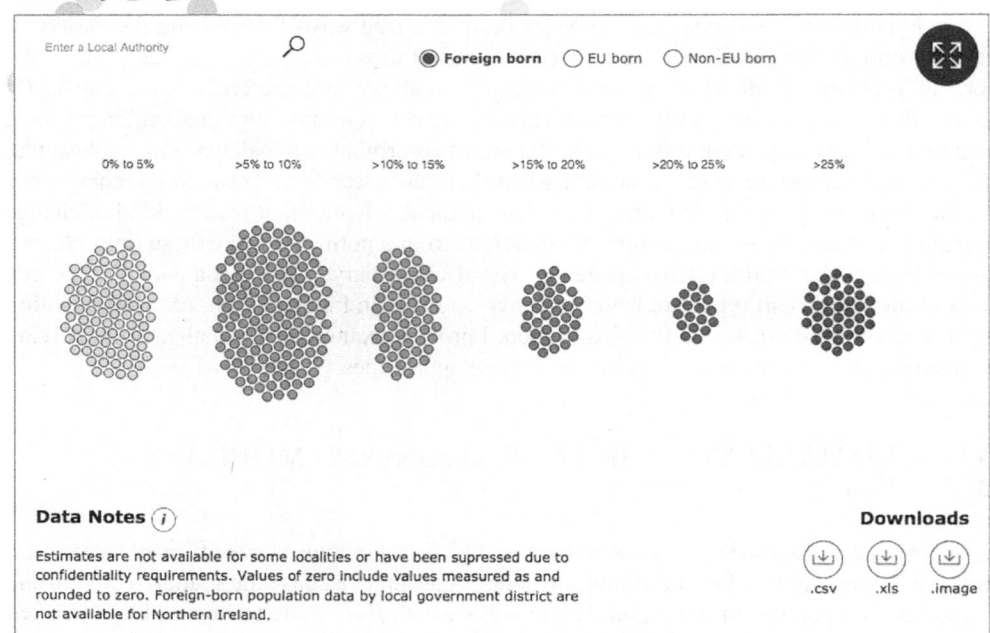

Figure 5.2   Screenshot of the Local Data Guide produced by The Migration Observatory

Another example involving static (i.e., non-interactive) visualizations comes from the World Migration Report, published by the International Organization for Migration (IOM). Across two chapters covering both global and regional migration data, a variety of charts comprehensively depict patterns of mobility subdivided into key categories including gender, reason for migration, and countries of origin (IOM 2019a, 2019b). They also present these figures as both gross levels and proportions to provide additional context for the statistics. Finally, in an effort to move beyond dominant themes and interests, the visualizations include data on emigration as well as how some countries receive and host large numbers of refugees. By providing details that might otherwise be overlooked or lumped together under broad categories of "migrants" and "refugees," these charts illustrate the complexities of mobility patterns.

Communicating Migration Data to Public Audiences

Another approach to data visualization involves using it to communicate key messages to non-expert audiences in settings where messengers might want to raise public awareness of particular migration issues or directly correct misperceptions and incorrect beliefs. In these situations, often but not exclusively in the realms of journalism and advocacy work, designers could turn to emotional appeals and figurative imagery rather than detailed data to make specific points—techniques that are closer to explaining and feeling data.[2]

Roopika Risam (2019) examines one such visualization, called "The Flow Towards Europe" and available at https://www.lucify.com/the-flow-towards-europe/. Relying on UN refugee data, it is a series of visualizations that aim "to clarify the scale of the crisis." Despite popular perceptions of huge numbers of refugees and migrants entering Europe—possibly sustained by media emphasizing the scale and pace of global migration (Allen, Blinder, and McNeil 2017; Allen and Blinder 2018)—this visualization tries to make the point that the actual arrival numbers are relatively low. For example, accompanying stylized icons of soccer fields as units, the visualization displays text saying "around a million Syrian refugees have sought asylum in Europe between April 2011 and December 2017. Standing very tightly together, they would fit on 15 soccer fields." The image of a playing field—probably recognizable to most public users—combined with a large main heading emphasizing "only a fraction makes it to Europe" demonstrates how this kind of visualization aims to send a clear (possibly counterintuitive) message that provokes reactions or re-thinking of held beliefs.

Informing and Enabling Decision-making in Policy and Practice

The third set of tasks for which some types of data visualization can be useful involves informing and enabling decision-making, particularly in the worlds of migration policymaking and practice. These areas often require specific forms of evidence delivered in timely and readily accessible ways (Oliver et al. 2014). However, specific to the migration domain, which is heavily politicized in some national contexts, quantitative evidence itself can be called upon to legitimize pre-determined government positions (Boswell 2009) or lend additional credibility to civil society organizations' agendas and claims of public impact (Allen 2017a). In these settings, data visualization may take a greater role in explaining the policy implications of trends and patterns while providing avenues for both feeling the aggregate "top-line" findings as well as reading values for specific dimensions of interest.

An example of this is the IOM's Global Migration Data Portal, available at https://migrationdataportal.org/. Launched in December 2017 by the IOM's Global Migration Data Analysis Centre with the financial support of the government of Germany and the UK's Department for International Development, the portal describes itself as "a unique access point to timely, comprehensive migration statistics and reliable information about migration data globally." It contains a wide range of visualizations based on data collated from a variety of sources, displayed in a "dashboard" format that combines separate charts into one overview pane (Froese and Tory 2016). Moreover, these dashboards are highly customizable: users can generate visualizations for different countries, regions, and thematic areas. Finally, text-based briefings under a tab labelled "themes" provide concise summaries of the available evidence and data for many migration-related themes, both in terms of their substantive content as well as their methodological rigor.

Although the visualizations themselves are similar in tone to the ones appearing in The Migration Observatory's Local Data Guide—exhibiting qualities that encourage users to "explore" the data by "reading" it in great detail—they take on more of an explanatory function when placed alongside the briefings that link migration data with key policy areas such as the Sustainable Development Goals (SDGs) or the Migration Governance Indicators (MGIs). This responds to the needs and priorities of users, in this case policymakers, whose interests are likely to lie in understanding the significance of national and regional trends in relation to global migration issues.

THE DATA POLITICS OF VISUALIZATION

The previous section presented a typology of data visualization to illustrate how different kinds of visualization can achieve different aims: there is no single "correct" visualization for all situations. Rather, as seen in the examples, visualization potentially speaks to multiple audiences and in multiple ways. However, this practical point has important implications for understanding the politics of data visualization—and, in relation to migration, both how human mobility is represented and which humans are important enough to have been "counted" as being mobile in the first place (Bigo, Isin, and Ruppert 2019). In the following sections, I outline the conceptual framework of data politics. Then, I apply this concept to migration data visualization, demonstrating how seemingly straightforward design choices can potentially reflect limited or incorrect assumptions about mobility. Finally, I place visualization in its wider relational context—within networks of researchers, designers, and intermediaries—to highlight how these visual outputs are actually the "brokered" products of many hands.

Data Politics and Visual Brokerage

Evelyn Ruppert and her colleagues have developed the concept of "data politics" to bring attention to the ways that data are "generative of new forms of power relations and politics at different and interconnected scales" (2017, 2). This concept has three key aspects: *worlds*, comprising the material infrastructures associated with the Internet and digital technologies; *subjects*, referring to the people and populations generated and governed by data; and *rights*, involving the struggles over who gets to generate and have authority over data as well as the ways those data are used (Bigo, Isin, and Ruppert 2019).

How are data politics relevant for migration studies generally, and data visualization specifically? First, migration as a subject domain is increasingly characterized by the presence and use of large-scale datasets. These include censuses (Ruppert 2011), mobile phones (Taylor 2016a), social media (Laczko 2015), and administrative data originally collected and held by governments for other purposes (Allen et al. 2018). Ambitions to more fully exploit these advances by linking data sources raise important questions about these new data subjects' privacy (Amoore 2009), as well as what kinds of responsibilities for sharing data with researchers should be borne by data collectors (Taylor 2016b).

Second, data politics draw attention to what and who is missing in the processes and outcomes of visualizing migration stocks and flows: "[w]ho, for example, has the ability—the privilege—to access, use, or speak with data? How are the uses of data the products of political, social, historical, and cultural contexts and values, and to what extent do they reinforce existing hierarchies of power" (Allen 2020, 188)? These kinds of questions are especially important for signaling how creating and communicating knowledge visually can on the one hand reinforce and widen existing divides between those who have the technological resources and skills to handle data and those who do not (Appadurai 2016), while on the other hand enabling critical and compelling counter-narratives to established ways of thinking.

To be sure, much of the data that eventually is visualized is underpinned, or otherwise informed by statistics and other kinds of quantitative measures. The uses (and abuses) of statistics are instructive for understanding why the politics of visualization matters. For example, survey poll results as snapshots of public attitudes are often used for political advantage by campaigning candidates (Herbst 1995), while quantitative evidence—especially on contentious topics with little agreement on a "correct" outcome—can be used to legitimize any number of positions (Boswell 2009). Meanwhile, even though bureaucratic statistical bodies themselves may strive for impartiality as a sort of "political observatory" (Schudson 2010) above the fray of party politics, a multitude of partisan actors may use and repurpose those statistical outputs for their own agendas (Allen and Blinder 2018; Spiegelhalter 2017). This is not to suggest that all normative uses of statistics are unethical or somehow inherently misleading. Rather, I am simply drawing attention to the ways that visualizations—and the data on which they are based—can be used to legitimate claims and persuade others of those claims.

Up to this point, I have analyzed data visualizations as objects that express data politics through representation: what (and who) features in these visual outputs matters for public understandings of consequential issues (see Allen 2021 for an extended discussion of this argument). However, visualizations also contribute to data politics in the ways that they are connected to other people, objects, and organizations—sometimes even before they take the forms of visualizations. I see them as part of a larger process of "visual brokerage" that involves "processes of conceiving, creating, interpreting, and responding to data visualisations as they occur in social, political, and cultural contexts" (Allen 2018, 907). Figure 5.3 illustrates how visual brokerage connects multiple actors as they contribute to visualization outputs. These include researchers who bring their own questions and interests (Boswell 2019), data collectors such as survey companies or national statistical offices who make technical decisions about question designs and response categories, and intermediaries such as journalists and representatives of non-governmental organizations who translate evidence for public audiences (Allen 2017b). All of these relationships exist within particular contexts that enable some possibilities while foreclosing others: for example, one aspect that is salient for migration involves the legal definitions surrounding who is counted as a "migrant" or "refugee" for the purposes of data collection and categorization. This model, read alongside the typology of visualizations, helps illustrate the wider implications of visualization beyond merely technical desires to communicate data more "effectively." Defining effectiveness, itself a multidimensional objective, is highly contingent on the audience and purpose (Kennedy, Hill, Allen, et al. 2016), while achieving effectiveness requires acknowledging how visualization involves many hands and steps.

Crucially for this chapter, each of these actors is using particular sets of technologies in the course of their work. Data collectors using satellite remote sensing, for instance, can take advantage of increasingly fine-grained imagery to trace and track human movement as well as its impacts on physical environments. Yet analyses based on these data are directly tied to what and how these satellites "see" and report as movement (Rothe 2017). In the case of user-generated data such as on social media, automatically or semi-automatically scraping this information using an Application Programming Interface (API) may involve some trade-offs in completeness or comprehensiveness owing to APIs' idiosyncrasies, which in turn can impact the scope of the conclusions (Rieder et al. 2015). Finally, when it comes to software, designers and intermediaries wanting to create and share data visualizations either for themselves or for clients in research and other sectors can now access a variety of tools, many of which are low-cost or free (Gatto 2015). Yet even as options have proliferated and arguably moved professional-quality visualization within the reach of many, there remains a need for greater understanding of how these software programs work at a technical level—particularly in terms of reshaping and transforming datasets for analysis.

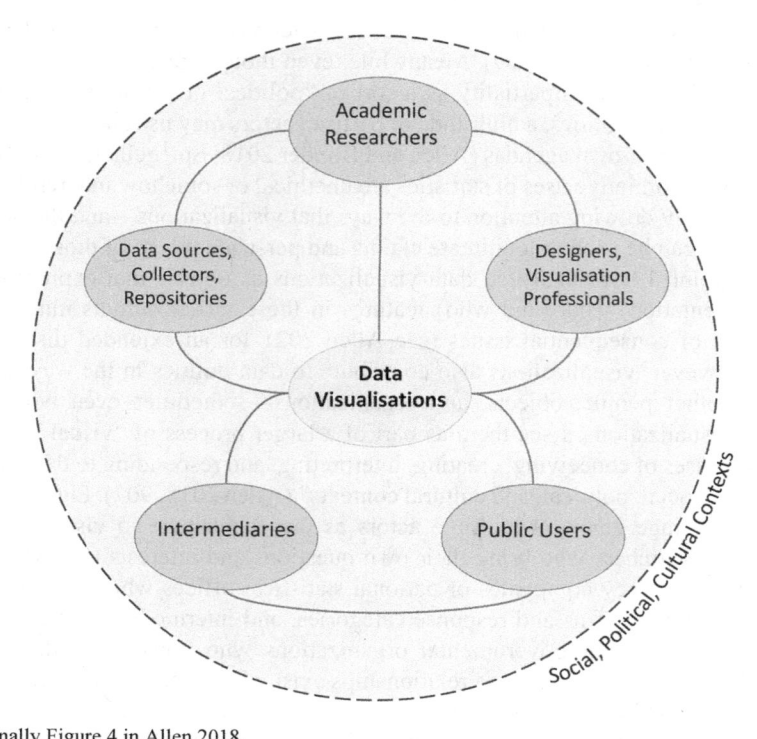

Source: Originally Figure 4 in Allen 2018.

Figure 5.3   A model of visual brokerage

EMERGING ISSUES AND FUTURE RESEARCH

This chapter outlined what data visualizations are, particularly highlighting how the forms they take (captured in a simple typology borrowed from design practice) depend on the audience and purpose of the final product. Then, it illustrated these forms by describing several exemplars within the domain of migration and mobility. Finally, it considered how both the visualizations as outputs as well as the processes surrounding their creation and communication necessarily involved many other actors, in a set of relationships called "visual brokerage" that gives rise to new forms of data politics.

The agenda for visualization research and practice continues to expand. How are these brokered objects perceived and understood by varied audiences who increasingly use a wider range of platforms to access them? What are the implications of these understandings for broader political behaviors such as voting, attitude formation, or expressing preferences for certain policies? How is the rise of crowdsourced data changing the norms of what is considered acceptable in terms of ethical practices when it comes to displaying and acting upon those data? More broadly, what kinds of assumptions, interests, and agendas are driving the sustained attention from policy and industry to data-based technologies and solutions? The existence of these kinds of questions illustrates the need for critically examining how, for and by whom, and in which circumstances technologies such as visualization are mobilized.

NOTES

1. Kirk includes "exhibiting" data as a third category between "explain" and "explore." In his words, exhibitory visualizations are "visual displays of data" where "the viewers have to do the work to interpret meaning relying on their own capacity to perceive and translate the features of a visualisation" (Kirk 2019, 86). In my summary, I have omitted this category for simplicity. Also, while he is not the first to make the explain–explore distinction in reference to data visualization (Healy and Moody 2014; Welles and Meirelles 2015), Kirk's addition of the read-feel dimension (what he calls the "tone" of a visualization's design) contributes important insight from visualization practice.
2. Of course, journalists may actually want readers to be able to read specific data to enhance the credibility of the story being crafted, by ascribing to perceptions that quantitative claims are more objective and scientific (Baele et al. 2017). Therefore, visualizations used in these data journalism contexts might contain elements that enable both reading and feeling data (Rall et al. 2016).

REFERENCES

Allen, William. 2017a. "Factors That Impact How Civil Society Intermediaries Perceive Evidence." *Evidence & Policy: A Journal of Research, Debate and Practice* 13 (2): 183–200. https://doi.org/10.1332/174426416X14538259555968.

Allen, William. 2017b. "Making Corpus Data Visible: Visualising Text with Research Intermediaries." *Corpora* 12 (3): 459–82. https://doi.org/10.3366/cor.2017.0128.

Allen, William. 2018. "Visual Brokerage: Communicating Data and Research through Visualisation." *Public Understanding of Science* 27 (8): 906–22. https://doi.org/10.1177/0963662518756853.

Allen, William. 2020. "Mobility, Media, and Data Politics." In *The SAGE Handbook of Media and Migration*, edited by Kevin Smets, Koen Leurs, Myria Georgiou, Saskia Witteborn, and Radhika Gajjala, 180–91. London: SAGE Publications Ltd.

Allen, William. 2021. "The Conventions and Politics of Migration Data Visualizations." *New Media & Society*. https://doi.org/10.1177/14614448211019300.

Allen, William, Bridget Anderson, Nicholas Van Hear, Madeleine Sumption, Lena Rose, Jennifer Hough, Rachel Humphris, and Sarah Walker. 2018. "Who Counts in Crises? The New Geopolitics of International Migration and Refugee Governance." *Geopolitics* 23 (1): 217–43. https://doi.org/10.1080/14650045.2017.1327740.

Allen, William, and Scott Blinder. 2018. "Media Independence through Routine Press–State Relations: Immigration and Government Statistics in the British Press." *The International Journal of Press/Politics* 23 (2): 202–26. https://doi.org/10.1177/1940161218771897.

Allen, William, Scott Blinder, and Robert McNeil. 2017. "Media Reporting of Migrants and Migration." World Migration Report 2018. Geneva: International Organization for Migration. At https://publications.iom.int/system/files/pdf/wmr_2018_en_chapter8.pdf (accessed on August 31, 2021).

Amoore, Louise. 2009. "Lines of Sight: On the Visualization of Unknown Futures." *Citizenship Studies* 13 (1): 17–30.

Appadurai, Arjun. 2016. "The Academic Digital Divide and Uneven Global Development." 4. CARGC Papers. At https://repository.upenn.edu/cargc_papers/4 (accessed on August 31, 2021).

Baele, Stephane J, Travis G. Coan, and Olivier C. Sterck. 2017. "Security through Numbers? Experimentally Assessing the Impact of Numerical Arguments in Security Communication." *The British Journal of Politics and International Relations* 20 (2): 459–76. https://doi.org/10.1177/1369148117734791.

Bigo, Didier, Engin Isin, and Evelyn Ruppert, eds. 2019. *Data Politics: Worlds, Subjects, Rights*. London and New York: Routledge.

Boswell, Christina. 2009. *The Political Uses of Expert Knowledge: Immigration Policy and Social Research*. Cambridge: Cambridge University Press.

Boswell, Christina. 2019. "Research, 'Experts', and the Politics of Migration." In *Bridging the Gaps: Linking Research to Public Debates and Policy-Making on Migration and Integration*, edited by Martin Ruhs, Kristof Tamas, and Joakim Palme, 21–33. Oxford: Oxford University Press.

boyd, danah, and Kate Crawford. 2012. "Critical Questions for Big Data." *Information, Communication & Society* 15 (5): 662–79. https://doi.org/10.1080/1369118X.2012.678878.

Bucchi, Massimiano, and Barbara Saracino. 2016. "'Visual Science Literacy': Images and Public Understanding of Science in the Digital Age." *Science Communication* 38 (6): 812–19. https://doi.org/10.1177/1075547016677833.

Chen, Min, Luciano Floridi, and Rita Borgo. 2014. "What Is Visualization Really For?" In *The Philosophy of Information Quality*, edited by Luciano Floridi and Phyllis Illari, 75–93. Cham: Springer International Publishing. http://dx.doi.org/10.1007/978-3-319-07121-3_5.

Dennett, Adam. 2015. "Visualising Migration: Online Tools for Taking Us beyond the Static Map." *Migration Studies* 3 (1): 143–52. https://doi.org/10.1093/migration/mnu073.

Froese, Maria-Elena, and Melanie Tory. 2016. "Lessons Learned from Designing Visualization Dashboards." *IEEE Computer Graphics and Applications* 36 (2): 83–89. https://doi.org/10.1109/MCG.2016.33.

Gatto, Malu A. C. 2015. "Making Research Useful: Current Challenges and Good Practices in Data Visualisation." University of Oxford: Reuters Institute for the Study of Journalism. At https://ora.ox.ac.uk/objects/uuid:526114c2-8266-4dee-b663-351119249fd5 (accessed on August 31, 2021).

Healy, Kieran. 2018. *Data Visualization: A Practical Introduction.* Princeton, NJ: Princeton University Press.

Healy, Kieran, and James Moody. 2014. "Data Visualization in Sociology." *Annual Review of Sociology* 40 (1): 105–28. https://doi.org/10.1146/annurev-soc-071312-145551.

Herbst, Susan. 1995. *Numbered Voices: How Opinion Polling Has Shaped American Politics.* Chicago, IL: University of Chicago Press.

Hullman, Jessica, and Nick Diakopoulos. 2011. "Visualization Rhetoric: Framing Effects in Narrative Visualization." *IEEE Transactions on Visualization and Computer Graphics* 17 (12): 2231–40. https://doi.org/10.1109/tvcg.2011.255.

IOM. 2019a. "Migration and Migrants: A Global Overview." In *World Migration Report 2020.* Geneva: International Organization for Migration. At https://publications.iom.int/system/files/pdf/wmr_2020_en_ch_2.pdf (accessed on August 31, 2021).

IOM. 2019b. "Migration and Migrants: Regional Dimensions and Developments." In *World Migration Report 2020.* Geneva: International Organization for Migration. At https://publications.iom.int/system/files/pdf/wmr_2020_en_ch_3_1.pdf (accessed on August 31, 2021).

Kelleher, Christa, and Thorsten Wagener. 2011. "Ten Guidelines for Effective Data Visualization in Scientific Publications." *Environmental Modelling & Software* 26 (6): 822–27. https://doi.org/10.1016/j.envsoft.2010.12.006.

Kennedy, Helen, and William Allen. 2017. "Data Visualisation as an Emerging Tool for Online Research." In *The SAGE Handbook of Online Research Methods*, edited by Nigel G Fielding, Raymond M Lee, and Grant Blank, 2nd Edition. London: SAGE Publications Ltd. https://uk.sagepub.com/en-gb/eur/the-sage-handbook-of-online-research-methods/book245027.

Kennedy, Helen, Rosemary Lucy Hill, Giorgia Aiello, and William Allen. 2016. "The Work That Visualisation Conventions Do." *Information, Communication & Society* 19 (6): 715–35. https://doi.org/10.1080/1369118X.2016.1153126.

Kennedy, Helen, Rosemary Lucy Hill, William Allen, and Andy Kirk. 2016. "Engaging with (Big) Data Visualizations: Factors That Affect Engagement and Resulting New Definitions of Effectiveness." *First Monday* 21 (11). https://doi.org/10.5210/fm.v21i11.6389.

Kirk, Andy. 2019. *Data Visualisation: A Handbook for Data Driven Design.* 2nd edn. London: SAGE.

Kitchin, Rob, and Tracey P. Lauriault. 2015. "Small Data in the Era of Big Data." *GeoJournal* 80 (4): 463–75. https://doi.org/10.1007/s10708-014-9601-7.

Krause, Kelly. 2017. "A Framework for Visual Communication at Nature." *Public Understanding of Science* 26 (1): 15–24.

Laczko, Frank. 2015. "Factoring Migration into the 'Development Data Revolution.'" *Journal of International Affairs* 68 (2): 1–17.

Oliver, Kathryn, Simon Innvar, Theo Lorenc, Jenny Woodman, and James Thomas. 2014. "A Systematic Review of Barriers to and Facilitators of the Use of Evidence by Policymakers." *BMC Health Services Research* 14 (1): 1–12.

Rall, Katharina, Margaret L. Satterthwaite, Anshul Vikram Pandey, John Emerson, Jeremy Boy, Oded Nov, and Enrico Bertini. 2016. "Data Visualization for Human Rights Advocacy." *Journal of Human Rights Practice* 8 (2): 171–97.

Rieder, Bernhard, Rasha Abdulla, Thomas Poell, Robbert Woltering, and Liesbeth Zack. 2015. "Data Critique and Analytical Opportunities for Very Large Facebook Pages: Lessons Learned from Exploring 'We Are All Khaled Said.'" *Big Data & Society* 2 (2): 2053951715614980. https://doi.org/10.1177/2053951715614980.

Risam, Roopika. 2019. "Beyond the Migrant 'Problem': Visualizing Global Migration." *Television & New Media* 20 (6): 566–80. https://doi.org/10.1177/1527476419857679.

Rothe, Delf. 2017. "Seeing like a Satellite: Remote Sensing and the Ontological Politics of Environmental Security." *Security Dialogue* 48 (4): 334–53. https://doi.org/10.1177/0967010617709399.

Ruppert, Evelyn. 2011. "Population Objects: Interpassive Subjects." *Sociology* 45 (2): 218–33. https://doi.org/10.1177/0038038510394027.

Ruppert, Evelyn, Engin Isin, and Didier Bigo. 2017. "Data Politics." *Big Data & Society* 4 (2): 2053951717717749. https://doi.org/10.1177/2053951717717749.

Schudson, Michael. 2010. "Political Observatories, Databases & News in the Emerging Ecology of Public Information." *Daedalus* 139 (2): 100–109. https://doi.org/10.1162/daed.2010.139.2.100.

Smith, Alan. 2019. *Visual Vocabulary.* At https://github.com/ft-interactive/chart-doctor/blob/master/visual-vocabulary/Visual-vocabulary.pdf (accessed on August 31, 2021).

Spiegelhalter, David. 2017. "Trust in Numbers." *Journal of the Royal Statistical Society: Series A (Statistics in Society).* https://doi.org/10.1111/rssa.12302.

Taylor, Linnet. 2016a. "No Place to Hide? The Ethics and Analytics of Tracking Mobility Using Mobile Phone Data." *Environment and Planning D: Society and Space* 34 (2): 319–36.

Taylor, Linnet. 2016b. "The Ethics of Big Data as a Public Good: Which Public? Whose Good?" *Philosophical Transactions of the Royal Society A: Mathematical, Physical and Engineering Sciences* 374 (2083): 20160126. https://doi.org/10.1098/rsta.2016.0126.

Tufte, Edward R. 1983. *The Visual Display of Quantitative Information.* Cheshire, CT: Graphics Press.

Tufte, Edward R. 2006. *Beautiful Evidence.* Vol. 1. Graphics Press Cheshire, CT.

Wainer, Howard. 1984. "How to Display Data Badly." *The American Statistician* 38 (2): 137–47. https://doi.org/10.1080/00031305.1984.10483186.

Welles, Brooke, and Isabel Meirelles. 2015. "Visualizing Computational Social Science." *Science Communication* 37 (1): 34–58. https://doi.org/10.1177/1075547014556540.

6. Migration networks: applications of network analysis to macroscale migration patterns[1]

Valentin Danchev and Mason A. Porter

INTRODUCTION

The complexity of international migration patterns has increased markedly over the last few decades (International Organization for Migration, 2003). Advancements in distance-shrinking technologies, such as transportation and digital communication, have interacted with other global processes—including accelerated international exchanges of goods, capital, and services; uneven global development (Wallerstein, 1974); and restrictive migration policies (Hatton and Williamson, 2005)—to yield complex migration patterns between previously disconnected places across the globe. To account for such tendencies, migration theories (e.g., Salt, 2013 [1986]; Kritz, Lim, and Zlotnik, 1992) have advanced the perspective that international migration is not merely a response to bilateral origin–destination forces, but also often takes place in a group (or in a network) of origins and destinations that are inter-connected through prior migration, trade, culture, and politics. Building on this idea, one can study migration by examining a network of nodes (which typically represent countries, but may alternatively represent rural areas, cities, provinces, or other geographically dispersed places) and edges (which encode migration ties that are based on movements of diverse populations, such as displaced individuals, low-skilled workers, or skilled professionals) between the nodes. Such 'migration networks' provide a useful perspective for the understanding of patterns of international migration.

In this chapter, we discuss how to leverage network analysis in concert with new technologies and data sources to study migration networks at several geographical scales. We focus in particular on migration networks at global scales. We first review recent work that has employed data from origin–destination matrices of bilateral migrant stocks (Özden *et al.*, 2011; UN DESA, 2019) and migration flows (Abel and Sander, 2014; Abel, 2018) to study various aspects of the structure and evolution of the global migration network[2] (e.g., Fagiolo and Mastrorillo, 2013; Danchev and Porter, 2018). We then discuss future research opportunities and ethical, methodological, data-availability, and reproducibility challenges that are associated with the use of digital geolocated data from online social networks, the World Wide Web, and mobile-phone networks (Lazer *et al.*, 2009; Salganik, 2018) to study migration networks. We consider possible ways of linking digital geolocated data to administrative and survey data for constructing and studying increasingly realistic migration networks. We also review recent developments in network methodology and modeling that can help advance network analysis in the field of migration.

CONCEPTS AND CONTEXT

Network analysis is a powerful tool for studying relationships between entities (e.g., individuals, countries, Web pages, computers, or neurons), rather than focusing solely on the entities themselves (Wasserman and Faust, 1994; Newman, 2018; Butts, 2009; Brandes *et al.*, 2013). Scholars from diverse fields—including sociology (Wasserman and Faust, 1994), economics (Jackson, 2008), political science (Maoz, 2011), applied mathematics (Porter, 2020), statistics (Kolaczyk, 2009), physics (Newman, 2018), social neuroscience (Baek, Porter, and Parkinson, in press (2020)), and others—have employed network analysis to study complex systems of interconnected entities. Taking into account system specificity, one can first carefully abstract a set of entities as a network's nodes and a set of relationships as its edges (e.g., encoding social ties, migration ties, or hyperlinks) between those nodes (Butts, 2009; Brandes *et al.*, 2013). One can then study patterns of relationships—in the form of network structure—that emerge from the interacting entities and examine how network structure affects the dynamics (and function) of a system and the features (e.g., importances or performance) of particular nodes, edges, and other substructures (Wasserman and Faust, 1994; Borgatti *et al.*, 2009; Newman, 2018).

It is convenient to categorize applications of network analysis of migration into two strands (Cushing and Poot, 2003; Maier and Vyborny, 2008; Bilecen, Gamper, and Lubbers, 2018; Lemercier, 2010). The first strand focuses on *microscale networks*, such as in the form of migrants' personal networks, which one constructs based on interpersonal relationships (e.g., kinship, friendship, or acquaintance) that link migrants and non-migrants (Massey *et al.*, 1998: 42; Boyd, 1989; Gurak and Caces, 1992). Migrants' personal networks can spread information about destination opportunities and provide initial employment, accommodation, and overall assistance, thereby reducing movement costs and risks (Palloni *et al.*, 2001; Liu, 2013; Blumenstock, Chi, and Tan, 2019). For example, recent work has used data from interviews and surveys (Lubbers *et al.*, 2010; Vacca *et al.*, 2018; Verdery *et al.*, 2018; Bilecen, Gamper, and Lubbers, 2018) and the online social network Facebook (Herdağdelen *et al.*, 2016) and employed what is often called 'social network analysis' (SNA) (Wasserman and Faust, 1994; Borgatti, Everett, and Johnson, 2018) to examine migrants' personal networks and the relative importances of migrants' ties to other migrants and their ties to non-migrants in home and host societies. The second strand, which uses aggregate data and focuses on *macroscale migration networks*, encompasses a multidisciplinary body of literature that leverages network methodology to study migration as a 'mechanism that connects "places"' (Maier and Vyborny, 2008). Among other phenomena, researchers have examined networks of internal (intra-country) movements that connect rural areas in Northern France (Lemercier and Rosental, 2010), movements that connect different states in the United States (Maier and Vyborny, 2008; Charyyev and Gunes, 2019), and global networks of international migration ties that connect different countries (Fagiolo and Mastrorillo, 2013; Davis *et al.*, 2013; Tranos, Gheasi, and Nijkamp, 2015; Danchev and Porter, 2018; Windzio, 2018).

In the present chapter, we focus on macroscale migration networks and especially on the global migration network. See Figure 6.1 for illustrations of global migration networks in two time periods. The figure shows global migration between countries and territories in 1960 and 2000 using the global database of bilateral migrant stocks of Özden *et al.* (2011). Each network in the figure consists of a set of nodes that encode countries (or territories) and a set of edges that encode migration ties between them. The positions of the nodes indicate the

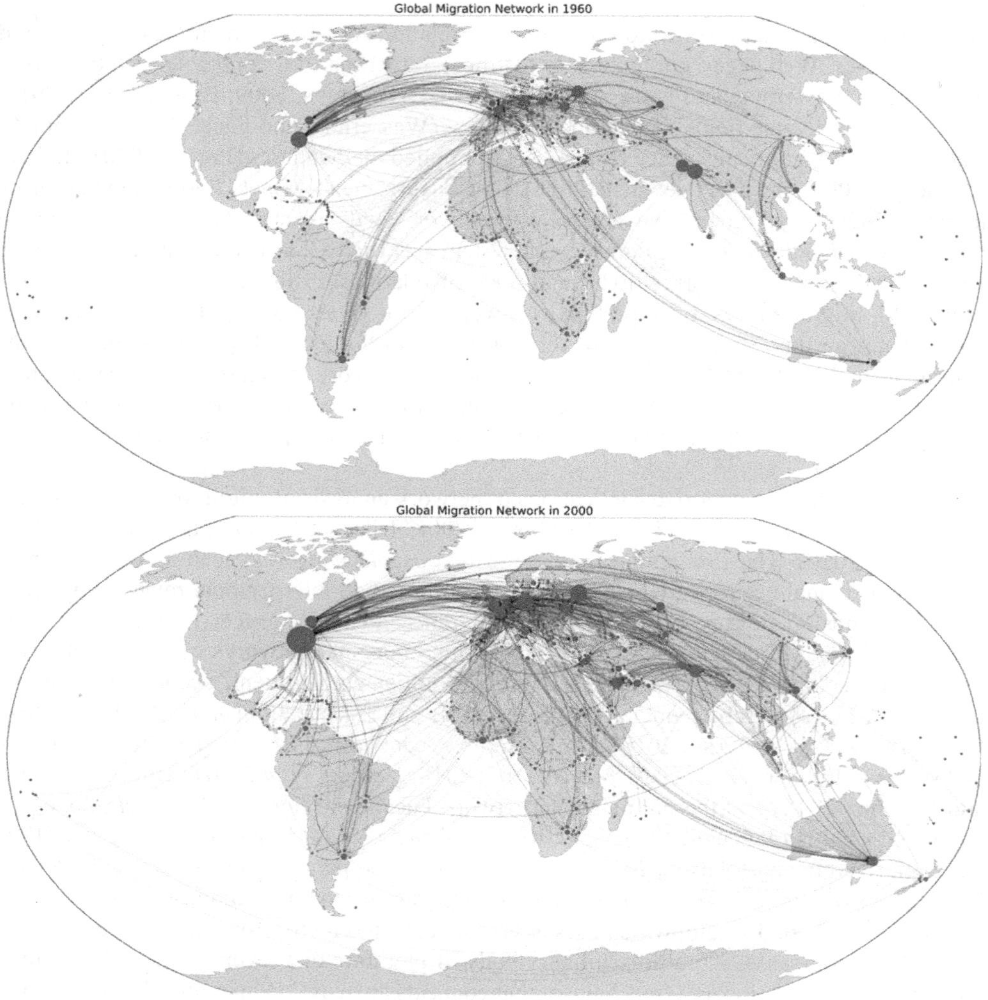

We created the visualization using PYTHON (version 3.7.10), NETWORKX (version 2.5) (Hagberg, Swart and Chult, 2008), and MATPLOTLIB (version 3.2.2) (Hunter, 2007), including the MATPLOTLIB 'Basemap' toolkit (version 1.2.2). For related visualizations of global migration networks, see Danchev (2015) and Danchev and Porter (2018). (The figure and the code that we used to generate it are available at https://osf.io/dgr6x/ under a CC-BY-4.0 license.)

Figure 6.1   *Global migration networks in (top) 1960 and (bottom) 2000*

geographical locations of the countries, and the sizes of the nodes encode weighted in-degrees. The weighted in-degree of a country (i.e., a node) is the sum of the weights (which encode migrant stocks) of the edges that are directed towards that country. Because of the heterogeneity in weighted in-degrees, we make nodes with minute weighted in-degrees visible in practice by setting their size to a small value. The arrow on an edge indicates the direction of migrant stock from an origin country to a destination country. The width of each directed edge encodes the volume of migrant stock that is associated with that edge. We use different widths for small (with 10,001–100,000 migrants), medium (with 100,001–1,000,000 migrants), and large (above 1,000,000 migrants) edges. We do not plot migration stocks with 10,000 or fewer migrants.

Until recently, networks were used in migration studies (Kritz and Zlotnik, 1992; Salt, 1989) primarily as a metaphor and without the application of explicit network-based methodology. Since the 2010s, however, various studies have employed concepts and methods from network analysis to study migration. A recent special issue of the journal *Social Networks* (Bilecen, Gamper, and Lubbers, 2018) included research at the intersection of social networks and migration. Apart from the dyadic assumptions that underlie many migration theories (Massey *et al.*, 1998), a major barrier to network analysis of macroscale international migration patterns was the lack of compatible migration data between world countries (Fagiolo and Mastrorillo, 2013; Özden *et al.*, 2011). However, over the last decade or so, researchers have had access both to (1) global and regional longitudinal migration data (including global bilateral migrant stocks) from national statistics and administrative sources (Özden *et al.*, 2011; UN DESA, 2019) and to (2) estimates of global bilateral migration flows (Abel and Sander, 2014; Abel, 2018). An important limitation of these migration data sets is the aggregate nature of the information that they provide, although recent data sets stratify migration by various characteristics, including age and sex (UN DESA, 2019; Abel, 2018). Computational and geospatial techniques have been used to infer migration trajectories from individual-level digital geolocated data from online platforms, such as Twitter (Zagheni *et al.*, 2014) and Facebook (Spyratos *et al.*, 2019). As we will discuss in the section 'Future Research Directions: Digital Technologies and Migration Networks', digital geolocated data can provide an opportunity for increasingly realistic network analyses of migration networks, but they have been used only rarely in this context (Messias *et al.*, 2016). Moreover, to facilitate future research opportunities, it is important to overcome several methodological, privacy, and proprietary limitations of digital geolocated data.

MIGRATION NETWORKS: DIRECTED, WEIGHTED, TEMPORAL, AND SPATIAL PROPERTIES

Danchev and Porter (2018) studied world migration in the form of a directed, weighted, temporal, and spatial network. In this section, we discuss these four features of world migration networks.

As Ravenstein observed long ago (1885), when thinking about migration, one needs to consider the directions of movements, as movement from place A to place B is distinct from movement in the opposite direction. To encode migration direction, one can construct a directed migration network, in which each edge has an associated direction that corresponds to either out-migration or in-migration. Migration movements also vary in terms of volume of migrant stocks or flows. One can represent migration volumes by constructing a weighted network (Newman, 2018), in which each edge has an associated weight that encodes the flow or volume of migrant stock from place A to place B for all origin–destination pairs (A,B). Migration networks are also longitudinal, as reflected in global bilateral migration databases, which typically report migrant stocks and flows at five-year or ten-year intervals. Instead of considering those time windows as separate snapshots, recent network methods (Holme and Saramäki, 2019) and statistical models (Krivitsky and Handcock, 2014) enable increasingly realistic investigations of network dynamics.

Migration networks are also spatial in nature (Danchev and Porter, 2018; Salt, 2013 [1986]). Spatial networks (Barthelemy, 2018) include both spatially embedded networks, whose nodes and edges are embedded in space in a literal sense (e.g., road networks), and networks that are 'merely' influenced by space, in the sense that space affects the probability of edge formation and/or the weights of the edges. Migration networks fall into the second category. Due to geographical constraints (Barthelemy, 2018), longer-distance migration edges are associated with costs, so they are less likely than shorter-distance edges to form and develop into strong edges.[3] Early migration theories treated the effects of distance on migration as a 'law' (Ravenstein, 1885; Zipf, 1946). However, because of significant advances in transportation, information, and communication technologies, geographical and cultural distances have shrunk since the 1970s in a phenomenon that is sometimes called 'distance-shrinking' or 'time–space compression' (Brunn and Leinbach, 1991; Harvey, 1989). Distance-shrinking technologies have increased the length and the spread of migratory movements (International Organization for Migration, 2003: 16; Vertovec, 2010; International Organization for Migration, 2017), although restrictive border control and global inequalities have caused distance-shrinking to have smaller effects on migration than it has on other types of cross-border exchanges (such as capital and goods) (Hatton and Williamson, 2005). The effects of distance on migration varies across the world, and it is important to take this spatial heterogeneity into account in network models and computations (Danchev, 2015; Danchev and Porter, 2018).

METHODS FOR INVESTIGATING MIGRATION NETWORKS

We now discuss some network diagnostics, techniques, and models that are useful for characterizing the structure of migration networks.

Network Diagnostics

Many diagnostics have been developed to measure the properties of nodes, edges, and other network structures (Newman, 2018; Wasserman and Faust, 1994).

Node degree is perhaps the simplest diagnostic for characterizing a country in a migration network. A node's degree is equal to the number of edges that are attached to it. In directed migration networks, one distinguishes a country's out-degree (i.e., the number of outgoing edges that originate at a node) from its in-degree (i.e., the number of incoming edges that terminate at a node) (Wasserman and Faust, 1994: 126). Out-degree and in-degree correspond to out-migration and in-migration, respectively. In weighted networks, it is also useful to consider a node's strength, which one can quantify as the total weight of the edges that are attached to a node (Barrat *et al.*, 2004: 2). The strength of a node is also sometimes called its weighted degree.

Node degree and node strength are basic measures of 'centrality' in networks. Centrality measures are ways to quantify the importances of nodes, edges, or other structures in a network (Newman, 2018). In network analysis, out-degree can be an indication of expansiveness, whereas in-degree is often a notion of popularity (Wasserman and Faust, 1994). From this perspective, one expects that potential migrants from a country with a large out-degree have more opportunities (in the form of potential destinations). Nodes that have disproportionately more connections than the other nodes in a network are sometimes called 'hubs' (Newman, 2018). In migration networks, hub countries are involved in the circulation of migrants to and from multiple countries. For discussions of other centrality measures (e.g., betweenness, closeness, and eigenvector centralities), see Borgatti, Everett, and Johnson (2018) and Newman (2018). One can evaluate the heterogeneity of the centrality values of the nodes of a network by calculating network centralization (Freeman, 1978; Borgatti, Everett, and Johnson, 2018).

An old idea in migration studies is that 'every migratory current has a counter-current' (Grigg, 1977: 48; Ravenstein, 1885: 199). To capture this intuition, consider the network notion of 'reciprocity', which measures whether an edge from node A to node B is matched by an edge in the opposite direction. One way to define reciprocity is as the number of pairs of mutually connected nodes (i.e., reciprocal relationships) divided by the total number of node pairs with any edge between them (Butts, 2008; Borgatti, Everett, and Johnson, 2018). One way to generalize reciprocity to weighted networks is to consider two weighted edges between a pair of nodes as reciprocated if the smaller edge weight divided by the larger edge weight is at least a certain threshold.

A characteristic property of many social and spatial networks is a tendency towards triadic closure (Davis, 1967; Wasserman and Faust, 1994; Barthelemy, 2018). In colloquial sociological terms, triadic closure refers to the tendency of friends of a person to themselves be friends. To study triadic closure, one examines the probability that two nodes with a connection to a common third node are themselves connected directly to each other (i.e., 'adjacent') via an edge (Wasserman and Faust, 1994; Newman, 2018). Both geographical proximity and social proximity (e.g., common language) facilitate triadic closure. Measures of the tendency of triadic closure in unweighted (i.e., binary) networks include clustering coefficients (Newman, 2018) and transitivity (Wasserman and Faust, 1994). There are also generalizations of clustering coefficients to directed (Fagiolo, 2007), weighted (Saramäki *et al.*, 2007), and multiplex (Cozzo *et al.*, 2015) networks.

Reciprocity, clustering coefficients, and related measures help characterize node neighborhoods, but they provide limited information about the overall structure of a network. One can use the mean shortest-path length, which is a measure of network distance between two nodes, as one indicator of global connectivity in a network. A path in a network is a sequence of adjacent nodes, so one can 'travel' from one node to another node along the edges between them. The length of a path in an unweighted network is equal to the number of edges that are traversed. A shortest path is a path that connects two nodes using the fewest possible number of edges (Newman, 2018: 132). In a directed network, one considers paths between origin nodes and destination nodes, and one calculates the length of a path by counting the number of edges from an origin to a destination. In weighted networks, one can transform from edge weights to edge costs to examine other notions of path lengths.

In the study of networks, it is also helpful to compute densities. The most common type of network density is edge density, which is the ratio of the actual edges in a network to the maximum possible number of edges in the network (Wasserman and Faust, 1994: 129; Borgatti, Everett, and Johnson, 2018). It takes values between 0 (if no edge is present) and 1 (if all edges are present).

Spatial Network Diagnostics

To examine spatial properties of migration networks, one can calculate diagnostics that combine network and geographical information. For example, a simple way of incorporating spatial information is to compute the probability of a migration edge in a given distance range, taking into account both actual and possible migration edges for that distance range. For a similar approach in the context of the 'geography' of online social networks, see Backstrom et al. (2010). See Barthelemy (2018) for a review of spatial networks.

Community Detection

The systems approach to international migration defines a 'migration system' as a set of countries with close historical, cultural, and economic linkages that exchange large numbers of migrants (Kritz, Lim, and Zlotnik, 1992; Fawcett, 1989; Salt, 1989). A major methodological difficulty is the demarcation of the boundaries of migration systems (Zlotnik, 1992). Techniques of community detection offer a large variety of algorithmic methods to delineate migration systems and other 'functional regions' (Ratti *et al.*, 2010; Farmer and Fotheringham, 2011) on the basis of empirical connectivity. A 'community' is a tightly knit subnetwork of densely connected nodes that are loosely connected to the rest of a network (Porter, Onnela, and Mucha, 2009; Fortunato and Hric, 2016). In Danchev and Porter (2018), we defined an international migration community as 'a tightly-knit group of countries with dense internal migration connections … but sparse connections to and from other countries in a network'.

There are many methods for algorithmic community detection in networks (Porter, Onnela, and Mucha, 2009; Fortunato and Hric, 2016). One popular method, which has been employed in many network studies of global migration (e.g., Fagiolo and Mastrorillo, 2013; Davis *et al.*, 2013; Tranos, Gheasi, and Nijkamp, 2015), is to maximize an objective function called 'modularity' (Newman, 2018). From a modularity-maximization perspective, an optimal division of a network into communities is one with the largest possible number (or the largest total weight, in a weighted network) of intra-community edges relative to the expected number of such edges in a specified null model (Newman, 2018). The purpose of a null model is to take into account 'statistically surprising' connectivity (Newman, 2006b: 8578).

The standard null model for modularity maximization (Newman, 2006a) works for (either unweighted or weighted) undirected networks. For migration networks, one can consider extensions of modularity that accommodate edge directionality and node attributes. For example, Leicht and Newman (2008) developed a null model for directed networks. Additionally, Expert *et al.* (2011) and Sarzynska *et al.* (2016) developed null models for spatial networks (with known node locations) and Mucha *et al.* (2010) extended modularity maximization to time-dependent and multiplex networks. The choice of community-detection method depends on the properties of available migration data and on one's research questions. When performing community detection, it is important to consider the parameter space, assumptions, and features of a method. For a review of community detection (including discussions of increasingly popular methods that are based on statistical inference), see Fortunato and Hric (2016).

Statistical Network Models

For testing network hypotheses, one can employ statistical models. The quadratic assignment procedure (QAP) of regression is appropriate for testing dyadic hypotheses (Dekker, Krackhardt, and Snijders, 2007). Additionally, a popular family of models called exponential random-graph models (ERGMs) allows both cross-sectional and longitudinal examination of higher-order network dependencies; these models can account for covariates that are encoded in node attributes. For a review of ERGMs, see Lusher, Koskinen, and Robins (2013). For a discussion of various statistical models for network data, see Kolaczyk (2009).

PRIOR RESEARCH AND KEY FINDINGS ABOUT MIGRATION NETWORKS

Probably one of the earliest works to integrate network analysis and migration studies is a paper by Vincent and Macleod (1974), who made analogies with physical networks (such as stream networks) to advance the argument that networks can influence migration patterns and therefore can inform the forecasting of migration rates. Vincent and Macleod (1974) examined patterns of migration by drawing on theories in regional science and on network methods in geography (Haggett and Chorley, 1969). In a pioneering study, Nogle (1994) used the systems approach to international migration, which was proposed by Fawcett and Arnold (1987) and Kritz and Zlotnik (1992), as a framework for studying migration flows within the European Union in the 1980s. By calculating centrality measures and applying techniques for detecting fully connected subgraphs (so-called 'cliques'), Nogle (1994) identified a tendency towards a 'Single Europe', in which more countries become interconnected via migration over time. More recently, Maier and Vyborny (2008) applied network analysis in an exploratory study of migration between states in the United States. An important contribution of theirs was to define migration as a 'mechanism that connects "places"' and to differentiate the analysis of migration patterns from individual-level analysis of migration decisions and motives of migrants (Maier and Vyborny, 2008). Slater (2008) also examined a network of internal migration in the United States, with a focus on the role of 'hubs' and 'functional regions'. Lemercier and Rosental (2010), inspired by the approach of Hägerstrand (1957) on migration fields, conducted an innovative study of migration patterns between rural areas in 19th century Northern France using an actor-oriented model of network dynamics (Snijders, van de Bunt, and Steglich, 2010).

In the last few years, many researchers have employed network approaches to study global migration (Tranos, Gheasi, and Nijkamp, 2015; Davis *et al.*, 2013; Fagiolo and Mastrorillo, 2013; Danchev, 2015; Novotný and Hasman, 2016; Peres, Xu, and Wu, 2016; Danchev and Porter, 2018; Windzio, 2018; Cerqueti, Clemente, and Grassi, 2019; Abel *et al.*, 2021). By calculating various network diagnostics and employing community detection, several of these papers have highlighted stylized observations about the network of international migration in the latter half of the twentieth century. For example, Davis *et al.* (2013) and Fagiolo and Mastrorillo (2013) concluded that interconnectivity and globalization of migration have increased over time based on increasing 'connectivity' (as reflected by the increase in value of various network diagnostics, such as the number of migration edges and migration weights between countries, countries' degrees and strengths, and clustering coefficients) and 'reachability' (as reflected by a decreasing mean shortest-path length). Davis *et al.* (2013) and Fagiolo and Mastrorillo (2013) also reported that these migration networks have a characteristic right-skewed edge-weight distribution, indicating that many edges have a small to moderate number of migrants and a small number of edges are responsible for many migrants. Additionally, Fagiolo and Mastrorillo (2013: 4) reported that the 'number of communities decreases across time' and concluded on this basis that 'globalization has made the architecture of the IMN [International Migration Network] less fragmented and modules more strongly interconnected between them'. Similarly, Davis et al. (2013: 6) argued their case of 'increasing globalization' by noting that 'the ratio between the internal and total fluxes slowly decreases in time: 0.8 in 1960; 0.8 in 1970; 0.76 in 1980; 0.75 in 1990 and 2000.' A major conclusion of these studies is that world migration has become more interconnected, in line with broader globalization tendencies.

Informed by the international-migration systems approach (Salt, 1989; Kritz, Lim, and Zlotnik, 1992), some research (DeWaard, Kim, and Raymer, 2012; Abel *et al.*, 2021) has employed community detection and other techniques to discover and characterize boundaries of migration systems. Abel *et al.* (2021) suggested that global migration from 1990 to 2015 consisted of 'geographically clustered' systems that bound neighboring countries in geographical regions. This line of research, which lies at the intersection of migration studies and geography, has emphasized localizing spatial tendencies in international migration.

Many of the above results rely on the structure and boundaries of migration communities that were obtained by maximizing the original modularity function (Newman and Girvan, 2004) or employing some other community-detection algorithm (Fortunato and Hric, 2016) that was designed for non-spatial networks. In a recent paper (Danchev and Porter, 2018), we employed a spatial modularity function (Expert *et al.*, 2011; Sarzynska *et al.*, 2016) that assigns countries to communities based both on migration ties and on distances between them. In Danchev and Porter (2018), we detected international migration communities using a generalized modularity function for spatial, temporal, directed, and weighted networks. We also leveraged properties of the detected communities to adjudicate among conflicting theoretical accounts. We concluded that, over the second half of the 20th century, 'world migration is neither regionally concentrated nor globally interconnected, but instead exhibits a heterogeneous connectivity pattern that channels unequal migration opportunities across the world.' Given appropriate data availability and quality, tailored community-detection techniques have the potential to help uncover the impact of recent events—including the enlargement of the European Union, the global financial crisis, and the withdrawal of the United Kingdom from the European Union ('Brexit')—on patterns of connectivity in global migration.

Windzio (2018) employed cross-sectional and longitudinal ERGMs to examine the impact of various factors—including geographical, demographic, economic, and linguistic ones—on migration between 202 countries during the period 1990–2013. He reported results that are consistent with migration theories, with (1) geographical distance tending to reduce the probability of migration edges between countries and (2) economic differences, shared geographical region, and similar religion and language tending to increase the probability of migration edges. Although they fit models using primarily binary data, the ERGM analysis of Windzio

(2018) provides a good foundation for sophisticated modeling of network dependencies and countries' attributes in global migration.

FUTURE RESEARCH DIRECTIONS: DIGITAL TECHNOLOGIES AND MIGRATION NETWORKS

Recent advances in digital technologies have created new opportunities to generate, store, and analyze unprecedented amounts of online geolocated data from the World Wide Web, social media, and mobile-phone networks (Lazer et al., 2009; Salganik, 2018). When online geolocated data are publicly available and accessible, they provide a valuable opportunity to responsibly study social interactions (Lazer *et al.*, 2009; Salganik, 2018) and human mobility (González, Hidalgo, and Barabási, 2008). In principle, one can exploit the availability of such digital geolocated data, along with computing infrastructure and scalable algorithms, to increase understanding of migration networks. However, the adoption of those digital innovations in research on migration networks has been relatively slow because of various methodological, ethical, data-availability, and reproducibility challenges that are associated with social data from digital sources (Ruths and Pfeffer, 2014; Salganik, 2018; Jasny et al., 2017). In this section, we discuss both the promise and the challenges of digital technologies for studying migration networks.

Digital, Geolocated, and 'Big' Migration Data

For a long time, research on macroscale patterns of international migration has relied exclusively on administrative records and national statistics (such as population censuses, surveys, and population registers). Although such data are useful in many respects (e.g., because of their wide geographical coverage and public availability), they have important limitations (Rango and Vespe, 2017). First, data from administrative records and national statistics are compiled and available only after a substantial time lag (Zagheni, Weber, and Gummadi, 2017; Rango and Vespe, 2017). Second, country-level aggregation may prove to be too coarse spatially for insightful network analysis. Because country-level aggregate data may obscure migration patterns, it is important to employ location-specific data about movements between actual settlements to conduct increasingly thorough and fine-grained analysis of international migration. Consider, for example, that over 95% of the Bangladeshis in Great Britain (estimated at 200,000 people in the mid 1980s) originated from specific villages in the urban area of Sylhet, which is located in the northeastern part of Bangladesh (Gardner, 1995: 2; Skeldon, 2006: 22; Skeldon, 2018).

Online geolocated data can provide granular and timely information about migration. Recent research has studied international migration trends using non-representative digital geolocated data of human mobility from the World Wide Web, social media, and mobile phones (Rango and Vespe, 2017; Spyratos *et al.*, 2019; Ahas, Silm, and Tiru, 2016; State *et al.*, 2014; Zagheni *et al.*, 2014; Böhme, Gröger, and Stöhr, 2020). Some studies have inferred geolocated information about migration from geographical (longitude and latitude) coordinates that are associated with data from Twitter (Zagheni *et al.*, 2014), Google Trends (Böhme, Gröger, and Stöhr, 2020), IP (Internet Protocol) addresses (Zagheni and Weber, 2012), and mobile-phone networks (Ahas, Silm, and Tiru, 2016). Other studies have used anonymized and publicly available self-reported geolocated data, such as country of employment in the professional-networking platform LinkedIn (State *et al.*, 2014), previous locations of residence in Facebook's advertising platform (Spyratos *et al.*, 2019), and geotagged photographs in the photograph-sharing platform Flickr (Barchiesi *et al.*, 2015).

Although not without limitations, which we detail later, digital geolocated data from social media are a promising resource that one can extend to the study of migration networks. For example, Spyratos et al. (2019; see also Zagheni, Weber, and Gummadi, 2017) used publicly available and aggregate information from Facebook's advertising platform to estimate stocks of migrants—specifically, using the number of Facebook users in each country who previously resided in a different country—for 119 destination countries. In principle, one can apply network analysis to such data (and similar data of migrant stocks from other social-media platforms), but one also has to somehow address the problem of missing nodes and edges (e.g., from heterogeneous and potentially limited coverage of countries). To take such limitations into account, it will be helpful to use methods for inferring missing nodes and edges in networks (Kossinets, 2006; Guimerà and Sales-Pardo, 2009). One also needs to consider other likely limitations—including selection biases and undocumented criteria for establishing previous country of residence (Spyratos *et al.*, 2019)—of data from Facebook and other social-media platforms.

Another promising research avenue is to leverage data from online activities (e.g., Web searches) to anticipate future movements in the context of broader patterns in migration networks. For example, Böhme et al. (2020) combined geolocated online search data from Google Trends with bilateral migration flows from the OECD International Migration Database (IMD) and survey data on migration intentions from the Gallup World Poll to estimate international migration flows. Along those lines, albeit at a smaller scale, a Pew Research Center report (Connor, 2017) combined traditional migration data with geolocated data of Web searches to track migration patterns of refugees from the Middle East to Europe. These studies did not use network analysis, but (to the extent that the selection of any destination depends in part on other destinations) one can construct migration networks and track their dynamics to anticipate changes in some of the origin and destination countries, migration ties, and other structures (such as migration communities).

Data sources from mobile-phone call-detail records (CDRs) provide another opportunity to harness new technologies to track patterns of migration (Rango and Vespe, 2017) and thereby construct migration networks. One example is the Data for Development (D4D) challenge that released anonymized CDRs (de Montjoye *et al.*, 2014), which were then used to study networks of internal migration patterns (Martin-Gutierrez *et al.*, 2016). One can also combine CDR data with census data. For example, Eagle, Macy, and Claxton (2010) linked communication networks with national census data to study associations between network structure and socio-economic opportunities. Anonymized geolocated data from CDRs have typically been useful for tracking internal migration patterns in a single country (Rango and Vespe, 2017), although some studies have used both domestic and roaming CDRs to construct star-like networks and study 'transnational' mobility between a seed country and other countries (e.g., Ahas, Silm, and Tiru, 2016).

Despite their immense potential promise, it is important to acknowledge the limitations of data from social-media platforms, CDRs, and other digital sources for the study of migration networks (International Organization for Migration's Global Migration Data Analysis Centre (IOM's GMDAC), 2017). In addition to the fact that geolocated data from online sources are typically non-representative, which is widely acknowledged, such data are often more restricted in their geographical scope than global migration databases. Importantly, because digital geolocated data typically include traces of individual human behavior, protection of privacy and confidentiality is a major concern (de Montjoye *et al.*, 2018). The problem is particularly pertinent to network data because dependencies between observations (a hallmark of network data) complicate efforts at anonymization. Furthermore, digital geolocated data are often proprietary and rarely available to openly reuse and/or reanalyze, making it impossible to evaluate both the computational workflow and the reproducibility of results (Stodden *et al.*, 2016; Jasny *et al.*, 2017). Recent initiatives, such as D4D-Senegal (de Montjoye *et al.*, 2014) and the Social Science One program (King and Persily, 2019), have offered secure solutions for leveraging proprietary and individual-level data for open research while simultaneously preserving privacy and data integrity. The success of such socio-technological innovations of data access and reuse can transform computational studies of migration networks and other areas of human science.

Advances in Network Analysis and Related Methodologies

Prior research on migration networks has largely employed standard techniques and diagnostics from network analysis. However, the analysis of migration networks that one constructs from complex and 'big' geolocated data requires the application and development of new computational methods and models. Examples of recent advances in network methodologies, which we expect to be valuable for research on migration networks, include work on machine learning and networks (Grover and Leskovec, 2016), new spatial null models for community detection (Sarzynska *et al.*, 2016), Bayesian network models (Peixoto, 2019), multilayer network analysis (Kivelä et al., 2014), and spatial applications of topological data analysis (Feng and Porter, 2021).

Multilayer Networks of Migration and Other Exchanges between Places

In this chapter, we have restricted our discussion thus far to ordinary ('monolayer') networks, which include a single type of edge (where, in our case, the edges encode bilateral migration stocks or flows). However, since the late 1980s, international migration has been viewed in the context of other spatial interactions, including economic ones (e.g., international trade), historical ones (e.g., previous colonial ties), cultural ones (e.g., language), and political ones (e.g., bilateral agreements) (Malmberg, 1997: 40; Kritz and Zlotnik, 1992; Portes and Böröcz, 1989; Fawcett, 1989). Additionally, research using data from social-media platforms indicates that there is a strong association between international social relationships and international migration (Takhteyev, Gruzd, and Wellman, 2012; Chi *et al.*, 2020). To encode multiple relationships between places, one can construct a multiplex network (Wasserman and Faust, 1994; Kivelä *et al.*, 2014; Aleta and Moreno, 2019), in which nodes can be linked to each other via more than one type of relationship. A multiplex network has multiple 'layers', which in the above description encode economic, historical, cultural, political, or communication relationships between countries. There have been few empirical studies of the interactions between multiple layers of international relationships; notable exceptions include Belyi *et al.* (2017) and network research on migration and trade (Fagiolo and Mastrorillo, 2014).

The analysis of multiplex networks can also facilitate the examination of multiple types of international migration. For example, different layers in a multiplex network can encode different policy categories by entry (e.g., work, free movements, family, and humanitarian (OECD, 2016)) or types of migrants (including asylum seekers, refugees, family members, low-skilled labor migrants, students, and highly skilled migrants). Unfortunately, current estimates of migration stocks and flows are either aggregated or stratified by general migrant attributes, such as age (UN DESA, 2019; Brücker, Capuano, and Marfouk, 2013), gender (Özden *et al.*, 2011; Abel, 2018), and educational attainment (Brücker, Capuano, and Marfouk, 2013; Docquier, Lowell, and Marfouk, 2009). Data on types of migrants are scattered and rarely available, except for OECD (Organisation for Economic Co-operation and Development) countries. For some types of migrants, it is possible to use estimates from digital geolocated data. For example, State et al. (2014) examined data from LinkedIn to track international migration flows of highly skilled migrants. It is desirable to extend such analysis to multiplex networks. Constructing a multiplex migration network requires the integration of comparable data from multiple sources, such as data for different types of migrants or from multiple social-media platforms.

With the availability of harmonized migration data for different types of migrants, multi-layer network analysis may provide both a theoretical framework and a practical approach for systematic investigations of multiple socio-economic relationships and/or multiple types of migration between world countries.

Relevance to Theory, Societal Issues, and Policy

To establish its relevance and impact, network-based research on macroscale migration patterns should target research questions of theoretical, societal, and policy importance. Research on macroscale migration networks thus far has primarily tested propositions from prior migration theories (e.g., Windzio, 2018; Danchev and Porter, 2018). Future research efforts should explore opportunities to develop new theoretical propositions. One important theoretical development is the conceptualization of the roles of migration networks in perpetuating, generating, or alleviating global inequalities in migration. For example, research on interpersonal networks has explored the conditions under which ethnic migrant networks have beneficial or adverse effects for economic integration of migrants (Portes, 1998; Martén, Hainmueller, and Hangartner, 2019).

Furthermore, to produce policy-relevant findings, research on migration networks should engage with concepts, categories, and legal-policy frameworks at subnational, national, and international levels (Betts, 2011; International Organization for Migration, 2019). One example of policy-relevant work is a recent paper by Bansak et al. (2018), who employed data-driven matching algorithms to assign refugees to jobs to improve refugee integration.

CONCLUSIONS

Advances in transportation, information, and communication technologies have combined with policy, economic, and social forces to reshape migration since the 1970s. The understanding of complex migration patterns cannot rely solely on bilateral-level investigations, and systematic network analysis can help shed light on unknown and unappreciated migration dynamics. For many decades, migration theories (Hägerstrand, 1957; Kritz and Zlotnik, 1992; Zlotnik, 1992; Mabogunje, 1970; Salt, 1989; Fawcett, 1989) developed the intuition of global migration as a network of places that are linked by movements, but these investigations lacked the formal language and techniques of network analysis to systematically encode migration connections between places in a network and study them explicitly from a network-based perspective. In this chapter, we reviewed ideas and techniques from network analysis that have been applied to international migration and have yielded fascinating insights into the structure and dynamics of world migration.

We discussed advances of digital technologies that have made it possible to generate, store, and analyze large quantities of geolocated data from online sources, mobile phones, and other resources. We outlined ways in which one can combine data from these new sources with more traditional data from national statistics, administrative records, and surveys to study the structure and dynamics of macroscale migration networks. There are many important challenges—including ones that are related to methodology, data privacy, geographical coverage, data availability, and research transparency—that have impacted the slow adoption of new data sources and data integration in the study of global migration networks.

For data sources to be available for reuse, it is crucial to implement socio-technological innovations to simultaneously improve privacy protection, ensure the secure use of proprietary data for research purposes, and promote research transparency and reproducibility. Making advances on issues such as data ethics and data availability will help investigations of migration networks to make better use of new developments in network analysis (e.g., new methods for community detection, multilayer network analysis, and statistical data analysis) and other promising areas of research, including machine learning, Bayesian statistics, and topological data analysis. Finally, it is important for theory-driven and problem-driven investigations that foster public debate and inform evidence-based policy to help guide future research efforts. Leveraging technological advances to make progress in such efforts requires interdisciplinary collaborations across many areas of research, including migration studies, geography, network science, computational social science, digital technologies, machine learning, and mathematics.

ACKNOWLEDGEMENTS

We thank Marie McAuliffe for the invitation to write this chapter and for helpful comments. The chapter draws on our prior work on these topics. We thank Marya Bazzi, Adam Dennett, Bernie Hogan, Lucas Jeub, Michael Keith, Mikko Kivelä, Renaud Lambiotte, Martin Ruhs, and Marta Sarzynska for helpful discussions.

NOTES

1. In this chapter, we draw on Danchev (2015) and Danchev and Porter (2018).
2. By convention, we refer to 'the' global migration network as a singular entity. However, there are many possible realizations of a global migration network, depending on data sources and on the ways in which one constructs such a network.

3.   Some networks of human mobility have more stringent spatial constraints than migration networks. For example, daily commuting depends on spatially embedded networks (e.g., transportation systems) (Montis, Caschili, and Chessa, 2013).

REFERENCES

Abel, G. J. (2018) 'Estimates of global bilateral migration flows by gender between 1960 and 2015', *International Migration Review*, **52**(3), 809–852.

Abel, G. J. and Sander, N. (2014) 'Quantifying global international migration flows', *Science*, **343**(6178), 1520–1522.

Abel, G. J., DeWaard, J., Ha, J. T., and Almquist, Z. W. (2021) 'The form and evolution of international migration networks, 1990–2015', *Population, Space and Place*, **27**(3), e2432.

Ahas, R., Silm, S., and Tiru, M. (2016) 'Tracking trans-nationalism with mobile telephone data', in *Estonian Human Development Report 2016/2017: Estonia at the Age of Migration*. Available at https://2017.inimareng.ee/en/open-to-the-world/tracking-trans-nationalism-with-mobile-telephone -data// (accessed on 18 August 2021).

Aleta, A. and Moreno, Y. (2019) 'Multilayer networks in a nutshell', *Annual Review of Condensed Matter Physics*, **10**(1), 45–62.

Backstrom, L., Sun, E., and Marlow, C. (2010) 'Find me if you can: improving geographical prediction with social and spatial proximity', *Proceedings of the 19th International Conference on World Wide Web*, Association for Computing Machinery, 61–70.

Baek, E. C., Porter, M. A., and Parkinson, C. (2021) 'Social network analysis for social neuroscientists', *Social Cognitive and Affective Neuroscience*, **16**(8), 883–901.

Bansak, K., Ferwerda, J., Hainmueller, J., Dillon, A., Hangartner, D., Lawrence, D., and Weinstein, J. (2018) 'Improving refugee integration through data-driven algorithmic assignment', *Science*, **359**(6373), 325–329.

Barchiesi, D., Moat, H. S., Alis, C., Bishop, S., and Preis, T. (2015) 'Quantifying international travel flows using Flickr', *PLoS ONE*, **10**(7), e0128470.

Barrat, A., Barthélemy, M., Pastor-Satorras, R., and Vespignani, A. (2004) 'The architecture of complex weighted networks', *Proceedings of the National Academy of Sciences of the United States of America*, **101**(11), 3747–3752.

Barthelemy, M. (2018) *Morphogenesis of spatial networks*. Cham, Switzerland: Springer International Publishing.

Belyi, A., Bojic, I., Sobolevsky, S., Sitko, I., Hawelka, B., Rudikova, L., Kurbatski, A., and Ratti, C. (2017) 'Global multi-layer network of human mobility', *International Journal of Geographical Information Science*, **31**(7), 1381–1402.

Betts, A. (2011) *Global migration governance*. Oxford, UK: Oxford University Press.

Bilecen, B., Gamper, M., and Lubbers, M. J. (2018) 'The missing link: social network analysis in migration and transnationalism', *Social Networks*, **53**, 1–3.

Blumenstock, J. E., Chi, G., and Tan, X. (2019) Migration and the value of social networks. *CEPR Discussion Paper No. DP13611*. Available at https://cepr.org/active/publications/discussion_papers/ dp.php?dpno=13611 (accessed on 18 August 2021).

Böhme, M. H., Gröger, A., and Stöhr, T. (2020) 'Searching for a better life: predicting international migration with online search keywords', *Journal of Development Economics*, **142**, 102347.

Borgatti, S. P., Everett, M. G., and Johnson, J. C. (2018) *Analyzing social networks*. London, UK: SAGE Publications.

Borgatti, S. P., Mehra, A., Brass, D. J., and Labianca, G. (2009) 'Network analysis in the social sciences', *Science*, **323**(5916), 892–895.

Boyd, M. (1989) 'Family and personal networks in international migration: recent developments and new agendas', *International Migration Review*, **23**(3), 638–670.

Brandes, U., Robins, G., McCranie, A. N. N., and Wasserman, S. (2013) 'What is network science?', *Network Science*, **1**(1), 1–15.

Brücker, H., Capuano, S., and Marfouk, A. (2013) 'Education, gender and international migration: insights from a panel-dataset 1980–2010'. Available at https://www.iab.de/en/daten/iab-brain-drain -data.aspx (accessed on 18 August 2021).

Brunn, S. D. and Leinbach, T. R. (1991) *Collapsing space and time: geographic aspects of communications and information*. London, UK: HarperCollins Publishers LLC.

Butts, C. T. (2008) 'Social network analysis: a methodological introduction', *Asian Journal of Social Psychology*, **11**(1), 13–41.

Butts, C. T. (2009) 'Revisiting the foundations of network analysis', *Science*, **325**(5939), 414–416.

Cerqueti, R., Clemente, G. P., and Grassi, R. (2019) 'A network-based measure of the socio-economic roots of the migration flows', *Social Indicators Research*, **146**(1), 187–204.

Charyyev, B. and Gunes, M. H. (2019) 'Complex network of United States migration', *Computational Social Networks*, **6**(1).

Chi, G., State, B., Blumenstock, J. E., and Adamic, L. (2020) 'Who ties the world together? Evidence from a large online social network', *Complex Networks and Their Applications VIII*. Cham, Switzerland: Springer International Publishing, 451–465.

Connor, P. (2017) *The digital footprint of Europe's refugees*, Washington, D.C. Available at https:// www.pewresearch.org/global/2017/06/08/digital-footprint-of-europes-refugees/ (accessed on 19 May 2020).

Cozzo, E., Kivelä, M., De Domenico, M., Solé-Ribalta, A., Arenas, A., Gómez, S., Porter, M. A., and Moreno, Y. (2015) 'Structure of triadic relations in multiplex networks', *New Journal of Physics*, **17**(7), 073029.

Cushing, B. and Poot, J. (2003) 'Crossing boundaries and borders: regional science advances in migration modelling', *Papers in Regional Science*, **83**(1), 317–338.

Danchev, V. (2015) *Spatial network structures of world migration: heterogeneity of global and local connectivity*. DPhil thesis, University of Oxford.

Danchev, V. and Porter, M. A. (2018) 'Neither global nor local: heterogeneous connectivity in spatial network structures of world migration', *Social Networks*, **53**, 4–19.

Davis, J. A. (1967) 'Clustering and structural balance in graphs', *Human Relations*, **20**(2), 181–187.

Davis, K. F., D'Odorico, P., Laio, F., and Ridolfi, L. (2013) 'Global spatio-temporal patterns in human migration: a complex network perspective', *PLoS ONE*, **8**(1), e53723.

de Montjoye, Y.-A., Gambs, S., Blondel, V., Canright, G., de Cordes, N., Deletaille, S., Engø-Monsen, K., Garcia-Herranz, M., Kendall, J., Kerry, C., Krings, G., Letouzé, E., Luengo-Oroz, M., Oliver, N., Rocher, L., Rutherford, A., Smoreda, Z., Steele, J., Wetter, E., Pentland, A. S., and Bengtsson, L. (2018) 'On the privacy-conscientious use of mobile phone data', *Scientific Data*, **5**(1), 180286.

de Montjoye, Y.-A., Smoreda, Z., Trinquart, R., Ziemlicki, C., and Blondel, V. D. (2014) 'D4D-Senegal: the second mobile phone data for development challenge', *arXiv:1407.4885*. A preprint is available at https://arxiv.org/abs/1407.4885.

Dekker, D., Krackhardt, D., and Snijders, T. (2007) 'Sensitivity of MRQAP tests to collinearity and autocorrelation conditions', *Psychometrika*, **72**(4), 563–581.

DeWaard, J., Kim, K., and Raymer, J. (2012) 'Migration systems in Europe: evidence from harmonized flow data', *Demography*, **49**(4), 1307–1333.

Docquier, F., Lowell, B. L., and Marfouk, A. (2009) 'A gendered assessment of highly skilled emigration', *Population and Development Review*, **35**(2), 297–321.

Eagle, N., Macy, M., and Claxton, R. (2010) 'Network diversity and economic development', *Science*, **328**(5981), 1029–1031.

Expert, P., Evans, T. S., Blondel, V. D., and Lambiotte, R. (2011) 'Uncovering space-independent communities in spatial networks', *Proceedings of the National Academy of Sciences of the United States of America*, **108**(19), 7663–7668.

Fagiolo, G. (2007) 'Clustering in complex directed networks', *Physical Review E*, **76**(2), 026107.

Fagiolo, G. and Mastrorillo, M. (2013) 'International migration network: topology and modeling', *Physical Review E*, **88**(1), 012812.

Fagiolo, G. and Mastrorillo, M. (2014) 'Does human migration affect international trade? A complex-network perspective', *PLoS ONE*, **9**(5), e97331.

Farmer, C. J. Q. and Fotheringham, A. S. (2011) 'Network-based functional regions', *Environment and Planning A: Economy and Space*, **43**(11), 2723–2741.

Fawcett, J. T. (1989) 'Networks, linkages, and migration systems', *International Migration Review*, **23**(3), 671–680.

Fawcett, J. T. and Arnold, F. (1987) 'The role of surveys in the study of international migration: an appraisal', *International Migration Review*, **21**(4), 1523–1540.

Feng, M. and Porter, M. A. (2021) 'Persistent homology of geospatial data: a case study with voting', *SIAM Review*, **63**(1), 67–99.

Fortunato, S. and Hric, D. (2016) 'Community detection in networks: a user guide', *Physics Reports*, **659**, 1–44.

Freeman, L. C. (1978) 'Centrality in social networks conceptual clarification', *Social Networks*, **1**(3), 215–239.

Gardner, K. (1995) *Global migrants, local lives: travel and transformation in rural Bangladesh*. Oxford, UK: Clarendon Press.

González, M. C., Hidalgo, C. A., and Barabási, A.-L. (2008) 'Understanding individual human mobility patterns', *Nature*, **453**(7196), 779–782.

Grigg, D. B. (1977) 'E. G. Ravenstein and the "laws of migration"', *Journal of Historical Geography*, **3**(1), 41–54.

Grover, A. and Leskovec, J. (2016) 'node2vec: scalable feature learning for networks', *Proceedings of the 22nd ACM SIGKDD International Conference on Knowledge Discovery and Data Mining*, San Francisco, CA, USA: Association for Computing Machinery, 855–864.

Guimerà, R. and Sales-Pardo, M. (2009) 'Missing and spurious interactions and the reconstruction of complex networks', *Proceedings of the National Academy of Sciences of the United States of America*, **106**(52), 22073–22078.

Gurak, D. T. and Caces, F. (1992) 'Migration networks and the shaping of migration systems', in Kritz, M.M., Lim, L.L., and Zlotnik, H. (eds), *International migration systems: a global approach*. Oxford, UK: Clarendon Press, pp. 150–176.

Hagberg, A., Swart, P., and S Chult, D. (2008) *Exploring network structure, dynamics, and function using NETWORKX*. Los Alamos National Laboratory (LANL).

Hägerstrand, T. (1957) 'Migration and area: survey of a sample of Swedish migration fields and hypothetical considerations on their genesis', in Hannerberg, D., Hagerstrand, T., and Odeving, B. (eds), *Migration in Sweden: a symposium*. Lund, Sweden: Lund Studies in Geography. Series B: Human Geography, pp. 26–158.

Haggett, P. and Chorley, R. J. (1969) *Network analysis in geography*. London, UK: Edward Arnold.

Harvey, D. (1989) *The condition of postmodernity: an enquiry into the origins of cultural change*. Cambridge, MA, USA: Blackwell Publishers.

Hatton, T. J. and Williamson, J. G. (2005) *Global migration and the world economy: two centuries of policy and performance*. Cambridge, MA, USA: MIT Press.

Herdağdelen, A., State, B., Adamic, L., and Mason, W. (2016) 'The social ties of immigrant communities in the United States', *Proceedings of the 8th ACM Conference on Web Science*, Hannover, Germany: Association for Computing Machinery, 78–84.

Holme, P. and Saramäki, J. (eds) (2019) *Temporal network theory*. Cham, Switzerland: Springer International Publishing.

Hunter, J. D. (2007) 'MATPLOTLIB: a 2D graphics environment', *Computing in Science & Engineering*, **9**(3), 90–95.

International Organization for Migration (2003) *Managing migration: challenges and responses for people on the move*. Geneva, Switzerland: IOM.

International Organization for Migration (2017) *World migration report 2018*. Geneva, Switzerland: IOM.

International Organization for Migration (2019) *World migration report 2020*. Geneva, Switzerland: IOM.

International Organization for Migration's Global Migration Data Analysis Centre (IOM's GMDAC) (2017) *Innovative data sources*, Washington, DC, USA: Global Knowledge Partnership for Migration and Development (KNOMAD), World Bank.

Jackson, M. O. (2008) *Social and economic networks*. Princeton, NJ, USA: Princeton University Press.

Jasny, B. R., Wigginton, N., McNutt, M., Bubela, T., Buck, S., Cook-Deegan, R., Gardner, T., Hanson, B., Hustad, C., Kiermer, V., Lazer, D., Lupia, A., Manrai, A., McConnell, L., Noonan, K., Phimister,

E., Simon, B., Strandburg, K., Summers, Z., and Watts, D. (2017) 'Fostering reproducibility in industry–academia research', *Science*, **357**(6353), 759–761.

King, G. and Persily, N. (2019) 'A new model for industry–academic partnerships', *PS: Political Science and Politics*, **53**(4), 703–709.

Kivelä, M., Arenas, A., Barthelemy, M., Gleeson, J. P., Moreno, Y., and Porter, M. A. (2014) 'Multilayer networks', *Journal of Complex Networks*, **2**(3), 203–271.

Kolaczyk, E. D. (2009) *Statistical analysis of network data: methods and models*. Heidelberg, Germany: Springer-Verlag.

Kossinets, G. (2006) 'Effects of missing data in social networks', *Social Networks*, **28**(3), 247–268.

Kritz, M. and Zlotnik, H. (1992) 'Global interactions: migration systems, processes, and policies', in Kritz, M.M., Lim, L.L., and Zlotnik, H. (eds), *International migration systems: a global approach*. Oxford, UK: Clarendon Press, pp. 1–16.

Kritz, M. M., Lim, L. L., and Zlotnik, H. (eds) (1992) *International migration systems: a global approach*. Oxford, UK: Clarendon Press.

Krivitsky, P. N. and Handcock, M. S. (2014) 'A separable model for dynamic networks', *Journal of the Royal Statistical Society: Series B (Statistical Methodology)*, **76**(1), 29–46.

Lazer, D., Pentland, A., Adamic, L., Aral, S., Barabási, A.-L., Brewer, D., Christakis, N., Contractor, N., Fowler, J., Gutmann, M., Jebara, T., King, G., Macy, M., Roy, D., and Van Alstyne, M. (2009) 'Computational social science', *Science*, **323**(5915), 721–723.

Leicht, E. A. and Newman, M. E. (2008) 'Community structure in directed networks', *Physical Review Letters*, **100**(11), 118703.

Lemercier, C. (2010) 'Formal network methods in history: why and how?', in Fertig, G. (ed.), *Social networks, political institutions, and rural societies*. Turnhout, Belgium: Brepols Publishers, pp. 281–310.

Lemercier, C. and Rosental, P.-A. (2010) *The structure and dynamics of migration patterns in 19th-century Northern France*. Available at https://hal-sciencespo.archives-ouvertes.fr/hal-01063603 (accessed on 26 July 2020).

Liu, M.-M. (2013) 'Migrant networks and international migration: testing weak ties', *Demography*, **50**(4), 1243–1277.

Lubbers, M. J., Molina, J. L., Lerner, J., Brandes, U., Ávila, J., and McCarty, C. (2010) 'Longitudinal analysis of personal networks: the case of Argentinean migrants in Spain', *Social Networks*, **32**(1), 91–104.

Lusher, D., Koskinen, J., and Robins, G. (2013) *Exponential random graph models for social networks: theory, methods, and applications: structural analysis in the social sciences*. Cambridge, UK: Cambridge University Press.

Mabogunje, A. L. (1970) 'Systems approach to a theory of rural–urban migration', *Geographical Analysis*, **2**(1), 1–18.

Maier, G. and Vyborny, M. (2008) 'Internal migration between US states: a social network analysis', in Poot, J., Waldorf, B., and van Wissen, L. (eds), *Migration and human capital*. Cheltenham, UK and Northampton, MA, USA: Edward Elgar Publishing, pp. 281–310.

Malmberg, G. (1997) 'Time and space in international migration', in Hammar, T., Brochmann, G., Tamas, K., and Faist, T. (eds), *International migration, immobility, and development: multidisciplinary perspectives*. Oxford, UK: Berg Publishers, pp. 21–48.

Maoz, Z. (2011) *Networks of nations: the evolution, structure, and impact of international networks, 1816–2001*. Cambridge, UK: Cambridge University Press.

Martén, L., Hainmueller, J., and Hangartner, D. (2019) 'Ethnic networks can foster the economic integration of refugees', *Proceedings of the National Academy of Sciences of the United States of America*, **116**(33), 16280–16285.

Martin-Gutierrez, S., Borondo, J., Morales, A. J., Losada, J. C., Tarquis, A. M., and Benito, R. M. (2016) 'Agricultural activity shapes the communication and migration patterns in Senegal', *Chaos*, **26**(6), 065305.

Massey, D. S., Arango, J., Hugo, G., Kouaouci, A., Pellegrino, A., and Taylor, J. E. (1998) *Worlds in motion: understanding international migration at the end of the millennium*. Oxford, UK: Oxford University Press.

Messias, J., Benevenuto, F., Weber, I., and Zagheni, E. (2016) 'From migration corridors to clusters: the value of Google+ data for migration studies', *2016 IEEE/ACM International Conference on Advances in Social Networks Analysis and Mining (ASONAM)*: Institute of Electrical and Electronics Engineers, pp. 421–428.

Montis, A., Caschili, S., and Chessa, A. (2013) 'Commuter networks and community detection: a method for planning sub regional areas', *The European Physical Journal Special Topics*, **215**(1), 75–91.

Mucha, P. J., Richardson, T., Macon, K., Porter, M. A., and Onnela, J.-P. (2010) 'Community structure in time-dependent, multiscale, and multiplex networks', *Science*, **328**(5980), 876–878.

Newman, M. E. J. (2018) *Networks*. 2nd edn. Oxford, UK: Oxford University Press.

Newman, M. E. J. (2006a) 'Finding community structure in networks using the eigenvectors of matrices', *Physical Review E*, **74**(3), 036104.

Newman, M. E. J. (2006b) 'Modularity and community structure in networks', *Proceedings of the National Academy of Sciences of the United States of America*, **103**(23), 8577–8582.

Newman, M. E. J. and Girvan, M. (2004) 'Finding and evaluating community structure in networks', *Physical Review E*, **69**(2), 026113.

Nogle, J. M. (1994) 'The systems approach to international migration: an application of network analysis method', *International Migration: Quarterly Review*, **32**, 329–342.

Novotný, J. and Hasman, J. (2016) 'Exploring the spatial relatedness network of the global system of international migration', *Journal of Maps*, **12**(sup1), 570–576.

OECD (2016) *OECD factbook 2015–2016: economic, environmental and social statistics*, Paris, France: OECD Publishing. Available at https://www.oecd-ilibrary.org/content/publication/factbook-2015-en (accessed on 18 August 2021).

Özden, C., Parsons, C. R., Schiff, M., and Walmsley, T. L. (2011) 'Where on Earth is everybody? The evolution of global bilateral migration 1960–2000', *World Bank Economic Review*, **25**(1), 12–56.

Palloni, A., Massey, Douglas S., Ceballos, M., Espinosa, K., and Spittel, M. (2001) 'Social capital and international migration: a test using information on family networks', *American Journal of Sociology*, **106**(5), 1262–1298.

Peixoto, T. P. (2019) 'Bayesian stochastic blockmodeling', in Doreian, P., Batagelj, V., and Ferligoj, A. (eds), *Advances in network clustering and blockmodeling*, Hoboken, NJ, USA: John Wiley & Sons, Inc., pp. 289–332.

Peres, M., Xu, H., and Wu, G. (2016) 'Community evolution in international migration top1 networks', *PLoS ONE*, **11**(2), e0148615.

Porter, M. A. (2020) 'Nonlinearity + networks: a 2020 vision', in Kevrekidis, P. G., Cuevas-Maraver, J., and Saxena, A. B. (eds), *Emerging Frontiers in Nonlinear Science*. Cham, Switzerland: Springer International Publishing, pp. 131–159.

Porter, M. A., Onnela, J.-P., and Mucha, P. J. (2009) 'Communities in networks', *Notices of the American Mathematical Society*, **56**(9), 1082–1097, 1164–1166.

Portes, A. (1998) 'Social capital: its origins and applications in modern sociology', *Annual Review of Sociology*, **24**(1), 1–24.

Portes, A. and Böröcz, J. (1989) 'Contemporary immigration: theoretical perspectives on its determinants and modes of incorporation', *International Migration Review*, **23**(3), 606–630.

Rango, M. and Vespe, M. (2017) *Big Data and alternative data sources on migration: from case-studies to policy support—Summary report*, Ispra, Italy. Available at https://knowledge4policy.ec.europa.eu/sites/default/files/BD4M-workshop-2017-summary-report.pdf (accessed on 18 August 2021).

Ratti, C., Sobolevsky, S., Calabrese, F., Andris, C., Reades, J., Martino, M., Claxton, R., and Strogatz, S. H. (2010) 'Redrawing the map of Great Britain from a network of human interactions', *PLoS ONE*, **5**(12), e14248.

Ravenstein, E. G. (1885) 'The laws of migration', *Journal of the Statistical Society of London*, **48**(2), 167–235.

Ruths, D. and Pfeffer, J. (2014) 'Social media for large studies of behavior', *Science*, **346**(6213), 1063–1064.

Salganik, M. J. (2018) *Bit by bit: social research in the digital age*. Princeton, NJ, USA: Princeton University Press.

Salt, J. (1989) 'A comparative overview of international trends and types, 1950–80', *International Migration Review*, **23**(3), 431–456.

Salt, J. (2013 [1986]) 'International migration: a spatial theoretical approach', in Pacione, M. (ed.), *Population geography: progress & prospect*. London, UK: Routledge, pp. 166–193.

Saramäki, J., Kivelä, M., Onnela, J.-P., Kaski, K., and Kertész, J. (2007) 'Generalizations of the cluster-ing coefficient to weighted complex networks', *Physical Review E*, **75**(2), 027105.

Sarzynska, M., Leicht, E. L., Chowell, G., and Porter, M. A. (2016) 'Null models for community detec-tion in spatially embedded, temporal networks', *Journal of Complex Networks*, **4**(3), 363–406.

Skeldon, R. (2006) 'Interlinkages between internal and international migration and development in the Asian region', *Population, Space and Place*, **12**(1), 15–30.

Skeldon, R. (2018) *International migration, internal migration, mobility and urbanization: towards more integrated approaches*. Geneva, Switzerland: International Organization for Migration. Available at: https://publications.iom.int/books/mrs-no-53-international-migration-internal-migration-mobility -and-urbanization-towards-more (accessed on 3 October 2021).

Slater, P. B. (2008) *Hubs and clusters in the evolving United States internal migration network*, arXiv: 0809.2768. A preprint is available at https://arxiv.org/abs/0809.2768.

Snijders, T., van de Bunt, G. G., and Steglich, C. E. G. (2010) 'Introduction to stochastic actor-based models for network dynamics', *Social Networks*, **32**(1), 44–60.

Spyratos, S., Vespe, M., Natale, F., Weber, I., Zagheni, E., and Rango, M. (2019) 'Quantifying interna-tional human mobility patterns using Facebook Network data', *PLoS ONE*, **14**(10), e0224134.

State, B., Rodriguez, M., Helbing, D., and Zagheni, E. (2014) 'Where does the US stand in the inter-national competition for talent? Evidence from LinkedIn data', *Population Association of America 2014 Annual Meeting*, Boston, MA, USA. Available at https://paa2014.princeton.edu/papers/141226 (accessed on 18 August 2021).

Stodden, V., McNutt, M., Bailey, D. H., Deelman, E., Gil, Y., Hanson, B., Heroux, M. A., Ioannidis, J. P. A., and Taufer, M. (2016) 'Enhancing reproducibility for computational methods', *Science*, **354**(6317), 1240–1241.

Takhteyev, Y., Gruzd, A., and Wellman, B. (2012) 'Geography of Twitter networks', *Social Networks*, **34**(1), 73–81.

Tranos, E., Gheasi, M., and Nijkamp, P. (2015) 'International migration: a global complex network', *Environment and Planning B: Urban Analytics and City Science*, **42**(1), 4–22.

UN DESA: Department of Economic and Social Affairs (2019) 'International migrant stock 2019 (United Nations database, POP/DB/MIG/Stock/Rev.2019)'. Available at: https://www.un.org/en/ development/desa/population/migration/data/estimates2/estimates19.asp (accessed on 18 August 2021).

Vacca, R., Solano, G., Lubbers, M. J., Molina, J. L., and McCarty, C. (2018) 'A personal network approach to the study of immigrant structural assimilation and transnationalism', *Social Networks*, **53**(1), 72–89.

Verdery, A. M., Mouw, T., Edelblute, H., and Chavez, S. (2018) 'Communication flows and the durabil-ity of a transnational social field', *Social Networks*, **53**, 57–71.

Vertovec, S. (2010) 'General Introduction', in Vertovec, S. (ed.), *Migration*. London, UK: Routledge, pp. 1–8.

Vincent, C. and Macleod, B. (1974) 'An application of network theory to migration and analysis', *Canadian Studies in Population*, **1**, 43–59.

Wallerstein, I. M. (1974) *The modern world-system I: capitalist agriculture and the origins of the European world-economy in the sixteenth century*. New York, NY, USA: Academic Press.

Wasserman, S. and Faust, K. (1994) *Social network analysis: methods and applications*. Cambridge, UK: Cambridge University Press.

Windzio, M. (2018) 'The network of global migration 1990–2013: using ERGMs to test theories of migration between countries', *Social Networks*, **53**, 20–29.

Zagheni, E., Garimella, V. R. K., Weber, I., and State, B. (2014) 'Inferring international and internal migration patterns from Twitter data', *Proceedings of the companion publication of the 23rd interna-tional conference on World Wide Web*. New York, NY, USA: Association for Computing Machinery, pp. 439–444.

Zagheni, E. and Weber, I. (2012) 'You are where you e-mail: using e-mail data to estimate international migration rates', *Proceedings of the 4th Annual ACM Web Science Conference*, Evanston, IL, USA: Association for Computing Machinery, pp. 348–351.

Zagheni, E., Weber, I., and Gummadi, K. (2017) 'Leveraging Facebook's advertising platform to monitor stocks of migrants', *Population and Development Review*, **43**(4), 721–734.

Zipf, G. K. (1946) 'The P_1P_2/D hypothesis: on the intercity movement of persons', *American Sociological Review*, **11**(6), 677–686.

Zlotnik, H. (1992) 'Empirical identification of international migration systems', in Kritz, M.M., Lim, L.L., and Zlotnik, H. (eds), *International migration systems: a global approach.* Oxford, UK: Clarendon Press, pp. 1–16.

PART II

DIGITAL TECHNOLOGY AND THE ACT OF MOVING: (IM)MOBILITY, BARRIERS AND BORDERS

7. Navigating borders/navigating networks: migration, technology and social capital

Farah Azhar, Sara Vannini, Bryce Clayton Newell and Ricardo Gomez

INTRODUCTION

In the 1990s, entry into the United States from Latin America became more difficult as the US–Mexico border became heavily militarized (Massey, Durand and Malone 2002). From 1986 to 2004, the budgetary increase in surveillance was tenfold, with more officers deployed at the border, and the number of deportations also expanded by a factor of ten (Massey 2005). Despite this, 10.3 million immigrants from Latin America were admitted legally from 1980 to 2006. However, there was also an increase in undocumented immigrants (Huntington 2004; Dobbs 2006). The act of migration is influenced by global media and its representation of foreign land. This representation plays a pivotal role in lives of individuals who are considering migration (King and Wood 2001). Research suggests that the use of information and communication technologies (ICTs) helps migrants get relevant information prior to migration and helps them comprehend the implications and consequences of migrating (Newell, Gomez and Guajardo 2016; Newell and Gomez 2015; Horst 2006). Panagakos and Horst (2006, p. 120) state that, "There is no doubt that new technologies have an impact on how transnational migrants imagine, negotiate and create their social worlds across broad transnational fields."

Social networks are integral to the process of migration (Fall 1998). People who have migratory experience become a valuable source of social capital to other people who are planning some form of migration, and those with friends with migration experience have a greater incentive to migrate themselves (Massey and Aysa-Lastra 2011). Social media has the potential to transform migration networks (Dekker and Engbersen 2014). The internet has thus given users an opportunity to be active producers as well as consumers of media content. Social media is not only a personal communication tool but a gateway for distributing information. This new generation of ICTs is significant for migrants and non-migrants who are geographically scattered but remain connected through transnational networks (Kissau 2012; Mahler 2001). The emergence of new technologies has lowered the communication- and travel-related costs of migration and has increased the "richness" of communication content. New technologies have led to the emergence of transnational identities as a new factor in traditional patterns of migration and assimilation into host societies (Bates and Komito 2012).

In their migration journeys, migrants have access to wider information services via different mobile technological devices like smartphones, global positioning apps, social media, and instant messaging systems (such as WhatsApp). The "polymedia" affordances of smartphones go beyond calling and texting (Madianou 2014) and provide more autonomy to migrants in certain contexts (e.g., in migration journeys to Europe; Schroeder 2015, as cited in Zijlstra and Liempt 2017). The portability of mobile phones allows migrants to adapt to changing circumstances during the journey (Schaub 2012). Information is often sought via online social networks. Migrants can stay connected to friends and family by exchanging pictures and text messages. Mobile phones may give migrants the feeling of security that they can always call for help if need arises (Germann Molz and Paris 2015).

However, depending on their circumstances and intended destinations, migrants might not be able or might not be willing to use ICTs. First, socio-economic status, level of education, urban/rural residence, gender and age can cause significant differences in people's ability to use ICTs, including migrants (Benitez 2006; Hamel 2009). Extensive literature indicates that ICTs may empower communities and groups that have been historically underserved, including women (Arun, Heeks and Morgan 2006). Yet, the literature also indicates how the groups with the least resources, including less-educated, lower-class women are not benefitted (Arun et al. 2006). Second, migrants might choose not to use ICTs during their journeys, and especially not mobile phones, as they perceive them as possible tools of surveillance. The perception that turning on a mobile phone will show their location to border agents is quite diffused among migrants at the US–Mexico border (Newell, Gomez and Guajardo 2016; 2017). Additionally, certain border regions, including along the US–Mexico border, do not always have the cellular infrastructure required to support consistent wireless connectivity.

In this chapter, we examine research on the role and use of online social media and other ICTs in the context of migration. We link the use of ICTs (including social media) by undocumented migrants—and the opportunities and risks enabled by these ICTs—to Bourdieu's (1986) definition of social capital. Our past research is particularly focused on undocumented migration into the United States across the country's southern border with Mexico and, to a lesser extent, migrant integration into various communities in the United States. The primary empirical research we reflect on here is drawn from three interrelated lines of research: the first was a study of the migration-related information practices of clandestine migrants in a shelter in Nogales, Sonora, Mexico (Newell, Gomez and Guajardo 2016; Newell, Gomez and Guajardo 2017; Newell and Gomez 2015; Yefimova et al. 2015); the second was a study of the information practices of Latino migrants, many of whom were undocumented, at the US–Mexico border, in Seattle, Washington, and in Cali, Colombia (Gomez and Vannini 2017; Gomez 2016; Gomez and Vannini 2015; Gomez, Gomez and Vannini 2017; Vannini, Gomez and Guajardo 2016); and the third was a study of the information practices of the humanitarian migrant-aid organizations that work to serve and support clandestine and undocumented migrants in and around Nogales, Sonora, and Nogales, Arizona (Gomez, Newell and Vannini 2020; Vannini, Gomez and Newell 2019; Newell, Vannini and Gomez 2020).

RELATED WORK

We discuss two areas of related work: the information practices, technology use and social networks of migrants, and the concept of social capital and how it relates to the migration experience.

Information Practices, Technology Use and Migrants' Social Networks

In some contexts, migration is viewed as a sense of achievement and the migrants are thought of as "national heroes" (Riccio 2006). Migration offers a unique opportunity to reinvent oneself. As Rouse (1995, p. 356) wrote, migration involves "asserting and organizing around either revalorized versions of ascribed identities or new ones that the (im)migrants develop for themselves." Chen and Choi (2011) found that computer-mediated social support is a precious supplement to the migrants' offline social support and that a growing number of Chinese migrants in Singapore go online to request and exchange information, counsel, companionship and even tangible assistance. Baron and Gomez (2017) examined how information practices affect the migration process by collecting stories from undocumented Latino migrants in the US. They argue that migration is not a linear process and show how ICTs facilitate the formation of new national–transnational identities, inculcate a sense of "in-betweenness," and generate understandings of nationhood.

Dekker and Engberson (2014) conducted a study to investigate how the use of online social media by migrants and non-migrants affects migration and the functioning of migrant networks. Brazilian, Ukrainian and Moroccan migrants were interviewed between January and June 2011 in the Dutch cities of Amsterdam and Rotterdam. Results revealed four main ways in which social media facilitated international migration: (1) promoting solid ties with loved ones, (2) addressing weak ties that are relevant to organizing the process of migration and integration, (3) building latent ties, and (4) offering discrete and unofficial migration information. Hence, social media changed the process of migration and lowered barriers.

Others, like Benítez (2012), analyzed how ICTs change the communication practices of transnational families and influence public policy. Based on survey research in El Salvador, Benítez (2012) found three ways in which ICTs offered new possibilities for communicative practices among immigrants and their relatives: (1) strengthening cultural values, (2) maintaining ties with home, and (3) providing affective support to family. Wilding (2006, p. 132) has argued that ICTs are important for transnational families in "constructing or imagining connected relationship and enabling them to overlook their physical separation by time and space even if only temporarily."

Migration within a country or beyond borders may entail risks that can be reduced to some extent (Tilly 1990) if international migrants maintain connections with social ties and have information about the conditions of migration (Ros et al. 2007). Schapendonk and Moppes (2007) interviewed migrants in Morocco, Spain and Senegal and found that migrants valued the connection—mediated by mobile phones—to friends and family members during and after migration. Many of their respondents used the internet to prepare for the journey, ranging from transferring money, using internet cafes, or looking for employment. Hence, the use of the internet mitigated risks for the migrants.

As the International Organization for Migration (IOM 2005) has noted, the rise of ICTs and their use for border control and national security now entails serious cost and investment in every country's information technology infrastructure coupled with information-sharing agreements among states. These technologies, in turn, make migration data more available but also produce concerns about privacy and state surveillance.

Smugglers and others may also use social media and spread rumors or misinformation and lure migrants into false deals or create unrealistic expectations (Frouws, Phillips, Hassan and Twigt 2016). In some countries, information campaigns to discourage migration address these issues. In Senegal, for instance, a state-initiated information campaign informed potential migrants of the dangers of crossing the Atlantic Ocean to the Canary Islands. However, trust in media and online information may also impact the effectiveness of such information campaigns. Often, migrants prefer to rely on information obtained through word-of-mouth from trusted sources (Guilmoto and Sandron 2001; Poot 1996).

From Social Networks to Social Capital

Many understand social capital as consisting of strong ties (Granovetter 1973)—in the context of migration, these are based on shared community, friendship, or kinship that connects both migrants and non-migrants (Massey et al. 1998). Perhaps the most straightforward approach to social networks is Granovetter's (1973) theory of the strength of weak ties, which asserts that weak ties are more likely to be bridges to outside networks than strong ties and that information flows through weak ties. Granovetter (1973) elaborates on the quality of connections between people as either strong or weak. Both types of links are crucial as the strong links provide support and are particularly important at the beginning and end of life, while weak links translate into new social and economic opportunities in adult life (Goodman 2003). Granovetter's theory enables us to answer questions like how individual and community social capital is affected by use of social media.

Social capital is a much-contested concept in the social sciences due to a variety of definitions (Castiglione 2008). Scholars have grappled over the concept's ambiguity, as evidenced by often vague explanations of social capital in academic literature (Ahn and Ostrom 2008; see, e.g., Solow 2000; Durlauf 1999; Manski 2000). In this chapter, we utilize Bourdieu's concept of social capital. Reference to Bourdieu's (1986) definition of social capital is quite common. According to Bourdieu (1986), social capital comprises social obligations that can be converted—in certain conditions—into economic capital, and maybe institutionalized, e.g., in the form of titles of nobility. The convertibility of these social, cultural and symbolic resources should have the effect of securing advantage or disadvantage particularly in terms of economics (Bourdieu 1986). Social ties are more meaningful when they result in access to those individuals who have more knowledge and resources (Bourdieu 1986, as cited by Ryan et al. 2008).

Inspired by Bourdieu, Cederberg (2012) interviewed Swedish refugees looking into their co-ethnic and non-ethnic networks. The concept of "social capital" is extremely significant for making sense of migrants' experience and position particularly with reference to inequality. Social groups are rarely homogenous and are marked by internal conflicts. Cederberg (2012) emphasized that it is imperative to look at a range of social networks and see the advantages as well as disadvantages they bring to members when studying different processes in the migration process.

Blanchard and Horan (2000) surveyed 342 people in a mid-sized Californian city that was about to get a "virtual community" (i.e., an online community network). The study was conducted to investigate whether virtual communities could compensate for a decrease in social capital as a consequence of decreased participation in face-to-face communities, and found that there was, indeed, an increase in social capital as people interacted in the new virtual space with neighbors, family, and friends. In an ethnography of a youth-oriented community technology center in Denver, Colorado, Clark (2003) examined how digital divide policy is actually practiced. Making use of Granovetter, she deduced that young people's online activities build their weak ties to a wider network. Using Bourdieu, she concludes that these networks create opportunities for them in the form of employment, housing, and other opportunities.

The social capital debate (Portes and Landolt 1996; Putnam 2000) has also been extended to include "network capital" (Larsen and Urry 2008). Larsen and Urry (2008, p. 93) explain the concept of network capital as

> access to communication technologies, transport, meeting places and the social and technical skills of networking. ... Network capital is the capacity to engender and sustain social relations with individuals who are not necessarily proximate, which generates emotional, financial and practical benefit. "Network capital" refers to a person's, or group's, or society's facility for "self-directed" corporeal movement and communication at-a-distance.

Migration research in some parts of the world (see, e.g., Garip 2008 [rural villages in Thailand]) suggest that individuals are more likely to migrate if their social capital is greater and more spread out by occupation rather than distance. Resources from weakly tied sources like an acquaintance had a higher effect on migration than from strongly tied sources like family. This result is in stark contrast to international migration from Mexico to the US, where strong ties like family members facilitate the decision to migrate (Davis, Stecklov and Winters 2002; Palloni et al. 2001). Massey and Aysa-Lastra (2011) conducted a study to find out the effects of social capital on international migration and how these effects differ based on contextual factors to estimate models predicting the probability of taking first and later trips to the US from five countries: Peru, Costa Rica, Nicaragua, Mexico and the Dominican Republic. The results confirmed that social capital had a universal and strong effect on migration, particularly on the first migration trip compared with later trips and interacts with cost of migration. On first trips, the effect of individual social capital (measuring strong ties) in encouraging migration increases with distance while the effect of community social capital (measuring weak ties) decreases with distance. On later trips, the direction of effects for both individual and community social capital is negative for long distances but positive for shorter ones.

Migration can be viewed as a special case of development of social networks (Eve 2010). The role of kinship and friendship networks is highly important in facilitating migration (Haug 2008; Heering, van Der Erf and van Wissen 2004). Bonding social capital is generally associated with technology that allows the migrants to maintain and strengthen strong ties with friends and family back home while bridging social capital is associated with migrants' use of technologies to "open up new perspectives" (Proulx 2008, p. 158). Bonding social capital and bridging social capital can serve during both the pre- and post-migration phase, and in the post-migration phase, to remain in contact with the source society as well as to integrate into the host society.

RESEARCH METHODS

The findings presented below come from three interrelated research projects active between 2014 and 2020, as discussed above. Additional details about methods for each of these studies have been published elsewhere. In the first line of research, we conducted semi-structured and informal interviews with 46 migrants and migrant-aid workers at a day shelter for migrants in Nogales, Sonora, Mexico, during multiple fieldwork trips in 2014. Interviews were conducted in either English or Spanish, depending on the language proficiency of our respondents, and all Spanish language interviews were translated into English prior to data analysis. Our respondents fell into the following three categories:

1. individuals who had been recently deported from the United States (generally within a few days of deportation, n=29),

2. migrants from Central America who had just arrived at the border with plans to cross into the United States in a clandestine fashion (n=4), and

3. migrant-aid workers affiliated with local and binational humanitarian organizations and who provide services at the shelter on a regular or recurring basis (n=13) (Newell, Gomez and Guajardo 2016, p. 181).

We generated a coding manual and engaged in multiple rounds of coding, beginning with pre-established codes derived from our interview protocols and expanding our code book through an iterative process of identifying additional concepts that emerged in the data.

Additionally, throughout this fieldwork in 2014, we also engaged the methodology of participatory photography, disseminating cameras and instructions to migrants at a day shelter in Nogales, Sonora, and conducting follow-up interviews the following day when they returned (see Yefimova et al. 2015; Gomez and Vannini 2015). After providing the participants with simple digital cameras, we asked them to take pictures of their daily lives in Nogales. The following day, when participants returned to the day shelter, we conducted semi-structured photo-elicitation interviews, using their captured photos as stimuli to guide our conversations. This methodology allowed "us to capture and understand the migrants' life experiences during some of their most vulnerable times: While receiving food and supplies at a migrant shelter just minutes from the border" (Yefimova et al. 2015, p. 3675).

In the third study, we conducted interviews and engaged in field observation of the work of five volunteer-led migrant-aid organizations active in the area surrounding Tucson and Nogales, Arizona, in the summer of 2018. Our focus was on understanding the information practices of these generally volunteer-led organizations and how they use ICTs within their work. We conducted twenty open-ended, semi-structured interviews with humanitarian volunteers and engaged in participant observation of their work (including ride-alongs on service trips into the desert and to a migrant shelter in Nogales, Sonora). We also attended weekly organizational meetings as well as immigration proceedings in federal court. As in the earlier research, we based our interviews (and subsequent coding) around our initial interview guide, expanding our set of basic questions throughout each interview based on the context, history, and roles of each participant in the organizations with which they were affiliated.

FINDINGS

Our findings show that migrants have different *uses of* and *feelings towards* ICTs at different stages of migration, reflecting not only their own migration stage, but also how they are intentional decisionmakers and active participants in their own lives (Favell, Brettell, and Hollifield 2008, as cited in Kozachenko 2013).

ICTs at the Border: Between Information Seeking and Risk Awareness

Our data show that, during their migration journeys to and across the US–Mexico border, migrants generally tend to distrust ICTs and prefer to rely on seeking information through word-of-mouth communication, and they place higher trust in such information. While, for example, the use of mobile phones and social media (such as Facebook) were important communication tools, migrants at the border are aware of the increased risks that they are facing—and if migrants were not previously aware of these risks, the staff at the day shelter consistently explained these issues to them through plenary announcements prior to serving meals. Leaking contact information, whether stored in cell phones or on pieces of paper, could lead to extortion and abuse by organized criminal traffickers, and even by the police, especially on the Mexico side of the border. Facebook was seen by at least one of our respondents as a way to secure contact information, at least decreasing the risk that a physical list might be stolen. One of the volunteers explained:

> Here and along the borders of the U.S. and Mexico, the migrant is ... just seen as a dollar sign. [...] So, the migrant who comes here, they have relatives on the other side who are going to help them, so what do they [criminals] do? They extort their family members; they get their phone numbers and try to extort the family. And their families, just to try to protect their relatives, they do whatever they can to send that money. (Volunteer 1)

One of our interviewees had first-hand experience with these kinds of threats:

> The other day they caught us, and I thought that it would be the last day of my life. [...] They took us, they took our shoes off, they took all our papers, they asked if we had any phone numbers of our friends, and that we had to give it to them. What I did was I took my wallet very carefully and took out the phone numbers and threw them out and [now] I cannot communicate with any of my family anymore; I only know my cousin's phone number but all the rest I lost, I don't have them anymore. (Migrant 5)

Migrants are also well aware of the increased employment of technology of surveillance by US Customs and Border Protection (CBP) and Immigration and Customs Enforcement (ICE), and they are afraid that using mobile phones will give away their location and facilitate their apprehension by US authorities.

Finding trustworthy information, especially regarding practical issues related to border crossing, is one of the biggest challenges they encounter. When possible, they rely on family members or friends to help them find a guide or trustworthy contacts. Statements such as "a neighbor helped me [arrange a smuggler]" (Migrant 1) and, "it's all through friends. You ask here and there [...]. And then you find somebody who knows someone. And they give you a phone number and you talk with someone, and that's the way you do it" (Migrant 2) were common explanations.

Migrant shelters across Mexico (like the one we used as a field site during much of our research), are commonly used by migrants as "information grounds" (Fisher and Naumer 2006)—places to find people who can help or point them to others who can. For example, one respondent explained, "That is why I'm here ... we're in the shelter to see if we can find information about who would be a good guide for us, to try to go back" (Migrant 3). More experienced migrants often share their knowledge with other migrants on the migrant trail in Mexico or in shelters at or near the border.

In this context, where migrants are often aware of the risks in storing information on physical devices, especially mobile phones, and where trust of information sources plays a very relevant role in terms of the kind of information that migrants will decide to believe and use, the use of Facebook seemed particularly relevant. Information shared through and stored into Facebook accounts was, in general, entrusted more than information stored into mobile phones. In a few cases, migrants indicated that social media could also be a good tool to share information about shelters and other safe spaces they could find during the journey:

> I could tell friends and migrants to come look for this place, for the shelter. Because when I was in Tijuana when they caught me, they mugged us on the mountains, and they took away my money. And [...] I had to ask, beg around to other people, and I did not know that there was this kind of place just like a shelter. Many people can learn about this. So that I could tell other people, like other migrants and other friends, to look for these kinds of places, so that they don't suffer like I suffered. Where to sleep, or [to find] clothes or food. (Migrant 6)

However, this use of Facebook raises concerns about the possible risks of mis- or dis-information and of accounts being compromised. First, migrants could erroneously trust or be fearful of information they find, which could influence their decisions on their migration journey. Second, migrants might not be aware of the risks connected to correctly logging off from public computers or that platforms may allow law enforcement access to profile information, and they may not always be knowledgeable about protective, privacy-preserving practices, such as adjusting privacy settings, using encrypted technologies, or managing multiple accounts.

Volunteers working with migrants on both sides of the US–Mexico border also recognize the importance of ICTs in migration journeys, and they try and offer migrants safer ways to access information and communicate with family and friends. Volunteers at shelters provide access to cell phones (and, in some cases, computers), allowing migrants to communicate with friends and family, minimizing migrants' exposure to fraud and extortion.

> Communication is important to them, we give them a chance to talk on the phone, on a secure phone. They can call whoever they want, it is important for them to remain in contact. Increasingly, over the years, more migrants have their own phones. Sometimes they don't have any airtime, sometimes they run out of battery, sometimes their phones are compromised and someone can listen in. Our phones give them a secure way to connect. (Volunteer 3)

These phones, owned by volunteers' organizations, are less likely to fall into the hands of smugglers and cartel members. Migrants are instructed to delete the last phone number dialed in pay phones, and to not accept free calls from people on the street. To increase their safety, organizations are also taking precautions, deleting the record of numbers called from their phones. More recently, organizations' smartphones have been set up to automatically delete the last number called.

On the US side of the border, in Arizona, things are different. Volunteers are wary of letting migrants use cell phones, as that might be construed as aiding and abetting unauthorized migration contrary to US law.

> We don't give them maps. If we gave them a map, the Border Patrol would get on us right away. [...] We don't let them use our cell phones. They want to call their mother... well, because you don't know who they're going to call, and they'll be calling a coyote or something, so we don't let them use cell phones. We don't get them in cars. All we do is, we provide food and water [...] and clothing, [...] and medical care. (Volunteer 4)

Using ICTs When Settling Down: Self-improvement, Ownership and Communication

When settling down in their new place, migrants show a more complex relationship with ICTs. Their decision to use ICTs is not dictated by the same immediate risks. As their lives become more and more integrated in the life of the city and community in which they settle, their use of ICTs also resembles general use throughout the community. Migrants in the US use ICTs not only to communicate with their family and friends, but also for learning, for working and for self-improvement:

> I call my mother... I don't tell her the bad things that happened to me. I just tell them the good things. I buy prepaid cards [...] to call long distance. I call her 4 or 5 times a week. I don't really know how to use the computer or the cell phone. I can just talk, and I like to hear their voice and talk to my family. I really miss my family. The cell phone I have is not a very fancy one. I can take pictures, but I can't send them to my family in Guatemala. I can send them to my nephew in New York and he then sends them to my family. (Migrant 7)

ICTs are used as tools to learn new things, including English, and relax:

> I have a laptop, but it doesn't work anymore. Now I use my phone. I have two cell phones. One has a phone line, and this other one is a smartphone, but it does not have any phone line. I put a Bluetooth keyboard on it, and I use it as a tablet. I watch programs in English, I play games, math games, and I watch things on YouTube, I listen to music. (Migrant 8)

They are also used as everyday tools to do their work:

> These are my laptop, my cell phone, and my desk phone. This is the place where I work, where I grow every day, where I discover my professional side. These tools push me toward success. [...] These tools have taught a lot of things throughout my job. I am open to learning because I am a human being. (Migrant 9)

Once they settle in the US, migrants' attitudes towards ICTs often exhibit ownership and appropriation of the technologies as tools for self-improvement:

> I started to come here [...] to the computer classes. Here they give you the very basics: how to use email, how to open an account, and how to use it. [...] I would like to learn how to repair cell phones or tablets because there's a lot of work I can no longer do because I don't have the strength. But my mind could do other jobs... I dream of doing something like repairing computers. [Migrant 10]

For those who do not have a stable income, libraries become a place of reference to learn, using ICTs and other technologies:

> This is the central library, but I also call it "the office" because for a lot of people, that's everyone's office. (Migrant 11)

> I always go to the library to use the translator; since I am an aficionado of English. I learn a lot of words and phrases there. I watch videos as well, but they have to have the closed captioning, otherwise I don't like it because I don't understand any of it. If the captioning comes up, and the person is speaking in English, that's what I like. I want to hear the pronunciation. That's why I go to the library. (Migrant 12)

Migrants refer often to the library as a safe space, a shelter, a place where they can learn, stay and be safe:

> I go to the library any day I have a chance. If I don't find work [...] I can go and spend all day at the library and improve myself. I can fill in all the knowledge that I'm missing and also it is a safe place to go hang out even if I have nothing to do. (Migrant 13)

I've always wanted to see the aurora borealis, but I've never seen them. And I like to go to the library and learn about all those things. I read the newspapers in the library. The library is really like a shelter for everybody who is on the street. Like a church, they cannot kick you out of there. (Migrant 14)

Libraries are information grounds and communication hubs for migrants in their new communities:

A friend asked me, "Do you want to speak with your family? I will show you how to get onto the computer," and that's how I got to know the library. He begged me, "Let's go to the library!" because I don't like to read books, but one day I agreed and up to now, I'm still going there. I use the computer for the translator, and I see movies with subtitles. And sometimes I speak to my family. (Migrant 15)

Once settled, migrants often see the use of electronic devices as an opportunity, rather than being associated with immediate risk.

Humanitarian organizations working with migrants on the US side of the border are aware of the possible privacy risks associated with sharing information about (especially undocumented) migrants publicly on social media. As recently brought to the attention of the public, ICE and other federal agencies use data and information from disparate databases: state and local governments, private data brokers, but also social networks (Funk 2019). Organizations, then, try to educate migrants about safer use of Facebook and social media (privacy settings, publishing and tagging photos, geo-location, etc.). They also adopt low-tech methods (e.g., visible signs people can wear to indicate they do not want to be photographed) to respect the privacy of people that might not want their photos to appear online.

However, social media platforms are perceived to have different levels of privacy and security risks. Email, iMessage and private Facebook groups were perceived by both migrants and those working with them as being safer than publicly accessible Facebook pages. Also, migrants are necessarily not discouraged to "be open about their story as undocumented individuals, if they should wish to." Young people and students, for example, may have chosen to "come out" as undocumented on public Facebook pages, using them as platforms for activism and peer support. Ultimately, organizations defer the decision to post personal information on social media to migrants themselves:

They have been living with their undocumented status their whole life. They understand the risks better than anyone and don't need me telling them what they should or shouldn't share. If this is something that is important for them to do, for themselves, I'm not going to try to stop them. We don't do any policing here. A few [...] choose to take on a more activist role and are open about their status. That is their decision to make, I am not going to try to stop them. (Staff 1)

This position can be problematic, as it possibly puts migrants at risk of being detained and deported. Importantly, "the failure to protect undocumented migrants' privacy in [humanitarian information activities] tends to exacerbate the migrants' vulnerability, whose legal status already places them at risk" (Vannini, Gomez and Newell 2020, p. 929). However, these practices do assign migrants a higher level of agency and ownership in their own use of ICTs.

CONCLUSION

ICTs are networking tools and sources of social capital providing opportunities to migrants to cross the border safely and integrate into the host society. This chapter attempts to examine how ICTs foster immigrants' social capital in the Mexico–US border crossing journey and after settling in the United States. It looks at how the use of social media simultaneously benefits and poses risks to the immigrant. The study rests on Bourdieu's concept of social capital that investigates size and type of social networks the migrant can access and draw upon.

The migrants displayed different uses and a mix of feelings towards ICTs. At the border, the migrants reported feeling scared of surveillance of their phones and invasion of their privacy. Facebook was more trusted than mobile phones for storage of their information though with the fear that their accounts might be compromised. Some migrant-aid workers provided access to cell phones to enable migrants to obtain accurate information. Before and during the border-crossing journey, the new migrants are constantly looking for experienced migrants' accounts of the journeys. Most new migrants place higher trust in information obtained from family members and close friends, rather than ICTs, during the migration process to help them find a guide or other valuable piece of information. This finding is supported by Massey and Aysa-Lastra's (2011) study in which migrants sought out and trusted prior migrants. When settling down in the US, ICTs were used for self-improvement, for communication and for work, consistent with Benítez's (2012) findings about how ICTs strengthen cultural values among immigrants and their relatives. Some migrants saw ICTs as an opportunity rather than a risk. ICTs provided the means to build social networks and find employment and resources, to integrate into their new society. Hence, ICTs remain an important tool for migrants to build their social capital. Some migrants did not have laptops and phones, and some had to seek support to learn to use these types of electronic devices and to access social networks. Thus, in line with Bourdieu's analysis, it is apparent that these networks can create opportunities in the form of employment, housing or other economic gains.

Mobile technologies and smartphones have the ability to build social capital. However, unequal access to ICTs, skill and experience in using technology and digital media may hinder this process. Differences in education, gender and foreign language skills are all factors that need to be considered when investigating the dynamics of the migration process. More empirical work is needed to better understand who benefits and who loses in this process. It is also worth investigating how migrants assess whether ICT-mediated information is trustworthy. As misinformation, disinformation and fake news are increasingly part of our daily informational environment, especially on digital media, it is important to assess the extent to which migration-related information will be affected and affect migrants. Future research should also examine how fake news and misinformation are perceived and whether they lower trust in the use of ICTs in the migration context, and how significant global events, such as the Coronavirus (COVID-19) pandemic may exacerbate these problems and perpetuate or exaggerate inequalities. Fake news and misinformation are becoming rampant, and in times of uncertainty and when lawful paths to international migration become limited, this may have serious negative implications for migrants. Whether ICTs prove a blessing or a bane in such circumstances needs further exploration.

REFERENCES

Ahn, T. K., and Ostrom, E. (2009). 'The meaning of Social Capital and its link to Collective Action', in D. Castiglione, R.F. Van Baumeister and M.R. Leary (eds), *Handbook of Social Capital: The Troika of Sociology, Political Science, and Economics*, Cheltenham, UK and Northampton, MA, USA: Edward Elgar Publishing, pp. 17–35.

Arun, S., R. Heeks and S. Morgan (2006), 'Improved livelihoods and empowerment for poor women through IT-sector intervention', in Nancy Hafkin and Sophia Huyer (eds), *Cinderella or Cyberella? Empowering Women in the Knowledge Society*, Bloomfield, CT: Kumarian Press, pp. 141–64.

Baron, L. F., and R. Gomez (2017), 'Living in the limits: Migration and information practices of undocumented Latino migrants', in *Proceedings of the International Conference on Social Implications of Computers in Developing Countries (ICT4D 2017)*, Cham: Springer, pp. 147–58.

Bates, J., and L. Komito (2012), 'Migration, community and social media', in G. Boucher, A. Grindsted and T. L. Vicente (eds), *Transnationalism in the Global City*, University of Deusto, pp. 97–112.

Benitez, J. L. (2006), 'Transnational dimensions of the digital divide among Salvadoran immigrants', *Global Networks*, 6 (2), 181–99.

Benítez, J. L. (2012), 'Salvadoran transnational families: ICT and communication practices in the network society', *Journal of Ethnic and Migration Studies*, 38 (9), 1439–49.

Blanchard, A., and T. Horan (2000), 'Virtual communities and social capital', in G. David Garson (ed.), *Social Dimensions of Information Technology: Issues for the New Millennium*, Hershey, PA: Idea Group Publishing, pp. 6–22.

Bourdieu, P. (1986), 'The forms of capital', in J. G. Richardson (ed.), *Handbook of Theory and Research for the Sociology of Education*, New York, NY: Greenwood Press, pp. 241–58.

Castiglione, D. (2008), 'Introduction: Conceptual issues in social capital theory', in D. Castiglione, Jan W. V. Deth and G. Wolleb (eds), *The Handbook of Social Capital*, Oxford University Press on Demand, pp. 13–21.

Cederberg, M. (2012), 'Migrant networks and beyond: Exploring the value of the notion of social capital for making sense of ethnic inequalities', *Acta Sociologica*, 55 (1), 59–72.

Chen, W., and A. S. K. Choi (2011), 'Internet and social support among Chinese migrants in Singapore', *New Media & Society*, 13 (7), 1067–84.

Clark, L. S. (2003), 'Challenges of social good in the world of Grand Theft Auto and Barbie: A case study of a community computer center for youth', *New Media & Society*, 5 (1), 95–116.

Davis, B., G. Stecklov and P. Winters (2002), 'Domestic and international migration from Rural Mexico: Disaggregating the effects of network structure and composition', *Population Studies*, 56 (3), 291–309.

Dekker, R., and G. Engbersen (2014), 'How social media transform migrant networks and facilitate migration', *Global Networks*, 14 (4), 401–18.

Dobbs, L. (2006), *War on the Middle Class: How the Government, Big Business, and Special Interest Groups are Waging War on the American Dream and How to Fight Back.* New York, NY: Viking.

Durlauf, S. N. (1999), *The Case 'against' Social Capital.* Social Systems Research Institute, University of Wisconsin.

Ericksen, E. and W. Yancey (1977), 'The Locus of Strong Ties', Unpublished manuscript, Department of Sociology, Temple University.

Eve, M. (2010), 'Integrating via networks: Foreigners and others', *Ethnic and Racial Studies*, 33 (7), 1231–48.

Fall, A. S. (1998), 'Migrants' long-distance relationships and social networks in Dakar', *Environment and Urbanization*, 10 (1), 135–46.

Favell, A. (2015), 'Migration Theory rebooted? Asymmetric challenges in a global agenda', in C. B. Brettell and J. F. Hollifield (eds), *Migration Theory: Talking across Disciplines*, London: Routledge, pp. 373–84.

Fisher, K. E., and C. M. Naumer (2006), 'Information grounds: theoretical basis and empirical findings on information flow in social settings', in A. Spink and C. Cole (eds), *New Directions in Human Information Behavior*, Dordrecht, Netherlands: Springer, pp. 93–111.

Frouws, B., M. Phillips, A. Hassan and M. Twigt (2016), 'Getting to Europe the WhatsApp way: the use of ICT in contemporary mixed migration flows to Europe', *Regional Mixed Migration Secretariat Briefing Paper*.

Funk, M. (2019), 'How ICE picks its targets in the surveillance age', *New York Times Magazine*, Oct. 2. At https://www.nytimes.com/2019/10/02/magazine/ice-surveillance-deportation.html (accessed on 13 August 2021).

Garip, F. (2008), 'Social capital and migration: How do similar resources lead to divergent outcomes?' *Demography*, 45 (3), 591–617.

Germann Molz, J. and C. M. Paris (2015), 'The social affordances of flashpacking: Exploring the mobility nexus of travel and communication', *Mobilities*, 10 (2), 173–92.

Gomez, L., R. Gomez and S. Vannini (2017), 'The power of Participatory Photography in ICTD programs: Freedom to explore beyond images', in *Proceedings of the 50th Hawaii International Conference on System Sciences*, pp. 2600–609.

Gomez, R. (2016), 'Vulnerability and information practices among (undocumented) Latino Migrants', *Electronic Journal of Information Systems in Developing Countries (EJISDC)*, 75(1), 1-43.

Gomez, R. and S. Vannini (2015), *Fotohistorias: Participatory Photography and the Experience of Migration*. Charleston, SC: CreateSpace.

Gomez, R. and S. Vannini (2017), 'Notions of home and sense of belonging in the context of migration in a journey through participatory photography', *Electronic Journal of Information Systems in Developing Countries (EJISDC)*, 78(1), 1–46.

Gomez, R., B. C. Newell and S. Vannini (2020), 'Empathic humanitarianism: Understanding the motivations behind humanitarian work with migrants at the US–Mexico border', *Journal on Migration and Human Security*. Early view online at https://doi.org/10.1177/2331502419900764.

Goodman, J. (2003), *Mobile Telephones and Social Capital in Poland: Summary of a Case Study with Vodafone*, Forum for the Future.

Granovetter, M. S. (1973), 'The strength of weak ties', *American Journal of Sociology*, 78 (6), 1360–80.

Guilmoto, C. and F. Sandron (2001), 'The internal dynamics of migration networks in developing countries', *Population: An English Selection*, 13 (2), 135–64.

Hamel, J. Y. (2009). Information and communication technologies and migration: Human Development Research Paper 2009/39, United Nations Development Programme. At https://mpra.ub.uni-muenchen.de/19175/1/MPRA_paper_19175.pdf (accessed on 13 September 2021).

Haug, S. (2008), 'Migration networks and decision making', *Journal of Ethnic and Migration Studies*, 34 (4), 585–605.

Heering, L., R. V. Erf and L. V. Wissen (2004), 'The role of family networks and migration culture in Moroccan emigration', *Journal of Ethnic and Migration Studies*, 30 (2), 323–37.

Horst, H. (2006), 'The blessings and burdens of communication: Cell phones in Jamaican transnational social fields', *Global Networks*, 6 (2), 143–59

Huntington, S. P. (2004), *Who Are We: The Challenges to America's National Identity*. New York, NY: Simon & Schuster.

International Organization for Migration [IOM] (2005), 'International migration, development and the information society', Geneva: World Summit on the Information Society. At http://www.itu.int (accessed on 5 January 2020).

King, R. and N. Wood (2001), *Media and Migration: Constructions of Mobility and Difference*, London: Routledge.

Kissau, K. (2012), 'Structuring migrants' political activities on the internet: A two-dimensional approach', *Journal of Ethnic and Migration Studies*, 38 (9), 1381–403.

Kozachenko, I. (2013), 'Horizon Scanning Report: ICT and Migration', *Working Papers of the Communities & Culture Network+*, 2 (October). At http://2plqyp1e0nbi44cllfr7pbor.wpengine.netdna-cdn.com/files/2013/01/ICT-and-Migration-Kozachenko.pdf (accessed on 13 September 2021).

Larsen, J. and J. Urry (2008), 'Networking in mobile societies', in J. O. Bærenholdt and B. Granas (eds), *Mobility and Place: Enacting Northern European Peripheries*, Aldershot: Ashgate, pp. 89–101.

Madianou, Mirca (2014), 'Smartphones as polymedia', *Journal of Computer-Mediated Communication*, 19, 667–80.

Mahler, S. J. (2001), 'Transnational relationships: The struggle to communicate across borders', *Identities*, 7 (4), 583–619.

Manski, C. F. (2000), 'Economic analysis of social interactions', *Journal of Economic Perspectives*, 14(3), 115–36.

Massey, D. S. (2005), 'Backfire at the border: Why enforcement without legalization cannot stop illegal immigration', Trade Policy Analyses no. 29, Washington, DC: Centre for Trade Policy Studies, Cato Institute, 13 June.

Massey, D. S. and M. Aysa-Lastra (2011), 'Social capital and international migration from Latin America', *International Journal of Population Research*, art. 834145, 1–18. https://doi.org/10.1155/2011/834145.

Massey, D. S., J. Durand and N. J. Malone (2002), *Beyond Smoke and Mirrors: Mexican Immigration in an Era of Economic Integration*, New York, NY: Russell Sage Foundation.

Massey, D. S., J. Arango, G. Hugo, A. Kouaouci, A. Pellegrino and J. E. Taylor, *Worlds in Motion: International Migration at the End of the Millennium*, Oxford: Oxford University Press.

Newell, B. C. and R. Gomez (2015), 'Informal networks, phones and Facebook: Information seeking and technology use by undocumented migrants at the U.S.-Mexico border', in *iConference 2015 Proceedings*, 1–11. At http://hdl.handle.net/2142/73434 (accessed on 13 September 2021).

Newell, B. C., R. Gomez and V. Guajardo (2016), 'Information seeking, technology use, and vulnerability among migrants at the United States–Mexico border', *The Information Society*, 32(3), 176–91.

Newell, B. C., R. Gomez and V. Guajardo (2017), 'Sensors, cameras, and the new "normal" in clandestine migration: How undocumented migrants experience surveillance at the US-Mexico border', *Surveillance & Society*, 15 (1), 21–41.

Newell, Bryce Clayton, Sara Vannini, and Ricardo Gomez (2020), 'The information practices and politics of migrant-aid work in the US–Mexico borderlands', *The Information Society*, 36(4), 199–213.

Palloni, A., D. Massey, M. Ceballos, K. Espinosa and M. Spittel (2001), 'Social capital and international migration: A test using information on family networks', *American Journal of Sociology*, 106, 1262–98.

Panagakos A. N. and Horst, H. A. (2006), 'Return to Cyberia: Technology and the social worlds of transnational migrants', *Global Networks*, 6 (2), 109–24.

Poot, J. (1996), 'Information, communication and networks in international migration systems', *The Annals of Regional Science*, 30 (1), 55–73.

Portes, A. and J. Sensenbrenner (1993), 'Embeddedness and immigration: Notes on the social determinants of economic action', *American Journal of Sociology*, 98(6), 1320–50.

Proulx, Serge (2008), 'Des nomades connectés: Vivre ensemble à distance', Hermès, La Revue 51 (2), 155–60.

Putnam, R. (2000), *Bowling Alone*, New York, NY: Simon & Schuster.

Riccio, B. (2006), '"Transmigrants" mais pas "nomads": Transnationalisme mouride en Italie' ("Transmigrants" but not "Nomads". Mouride transnationalism in Italy), *Cahiers d'études Africaines*, 46 (181), 95–114.

Ros, A., E. Gonzalez, A. Marin and P. Sow (2007), 'Migration and information flows: A new lens for the study of contemporary international migration', Barcelona: Internet Interdisciplinary Institute.

Rouse, R (1995), 'Questions of identity: Personhood and collectivity in transnational migration to the United States,' *Critique of Anthropology*, 4, 351–80.

Ryan, L., R. Sales, M. Tilki and B. Siara (2008), 'Social networks, social support and social capital: The experiences of recent Polish migrants in London', *Sociology*, 42 (4), 672–90.

Schapendonk J, and Van Moppes D (2007), *Migration and Information: Images of Europe, migration encouraging factors and en route information sharing*. Nijmegen: Radboud University Nijmegen.

Schaub, M. L. (2012), 'Lines across the desert: Mobile phone use and mobility in the context of trans-Saharan migration', *Information Technology for Development*, 18 (2), 126–44.

Schroeder, S. (2015), 'Refugees in Croatia can't get to the internet, so the internet comes to them', *Mashable*, 21 September. At https://mashable.com/archive/mobile-free-internet-refugees (accessed on 8 August 2021).

Solow, R. M. (2000), 'Notes on social capital and economic performance', in P. Dasgupta and I. Serageldin (eds), *Social Capital: A Multifaceted Perspective* (pp. 3–5). Washington, DC: The World Bank.

Tilly C. (1990), 'Transplanted networks', in V. Yans-MacLoughlin (ed.), *Immigration Reconsidered*, New York, NY: Oxford University Press, pp. 79–95.

Vannini, S., R. Gomez and B. C. Newell (2016), "'Mind the Five': Guidelines for data privacy and security in humanitarian work with undocumented migrants and other vulnerable populations', *Journal of the Association for Information Science and Technology* (JASIS&T). Early view online at https://doi.org/10.1002/asi.24317.

Vannini, Sara, Ricardo Gomez and Bryce Clayton Newell (2020). "Mind the five": Guidelines for data privacy and security in humanitarian work with undocumented migrants and other vulnerable populations, *Journal of the Association for Information Science & Technology* 71 (8), 927–38.

Vannini, Sara, Ricardo Gomez and Veronica Guajardo (2016), 'Security and activism: Using participatory photography to elicit perceptions of information and authority among Hispanic migrants in the U.S.', in *Proceedings of 2016 iConference*, 1–10. doi: 10.9776/16279. At http://hdl.handle.net/2142/89319 (accessed on 13 September 2021).

Wilding, R. (2006), '"Virtual" intimacies? Families communicating across transnational contexts', *Global Networks*, 6 (2), 125–42.

Yefimova, K., M. Neils, B. C. Newell and R. Gomez (2015), 'Fotohistorias: Participatory photography as a methodology to elicit the life experiences of migrants', in *Proceedings of the 48th Annual Hawaii International Conference on System Sciences (HICSS)*, 3672–81.

Zijlstra, J. and Liempt, I. V. (2017), 'Smart (phone) travelling: Understanding the use and impact of mobile technology on irregular migration journeys', *International Journal of Migration and Border Studies*, 3 (2–3), 174–91.

8. Mobile data challenges for human mobility analysis and humanitarian response

Albert Ali Salah

INTRODUCTION

The capabilities of computer systems expand every day and new approaches are being developed for computer analysis of human behaviour (Salah and Gevers, 2011). At the largest scale of modeling, data collected over long periods of time from millions of people living in a region can be visualized, analyzed, and interpreted with the help of computer systems. Especially during the COVID-19 pandemic, the value of large-scale behaviour modeling and analysis became apparent, both for prediction of the spread of the disease, as well as in assessing measures in its control (Oliver et al., 2020).

A common processing pipeline starts with "sensing" the behaviours in question, using physical sensors like cameras, or virtual sensors like user traces left on the Internet and on phone-based applications. The analysis of spatio-temporal data generated by these sensors can use models developed by social scientists, but also be accomplished in a data-driven way, with the help of statistics, pattern recognition, and machine learning. Finally, the results of the analysis, which can include prediction of behaviours, or explanation of their origin or dynamics, are put to some use. They can feed applications that provide better services, or help them accomplish their tasks more effectively.

Computational social science is a new field that "leverages the capacity to collect and analyze data at a scale that may reveal patterns of individual and group behaviours" (Lazer et al., 2009, p. 72). Using social media or mobile phones, it is possible to get data from millions of people in a very short amount of time. It is also possible to observe longitudinal changes as reflected on such data sources, because the measurements can be repeated regularly. For large-scale human behaviour analysis, traditional methods of data collection, such as surveys, interviews, and focus groups have limitations, including the high cost of data collection, response biases, and the difficulty of ensuring the same conditions over longitudinal designs (Margolin et al., 2013). The difficulties are exacerbated in the case of studying migration and human mobility, which can greatly benefit from the analysis of new data sources (Zagheni and Weber, 2015).

In this chapter, we discuss a specific sensing modality for large-scale human mobility analysis, namely, mobile phone data, that may overcome some of these limitations. In particular, mobile phones are ubiquitous in many countries (and increasingly used, thus implying less bias by wealth),[1] and offer very high temporal resolution in data acquisition. The potential of mobile phone data for the analysis of human mobility is perhaps not very surprising, as the location of a mobile phone can be tracked with great accuracy. Additional information from a smartphone and its applications can greatly increase this potential.

This power comes with a range of ethical and privacy issues that need to be considered when dealing with large-scale, sensitive data analysis. In particular, proprietary data sources, such as call detail records of mobile telecommunications companies, are extremely valuable, but also very sensitive data collections. In this chapter we discuss some mechanisms to share and analyze such collections responsibly.

This chapter starts with describing how mobile data are used for behaviour analysis in general, and mobile call data records in particular. We then explain the concept of data collaboratives, and summarize efforts in several large-scale mobile data challenges, including the recent Data for Refugees Challenge (Salah et al., 2019). We discuss mobile data in the context of mobility, migration, displacement, and for humanitarian response. We deal with the ethical and privacy issues at some length, and we conclude the chapter by discussing key implications for research and governance.

MOBILE PHONE DATA FOR BEHAVIOUR ANALYSIS

The potential of using mobile phone data in gaining in-depth insight into movements of people was revealed forcefully during the COVID-19 pandemic (Oliver et al., 2020), but investigations of the potentials and limits started much earlier. In a landmark study, Eagle and Pentland (2006) provided 100 smartphones to students at MIT, and just by tracking simple indicators, such as call logs, Bluetooth devices in proximity, cell tower IDs, application usage, and phone status (such as charging and idle), were able to obtain interesting insights into the social behaviours of the students.[2] Their goal was to explore the capabilities of smartphones for enabling social scientists to investigate human interactions beyond the traditional survey- or simulation-based methodologies. Indeed, applying well-known pattern recognition techniques on these types of data revealed most-frequented locations of individuals, their social contacts (by using regular proximity with other individuals), networks and groups, and even organizational dynamics. Given data from a new student, they were able to classify whether the student was a first-year student or a graduate student (with over 95% accuracy), and given half a day of observing a student's behaviour, they could predict the rest of the day's behaviour with almost 80% accuracy (Eagle and Pentland, 2009). Bayir et al. (2009) found that users in this dataset spent 85% of their time in three to five favourite locations. One can argue that a small number of MIT students may not have a lot of variability in their daily routines, yet daily routines and social activities are important to many people, and these studies predicted that such methods will play an increasingly important role in behaviour research.

What was also initially surprising was how much information could be extrapolated by combining such mobile phone data with other data sources. In 2005, Ahas and Mark combined mobile phone data with demographics and survey data in what they termed the "Social Positioning Method", for the analysis of social spaces in Estonia. They remarked that one can study social flows in space and time through such an approach, but questions of privacy and freedom of the individual should be asked.

Mobile applications can collect a lot of information from their users, but they are restricted to the people who use the application, whereas data available to telecommunication operators encompass all cellphone users. In the remainder of this section, we focus on such data.

Mobile Call Detail Records

A "call detail record" (CDR) is a data structure that contains the exact time, duration, and location (antenna) of a call, as well as the calling number and callee number. If within the same telecommunications operator, antenna IDs of both sides of the call can be known. A similar record will be created for the SMS messages. Naturally, the actual content of the messages or calls is not stored; this is mostly a record that the telecommunication company (or, equivalently, the network operator) creates for billing and accounting purposes. They may also be internally used to optimize resources and infrastructure.

What would count as a personal information in a CDR database is the phone number itself, and removing this number, or replacing it with a random indicator could be a way of anonymizing such data. However, this by itself is not nearly enough, and there are additional considerations to be taken into account. If there are too few people around an antenna at a given time, the CDR can be associated to a known person by a process of elimination. Subsequently, spatial and temporal aggregation is used in addition to anonymization (see Figure 8.1). Furthermore, external data sources can help de-anonymize such a dataset. For example, if precise information for a number of travels of a person is known, that person can be identified in such a database. One solution is to track individuals for very short periods, such as for one or two weeks at a time. These records are called *tracklets*, and each time a person is sampled from the pool of users, they are given a new random identifier. When we just look at tracklets from thousands of people, we can see the dynamics of mobility between connected locations without jeopardizing the anonymity of individuals from which the tracklets were obtained.

Regularities in people's living allow getting semantically rich information out of CDR records. By determining the most frequently used locations for phone calls during work hours, it is possible to estimate the location(s) where a person works. The same can be done for a home location by looking at early morning or late-night calls. Once these landmarks are known, they can serve as anchors for classifying behaviour.

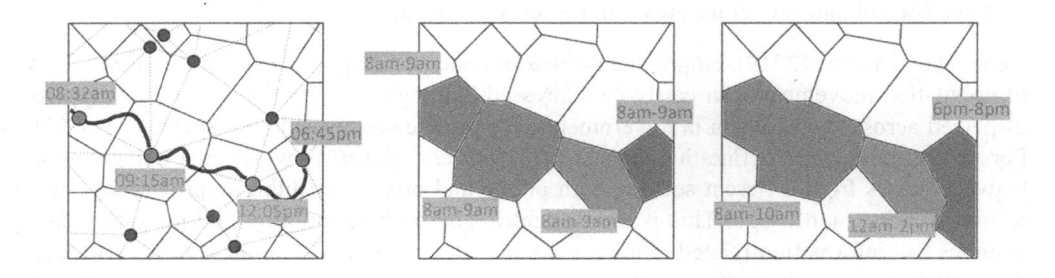

Notes: (a) A user has four recorded calls in the telecom database, with precise call times, call duration, source, and target base stations. The phone number is first replaced with a pseudorandom number. (b) The temporal granularity is decreased by replacing exact times with time intervals. The larger the interval, the coarser the granularity. Choosing one or two hours is typically enough. (c) Spatial granularity is decreased by joining adjacent cells. For tracks that are longer than a few weeks, such a step is a necessary precaution. Figure adapted from de Montjoye et al., 2013.

Figure 8.1   Anonymization and spatiotemporal aggregation of CDR data

Sharing Mobile CDR

Detailed mobile CDR is a rich data source for looking at mobility, but it is generated and used internally by telecommunications operators. Since it is highly sensitive, it is not released to the public. These datasets are analyzed from time to time by a small group of researchers who are granted access through project collaborations (Deville et al., 2014; Lu et al., 2016) or by governments, under special regulations. According to Maxmen (2019), more than 20 mobile phone companies participated in such efforts. This includes operators in 100 countries that backed the Big Data for Social Good initiative, sponsored by the Global System for Mobile Communications Association (GSMA), which also developed an online resource to facilitate further use cases.[3]

There are in fact several data sharing models for CDR, each with different strengths and weaknesses. Letouzé and Oliver (2019) list five models, based on an analysis of scale and protection level (de Montjoye et al., 2018):

1. Limited data release: Private companies package the data and release it to a small number of groups. Scale and protection are both low.
2. Remote access: Company grants access for limited data to a few authorized groups. The scale is similarly low, but protection level is higher.
3. Access through application programming interfaces (APIs): This is a question-and-answer model, where data users send a query to an API, and data are aggregated within the private company, never leaving the premises. Only the "answer" is sent out. It is both scalable and secure, but difficult to set up and maintain. A good example is the Open Algorithms (OPAL) framework, which is currently being tested in several countries (Letouzé, 2019).
4. Pre-computed indicators: Data are not shared; only indicators are computed and shared. This limits the analysis and precludes innovative approaches for using the data. Similarly, deriving statistics from the data and producing a synthetic dataset for sharing will miss aspects not included in the statistical modeling already.
5. Data collaboratives or cooperatives: Data cooperatives are institutional agreements for incentivizing data sharing, where individuals will pool their data under one umbrella, and have control (and sometimes revenue) when sharing data.

Verhulst and Young (2019) remark that the task of understanding the causes and consequences of population movements can partly be addressed "through the targeted analysis of datasets dispersed across stakeholders in governments, the private sector, and civil society" (p. 465). For this purpose, they define the concept of a *data collaborative*, which is a collaboration between actors from different sectors, both public and private, enabling improved data collection, sharing, and usage. This is a data-centric approach to mobility and migration, and assumes that information-related problems are of utmost importance for these fields. Through data collaboratives, it becomes possible for the stakeholders to access more relevant, real-time and detailed data, which in turn guides decision and policy making. For migration and mobility specifically, there are currently several projects underway aiming to use new data sources via such collaboratives. For example, the SoBigData and HummingBird projects funded by European Union's Horizon2020 program are investigating data collaborations (including mobile CDR) related to migration policy and migration indicators.[4]

Mobile Data Challenges

In the rest of this section, we describe several initiatives to open CDR data to the scrutiny of a large number of researchers as data collaboratives with telecommunication operators. Each of these initiatives required significant effort by a large number of people addressing technical, organizational, and ethical aspects of such a data collaborative, as well as the cooperation and commitment of the telecommunication operator in question. There is also a strong, inter-disciplinary community building effort around these topics. The Analysis of Mobile Phone Networks Conference (NETMOB), initiated in 2010, continues to be a gathering place for scholars who explore the potential of mobile phone data in various settings.[5]

The first large initiative involving a data collaborative with mobile data was the Data for Development (D4D) Challenge, which was initiated by Orange Cote D'Ivoire in 2012 (Blondel et al., 2012). This initiative opened a big dataset of mobile CDR for the first time to a large group of scientists, in the form of a data challenge. The data consisted of five months of anonymized CDR records and antenna positions, and the aim of the challenge was "to help address society development questions in novel ways by contributing to the socio-economic development and well-being of the Ivory Coast population" (Blondel et al., 2012, p. 1).

The CDR data for D4D contained:

- antenna-to-antenna traffic on an hourly basis,
- individual trajectories for 50,000 randomly sampled users for two-week time windows with antenna location information,
- individual trajectories for 500,000 randomly sampled users over the entire observation with sub-prefecture location information,
- a sample of communication graphs for 5,000 customers.

The scientific challenge, which ran for eight months, required institutions interested in the data to sign a legal agreement with binding terms and conditions, and the proposals were screened by an evaluation committee. A total of 263 project proposals was submitted to the challenge, and over 80 reports were received.[6]

Some of the outcomes of the projects were potentially useful in improving the infrastructure in Ivory Coast. For example, an analysis of the transit network jointly with CDR revealed the differences of Origin-Destination flows indicated by the CDR with the transit network, and four new routes were suggested for addition to the city transit network, decreasing the citywide travel time by 10% (Di Lorenzo et al., 2015). Other project outcomes suggested improvements for the healthcare system through modeling of disease dynamics, provided better population statistics and economic indicators, and provided insights into the social dynamics of Ivory Coast.

In 2014, Telecom Italia opened a dataset of its own CDR records from two cities as a challenge.[7] In the same year, Sonatel Senegal and Orange Labs embarked on a second edition of D4D, by providing CDR collected over an entire year from over nine million unique mobile phones (de Montjoye et al., 2014). By that time, it was already known that four spatio-temporal points are sufficient to find a person in a mobility database of 1.5 million users with 95% accuracy (de Montjoye et al., 2013), so the data were aggregated and anonymized to prevent such re-identification. The same data aggregation scheme was used in 2018, during the Data for Refugees (D4R) Challenge, which we describe in Section 3.3 in more detail.

The ethical safeguards and privacy measures used in all these challenges focused on the principle of "do no harm", where responsible data practices are promoted, a broad set of stakeholders involved in decision process (e.g., in the Data for Refugees Challenge, this included refugees and institutions protecting the rights of refugees), and measures are taken to minimize the risk to individuals (Salah et al., 2018, Vinck et al., 2019). We discuss these issues in Section 4 on ethics and privacy, and broadly comment on existing frameworks and concerns.

MOBILITY, MIGRATION, DISPLACEMENT ANALYSIS AND HUMANITARIAN RESPONSE WITH CDR

In this section, we provide examples on how mobile CDR can be used for the analysis of mobility, migration and displacement, discussing specific applications, including analysis of refugee movements and humanitarian response.

Movement Patterns

Movement patterns of the masses show great regularity, and inspecting regularities and breaks from regularity is a great way to start analysis of mobile CDR. Dobra et al. (2015) processed mobile phone records from Rwanda with an unsupervised learning approach to detect deviations from expected behaviour, and connected the identified anomalous days and locations with records of violent and political events such as protests, and natural disasters like earthquakes. In a similar approach, Gundogdu et al. (2016) analyzed CDR data from Ivory Coast and showed that movements of different ethnic groups on religious holidays were different. It was possible to use regular visits to a basilica during Christian holidays as a proxy indicator for religion, allowing the profiling of users in terms of this variable.[8] Even though such a classification will be noisy, when hundreds of thousands of people are used in the analysis, it can produce meaningful patterns. In a sense, computational social science replaces the carefully controlled experimental settings of traditional social science with a much less controlled setting but relies on orders of magnitude larger sample sizes to reduce the effect of noise and the uncontrolled variables (Giannotti et al., 2016).

The CDR collected from an entire country gives a detailed picture about internal displacement and mobility, not only connecting cities to cities, but potentially giving prefecture-level mobility patterns. But how can we deduce cross-border movements? One obvious solution is to work with multiple telecom operators at the same time, and link roaming and cell phone usage across operators. This, however, may create serious privacy concerns. In some cases, the context of mobility may provide cues. In the Ivory Coast study mentioned above, the data were collected during a severe civil conflict, and many people escaped the country over borders. When CDR is tracked longitudinally, it is possible to see the movement tracks leading to a border and the cessation of activity for the remainder of the data collection period, for a number of users. These tracks, with a high probability, illustrate displacement of people across borders.

Gonzáles et al. (2008) tracked trajectories of 100,000 anonymized mobile phone users, and observed a high degree of temporal and spatial regularity. Mathematical models were used to characterize the distribution of movements of individuals. Most individuals moved within a small radius of gyration, which denotes the distance traveled by a person when observed for a specific time. It is clear that internal (or external) migration and displacement will create quite different trajectories, which can be modeled similarly.

Ahas et al. (2018) classified people who were traveling abroad from Estonia into groups of tourists, commuters, transnationals, and foreign workers, using roaming data from mobile operators, depending on the number of visits and the number of days spent in the destination. Since register-based data are not very suitable to describe the transnational community, mobile data provide a good alternative.

Generalized radiation models are the state of the art in modeling diverse mobility scenarios (Kang et al., 2015). Isaacman et al. (2018) incorporated weather conditions into an extended radiation model to predict mobility in La Guajira, Colombia during a drought, and used mobile CDR as a ground truth to contrast different models. They concluded that mobile CDR is powerful for modeling climate change-related migration. Both gravity and radiation models were developed with geospatial population information as their input sources. If one has access to mobile CDR, models can use additional information. For example, Palchykov et al. (2014) developed a simple communication model based on the frequency of mobile phone calls between two locations and their geographical distance.

Indicators

The use of mobile CDR is not only a good source for predicting movements, but also for linking mobility behaviours to other indicators, which are potentially even more powerful. It is possible to derive proxies from mobile CDR for many indicators, including population density (Deville et al., 2014), urban activity (Reades et al., 2007), travel behaviour (Wang et al., 2019), civic engagement and political participation (Campbell and Kwak, 2010), economic status and poverty (Šćepanović et al., 2015), social integration (Bakker et al., 2019), and undeclared employment (Bruckschen et al., 2019). The raw data are first aggregated and converted into more informative representations, such as origin-destination matrices, total traveled distance for individuals or geographical areas, as well as spatial and temporal properties of routine behaviours (Hughes et al., 2016). Regularities of movement lead to tagging locations of "home" and "work" for each individual (Isaacman et al., 2011). Further analysis on social network, travel behaviour and frequently visited places, makes behavioural profiling very powerful.

An example is predicting poverty and wealth from mobile data, for example by measuring commuting patterns (Šćepanović et al., 2015). Some qualitative information about the socio-economic status of different regions needs to be combined with mobile CDR in this case. Blumenstock et al. (2015) used data from Rwanda's largest mobile phone network for wealth analysis, and verified their findings with follow-up phone surveys of a geographically stratified random sample of 856 individual subscribers. When they aggregated the results, the predicted composite wealth index, computed from 2009 call data and aggregated by administrative districts showed remarkable similarity with the actual composite wealth index as computed from a 2010 government Demographic and Health Survey (DHS). The latter was collected from about 12,800 households, whereas the call data was obtained from 15 million individuals over one year.

The most alluring benefit of using mobile CDR is arguably the cost of obtaining information. The cost of a national household survey is estimated at over $1 million, taking 12 to 18 months to complete (Jerven, 2014), but the CDR can be accumulated and analyzed much faster, and cheaply. One additional benefit was that the phone data could provide a much higher granularity in providing characteristics. Indeed, it was difficult to verify the quality of the insights provided by the phone data in Rwanda, because no other data source was able to provide wealth information with such geographical resolution.

Refugees

Mobility data are particularly useful for studying groups that are not easily covered in the national surveys, such as refugees. An example initiative is the Data for Refugees (D4R) Challenge, which was a data collaborative initiated by Turk Telekom, Boğaziçi University and TUBITAK and in collaboration with several academic and non-governmental or intra-governmental organizations, including UNHCR Turkey, UNICEF, and IOM, to gain insights for improving the living conditions of over (then) three million Syrian refugees[9] living in Turkey (Salah et al. 2018; Salah et al. 2019). In D4R, an additional flag was used to tag each mobile CDR as possibly originating from a refugee, in case the phone line was (1) obtained with a subsidized, cheaper tariff for Syrian refugees, (2) registered with a Syrian passport, (3) registered with a special ID number that the Turkish government was providing the Syrian refugees. Each of these conditions contained some unspecified noise. Consequently, it was not possible to associate a CDR with a refugee with certainty. However, it was possible to derive conclusions from aggregated data on internal migration. For example, it was possible to see how many refugees were coming to the city of Ordu to work at the hazelnut harvest, which cities they were coming from, how long they were staying, and how they were distributed in and around Ordu during the harvest (Bruckschen et al., 2019; Turper Alisik et al., 2019). The dataset opened for the challenge contained data from 200K refugees and 800K Turkish citizens, and CDR were collected over one year (2017). Thirty groups completed the challenge successfully, and submitted project reports.[10]

As in all research involving mobile CDR, additional data sources or indicators must be combined with such data to investigate more complex aspects of mobility. For example, Beine et al. (2019) employed a gravity model to empirically estimate a series of determinants of refugee movements using the D4R challenge data, in order to evaluate how policy can facilitate mobility and integration of refugees. They considered standard determinants such as province characteristics, distances across provinces, levels of income, network effects, as well as refugee-specific determinants and the effect of certain categories of news events, protests, violence, and asylum grants.

Humanitarian Response

Humanitarian response requires up-to-date data processing for operational capabilities and informed decision making. Mobile CDR has been used by humanitarian organizations like UNHCR (Earney and Jimenez, 2019), UN Global Pulse (Boy et al., 2019), and UNICEF (Sekara et al., 2019) in refugee scenarios, but there are other crisis response settings, where its use has been demonstrated as well. One of the first applications was the analysis of people's movements in the aftermath of the 2010 earthquake in Haiti (Bengtsson et al., 2011). Mobility during floods, and the spread of epidemics, such as cholera, malaria, Ebola, and COVID-19, were also analysed and modeled with mobile CDR (Sandvik et al., 2014).

An important use of CDR is conducting pre-analysis to develop crisis response strategies and information preparedness. However, most research demonstrating the potential of mobile CDR analysis for humanitarian response is typically performed in the aftermath of emergencies, involving a significant delay. A major reason for this is that the sharing and processing of mobile CDR requires technical capabilities and resources, carefully prepared legal agreements, and ethical committee approvals that take time. However, researchers have discussed the possibility of preparing data processing pipelines and technical infrastructures well in advance of crises. A wake-up call was the COVID-19 pandemic, which will surely not be the last pandemic the world is facing. In Oliver et al. (2020), we have stressed the importance of being ready for the next one, by creating data collaboratives that can be activated rapidly when a new epidemic is in its initial stages.

The need for actionable data is widely acknowledged, and there are important initiatives in this area. UN Global Pulse was established in 2009 to leverage big data sources in studying population behaviours for understanding and responding to global crises. A key role is played by coordinating agencies such as UN Office for the Coordination of Humanitarian Affairs (OCHA), and by data holders (such as telecommunication operators) willing to act rapidly for humanitarian response.

Processing Mobile CDR

A detailed exposition of how mobile CDR data (or the higher-frequency x-Detail Record xDR, which is obtained from data package exchanges between phones and operators) can be processed is beyond the scope of this chapter. Tools specifically developed for analyzing mobility data include algorithmic packages like *scikit-mobility* (Pappalardo et al., 2019), and visualization tools such as *Urban Mobility Atlas*, which visually summarizes a number of mobility indicators over a geographical area (Giannotti et al., 2016). A good starting point that surveys advances made recently in the study of mobile phone datasets is Blondel et al. (2015).

There are numerous factors and potential biases one needs to consider when analyzing mobile CDR data. Calabrese et al. (2013) list four of these as (1) the market share of the mobile phone operator from which the dataset is obtained; (2) the potential non-randomness of the mobile phone users; (3) calling plans that can limit the number of samples acquired at each hour or day; and (4) the number of devices that each person carries. Additional factors include uneven gender distribution in phone ownership (e.g., in the D4R Challenge in Turkey, over 75% of phones were registered to male users, even though the proportion of females using the phones was probably much higher than 25%), and the lack of children, as they cannot legally own a phone line before a certain age. These are called "data gaps", and they should be carefully integrated into the models of analysis. Zagheni and Weber (2015) propose calibration of data with reliable official statistics, when they are available, and evaluating relative trends, when such data are lacking.

ETHICS AND PRIVACY

Using big data technologies that provide a granular and detailed view into human behaviour raises a number of important concerns that need to be addressed carefully. This is not only true of mobile phone data, but also of, for example, satellite imaging to detect human movements, or of using social media to chart out opinions and influences in social networks. These technologies, in the hands of controlling governments, can easily usher in an unprecedented degree of surveillance. As an example, in February 2020, the press reported that the U.S. Immigration and Customs Enforcement agency (ICE) has purchased commercial mobile phone datasets and used the information to arrest undocumented migrants (Tau and Hackman, 2020). Furthermore, irresponsible handling of data can harm the "data subjects", turning a social good application into a source of harm for the population it is intended to help (Berens et al., 2018). Subsequently, it is important to ask the question of what the risks are in designing technology and algorithms for such data analysis, and whether they are worth the potential benefits (Maxmen, 2019).

For each of the data collaboratives described in Section 2.3, a special committee was formed to screen proposals and reports, observing the "Do No Harm" principle for the target populations (Vinck et al., 2019). In D4D, the D4D External Ethics Panel (DEEP) assessed applications for business ethics and intended applications, to balance risks and opportunities. This goes beyond what is permissible legally, and the panel was able to request amendments to reports or papers when necessary. Later, the Institute of Business Ethics (IBE) penned a report called "Data for Development Senegal: Report of the External Review Panel" to share this framework and the findings of DEEP. The report stressed the importance of group privacy, as well as the importance of cultural factors, stating that "a particularly sensitive challenge is related to the lack of a regulatory framework or widely accepted benchmark on the privacy of data for individuals and for groups and its subsequent use or sharing. Another is the consideration for the publication of scientific results, the application of which might be considered more sensitive in certain cultures."

The D4R Challenge followed the recommendations of DEEP, and a project examination committee (PEC) was formed with members from academia, IGOs/NGOs, the telecommunications company initiating the challenge, representatives from two ministries of Turkey related to the challenge topics (i.e., education and health), as well as Syrian refugees. Together with the scientific committee, PEC evaluated all project proposals to determine who gets access to the data, and later evaluated all project reports and publications for potential risks. Once the challenge was over, PEC was disbanded, and it was not possible to screen later publications based on the D4R data.

The most important principle used in D4R was "data protection by design and default", where any name, real phone number, or other identifying information was excluded from the design of the database. The pseudo-random numbers representing customers were not stored anywhere along with actual phone numbers. Subsequently, the anonymization worked only one way. Since the refugee status was indicated by purposefully noisy indicators, and without any effort spent to ensure its validity, person-level conclusions about refugees could not be drawn from the data.

In all these challenges, the legal team of the data owner prepared an extensive license agreement for the challenge participants. As the datasets did not contain personally identifying information, they were not considered personal information by definition, and did not legally require subscriber consent beyond what was specified in the mobile user agreement. However, the preparation of a legal document that maintains a high standard of accountability for data users is essential. To better understand the legal conditions that can enable effective data collaboration, GovLab, SDSN TReNDS, University of Washington's Information Risk Research Initiative, and the World Economic Forum have recently started an initiative called Contracts for Data Collaboration (C4DC),[11] and created an online repository of data-sharing agreements that have facilitated data sharing in a variety of settings and countries. These agreements cover a range of data applications and consider a host of legal issues that differ from country to country. The repository also contains associated use cases, guides, and other resources.

The ICRC Data Protection Impact Assessment (DPIA) template[12] provides a good framework that addresses many issues in mobile CDR processing in a systematic manner, providing a number of data protection issues together with related code of conduct, an example and practice-oriented assessment of risks, potential mitigation measures, and potential outcomes for each issue, whereby a given scenario can be assessed. We provide a summary of the highlighted issues in Box 8.1. This, of course, is only one potential framework that can be used to address the issues that arise in using mobile CDR for migration and mobility. Several other frameworks are contrasted and detailed (see Berens et al., 2016).

BOX 8.1 KEY ISSUES IN ICRC DATA PROTECTION IMPACT ASSESSMENT (DPIA) TEMPLATE

- Purpose specification: The data should be collected only for a specific purpose.
- Data limitation: There should be no personal data items that are not required for analysis.
- Right to information: The individuals should be informed about collection and use of personal information.
- Legal basis for data processing / transfer: Informed consent should be obtained, under conditions where individuals giving consent are able to assess positive or negative outcomes.
- Right to access / rectification / deletion: Individuals should be able to access, correct, and delete their personal information.
- Information quality and accuracy: Processes should be in place to ensure information quality and accuracy.
- Appropriate security measures: The information should not jeopardize people, and must be appropriately protected against malicious uses, including surveillance.
- Data sharing, disclosure/publication and/or transfer: Data sharing with third parties and national societies should be transparent, meaningful, and accountable.
- Data retention: Data should not be retained longer than necessary.
- Risks to individuals: Additional risks, such as risks to physical or moral integrity of individuals, should be assessed.
- Accountability / oversight mechanism: Mechanisms should be in place to ensure responsible code of conduct and data protection standards.

CONCLUSIONS AND KEY IMPLICATIONS FOR RESEARCH AND GOVERNANCE

The effort in data challenges with mobile information illustrated that it takes a long time until research results can even begin to influence policy making. However, mobile CDR can potentially be processed very fast, much faster than traditional data acquisition mechanisms used by states. It is clear that the state would benefit from such faster decision making for informed policy decisions, but the potential issues of such processing are not yet tested, not even with the recent data challenges.

Many scientific papers were published on the findings of the large data challenges mentioned in this chapter, international media and NGOs showed great interest in the results, and several other initiatives were inspired. However, in many cases policy implications were limited, as the projects were exploratory. One proposed solution to increase impact for governance was to secure government funding for pilot projects following the challenge. Other strategies were to involve representatives from related ministries in the organization of the challenges, and preparing white papers for informing policy makers. What is still lacking at the time of writing this chapter is a data processing pipeline involving private data sources, designed in a privacy-aware way to inform authorities on urgent issues and help policy making (Letouzé and Oliver, 2019; Oliver et al., 2020).

The value of mobile CDR for studying displaced populations is mainly in providing a rich data source that complements official statistics, and under certain safeguards, one that can give insights in real time. This is particularly important for crisis management, where unexpected and large-scale displacement happens, or for studying challenging questions, such as the integration prospects of arriving migrants. However, the safeguards of using mobile phones as a data source without compromising the privacy of individuals are just as important as the potential benefits.

It is clear that policy makers should continue an interdisciplinary dialogue to evaluate and discuss the positive and potentially negative consequences of mobile data processing. A clear understanding of how these technologies operate and what they can provide is essential for enabling the potential of using mobile data for social good.

ACKNOWLEDGMENTS

We thank Almıla Akdağ, Marie McAuliffe, Bilgeçağ Aydoğdu, and Metehan Doyran for their constructive comments. This work is supported by the HumMingBird project, which has received funding from the European Union's Horizon 2020 research and innovation programme under Grant Agreement No 870661.

NOTES

1. The International Telecommunications Union (ITU) estimates the percentage of the population covered by a mobile-cellular network to be 98.8% in the developed world and 96.2% in the developing world, according to 2019 data. Detailed statistics are available at: https://www.itu.int/en/ITU -D/Statistics/Pages/stat/default.aspx (all webpages in endnotes accessed 23 August 2021).
2. The anonymized Reality Mining dataset, collected over the course of nine months, can be downloaded from http://realitycommons.media.mit.edu/realitymining.html.
3. The GSMA Digital Toolkit includes detailed information on policy and regulations, business models, as well as a portfolio of case studies, and can be accessed at: https://aiforimpacttoolkit.gsma .com/.
4. See more on the SoBigData project at: http://project.sobigdata.eu/ and the HummingBird project at: https://cordis.europa.eu/project/rcn/225807/en.
5. Contributions to all past editions of NETMOB can be accessed at: https://netmob.org/#pasteditions.

6. The scientific reports of all D4D projects can be accessed at: https://perso.uclouvain.be/vincent .blondel/netmob/2013/D4D-book.pdf.
7. Unlike the D4D and D4R Challenges, Telecom Italia Challenge made its data public for a limited time: http://theodi.fbk.eu/openbigdata/
8. It is important to note here that when the database is appropriately anonymized, such profiling can indicate a group membership with a high probability, but is not connected to a specific identity. Nonetheless, if sufficiently many indicators are added to the analysis, it becomes possible to pinpoint to an individual about which a lot of information is externally obtained. For this reason, both spatial and temporal resolution of the data are reduced on purpose.
9. Turkey is party to the 1951 Geneva Refugee Convention but only acknowledges "refugee" status for people originating from Europe. Syrian refugees are officially and legally considered "temporarily protected foreign individuals".
10. The scientific reports of all D4R projects can be accessed at: https://webspace.science.uu.nl/ ~salah006/d4r-proceedings.pdf.
11. Contracts for Data Collaboration (C4DC) website contains more information at: https://contracts fordatacollaboration.org/.
12. ICRC Data Protection Impact Assessment template can be accessed at: https://www.icrc.org/en/ download/file/18149/dpia-template.pdf.

REFERENCES

Ahas, R. and Ü. Mark (2005), 'Location based services: new challenges for planning and public administration?', *Futures*, **37**(6), 547–561.

Ahas, R., S. Silm, and M. Tiru (2018), 'Measuring transnational migration with roaming datasets', paper in *Adjunct Proceedings of the 14th International Conference on Location based services*, ETH Zurich, 105–108.

Bakker, M.A., D.A. Piracha, P.J. Lu, K. Bejgo, M. Bahrami, Y. Leng, J. Balsa-Barreiro, J. Ricard, A.J. Morales, V.K. Singh, and B. Bozkaya (2019), 'Measuring fine-grained multidimensional integration using mobile phone metadata: the case of Syrian refugees in Turkey', in A.A. Salah, A. Pentland, B. Lepri, and E. Letouzé (eds), *Guide to Mobile Data Analytics in Refugee Scenarios*, Cham, Switzerland: Springer Nature Switzerland AG, 123–140.

Bayir, M.A., M. Demirbas, and N. Eagle (2009), 'Discovering spatiotemporal mobility profiles of cellphone users', paper presented at the *IEEE International Symposium on a World of Wireless, Mobile and Multimedia Networks & Workshops,* IEEE.

Beine, M., L. Bertinelli, R. Cömertpay, A. Litina, J.F. Maystadt, and B. Zou (2019), 'Refugee mobility: evidence from phone data in Turkey', in A.A. Salah, A. Pentland, B. Lepri, and E. Letouzé (eds), *Guide to Mobile Data Analytics in Refugee Scenarios*, Cham, Switzerland: Springer Nature Switzerland AG, 433–449.

Bengtsson, L., X. Lu, A. Thorson, R. Garfield, and J. Von Schreeb (2011), 'Improved response to disasters and outbreaks by tracking population movements with mobile phone network data: a post-earthquake geospatial study in Haiti', *PLoS Med*, **8**(8), e1001083.

Berens, J., U. Mans, and S. Verhulst (2016), 'Mapping and comparing responsible data approaches', available at *SSRN*. Accessed 2 October 2019: http://dx.doi.org/10.2139/ssrn.3141453

Blondel, V. D., A. Decuyper, and G. Krings (2015), 'A survey of results on mobile phone datasets analysis', *EPJ data science*, **4**(1), 10.

Blondel, V. D., M. Esch, C. Chan, F. Clérot, P. Deville, E. Huens, F. Morlot, Z. Smoreda, and C. Ziemlicki (2012), 'Data for development: the D4D challenge on mobile phone data', *arXiv preprint arXiv:1210.0137*.

Blumenstock, J., G. Cadamuro, and R. On (2015), 'Predicting poverty and wealth from mobile phone metadata', *Science*, *350*(6264), 1073–1076.

Boy, J., D. Pastor-Escuredo, D. Macguire, R.M. Jimenez, and M. Luengo-Oroz (2019), 'Towards an understanding of refugee segregation, isolation, homophily and ultimately integration in Turkey using call detail records', in A.A. Salah, A. Pentland, B. Lepri, and E. Letouzé (eds), *Guide to Mobile Data Analytics in Refugee Scenarios*, Cham, Switzerland: Springer Nature Switzerland AG, 141–164.

Bruckschen, F., T. Koebe, M. Ludolph, M.F. Marino, and T. Schmid (2019), 'Refugees in undeclared employment: a case study in Turkey', in A.A. Salah, A. Pentland, B. Lepri, and E. Letouzé (eds), *Guide to Mobile Data Analytics in Refugee Scenarios*, Cham, Switzerland: Springer Nature Switzerland AG, 329–346.

Calabrese, F., M. Diao, G. Di Lorenzo, J. Ferreira Jr, and C. Ratti (2013), 'Understanding individual mobility patterns from urban sensing data: a mobile phone trace example', *Transportation Research Part C: Emerging Technologies*, **26**, 301–313.

Campbell, S. W. and N. Kwak (2010), 'Mobile communication and civic life: linking patterns of use to civic and political engagement', *Journal of Communication*, **60**(3), 536–555.

de Montjoye, Y.-A., C.A. Hidalgo, M. Verleysen, and V.D. Blondel (2013), 'Unique in the crowd: The privacy bounds of human mobility', *Scientific Reports*, **3**:1376.

de Montjoye, Y.-A., Z. Smoreda, R. Trinquart, C. Ziemlicki, and V.D. Blondel (2014), 'D4D-Senegal: the second mobile phone data for development challenge', *arXiv preprint arXiv:1407.4885*.

de Montjoye, Y.A., S. Gambs, V. Blondel, G. Canright, N. De Cordes, S. Deletaille, K. Engø-Monsen, M. Garcia-Herranz, J. Kendall, C. Kerry, and G. Krings (2018), 'On the privacy-conscientious use of mobile phone data', *Scientific data*, 5.

Deville, P., C. Linard, S. Martin, M. Gilbert, F.R. Stevens, A.E. Gaughan, V.D. Blondel, and A.J. Tatem (2014), 'Dynamic population mapping using mobile phone data', *Proceedings of the National Academy of Sciences*, **111**(45), 15888–15893.

Di Lorenzo, G., M. Sbodio, F. Calabrese, M. Berlingerio, F. Pinelli, and R. Nair (2015), 'AllAboard: visual exploration of cellphone mobility data to optimise public transport', *IEEE Transactions on Visualization and Computer Graphics*, **22**(2), 1036–1050.

Dobra, A., N.E. Williams, and N. Eagle (2015), 'Spatiotemporal detection of unusual human population behavior using mobile phone data', *PloS One*, **10**(3), e0120449.

Eagle, N. and A.S. Pentland (2006), 'Reality mining: sensing complex social systems', *Personal and Ubiquitous Computing*, **10**(4), 255–268.

Eagle, N. and A.S. Pentland (2009), 'Eigenbehaviors: identifying structure in routine', *Behavioral Ecology and Sociobiology*, **63**(7), 1057–1066.

Earney, C. and R.M. Jimenez (2019), 'Pioneering predictive analytics for decision-making in forced displacement contexts', in A.A. Salah, A. Pentland, B. Lepri, and E. Letouzé (eds), *Guide to Mobile Data Analytics in Refugee Scenarios*, Cham, Switzerland: Springer Nature Switzerland AG, 101–119.

Giannotti, F., L. Gabrielli, D. Pedreschi, and S. Rinzivillo (2016), 'Understanding human mobility with big data', in *Solving Large Scale Learning Tasks: Challenges and Algorithms*, Cham: Springer, 208–220.

González, M.C., C.A. Hidalgo, and A.L. Barabasi (2008), 'Understanding individual human mobility patterns', *Nature*, **453**(7196), 779.

Gundogdu, D., O. Durmaz Incel, A.A. Salah, and B. Lepri (2016), 'Countrywide arrhythmia: emergency event detection using mobile phone data', *EPJ Data Science*, **5**(1), 25.

Hughes, C., E. Zagheni, G.J. Abel, A. Sorichetta, A. Wi'sniowski, I. Weber, and A.J. Tatem (2016), 'Inferring migrations: traditional methods and new approaches based on mobile phone, social media, and other big data: Feasibility study on Inferring (labour) mobility and migration in the European Union from big data and social media data', report prepared for the European Commission project #VT/2014/093.

Isaacman, S., V. Frias-Martinez, and E. Frias-Martinez (2018), 'Modeling human migration patterns during drought conditions in La Guajira, Colombia', paper presented at *1st ACM SIGCAS Conference on Computing and Sustainable Societies*, 1–9.

Isaacman, S., R. Becker, R. Cáceres, S. Kobourov, M. Martonosi, J. Rowland, and A. Varshavsky (2011), 'Identifying important places in people's lives from cellular network data', in K. Lyons, J. Hightower, and E.M. Huang (eds), *Pervasive Computing*, 6696. Springer, Berlin, 133–151.

Jerven, M. (2014), 'Benefits and costs of the data for development targets for the post-2015 development agenda', *Data for Development Assessment Paper*, Copenhagen Consensus Center.

Kang, C., Y. Liu, D. Guo, and K. Qin (2015), 'A generalized radiation model for human mobility: spatial scale, searching direction and trip constraint', *PloS one*, **10**(11).

Lazer, D., A.S. Pentland, L. Adamic, S. Aral, A.L. Barabási, D. Brewer, ... and M. Van Alstyne (2009), "Life in the network: the coming age of computational social science', *Science*, **323**(5915), 721–723.

Letouzé, E (2019), 'Leveraging Open Algorithms (OPAL) for the safe, ethical, and scalable use of private sector data in crisis contexts', in A.A. Salah, A. Pentland, B. Lepri, and E. Letouzé (eds), *Guide to Mobile Data Analytics in Refugee Scenarios*, Cham, Switzerland: Springer Nature Switzerland AG, 453–464.

Letouzé, E. and N. Oliver (2019), 'Sharing is Caring: Four Key Requirements for Sustainable Private Data Sharing and Use for Public Good', *DataPop Alliance: Vodafone Institute for Society and Communications White Paper.* Accessed on September 2021: http://datapopalliance.org/wp-content/uploads/2019/11/DPA_VFI-SHARING-IS-CARING.pdf

Lu, X., D.J. Wrathall, P.R. Sundsøy, M. Nadiruzzaman, E. Wetter, A. Iqbal, and L. Bengtsson (2016), 'Unveiling hidden migration and mobility patterns in climate stressed regions: A longitudinal study of six million anonymous mobile phone users in Bangladesh', *Global Environmental Change*, **38**, 1–7.

Margolin, D., Y.R. Lin, D. Brewer, and D. Lazer (2013), 'Matching data and interpretation: Towards a rosetta stone joining behavioral and survey data', paper presented at the *Seventh International AAAI Conference on Weblogs and Social Media.*

Maxmen, A. (2019), 'Can tracking people through phone-call data improve lives?' *Nature News Feature*, 29 May. Accessed 23 August 2021: https://www.nature.com/articles/d41586-019-01679-5

Oliver, N., B. Lepri, H. Sterly, R. Lambiotte, S. Deletaille, M. De Nadai, E. Letouzé, A.A. Salah, R. Benjamins, C. Cattuto, and V. Colizza (2020), 'Mobile phone data for informing public health actions across the COVID-19 pandemic life cycle', *Science Advances*, **6**(23), eabc0764.

Palchykov, V., M. Mitrović, H.H. Jo, J. Saramäki and R.K. Pan (2014), 'Inferring human mobility using communication patterns', *Scientific Reports*, **4**(1), 1–6.

Pappalardo, L., G. Barlacchi, F. Simini, and R. Pellungrini (2019), 'scikit-mobility: An open-source Python library for human mobility analysis and simulation', *arXiv preprint arXiv:1907.07062.*

Reades, J., F. Calabrese, A. Sevtsuk, and C. Ratti (2007), 'Cellular census: Explorations in urban data collection', *IEEE Pervasive Computing*, **6**(3), 30–38.

Salah, A.A. and T. Gevers (2011), *Computer Analysis of Human Behavior.* London: Springer.

Salah, A.A., A. Pentland, B. Lepri, E. Letouzé, P. Vinck, Y.-A. de Montjoye, X. Dong, and Ö. Dağdelen (2018), 'Data for refugees: The D4R challenge on mobility of Syrian refugees in Turkey', *arXiv preprint arXiv:1807.00523.*

Salah, A.A., A. Pentland, B. Lepri, E. Letouzé, Y.-A. de Montjoye, X. Dong, Ö. Dağdelen, and P. Vinck (2019), 'Introduction to the data for refugees challenge on mobility of Syrian refugees in Turkey', in A.A. Salah, A. Pentland, B. Lepri, and E. Letouzé (eds), *Guide to Mobile Data Analytics in Refugee Scenarios*, Cham, Switzerland: Springer Nature Switzerland AG, 3–27.

Sandvik, K.B., M.G. Jumbert, J. Karlsrud, and M. Kaufmann (2014), 'Humanitarian technology: A critical research agenda', *International Review of the Red Cross*, **96**(893), 219–242.

Šćepanović, S., I. Mishkovski, P. Hui, J.K. Nurminen, and A. Ylä-Jääski (2015), 'Mobile phone call data as a regional socio-economic proxy indicator', *PloS one*, **10**(4).

Sekara, V., E. Omodei, L. Healy, J. Beise, C. Hansen, D. You, ... and M. Garcia-Herranz (2019), 'Mobile phone data for children on the move: Challenges and opportunities' in A.A. Salah, A. Pentland, B. Lepri, and E. Letouzé (eds), *Guide to Mobile Data Analytics in Refugee Scenarios*, Cham, Switzerland: Springer Nature Switzerland AG, 53–66.

Tau B. and M. Hackman (2020), 'Federal agencies use cellphone location data for immigration enforcement', *The Wall Street Journal*, Feb 7, 2020. Accessed 23 August 2021: https://www.wsj.com/articles/federal-agencies-use-cellphone-location-data-for-immigration-enforcement-11581078600.

Turper Alisik, S., D.B. Aksel, A.E. Yantac, I. Kayi, S. Salman, A. Icduygu, D. Cay, L. Baruh, and I. Bensason (2019), 'Seasonal labor migration among Syrian refugees and urban deep map for integration in Turkey', in A.A. Salah, A. Pentland, B. Lepri, and E. Letouzé (eds), *Guide to Mobile Data Analytics in Refugee Scenarios*, Cham, Switzerland: Springer Nature Switzerland AG, 305–328.

Verhulst, S.G. and A. Young (2019), 'The potential and practice of data collaboratives for migration', in A.A. Salah, A. Pentland, B. Lepri, and E. Letouzé (eds), *Guide to Mobile Data Analytics in Refugee Scenarios*, Cham, Switzerland: Springer Nature Switzerland AG, 465–476.

Vinck, P., P.N. Pham, and A.A. Salah (2019), '"Do No Harm" in the Age of Big Data: Data, ethics, and the refugees', in A.A. Salah, A. Pentland, B. Lepri, and E. Letouzé (eds), *Guide to Mobile Data Analytics in Refugee Scenarios*, Cham, Switzerland: Springer Nature Switzerland AG, 87–99.

Wang, Y., L. Dong, Y. Liu, Z. Huang, and Y. Liu (2019), 'Migration patterns in China extracted from mobile positioning data', *Habitat International*, **86**, 71-80.

Zagheni, E. and I. Weber (2015), 'Demographic research with non-representative internet data', *International Journal of Manpower*, **36**(1), 13–25.

9. Migrant smuggling and ICT: research advances, prospects and challenges

Georgios Papanicolaou, Parisa Diba and Georgios A. Antonopoulos

INTRODUCTION

This chapter explores the question of the role of ICT (Information and Communications Technology) in human smuggling and provides a critical appreciation of the emerging research literature around it. There is no lack of alarmism or even sensationalism in representations of human smuggling as a form of transnational organised crime in official and, particularly, media accounts of the phenomenon. Cross-border or transnational organised crime has emerged as a key policy area in our era of globalisation, one that is backed up by a new international policy regime featuring international instruments of major significance, such as the 2000 UN Convention on Transnational Organised Crime (UNTOC) (United Nations, 2004). These, in turn, have engendered a new law enforcement alertness, infrastructure, and capacity to monitor and address the issue at national, regional, and global level. In the past 25 years or so, there has emerged a substantial body of knowledge exploring and documenting the possible threats and real harms involved in the business of human smuggling (see McAuliffe and Laczko, 2016), and it continues to grow in the context of the booming of irregular migration flows from Asia and Africa to Europe. Much of the literature focuses on the link between the predatory intention and practices of human smugglers and the exposure of irregular migrants to conditions of vulnerability and abuse, up to and including the possibility of death *en route* (see, e.g., Clarke-Billings, 2017). Official accounts typically emphasise the role of members of criminal networks in irregular migration (see Europol and Interpol, 2016) as well as the gravity of the threat this connection engenders. Since the beginning of 2014, IOM's *Missing Migrants Project* has reported the deaths of over 35,000 people, noting that even this figure should be treated as 'indicative of the risks associated with migration, rather than representative of the true number of deaths across time or geography' (IOM, 2020).

It is unsurprising that the increasing use of ICT by both human smugglers and irregular migrants amplifies these concerns. In light of the typical perception of human smuggling and the conditions in which it occurs, the increasing use of (mobile) telephony, the internet, social media and other applications is understood as an enhancement in the sophistication, versatility and reach of the human smuggling business. According to this narrative, smugglers use the Internet, social media and also widely popular software applications to recruit customers, to arrange facilitation services, including transportation, accommodation and the provision of fraudulent travel documents, and also to make and receive payments. Conversely, the use of ICT by migrants exacerbates the risk that they are exposed to the deceptive and predatory activity of the smugglers; equally, other users inadvertently become facilitators of the latter's business (Europol, 2017b).

While the official and mainstream narrative is very much in harmony with the escalating political controversy around irregular migration and the emergence of transnational organised crime as the folk devil of the globalisation era (see, e.g., Shelley, 1999; 2014), it far from settles the matter. There is now an emerging stream of research literature on irregular migration and human smuggling, which has presented a much more nuanced view of the phenomenon (e.g., Zhang et al., 2018; Achilli, 2018a; Baird, 2016; Sanchez, 2016; Siegel, 2019). Similarly, the global diffusion of ICT among populations (see, e.g., Lechman, 2019) entails that its use cannot tenably be construed to be monopolised by organised crime. A consideration, therefore, of the connection between human smuggling and ICT critically depends on a critical reappreciation of the key messages of the research literature on the phenomenon as an instance of organised crime. Equally, a consideration of the possible role and use of ICT depends on some conceptual clarity about ICT itself—particularly in light of the fact that assessing the implications of accelerated development of information technology in the 21st century remains an uncertain exercise (Europol, 2017a; 2017c), which may increase anxieties and the sense of threat this technology is known to cause (Brosnan, 1998).

These aspects, alongside some definitional issues, will be explored in the first part of the remainder of this chapter. In the second part we proceed to provide an overview of more specific aspects of the question of the role and use of ICT in human smuggling as well as the methodological approaches through which knowledge of the role of ICT in human smuggling can be gained. Finally, we raise some questions about the challenges future research should address.

ICT, CRIME, ORGANISED CRIME AND HUMAN SMUGGLING: IN SEARCH OF CONCEPTUAL CLARITY

As already suggested in the introduction, the study of the role of ICT in human smuggling is a relatively new focal point for research, situated at the intersection of wider—parent, as it were—fields of study. This may have the implication that some of the general insights emerging from those fields may be impairing the capacity to define the specific issue in hand. For example, a general knowledge of the role of ICT in crime may not necessarily translate into a specific knowledge of the role of ICT in the specific crime of human smuggling; similarly, rather than rely on the (readily available) general insights about the role of ICT in migration, an adequate conceptual apparatus, and corresponding research strategies potentially, should be sought at the level of the social organisation, practices and experiences of facilitated illicit (international) migration. The discussion in this section explores this possibility, with the overall direction of the argument being that certain precautions are necessary when approaching the question of the role of ICT in the business of human smuggling.

It is useful to begin with a very brief reminder about ICT itself. While mentions of ICT are ubiquitous today, a universal definition is notoriously hard to find, since ICT can be approached from the perspective of any of the various components that it encompasses as a reality. Broadly defined, ICT can include older technologies, such as fax machines and television, to encompass all the electronic technologies that people use to share, distribute, and stock all sorts of information and knowledge (see Lechman, 2019). Typically, the emphasis is on the wireless telecommunication technologies that, since the beginning of the 21st century, have evolved from simple voice and text services to diversified innovative applications and mobile broadband Internet. UNESCO (2002) has defined ICT as the combination of informatics technology with other, related technologies, specifically communication technology, whereas the World Bank measures the diffusion of ICT on the basis of such indicators as fixed-telephone, mobile-cellular telephone, fixed broadband subscriptions, internet access at home or the affordability of mobile broadband (see World Bank, 2017; 2016). The above suggest that, while it is possible to approach ICT with an emphasis that is not uncommon in either lay or expert (e.g., Hargittai, 1999) discourse, ICT is not the internet or access to the internet alone.

The implications of how broadly ICT can be approached are visible in both the 'parent' scholarly fields that claim an interest in human smuggling. Firstly, there now exists substantial and well-established research literature on the use of ICT for criminal purposes (e.g., Wall, 2007; McQuade, 2009, 2011; Jaishankar, 2011; Jewkes and Yar, 2010; Martellozzo and Jane, 2017; Yar and Steinmetz, 2019; 2006; Lavorgna, 2020; Whittle et al., 2013). In this context, it has been shown quite effectively how ICT may either expand the opportunity for known, conventional, as it were, types of crime or open up an opportunity for new types of crime enabled by new conditions that the existence of these technologies creates. On the other hand, the question of the use of ICT in the context of migration and the migrants' experience has been picked up by researchers in the 2000s (e.g., Hamel, 2009), and it has naturally gained a renewed relevance in the context of the irregular migration flows and refugee crisis of the past decade (e.g., McAuliffe et al., 2017; Latonero and Kift, 2018; Zijstra and van Liemt, 2017; Frouws et al., 2016; Gillespie et al., 2016; Dekker and Engbersen, 2012; Dekker et al., 2018). What emerges from the research literature is the robust finding that ICT does play an increasingly important role not only in the experience of migrants and refugees at all stages of the migratory process but also in shaping the patterns of international migration more generally (see McAuliffe et al., 2017).

The above observations, however, do not settle the specific question regarding the role of ICT in human smuggling. Within the wide context of (international) migration and the cross-border flows of populations, human smuggling is a specific phenomenon and, equally, a *specific legal construct*. The UNTOC in Article 3 of the Protocol against Human Smuggling has provided a definition of human smuggling as:

> the procurement, in order to obtain, directly or indirectly, a financial or other material benefit, of the illegal entry of a person into a State Party of which the person is not a national or a permanent resident.

The Protocol identifies the criminalised activity as having to do, as Latonero and Kift (2018) correctly remark, with transportation and, more generally, with the facilitation of cross-border mobility, whether it involves the entry, transit and residence of (irregular) migrants (see European Commission, 2015). It follows that the question of the role of ICT in human smuggling can emerge in two related and often overlapping but distinct areas of research, namely the study of (irregular) migration on one hand, and on the other, the study of transnational organised crime, within which it must be explicitly problematised or raised as an issue, since it pertains to a subset of patterns and practices involved in those two areas respectively.

The particularity of the topic of ICT in human smuggling is clearly reflected on the state of play in those respective fields. Firstly, while, as noted above, the study of the use and implications of communication technologies within migration studies is well established, studies of its role in specific connection with human smuggling have only begun to appear recently (e.g., Di Nicola et al., 2017, Diba et al., 2019; Latonero and Kift, 2018). The question has, of course, naturally emerged in studies of migrants and refugees, particularly in light of the higher visibility of the abusive and harmful practices of smugglers at key points of contemporary migration routes, such as the Mediterranean (UNODC, 2018). The body of empirical data and other information cumulated by these studies is becoming substantial; however, one must not fail to note that in this specific context human smugglers are only one group of actors populating the landscape of 'information precarity' (Wall et al., 2017; see also McAuliffe, 2016; McAuliffe et al., 2017) that migrants and refugees negotiate while planning or carrying out their journeys.

On the other hand, in studies of transnational organised crime, human smuggling has been more often addressed in conjunction with human trafficking (Laczko and Thompson, 2000; UNODC, 2010; Aronowitz, 2001), in light of the socio-legal fact that these two types of crime constitute a prime focus of the international regime pertaining to irregular cross-border mobility that was instituted by the UNTOC. There is also a real possibility that they may practically overlap and that smuggled migrants are exposed to the risk of trafficking during and after their journey, depending on the conditions they may be facing or the organisation of the criminal business (UNODC, 2018: 41–42). While the two should not be confused, as the different goals and processes of each activity can be clearly differentiated in both theory, law and reality (Salt and Hogarth, 2000), human trafficking has received more attention and as a result more is arguably known about the role and use of ICT in the human trafficking business, particularly sex trafficking. While the literature on ICT in human trafficking grows (e.g., Di Nicola, Cauduro and Falletta, 2013; Latonero et al., 2011; Latonero et al., 2012; Sykiotou, 2007; Myria, 2017; Sarkar, 2015), the study of ICT in human smuggling has rarely gone beyond the observation that these technologies are used by smugglers and (prospective) irregular migrants.

Overall, the outlook of extant research emerging from the general fields of study maintaining an interest on human smuggling underscores the need for a more precise knowledge of how and to what extent ICT is integrated in the process and business of smuggling. The implications of the particular question of the role of ICT in the process are threefold. Firstly, not every aspect of ICT is equally relevant in the process of smuggling; in so far as the smuggling of migrants pertains primarily to the journey, mobile communications and the ability to process and exchange information on the move are arguably far more critical to the process than fixed forms of access and use of ICT. Secondly, the particular characteristics of the smuggling business must be taken seriously and specifically into account when considering the specific utility of ICT infrastructures and applications, such as mobile phones, VoIP (Voice over Internet Protocol) applications such as Skype or Viber, social media platforms or GPS, mapping or route planning applications (such as Google Maps). Finally, in light of the diffusion of ICT globally, including among the populations that take the journal and possibly seek the services of smuggling entrepreneurs, the particular use patterns and experiences of the users are key to an adequate knowledge of the issue (see Whittle and Antonopoulos, 2020).

THE ROLE OF ICT IN HUMAN SMUGGLING: WHAT DO WE KNOW? HOW DO WE KNOW IT?

The body of empirically driven research that has only recently begun to develop about the role of ICT in human smuggling appears to be largely congruent across key aspects of the issue and its study. This section provides an overview of the knowledge that has begun to emerge across key dimensions of the issue—also noting some key differentiations.

ICT Infrastructures and Applications

A key fact emerging from research specifically interrogating the role of ICT in human smuggling is congruent with the established insight from the wider field of migration and migrant studies, namely that communication technologies are firmly embedded in the experiences and life trajectories of migrant populations, to an extent that, as McAuliffe (2016) has remarked with reference to mobile phone technology, they have become the norm. Earlier research into the use of ICT had documented it in no uncertain terms (e.g., Schapendonk and van Moppes, 2007), and more recent research has equally captured the significance of an increasingly wide range of infrastructures and applications used in and even shaping the experience of the journey across contexts and routes (e.g., Frouws et al., 2016; Dekker et al., 2018). The advent of the smartphone has meant that the widespread use of mobile telephony is nowadays combined with internet access and the use of internet-dependent applications, such as social media or mapping applications.

For example, Diba et al.'s (2019) research into the smuggling of migrants into the UK has highlighted the significance of Facebook, a social media platform, for the overall organisation and experience of the journey for migrants. With specific reference to the connection of these platforms to migrant smuggling, it emerges that Facebook pages may be used to access and distribute information about transportation services, the sale of counterfeit travel documents such as passports, visas and identification papers including driving licences, as well as general discussions of the navigation of paths into the transit and destination countries. Individual users may also share their personal experiences (see Antonopoulos et al., 2020).

It is important to note that not all use of ICT in the process of irregular migration points to a connection with human smuggling as a particular type of criminal activity facilitating cross border mobility. As Frouws et al.'s (2016) report on the use of ICT in migration flows shows, the information sharing that is a core characteristic of online migration communities may in fact be used by migrants wishing to avoid relying on smugglers for their journey. The extant evidence suggests that the possibility of smuggling entrepreneurs becoming a part of such emerging online communities to promote their business is real, but this is something to be investigated by further research rather than a fact to be taken for granted.

The latter point is of particular significance in light of the fact that studies approaching the issue from the viewpoint of the smugglers are few and far apart—in fact, studies problematising entrepreneurs as complex and reflexive actors in the process of human smuggling are a relatively recent development (see Achilli, 2018a and 2018b; 2015). The available evidence suggests that smugglers appear to take a utilitarian approach to the use of ICT depending on how they approach their potential clientele and on how they conduct their business in the particular locales and contexts they operate. Equally, the potential threat that the use of ICT presents to their illicit business, for fear of detection by increasingly alert law enforcement may entail that smugglers may be resistant to the adoption of the latest communicational fads, such as social media, and they may retain the use of earlier technologies, such as (traditional) mobile telephony as means to reinforce security and minimise exposure in communications with clients (Antonopoulos et al., 2020).

Methodologies

The question 'What type of ICT?' in the smuggling process immediately raises issues of research methodology. Extant research clearly showcases the relevance of qualitative inquiry, ethnography in particular, including online ethnography which is a possibility enabled by the growth of the use of digital platforms. The key point is that, as Achilli remarks, 'human smuggling cannot be understood without attending to the interactions between migrants and smugglers' (Achilli, 2018b: 80). In both the offline and online context obtaining first-hand direct accounts of smugglers' perception and use of ICT is equally indispensable as investigating those interactions from the migrants' perspective.

Explicitly problematising the use of ICT opens up fascinating possibilities for online ethnography and an immersion into the cyber-universe of social media relating to the information-sharing and real-word experiences of migrants. The online environment enables research strategies that may range from the type of study presented above and also found in Frouws et al. (2016) up to the harnessing of big data from media such as Twitter or Facebook and even network statistical data (see, e.g., UN Global Pulse & UNCHR Innovation Service, 2017; Spyratos et al., 2018).

Nevertheless, assigning methodological primacy to digital environment involves inherent limitations. Firstly, while information and data on the migrants' end of the process may be overwhelmingly abundant on the internet, this is not necessarily the case with smugglers, who may not only take an opportunistic and inconsistent approach, but also operate under an increasingly vigilant eye of law enforcement agencies (see, e.g., European Migration Network, 2016; Latonero and Kift, 2018). Secondly, there exists the wider issue of access to ICT infrastructure and media under the practical conditions experienced by the actors of the human smuggling process either en route or at the locations where the process is taking place, and, more generally, the wider conditions of 'information precarity' (Wall et al., 2017; Whittle and Antonopoulos, 2020).

The Profile of Actors

Raising the latter issue is not intended as a return to a wholesale questioning of the availability and diffusion of ICT among irregular migrants, as mobile phone use and internet access has increased dramatically across the globe in recent years. As a result, the issue of 'digital infrastructure' (Gillespie et al., 2016; 2018) is decided more by social factors, such as class, gender, and ethnicity (McAuliffe et al., 2017), rather than the (increasing) infrastructural availability of these means in the developing countries from which irregular migrants typically originate. These social dimensions appear to mediate the consequences of the use of ICT in the facilitated migration process, in light of the fact that infrastructures and applications are available to and used by all the actors involved in the process (Antonopoulos et al., 2020; Frouws et al., 2016; McAuliffe, 2016).

This is to say that the information-sharing practices enabled by phones, the internet and social media are conditioned by the backgrounds and social ties of those participating in the process. As Sanchez (2017) observes, prospective migrants explore smuggling options usually by consulting their friends and relatives who have made and undertaken successful smuggling journeys, often meeting with smugglers in person or chatting via telephone and may meet several smugglers until a suitable option is chosen. Moreover, a small number of smugglers manage to form a client base and positive reputation attesting to the reliability and consistency of communication throughout the journey and the quality of the transportation, all of which contribute to sustaining the smuggling enterprise and allow smugglers to generate and conduct business with prospective customers. It is not the availability or the use of ICT that enables or technologically revolutionises the smuggling process, but rather these technologies are attached to the social organisation of the latter, whose small-scale operation—building on pre-existing social ties, requiring practical skills, knowledge of locations and routes, and flexibility in decision making—have been well established in the extant research literature (Andreas, 2000; Carrera and Guild, 2016; Içduygu & Toktas, 2002; Sanchez, 2016; Spener, 2004, 2009; Zhang, 2007). It does appear that the availability of and use of ICT may have a positive, protective effect overall in so far as it empowers, on one hand, migrants by allowing information to circulate and be shared more easily and in larger volume (see e.g., Achilli, 2015; Sanchez, 2017; Triandafyllidou, 2018), and, on the other, it may expose the more opportunistic and unscrupulous illicit entrepreneurs in the smuggling business.

CONCLUDING REFLECTIONS

Overall, the emerging research literature on the use of ICT in human smuggling documents the important role of ICT in the human smuggling process, making it clear that ICTs are available and are being used by all the actors in the process. Ironically, this particular achievement humanises the knowledge of the phenomenon in two intriguing ways. Firstly, it provides an important corrective approach to official and law enforcement representations of ICT as an unqualified, novel threat, extending the established 'predator–victim binary' of those representations. Much of recent research on the human smuggling process has challenged successfully the quasi-stereotypical representation of the issue by focusing on the practices and experiences of the people involved in this process (e.g., Zhang et al., 2018; Achilli, 2018a; Baird, 2016; Sanchez, 2016), and a closer interrogation of the role of ICT specifically points to a similar direction. Secondly, by pointing to the socially embedded, and socially (e.g., by class- or gender-) mediated uses of ICT, the emerging body of knowledge on ICT also reinforces the message that much of what the smuggling process involves, crucially relies on trust, trust-building and personal reassurance, all of which are the products of real-world, 'flesh and blood' human contact and interaction. More generally speaking, research on the use of ICT in human smuggling is consonant with a well-established stream of research on organised crime, which has documented the mundane, pragmatic, this-worldly nature of illegal entrepreneurship (van Duyne, 1996; see also Antonopoulos and Papanicolaou, 2018).

Nevertheless, the above corrective still entails that more research is needed on how these technologies are embedded and filtered by the social organisation of the process. In so far as human smuggling is approached abstractly as crime and specific legal construct, it possesses an irreducible physical quality that pertains to the facilitation of the cross-border movement of people. ICT cannot transform or revolutionise radically this activity, simply by virtue of the plain fact that smuggling cannot take place in the virtual spaces that communication technologies open up. From a criminological perspective, this characteristic places a limit to the direct applicability of analyses and assessments of the transformative effect of ICT in contemporary criminality (see, e.g., McCusker, 2006; Naylor, 2000; Levi, 2001; Savona and Mignone, 2004). The matter, however, appears differently from the viewpoint of actual social life, in which human smuggling can be found in a continuum with human trafficking and modern slavery. The latter's core element—exploitation— grows on real social conditions that also shape the irregular migration process, and there is an established and ever developing pool of evidence regarding the potentially blurred lines between human smuggling and trafficking (e.g., Carling et al., 2015; Reitano et al., 2018). Since exploitation can unfold in new ways in digital environments, the challenge for research is to remain alert to the possibility that ICT will have a novel transformative effect on this nexus.

REFERENCES

Achilli, L. (2015), *The Smuggler: Hero or Felon?* Florence: Migration Policy Centre.
Achilli, L. (2018a), 'Smuggling and trafficking in human beings at the time of the Syrian conflict'. In Gebrewold, B. Kostenzer, J. & Müller, A. (eds) *Human Trafficking and Exploitation: Lessons from Europe*. London: Routledge, pp.129–146.
Achilli, L. (2018b), 'The "good" smuggler: The ethics and morals of human smuggling among Syrians'. *Annals of the American Academy of Political and Social Science*, **676**(1), 77–96.
Andreas, P. (2000), *Border Games: Policing the US–Mexico Divide*. Ithaca, NY and London: Cornell University Press.
Antonopoulos, G.A. & Papanicolaou, G. (2018), *Organised Crime: A Very Short Introduction*. Oxford: Oxford University Press.
Antonopoulos, G.A., Baratto, G., Di Nicola, A., Diba, P., Martini, E., Papanicolaou, G., & Terenghi, F. (2020), *Technology in Human Smuggling and Trafficking: Case Studies from Italy and the United Kingdom*. Cham: Springer.
Aronowitz, A.A. (2001), 'Smuggling and trafficking in human beings: The phenomenon, the markets that drive it and the organisations that promote it'. *European Journal on Criminal Policy and Research* **9**(2), 163–195.
Baird, T. (2016), *Human Smuggling in the Eastern Mediterranean*. London: Routledge.
Brosnan, M. (1998), *Technophobia: The Psychological Impact of Information Technology*. London: Routledge.
Carling, J., Gallagher, A., & Horwood, C. (2015), *Beyond Definitions: Global Migration and the Smuggling–Trafficking Nexus*. Nairobi: Regional Mixed Migration Secretariat.
Carrera, S., and Guild, E. (eds) (2016), *Irregular Migration, Trafficking and Smuggling of Human Beings: Policy Dilemmas in the EU*. Brussels: Centre for European Policy Studies.
Clarke-Billings, L. (2017), 'Dozens of desperate migrants feared dead as smugglers force hundreds into rough sea hours after 29 teenagers drowned', *Mirror*, August 17, accessed 17 August 2017 at https://www.mirror.co.uk/news/world-news/dozens-desperate-migrants-feared-dead-10963244.
Dekker, R. & Engbersen, G. (2012), *How Social Media Transform Migrant Networks and Facilitate Migration*. Oxford: International Migration Institute.
Dekker, R., Engbersen, G., Klaver, J., & Vonk, H. (2018), 'Smart refugees: How Syrian asylum migrants use social media information on migration and decision-making'. *Social Media and Society*, **4**(1), 1–11.
Di Nicola, A., Cauduro, A., & Falletta, V. (2013), 'Dal marciapiede all'autostrada digitale: uno studio sul web come fonte di informazioni su prostituzione e vittime di tratta in Italia'. *Rassegna Italiana di Criminologia*, **7**(3), 219–228.

Di Nicola, A., Baratto, G., Martini, E., Antonopoulos, G.A., Cicaloni, M., Damian, A., Diba, P., Ivanova, S., Papanicolaou, G., & Rusev, A. (2017), *Surf and Sound: The Role of the Internet in People Smuggling and Human Trafficking*. eCrime Research Report no.3. Middlesbrough/Trento: eCrime.

Diba, P., Papanicolaou, G., & Antonopoulos, G.A. (2019), 'The digital routes of human smuggling? Evidence from the UK'. *Crime Prevention and Community Safety*, 21(2), 159–175.

European Commission (2015), *A Study on Smuggling of Migrants: Characteristics, Responses and Cooperation with Third Countries. Final Report*. Brussels: European Commission, DG Migration and Home Affairs.

European Migration Network (2016), *The Use of Social Media in the Fight against Migrant Smuggling*. Brussels: European Commission, Directorate General Migration and Home Affairs.

Europol (2017a), *Crime in the Age of Technology*. The Hague: Europol.

Europol (2017b), *European Union Serious and Organised Crime Threat Assessment 2017*. The Hague: Europol.

Europol (2017c), *Internet Organised Crime Threat Assessment 2017*. The Hague: Europol.

Europol and Interpol (2016), *Migrant smuggling networks: joint Europol-INTERPOL report. Executive summary*. Accessed 20 February 2020 at https://www.europol.europa.eu/newsroom/news/europol -and-interpol-issue-comprehensive-review-of-migrant-smuggling-networks.

Frouws, B., Phillips, M., Hassan, A., & Twigt, M. (2016), *Getting to Europe the 'WhatsApp' Way: The Use of ICT in Contemporary Mixed Migration Flows to Europe*. Nairobi: Regional Mixed Migration Secretariat.

Gillespie, M., Osseiran, S., & Cheesman, M. (2018), 'Syrian refugees and the digital passage to Europe: smartphone infrastructures and affordances'. *Social Media and Society*, 4(1), 1–12.

Gillespie, M., Ampofo, L., Cheesman, B.F., Illiadou, E., Issa, A., Osseiran, S., & Skleparis, D. (2016), *Mapping Refugee Media Journeys: Smart Phones and Social Media Networks. Research Report*. Milton Keynes: The Open University.

Hamel, J.-Y. (2009), *Information and Communication Technologies and Migration*. New York, NY: United Nations Development Programme.

Hargittai, E. (1999), 'Weaving the western web: explaining differences in Internet connectivity among OECD countries'. *Telecommunications Policy*, 23(10), 701–718.

Içduygu, A., & Toktas, Ş. (2002), 'How do smuggling and trafficking operate via irregular border crossings in the middle east? Evidence from fieldwork in Turkey'. *International Migration*, 40, 25–54.

IOM (2020), 'Mediterranean monthly report: migrant arrivals reach 7,168 in 2020; deaths reach 77'. Accessed 20 February 2020 at https://www.iom.int/news/mediterranean-monthly-report-migrant -arrivals-reach-7168-2020-deaths-reach-77.

Jaishankar, J. (ed.) (2011), *Cyber Criminology: Exploring Internet Crimes and Criminal Behaviour*. Boca Raton: CRC Press.

Jewkes, Y., & Yar, M. (eds) (2010), *Handbook of Internet Crime*. Cullompton: Willan Publishing.

Laczko, F., & Thompson, D. (eds) (2000), *Migrant Trafficking and Human Smuggling in Europe: A Review of the Evidence with Case Studies from Hungary, Poland and Ukraine*. Geneva: IOM.

Latonero, M., Berhane, G., Hernandez, A., Mohebi, T., & Movius, L. (2011), *Human Trafficking Online: The Role of Social Networking Sites and Online Classifieds*. Los Angeles, CA: Centre on Communication Leadership and Policy.

Latonero, M., Musto, J., Boyd, Z., Boyle, E., Bissel, A., Gobson, K., & Kim, J. (2012), *The Rise of Mobile and the Diffusion of Technology-Facilitated Trafficking*. Los Angeles, CA: Centre on Communication Leadership and Policy.

Latonero, M., & Kift, P. (2018), 'On digital passages and borders: refugees and the new infrastructure for movement and control'. *Social Media and Society*, 4(1), 1–11.

Lavorgna, A. (2020), *Cybercrimes: Critical Issues in a Global Context*. London: Red Globe Press.

Lechman, E. (2019), *The Diffusion of Information and Communication Technologies*. London: Routledge.

Levi, M. (2001), 'Between the risk and the reality falls the shadow: Evidence and urban legends in computer fraud'. In Wall, D. (ed.), *Crime and the Internet*, Abingdon: Routledge, pp. 44–58.

Martellozzo, E., & Jane, E.A. (eds) (2017), *Cybercrime and Its Victims*. Oxon: Routledge.

McAuliffe, M. (2016), 'The appification of migration: A million migrants? There are apps for that'. *Policy Forum.* Accessed 15 January 2020 at https://www.policyforum.net/the-appification-of -migration.

McAuliffe, M., & Laczko, F. (eds) (2016), *Migrant Smuggling Data and Research: A Global Review of the Emerging Evidence Base.* Geneva: International Organisation for Migration.

McAuliffe, M., Goossens, A.M., & Sengupta, A. (2017), 'Mobility, migration and transnational connectivity'. In IOM (ed.), *World Migration Report 2018*, Geneva: IOM, pp. 149–169.

McCusker, R. (2006), 'Transnational organised cyber-crime: distinguishing threat from reality'. *Crime, Law & Social Change*, **46**, 257–273.

McQuade, S. (ed.) (2009), *Encyclopedia of Cybercrime.* Westport, CT: Greenwood Press.

McQuade, S. (2011), 'Technology-enabled Crime, Policing and Security'. *The Journal of Technology Studies*, **32**(1), 1–9.

Myria (2017), *Trafficking and Smuggling of Human Beings Online: 2017 Annual Report.* Brussels: Myria, Federal Migration Centre.

Naylor, R.T. (2000), *Economic and Organised Crime: Challenges for Criminal Justice.* Ottawa: Research and Statistics Division

Reitano, T., McCormack, S., Micallef, M., & Shaw, M. (2018), *Responding to the Human Trafficking– Migrant Smuggling Nexus, with a Focus on the Situation in Libya.* Geneva: The Global Initiative against Transnational Organised Crime.

Salt, J., & Hogarth, J. (2000), 'Migrant trafficking and human smuggling in Europe: A review of the evidence'. In F. Laczko and D. Thompson (eds), *Migrant Trafficking and Human Smuggling in Europe: A Review of the Evidence with Case Studies from Hungary, Poland and Ukraine*, Geneva: IOM, pp. 11–232.

Sanchez, G. (2016), *Human Smuggling and Border Crossings.* London: Routledge.

Sanchez, G. (2017), 'Critical perspectives on clandestine migration facilitation: An overview of migrant smuggling research'. *Journal on Migration and Human Security*, **5**(1), 9–27.

Sarkar, S. (2015), 'Use of technology in human trafficking networks and sexual exploitation: A cross sectional multi-country study'. *Transnational Social Review*, **5**(1), 55–68.

Savona, E.U., & Mignone, M. (2004), 'The fox and the hunter: How ICT changes the crime race'. *European Journal on Criminal Policy and Research*, **10**, 3–26.

Schapendonk, J., & van Moppes, D. (2007), *Migration and Information: Images of Europe, Migration Encouraging Factors and En Route Information Sharing.* Nijmegen: Radboud University.

Shelley, L. (1999), 'Transnational Organised Crime: The New Authoritarianism'. In H.R. Freeman & P. Andreas (eds), *The Illicit Global Economy and State Power*, Lanham, MD: Rowman & Littlefield Publishers, pp. 25–51.

Shelley, L. (2014), *Human Smuggling and Trafficking into Europe: A Comparative Perspective.* Washington DC: Migration Policy Institute.

Siegel, D. (2019), 'Human smuggling reconsidered: the case of Lesbos'. In P.C. van Duyne, A. Serdyuk, G.A. Antonopoulos, J. Harvey, and K. von Lampe (eds), *Constructing and Organising Crime in Europe*, The Hague: Eleven International Publishing, pp. 105–123.

Spener, D. (2004), 'Mexican migrant-smuggling: A cross-border cottage industry'. *Journal of International Migration and Integration*, **5**(3), 295–320.

Spener, D. (2009), *Clandestine Crossings: Migrants and Coyotes on the Texas–Mexico Border.* Ithaca, NY: Cornell University Press.

Spyratos, S., Vespe, M., Natale, F., Weber, I., Zagheni, E., & Rango, M. (2018), *Migration Data Using Social Media.* Brussels: European Commission.

Sykiotou, A.P. (2007), *Trafficking in Human Beings: Internet Recruitment. Misuse of the Internet for the Recruitment of Victims of Trafficking in Human Beings.* Strasbourg: Council of Europe, Directorate General of Human Rights and Legal Affairs.

Triandafyllidou, A. (2018), 'Migrant smuggling: Novel insights and implications for migration control policies'. *Annals of the American Academy of Political and Social Science*, **676**(1), 212–221.

UN Global Pulse & UNCHR Innovation Service (2017), *Social media and Force Displacement: Big Data Analytics and Machine Learning. White Paper.* New York: UN Global Pulse and UNCHR Innovation Service.

UNESCO (2002), *Information and Communication Technology in Education: A Curriculum for Schools and Programme for Teacher Development*. Paris: UNESCO.

United Nations (2004), *United Nations Convention against Transnational Organised Crime and the Protocols Thereto*. New York, NY: United Nations.

UNODC (2010), *Organised Crime Involvement in Trafficking in Persons and Smuggling Of Migrants*. Vienna: United Nations Office on Drugs and Crime.

UNODC (2018), *Global Study on Smuggling of Migrants*. New York, NY: United Nations.

van Duyne, P. (1996), 'The phantom and threat of organised crime'. *Crime, Law and Social Change*, **24**, 341–377.

Wall, D.S. (2007), *Cybercrime: The Transformation of Crime in The Information Age*. Malden, MA: Polity Press.

Wall, M., Campbell, M.O., & Janbek, D. (2017), 'Syrian refugees and information precarity'. *New Media and Society*, 19(2), 240–254.

Whittle, H., Hamilton-Giachritsis, C., Beech, A., & Collings, G. (2013), 'A review of online grooming: Characteristics and concerns'. *Aggression and Violent Behaviour*, **18**, 62–70.

Whittle, J. & Antonopoulos, G.A. (2020), 'How Eritreans plan, fund and manage irregular migration, and the extent of involvement of "organised crime"'. *Crime Prevention and Community Safety*, **22**(2), 173–190.

World Bank (2017), *The Little Data Book on Information and Communication Technology 2017*. Washington DC: World Bank Group and ITU.

World Bank (2016), *Digital Dividends Overview: World Development Report 2016*. Washington, D.C.: World Bank Group.

Yar, M., & Steinmetz, K.F. (2019), *Cybercrime and Society* (3rd edn). London: Sage Publications.

Yar, M., & Steinmetz, K.F. (2006), *Cybercrime and Society*. London: Sage Publications.

Zhang, S. (2007), *Smuggling and Trafficking in Human Beings: All Roads Lead to America*. Westport, CT: Praeger.

Zhang, S., Sanchez, G., & Achilli, L. (2018), 'Crimes of solidarity in mobility: alternative views on migrant smuggling'. *Annals of the American Academy of Political and Social Science*, **676**, 6–15.

Zijstra, J., & van Liemt, I. (2017), 'Smart(phone) travelling: understanding the use and impact of mobile technology on irregular migration journeys'. *International Journal of Migration and Border Studies*, **3**(2/3), 174–191.

10. Robots and refugees: the human rights impacts of artificial intelligence and automated decision-making in migration

Petra Molnar

INTRODUCTION: TECHNOLOGIES OF MIGRATION

Millions of people are on the move due to conflict, instability, environmental factors and economic reasons. As a result, many states and international organizations involved in migration control are exploring various technological experiments to strengthen border enforcement and improve decision-making. These experiments range from Big Data predictions about population movements in the Mediterranean, to Canada's use of automated decision-making in immigration, to Artificial Intelligence (AI) lie detectors at European borders. However, these technological innovations often fail to account for profound human rights ramifications and real impacts on human lives. Now, with the COVID-19 pandemic sweeping the globe, the use of new technologies is increasing. As governments move towards biosurveillance to contain the spread of the pandemic, there has been an increase in the use of tracking, automated drones and other types of technologies that purport to help manage migration, exacerbating potential human rights concerns (Cliffe, 2020; Lewis & Mok, 2020; Molnar & Naranjo, 2020).

Emerging research[1] is beginning to highlight how new technologies such as biometrics, Big Data and airport AI lie detectors by private companies such as iBorderCtrl are used in the management of migration, but there is a gap in the conversation around the disproportionate impact of technological experimentation on migrants and refugees without appropriate mechanisms of accountability and oversight (Picheta, 2018). New technologies challenge our understanding of decision-making and procedural protections, with the risk of creating legal black holes in which States seek to leave migrants beyond the duties and responsibilities enshrined in law. Rendering certain populations such as migrants more trackable and intelligible justifies more technology and data collection under the guise of national security, or even under tropes of humanitarianism and development.[2]

Technological development does not occur in a vacuum, but in fact replicates existing power hierarchies and differentials. Technology is not inherently democratic and issues of informed consent and right of refusal are particularly salient in humanitarian and forced migration contexts when, for example, refugees in Jordan have their irises scanned in order to receive their weekly rations under the justification of efficiency, while not being able to refuse biometric registration (Staton, 2016). Technologies of migration management also operate in an inherently global context. They reinforce institutions, cultures, policies and laws, and exacerbate the gap between the public and the private sector, where the power to design and deploy innovation comes at the expense of oversight and accountability. Technologies have the power to shape democracy and influence elections,[3] through which they can reinforce the politics of exclusion (Hao, 2019). Technologies also reinforce the North–South power asymmetries and concretise which locations are seen as innovation centres, while spaces like conflict zones and refugee camps become sites of experimentation under the guise of 'humanitarianism' and 'empowerment of migrants' through innovation.[4] Technological innovations exude the promises of increased fairness and efficiency. However, these advances also reveal the fissures of imbalanced power relations in society. International human rights law is useful for codifying potential harms because technology and its development is inherently global and transnational. Currently, new technologies in migration are insufficiently regulated at the international level. More global oversight and accountability mechanisms are urgently needed to safeguard fundamental rights such as freedom from discrimination, privacy rights and procedural justice safeguards such as the right to a fair decision-maker and the rights of appeal.

This chapter canvasses some of the uses of AI and automated decision-making used in migration control globally and highlights existing and emerging human rights implications that have far-reaching impacts on the lives of migrants. This analysis then explores the growing privatisation of migration control and the increasing incursion of 'Big Tech' solutions without appropriate methods of oversight. This chapter ultimately argues that, before developing and deploying new technologies in migration control, we need governance and oversight mechanisms, leading to greater transparency and accountability regarding the development and use of AI in the complex, opaque, and discretionary realm of migration.

TECHNOLOGIES OF MIGRATION CONTROL

The introduction of new technologies impacts both the processes and outcomes associated with decisions that would otherwise be made by administrative tribunals, immigration officers, border agents, legal analysts and other officials responsible for the administration of immigration and refugee systems, border enforcement and refugee response management.

What Constitutes Artificial Intelligence?

AI, machine learning, automated decision systems and predictive analytics are a series of overlapping terms and refer to a class of technologies that assist or replace the judgment of human decision-makers. These systems, which can be taught and can learn, will behave in various ways through various techniques. Different disciplines or regulatory mechanisms also use different definitions.[5] As such, delineating the limits of what constitutes AI can be difficult. For clarity, this chapter will refer to all AI technologies discussed as automated decision-making. This can include technologies that automate the mining of vast stores of data as well as processes that mimic human cognition and come up with novel decisions about outcomes.

Automated decision systems process information in the form of input data, using an algorithm to generate an output. An algorithm can be thought of as a set of instructions, like 'a recipe composed in programmable steps ... organizing and acting on a body of data to quickly achieve a desired outcome' (Gillespie, 2016, p.19). Certain algorithms are 'trained' using a large, existing corpus of data, which allows the algorithm to classify and 'generalize beyond the examples in the training set' (Gillespie, 2016, p.21). Training data can be a body of case law, a collection of photographs or a database of statistics, some or all of which have been pre-categorised or labelled based on the designer's criteria.

Algorithms have been criticised for being 'black boxes'.[6] This is because an algorithm's source code, its training data or other inputs may be proprietary, and can be shielded from public scrutiny on the basis of intellectual property legislation or as confidential business assets. Moreover, when algorithms are used in immigration and refugee matters, and form a nexus with issues of national security, both input data and source code may also be classified (Molnar & Gill, 2018, p.18). However, without being able to scrutinise input data to understand how the algorithm starts to make decisions, iterate and improve upon itself in unpredictable or unintelligible ways, their logic becomes less and less intuitive to human oversight. One of the main concerns with not being able to scrutinise and critique automated decision-making is the introduction of bias. Training data can be coloured by direct or indirect human agency and pre-existing bias, or when seemingly non-discriminatory variables become 'proxies' for other categories, such as postal codes serving as proxies of race (Council of Europe, 2018). In the United States, the COMPAS algorithm used to predict the likelihood of crime recidivism has been widely criticised for falsely recommending racialised individuals for higher pre-custodial sentences than white offenders (Larson et al., 2016).

Algorithms are vulnerable to the same decision-making concerns that plague human decision-makers: transparency, accountability, discrimination, bias and error, factors that are incredibly relevant to the opaque and discretionary decision-making inherent in various facets of migration control, as explored below (Tufekci, 2015).

An Ecosystem of Automation in Migration

Automated decision-making technologies require vast amounts of data on which to learn. Various projects use Big Data, or extremely large data sets, to predict population movements during conflicts and make the delivery of aid more efficient (Verity & Whipkey, 2015). However, data collection is not an apolitical exercise, when powerful actors collect information on vulnerable populations with few regulated methods of oversights and accountability. In an increasingly anti-immigrant global landscape, migration data have also been misrepresented for political ends, to affect the distribution of aid dollars and resources and support hardline anti-immigration policies (Nature Editorial Team, 2017). The use of new technologies also raises issues of informed consent, particularly in the increasing reliance on biometric data. In Jordan, refugees now have their irises scanned in lieu of identification to receive their food rations. However, are they able to meaningfully opt out from having their data collected? Most refugees reported being uncomfortable with this collection but felt that they could not refuse to get scanned if they wanted to eat that week (Staton, 2016). Consent is not free if it is given under coercion, even if the coercive circumstances masquerade as efficiency and better service delivery.

Autonomous technologies are also increasingly used in securing border spaces. For example, FRONTEX, the European Border and Coast Guard Agency, has been testing unpiloted military-grade drones in the Mediterranean for the surveillance and interdiction of migrant vessels hoping to reach European shores to file asylum applications (Csernatoni, 2018). These technologies can have drastic results. While 'smart-border' technologies have been called a more 'humane' alternative to the Trump Administration's physical wall, using new surveillance technologies along the US–Mexico border have more than tripled migrant deaths and pushed migration routes towards more dangerous terrains in the Arizona dessert (Chambers et al., 2019). This phenomenon has created what anthropologist De León has called the 'land of open graves', echoing the rising numbers of deaths in the Mediterranean (De Leon, 2015). The use of these technologies by border enforcement is only likely to increase in the 'militarized technological regime' of border spaces, without appropriate public consultation, accountability frameworks, and oversights mechanisms (Csernatoni, 2018).

States receiving large numbers of migrants have been experimenting with automated decision-making. A 2018 report (co-written by the author) explored the human rights risks of using AI to replace or augment immigration decisions in Canada (Molnar & Gill, 2018). In other jurisdictions, these experiments are in full force. Following the Trump Administration's executive orders cracking down on migration, Immigration and Customs Enforcement (ICE) used an algorithm at the USA–Mexico border to justify detention of migrants in every single case (Oberhaus, 2018). Instances of bias in automated decision-making are widely documented. These biases have far-reaching results if embedded in emerging migration technologies. In airports in Hungary and Greece, a new project by a company called iBorderCtrl has introduced AI-powered lie detectors at border checkpoints (Picheta, 2018). Passengers' faces will be monitored for signs of lying, and if the system becomes more 'skeptical' through a series of increasingly complicated questions, the person will be selected for further screening by a human officer. However, what happens if a refugee claimant interacts with these systems? Can this system account for trauma and its effects on memory, or for cultural differences in communication? This use of AI again raises concerns about information sharing without people's consent, as well as about bias in identification through facial recognition, as facial recognition technologies struggle when analysing women or people with darker skin tones.

What happens when an algorithm like this makes a mistake? An algorithm already wrongfully deported over 7,000 students from the UK after accusing them of cheating on a language acquisition test (Baynes, 2019). Where does liability lie in a decision like this – with the designer, the coder, the immigration officer, or the algorithm itself? Should algorithms have legal personality? Much of immigration and refugee decision-making already sits at an uncomfortable legal nexus: the impact on the rights of individuals is very significant, even where procedural safeguards are weak. It is unclear how a whole new system of decision-making will impact mechanisms of redress and how courts will interpret algorithmic decision-making and relevant administrative law principles like procedural fairness and the right to an impartial decision-maker.

However, none of these technologies is developed in a vacuum. What is particularly concerning is the growing role of the private sector in the collection, use and storage of these data and the increasing power over the development and deployment of new technological interventions.

PRIVATE SECTOR INCURSIONS

Increasingly, private companies are setting the migration control agenda. The private sector is an integral player in the management of migration. States and government departments over-rely on private actors to develop and deploy technologies used to control migration. As a result, government liability and accountability get watered down and shifted to the private sector. People on the move get caught in the middle, leading to grave human rights abuses and infringements on fundamental freedoms. As recent research by European Digital Rights

(EDRi) suggests, the opaque, private technologies deployed at border zones to control migration desperately need to be regulated (Molnar, 2020).

Private Surveillance of People on the Move

Private companies have long played a central role in surveillance, but today, they're making inroads into migration control. For example, Privacy International reported that the Israeli company Cellebrite, a surveillance firm marketing itself as the 'global leader in digital intelligence,' is advertising its devices used to extract data from mobile extraction devices towards a new target: authorities interrogating people seeking asylum (Privacy International, 2019). Israel-based Cellebrite, a subsidiary of Japan's Sun Corporation, markets forensic tools that empower authorities to bypass passwords on digital devices, allowing them to download, analyse and visualise data. The Israeli arms company Elbit Systems has been contracted by the US Border Patrol to put Native American reservations along the Arizona border under constant surveillance, including surveillance cameras and drones (Parrish, 2019). This technology was first rolled out in the surveillance sensors for Israel's separation barrier through the West Bank, a barrier the International Court of Justice ruled to be illegal 15 years ago but still stands today (Parrish, 2019). *The Observer* and Statewatch also uncovered three contracts to track migrant boats reaching the shores of Europe between Elbit Systems, the European Union's Frontex (the border and coast guard agency) and the European Maritime Safety Agency that total $115 million (Privacy International, 2019).

While 'smart border' technologies have been called a more humane alternative to the Trump Administration's physical wall, using new surveillance technologies along the US–Mexico border has more than tripled migrant deaths and pushed migration routes towards more dangerous terrains through the Arizona desert (European Commission, 2014; Boyce et al., 2019). This toll echoes the rising deaths in the Mediterranean as more migrant boats are intercepted and as dubious innovations like floating walls are erected to prevent people reaching the shores of Europe (Howden et al., 2019; Keady-Tabbal & Mann, 2020; Block, 2020). The European Union is also increasingly pushing its borders beyond the physical corridor of the Mediterranean through its border externalization policies, employing various actors in North and Sub-Saharan Africa through private contracts to keep migrants from reaching the shores of Europe. Through technology, the state control over the management of migration has extended beyond traditional sovereignty. The externalization of borders, opacity of border zones, and transnational surveillance all work to transform migration into a site of criminality that must be surveilled and managed to root out the ever-present spectre of terrorism and irregular migration.[7]

Various private sector interventions in migration have come under fire, sometimes by the companies' own workforce. For example, Amazon workers demanded that the company cut its ties with Palantir Technologies over its use of facial recognition technology to support the detention and deportation programs run by the US Immigration and Customs Enforcement (ICE) and the Department of Homeland Security, which Amazon workers have decried as an 'immoral U.S. policy', and part of 'the U.S.'s increasingly inhumane treatment of refugees and immigrants' (Chan, 2019). Microsoft employees have also taken a stand against the use of private sector technology to facilitate abuses conducted by the US government, calling on CEO Satya Nadella to cancel Microsoft's $19.4 million contract with ICE, as well as contracts with clients directly related to ICE (Frenkel, 2018). Civil society has also been pushing back.

Groups such as Mijente have launched initiatives such as the No Tech for Tyrants and No Tech for ICE campaigns (Mijente, 2018). The Canadian government also received criticism for partnering with Palantir on a variety of projects, including working with the Calgary Police Department to integrate their databases and the Canadian Department of National Defence to provide data analytics software for the Canadian Forces Special Operations Command (Braga, 2017; Hemmadi, 2019; Ling, 2019). Internationally, Palantir partnered with the United Nations' World Food Programme and now has access to the data of over 90 million humanitarian aid recipients (Privacy International, 2019). While the World Food Programme call this partnership a way to 'help transform global humanitarian delivery,' concerns over unauthorized use of data remain and it is not clear what role Palantir will actually play in the humanitarian space (WFP, 2019).

Human Rights Responsibilities in the Era of Privatization

The private sector already has an independent responsibility to ensure that technologies do not violate international human rights, such as the United Nations Guiding Principles on Businesses and Human Rights (OHCHR, 2011). In the development of products and services, private entities also have clear legal obligations to comply with domestic law, including privacy and human rights legislation. Technologists, developers and engineers responsible for building this technology also have special ethical obligations to ensure that their work does not facilitate human rights violations (Martin, 2018). However, the tension between private and public regulation highlights an overall lack of institutional capacity to effectively regulate technology and a disjuncture between those who develop migration-related technology in the private sector, and those in the public sector who deploy it on specific populations. The so-called AI divide, or the gap between those who are able to design AI and those who do not is broadening and highlights problematic power dynamics in participation and agency when it comes to the roll out of new technologies. Most often, the viewpoints of those most affected are excluded from the discussion, particularly around areas of no-go zones or ethically fraught usages. There is a lack of contextual analysis when thinking through the impact of new technologies resulting in great ethical, social and political harm.

A lack of technical capacity within government can lead to a potentially inappropriate reliance on the private sector. As with many large software projects, in the case of automated decision system technology there may be strong commercial incentives for vendors to impose licensing conditions that keep source code closed, opaque and unavailable for public scrutiny. Similarly, there may be vendor interest in using input or training data provided by government for other commercial purposes beyond the government's own use case (for example, to improve and train the vendor's unrelated products and services), raising privacy concerns. The not-for-profit sector working with refugees and people on the move can also over-rely on the private sector. A growing number of connectivity interventions have been developed to provide Internet access to refugees, financed by both public and private sector actors. These initiatives grew rapidly and without central coordination in response to the 2015 refugee crisis in Europe, using broadband, mobile connectivity, Wi-Fi, GSM, mesh networks, as well as satellites, drones, white space and balloons. Lacking sustainable funding and a framework for these interventions led many nonprofit organizations to turn to private sector actors for their resources, support and funding, despite having very different values and interests in supporting this service for refugees (Kaurin, 2020).

Adopting emerging and complex tools at the bleeding edge of scientific development without first analysing their necessity, proportionality and impact on fundamental rights while lacking at the same time in-house talent capable of understanding, evaluating and managing these technologies is irresponsible from not only an engineering perspective, but also a legal and ethical one. In the case of automated decision systems that impact individuals' immigration or refugee status, error and system failure also pose a threat to human rights. The impacted rights can be extensive, including the right to life, liberty and security of the person, freedom from discrimination, freedom of association, privacy rights and various due process rights, all of which are fundamental and internationally protected (Molnar & Gill, 2018).

Politics of Privatization

Politics also cannot be discounted, as migration management is inherently a political exercise. Migration data is already being politicised to justify greater interventions in support of threatened national sovereignty (Scheel & Ustek-Spilda, 2018). The state's ultimate power to decide who is allowed to enter and under what conditions is bolstered by ongoing beliefs in technological impartiality (Scheel & Ustek-Spilda, 2018). However, there is an inherent tension between the claimed prerogative of nation states over sovereignty and the malleable nature of technology. In its fluidity, technology is inherently oppositional to borders, and by extension sovereignty. Indeed, oftentimes it impinges on the very definition of 'humanness' in the digital era (Zureik & Hindle, 2004, p.113). Ultimately, the primary purpose of the technologies used in migration management is to track, identify and control those crossing borders. The unequal distribution of benefits that accrue in technological development work to create monopolies of knowledge and consolidate power and authority vested in the sovereign state. These monopolies are allowed to exist because there is no unified global regulatory regime governing the use of new technologies, creating laboratories for high risk experiments with profound impacts on people's lives.

Immigration and refugee law is a useful lens through which to examine the far-reaching impacts of technological experimentation. Particularly in times of greater border control security and screening measures, complex systems of global migration management, the increasingly widespread criminalisation of migration, and rising xenophobia. Immigration law operates at the nexus of domestic and international law and draws upon global norms of international human rights and the rule of law. States have clear domestic and international legal obligations to respect and protect human rights when it comes to the use of these technologies, particularly given the growing incursions of the private sector in this space. It is therefore incumbent upon policy makers, government officials, technologists, engineers, lawyers, civil society and academia to take a broad and critical look at the very real impacts of these technologies on human lives.

EXISTING REGULATORY FRAMEWORKS AND HUMAN RIGHTS IMPACTS

As the discussion above shows, a number of internationally protected rights are already engaged in the increasingly widespread use of new technologies that manage migration. However, currently there is no integrated regulatory global governance framework for the use of automated technologies, and no specific regulations in the context of migration management. Much of the global conversation centres on ethics without clear enforceability mechanisms.

Various countries have begun the process of setting up piecemeal guidelines on the use of AI. For example, in April 2018, the European Commission adopted the Communication on Artificial Intelligence, creating the 'European AI Alliance', and a new set of AI ethics guidelines to address issues such as fairness, safety and transparency, with its High-Level Expert Group on Artificial Intelligence tasked with establishing recommendations on future-related policy development and on ethical, legal and societal issues related to AI (European Commission, 2018).[8] Countries like the Netherlands and Canada are leading in the establishment of certification and assessments for responsible and ethical AI (Consultancy EU, 2018; Government of Canada, 2019). Outside of North America and Europe, India has established its inclusive AI strategy 'AIforAll,' while Kenya recently announced the formation of a taskforce to provide guidance on Blockchain and AI-related technologies' application in the areas of financial inclusion, cybersecurity, election processes and public-service delivery (Government of India, 2018; Kenyan Wall Street, 2018).

While binding regional mechanisms that touch on the use of automated decision-making, such as Article 22 of the European Union's (EU) General Data Protection Regulation (GDPR), are being implemented and guiding principles for the development of ethical standards in engineering and design are also being explored, there are currently no legally binding international legal documents to regulate these technologies and limit their risks. However, states are already bound by the rules of customary international law that can also apply to the development of new technologies (Rayfuse, 2018).

This section will analyse some of these principles and rules, and argue that more specific regulation is required to protect against human rights infringements in the use of new technologies, particularly given how far-reaching the ramifications can be in immigration and refugee implementations. An international human rights law (IHRL) framework is particularly useful for codifying and recognising potential harms, because technology and its development is inherently global and translational. Under IHRL, States must commit to preventing violations from occurring, establish monitoring and oversight, and provide remedy and redress for rights violations to hold violators accountable (HRC, 2004, paras 3–8). This also includes the obligations of a state to protect individuals from harms perpetrated by third parties, including private entities (UN Human Rights Council, 2011).

However, States are willing to experiment with these new unregulated technologies in the space of migration precisely because it is a discretionary space of opaque decision-making. Moreover, much of migration management is also enacted by international organisations such as the UNHCR and various other bodies. As non-state actors operating under various legal and quasi-legal authorities and regulations globally, international organisations are 'arenas for acting out power relationships' without being beholden to the responsibilities that States have to protect human rights (Evans & Wilson, 1992, pp.329–330). States that operate through international organisations can also 'launder' their legal responsibility for acts or omissions that are attributed to the organisation (Benvenisti, 2018a). With the proliferation of migration control technologies, international organisations are overly empowered to administer technology without being beholden to rights-protecting laws and principles, resulting in problems with compliance.[9]

Life and Liberty

The far-reaching impact of new technologies on the lives and security of persons affected should not be underestimated. The right to life and liberty is one of the most fundamental internationally protected rights and highly relevant to migration and refugee contexts. The Convention Relating to the Status of Refugees,[10] with its accompanying Protocol,[11] enshrines the right to seek protection from persecution when life, liberty and security are threatened, including the right not to be returned to a country where persecution and risk to life is likely, under the principle of *nonrefoulement* (Convention Relating to the Status of Refugees, 1951; Protocal Relating to the Status of Refugees, 1967). Numerous other specific legal instruments, such as the International Covenant on Civil and Political Rights (ICCPR),[12] Convention Against Torture and Other Cruel, Inhumane or Degrading Treatment or Punishment (CAT),[13] Convention on the Rights of Persons with Disabilities (CRPD),[14] Convention on the Elimination of All Forms of Discrimination Against Women (CEDAW),[15] Convention on the Rights of the Child (CRC),[16] and the International Convention on the Elimination of All Forms of Racial Discrimination (CERD)[17] also recognise the right to liberty and security for all persons.

Multiple technological experiments already impinge on the right to life and liberty. The starkest example is the denial of liberty when migrants are placed in administrative detention at the US–Mexico border. Immigration detention is a highly discretionary phenomenon, and the justification of increased incarceration on the basis of algorithms that have been tampered with shows just how far the State is willing to justify incursions on basic human rights under the guise of national security and border enforcement (Silverman & Molnar, 2016, p.109). Errors, mis-calibrations and deficiencies in training data can result in rights-infringing outcomes. For example, aspects of training data that are mere coincidences in reality may be treated as relevant patterns by a machine-learning system, leading to outcomes that are considered arbitrary when examined against the purpose of the governing statute. This is one reason why the GDPR requires the ability to demonstrate that the correlations applied in algorithmic decision-making are 'legitimate justifications for the automated decisions' (Moerel & Storm, 2018).

Equality Rights and Freedom from Discrimination

Given the problematic track record that automated technologies have on race and gender, it is very plausible that similar issues will occur, or have already occurred, in migration. Proxies for discrimination, such as country of origin, can be used to make problematic inferences leading to discriminatory outcomes. Freedom from discrimination and equality rights are widely protected by the International Covenant on Economic, Social and Cultural Rights (ICESCR),[18] the ICCPR,[19] CERD, CEDAW, CRPD,[20] CRC[21] and the Refugee Convention.[22] The Inter-American Commission on Human Rights' (IACHR) American Declaration of the Rights and Duties of Man[23] and the Universal Declaration of Human Rights (UDHR)[24] also enshrine equality. The UNHCR has also issued guidelines specific to countering discrimination against refugees.[25]

As discussed above, algorithms are vulnerable to the same decision-making concerns that plague human decision-makers: transparency, accountability, discrimination, bias and error.[26] The opaque nature of immigration and refugee decision-making creates an environment ripe for algorithmic discrimination. Decisions in this system – from whether a refugee's life story is 'truthful' to whether a prospective immigrant's marriage is 'genuine' – are highly discretionary, and often hinge on assessment of a person's credibility.[27] To the extent that these technologies will be used to assess 'red flags', 'risk' and 'fraud', they also raise definitional issues, as it remains unclear what the parameters of these markers will be. For example, in the experimental use of AI lie detectors at EU airports, it is unclear what will constitute truthfulness and how differences in cross-cultural communication will be dealt with in order to ensure that problematic inferences are not encoded and reinforced into the system. The complexity of human migration is not easily reducible to an algorithm. The US Government's proposed Extreme Vetting Initiative is also rife with potential discriminatory inferences, based on travel patterns, countries of origin and various other markers that can flag an individual for further surveillance and even deny them access into the country.[28] The heavy monitoring of social media sites is also contentious, as the information can often be misleading to a non-human analyst (Laperruque, 2018). As with predictive policing, the system 'risks hiding biased, politicised and discriminatory decision-making behind the scientific objectivity of algorithms and machine learning' (Root, 2018). Further, 'once applicants or visa holders know they are being monitored, a chilling effect could occur on their freedom of speech, forcing them to censor themselves online to avoid scrutiny,' as well as curtail their freedom of movement,[29] association,[30] and religion[31] by limiting their time in places of worship (Root, 2018).

Privacy Rights

Privacy is not only a consumer or property interest: it is a human right, rooted in foundational democratic principles of dignity and autonomy.[32] It is protected under Article 17 of the ICCPR. The UN has also explicitly recognized the impact of digital technologies on the right to privacy (OHCHR, n.d.). For example, the High Commissioner for Human Rights issued various statements on the risk surveillance poses to individuals' rights, particularly privacy and freedom of expression and association (Pillay, 2013; Pillay, 2014). The General Assembly also adopted Resolution 68/167, which expressed concerns regarding the potential negative impacts that surveillance may have on international human rights (UNGA, 2014). With the adoption of this Resolution, the General Assembly also requested that the High Commissioner for Human Rights prepare a report on the right to privacy in the digital age.[33] In 2015, the Human Rights Council adopted Resolution 28/16, which saw the appointment of a Special Rapporteur on the right to privacy.[34] In 2016, the UN Special Committee on Social, Humanitarian and Cultural Issues adopted a new resolution on the right to privacy in the digital age, which recognises the importance of respecting pre-existing international commitments regarding privacy rights and calls on States to develop appropriate remedies (UNGA, 2016).[35] It emphasises that States must address legitimate concerns regarding their national security in a manner that is consistent with these obligations and that personal data are increasingly susceptible to being sold without the individuals' consent or knowledge. The resolution also highlights the increased vulnerability of women, children and communities made marginalised to these privacy right violations, and links the right to privacy with other human rights such as freedom of expression (Brown, 2016). Increasingly, there is a recognition that 'the same rights that people have offline must also be protected online' and that technologies should be designed and operated in a way that respects and fulfils human rights, freedoms, human dignity and cultural diversity (Brown, 2016; UNHRC, 2012, para 1).

However, the differential impacts of privacy infringements must be considered when analysing the experiences of migrants.[36] If collected information is shared with repressive governments or actors from which refugees are fleeing, the ramifications can be life-threatening. Or, if automated decision-making systems designed to predict a person's sexual orientation are infiltrated by States targeting the LGBTQ community, discrimination and threats to life and liberty are likely outcomes (Murphy, 2017). It is the power of pattern recognition to extract personal details from available data that is concerning, particularly given the proliferation of surveillance by authoritarian regimes (Privacy International, 2017, para 12–16).

Efforts like the GDPR enshrine certain protections, particularly around the use of automated individual decision-making and the protection of personal data.[37] This is hopefully a starting point for broader global standards, but when analysing the protection of rights for migrants, the GDPR does not go far enough, as its protections are only available for EU citizens. The United Kingdom's data protection legislation actually goes further and protects the privacy of 'anybody' in its jurisdiction,[38] presumably including migrants.

Administrative Law Frameworks and Principles of Natural Justice

Any discussion of pertinent human rights in migration must also include an analysis of administrative legal frameworks and principles of natural justice that are inherent in migration management. For example, in immigration and refugee decision-making, procedural fairness dictates that the person affected by administrative processes has a right to be heard, the right to a fair, impartial and independent decision-maker, the right to reasons – also known as the right to an explanation – and the right to appeal an unfavourable decision. However, it is unclear how administrative law will handle the augmentation or even replacement of human decision-makers by algorithms. While these technologies are often presented as tools to be used by human decision-makers, the line between machine-made and human-made decision-making is not often clear. Given the persistence of automation bias, or the predisposition towards considering automated decisions as more accurate and fair, it remains unclear what rubric human decision-makers will use to determine how much weight to place on the algorithmic predictions, as opposed to any other information available to them, including their own judgment and intuition. Furthermore, when a person wishes to challenge an algorithmic decision, what will the appropriate standard of review look like? Inappropriate deference given to algorithmic decision-making has been widely documented.[39] It is unclear how tribunals and courts will assign reasonableness to automated decision-making, what standards of review will be used, and what mechanisms of redress will look like.

The unique context of migration should be the central consideration when analysing which human rights should be taken into consideration when exploring new technologies, given the very real risks to life, liberty and security, as well as heightened privacy considerations. Analysing the ramifications of new technologies on internationally protected human rights is a principled way to codify harms and think through mechanisms of redress.[40] Yet why are States reluctant to engage with this, seeing technology through the lens of IHRL? States are able to justify technological experiments in migration control precisely because migrants are not able to exercise the same rights as citizens, and because they are seen as a useful tool through which to exercise powers of sovereignty in an increasingly destabilised world.

WAYS FORWARD: CONTEXT-SPECIFIC GOVERNANCE OF MIGRATION TECHNOLOGIES

Technology replicated power in society and its benefits do not accrue equally. Yet no global regulatory framework exists to oversee the use of new technologies in the management of migration. Much of technological development occurs in so-called 'black boxes,' where intellectual property laws and proprietary considerations shield the public fully understanding how the technology operates. States are willing to experiment with these new unregulated technologies in the space of migration precisely because it is a discretionary space of opaque decision-making. States are able to justify increasing technological experiments in migration because migrants have been historically rendered as a population which is intelligible, trackable and manageable.[41] Indeed, the very rhetoric of migration 'management' implies that refugees and migrants must be presided over and controlled, as they are construed to be a threat to national sovereignty, particularly in times when more and more states are turning inward and reifying their sovereign power.

Moreover, much of migration management is also enacted by international organizations such as the UNHCR and various other bodies. As non-state actors operating under various legal and quasi-legal authorities and regulations globally, international organizations are 'arenas for acting out power relationships' without being beholden to the responsibilities that states have to protect human rights (Evans & Wilson, 1992, pp.329–330). States that operate through international organizations can also 'launder' their legal responsibility for acts or omissions that are attributed to the organization (Benvenisti, 2018b). Affected communities must also be involved in technological development. While conversations around the ethics of AI are taking place, ethics do not go far enough. What is needed is a sharper focus on oversight mechanisms grounded in fundamental human rights and context-specific accountability that recognizes the particular lived experiences of people on the move.

Yet there is also a hopeful promise to the proliferation of new technologies. Policy-makers, academia and the public are being forced to reckon with fundamental normative ideas around what constitutes intelligence, how to manage and regulate new systems of cognition, and who should be at the table when designing and deploying new tools that can be used to either dismantle or reinforce the status quo. Culture, institutions and technology all iteratively shape one another. Ultimately, technology is a social construct,[42] a mirror to reflect the positives and negatives inherent in our societies, forcing us to rethink ideas of privilege and power. It remains to be seen whether the current global push towards fervent technological innovation will result in robust global governance, centred on the experiences of people on the move.

NOTES

1. See Molnar 2019; Johns 2017; Csernatoni 2018. On broader theories around surveillance of migrants see Wickins 2007; Thomas 2005; Magnet 2011; Farraj 2011. On data collection, biometrics and power relations, see e.g. McFarland and McFarland 2015; Edwards 2012.
2. For discussion on the political economy of rendering migrants trackable, see for example Macklin 2005; De Genova 2002; Blommaert 2009; Inda 2004; Ahmed 2004; Appadurai 2006.
3. See e.g. Russell and Zamfir 2018.
4. See e.g. initiatives such as 'Techfugees: Empowering the Displaced through Technology', accessed 17 March 2019 at https://techfugees.com/

5.   See e.g. Amnesty International & Access Now 2018; Université de Montréal 2017; or Regulation 2016/679 of the European Parliament and of the Council of 27 April 2016 on the Protection of Natural Persons with regard to the Processing of Personal Data and on the Free Movement of Such Data, and Repealing Directive 95/46/EC [2016] OJ L119/1 (General Data Protection Regulation or GDPR).
6.   See e.g. Pasquale 2015.
7.   See also SITA 2019.
8.   See also Select Committee on Artificial Intelligence, 'AI in the UK: Ready, Willing and Able?' (HL 2017–2019, 100).
9.   See e.g. Raustiala and Slaughter 2002.
10. Convention Relating to the Status of Refugees (adopted 28 July 1951, entered into force 22 April 1954) 189 UNTS 150 (Refugee Convention).
11. Protocol Relating to the Status of Refugees (adopted 31 January 1967, entered into force 4 October 1967) 606 UNTS 267.
12. International Covenant on Civil and Political Rights (adopted 16 December 1966, entered into force 23 March 1976) 999 UNTS 171 (ICCPR) art 9. See also Universal Declaration of Human Rights (adopted 10 December 1948 UNGA Res 217 A(III)) (UDHR) art 3.
13. Convention Against Torture and Other Cruel, Inhuman or Degrading Treatment or Punishment (adopted 10 December 1984, entered into force 26 June 1987) 1465 UNTS 85 (CAT).
14. Convention on the Rights of Persons with Disabilities (adopted 13 December 2006, entered into force 3 May 2008) 2515 UNTS 3 (CRPD) art 14.
15. Convention on the Elimination of All Forms of Discrimination Against Women (adopted 18 December 1979, entered into force 3 September 1981) 1249 UNTS 13 (CEDAW).
16. Convention on the Rights of the Child (adopted 20 November 1989, entered into force 2 September 1990) 1577 UNTS 3 (CRC).
17. International Convention on the Elimination of All Forms of Racial Discrimination (adopted 7 March 1966, entered into force 4 January 1969) 66 UNTS 195 (CERD) art 5(b), for example.
18. International Covenant on Economic, Social and Cultural Rights (adopted 16 December 1966, entered into force 3 January 1976) 993 UNTS 3 (ICESCR) arts 2, 10.
19. ICCPR arts 4, 24, 26.
20. CRPD arts 3, 5.
21. CRC art 2.
22. Refugee Convention art 3.
23. Inter-American Commission on Human Rights (IACHR), American Declaration of the Rights and Duties of Man, 2 May 1948 (Bogota Declaration) art 2.
24. UDHR arts 7, 10.
25. United Nations High Commissioner for Refugees, 'Guidelines on International Protection No 9: Claims to Refugee Status Based on Sexual Orientation and/or Gender Identity within the Context of Article 1A(2) of the 1951 Convention and/or its 1967 Protocol Relating to the Status of Refugees' (23 October 2012) UN Doc HCR/GIP/12/09.
26. Tufekci (n 29) 216–217.
27. See e.g. Satzewich 2015.
28. Glaser 2017.
29. ICCPR art 12; UDHR art 13; CRC art 10(2).
30. ICCPR art 22. This right is also protected by art 20 of the UDHR and art 22 of the Bogota Declaration.
31. ICCPR art 18; UDHR art 18; Bogota Declaration art 3.
32. See also Austin 2018.
33. Pillay 2013.
34. UNHRC Res A/HRC/RES/28/16 ,'The Right to Privacy in the Digital Age', 26 March 2015.
35. See also Brown 2016.
36. See also Johns 2017.
37. GDPR art 22. The GDPR also includes mandatory provisions to make machine-made decisions explainable and transparent: see art 5(1).
38. See Data Protection Act 2018.

39. See Koliska and Diakopoulos 2018; see also Wilson 2017, 137, 141, 143–144, 147.
40. See also McGregor, Murray and Ng 2019.
41. See for example Macklin 2005; De Genova 2002; Blommaert 2009; Inda 2004; Ahmed; Appadurai 2006. See also Csernatoni 2018.
42. Franklin 1990.

REFERENCES

Ahmed, S. (2004), 'Affective Economies', *Social Text*, 22(2), 117–39.

Amnesty International & Access Now (2018), 'The Toronto Declaration: Protecting the Rights to Equality and Non- Discrimination in Machine Learning Systems', accessed on 23 July 2019 at www.accessnow.org/cms/assets/uploads/2018/05/Toronto-Declaration-D0V2.pdf.

Appadurai, A. (2006), *Fear of Small Numbers: An Essay on the Geography of Anger*. Durham, NC: Duke University Press.

Austin, L. (2018), 'We Must Not Treat Data Like a Natural Resource', accessed on 23 July 2019 at https://www.theglobeandmail.com/opinion/article-we-must-not-treat-data-like-a-natural-resource/.

Baynes, C. (2019), 'Government "deported 7,000 foreign students after falsely accusing them of cheating in English language tests"', accessed on 10 September 2021 at https://www.independent.co.uk/news/uk/politics/home-office-mistakenly-deported-thousands-foreign-students-cheating-language-tests-theresa-may-a8331906.html.

Benvenisti, E. (2018a), 'EJIL Foreword: Upholding Democracy amid the Challenges of New Technology: What Role for the Law of Global Governance?', *GlobalTrust Working Paper*, accessed on 10 September 2021 at http://globaltrust.tau.ac.il/publications.

Benvenisti, E. (2018b), 'Upholding Democracy amid the Challenges of New Technology: What Role for the Law of Global Governance?', *European Journal of International Law*, 29(1), 1–18.

Block, I. (2020), 'Greece Plans Floating Sea Barrier to Keep Out Refugees', accessed on 10 September 2021 at https://www.dezeen.com/2020/02/10/greece-floating-sea-border-wall-news/https://www.dezeen.com/2020/02/10/greece-floating-sea-border-wall-news/.

Blommaert, J. (2009), 'Language, Asylum, and the National Order', *Current Anthropology*, 50(4), 415–41.

Boyce, G.A., Chambers, S. & Launius, S. (2019), 'Democrats' "Smart Border" Technology is not a "Humane" Alternative to Trump's Wall', accessed on 10 September 2021 at https://thehill.com/opinion/immigration/429454-democrats-smart-border-technology-is-not-a-humane-alternative-to-trumps.

Braga, M. (2017), 'A Secretive Silicon Valley Tech Giant Set Up Shop in Canada: But What Does It Do?', accessed on 10 September 2021 at https://www.cbc.ca/news/technology/palantir-silicon-valley-technology-giant-data-canada-1.4111163.

Brown, D. (2016), 'New UN Resolution on the Right to Privacy in the Digital Age: Crucial and Timely', accessed 23 July 2019 at https://policyreview.info/articles/news/new-un-resolution-right-privacy-digital-age-crucial-and-timely/436.

Chambers, S. N., Boyce, G. A., Launius, S., & Dinsmore, A. (2019), 'Mortality, Surveillance and the Tertiary "Funnel Effect" on the US–Mexico Border: A Geospatial Modeling of the Geography of Deterrence, *Journal of Borderlands Studies*, 25, 1–26.

Chan, R. (2019), 'Read the Internal Letter Sent By a Group of Amazon Employees Asking the Company to Take a Stand against ICE', accessed on 10 September 2021 at https://www.businessinsider.com/amazon-employees-letter-protest-palantir-ice-camps-2019-7?r=US&IR=T.

Cliffe, J, (2020), 'The Rise of the Bio-Surveillance State', accessed on 10 September 2021 at https://www.newstatesman.com/science-tech/2020/03/rise-bio-surveillance-state.

Consultancy EU (2018), 'Deloitte Launches Certified Quality Mark for AI and Robotics', accessed 23 July 2019 at www.consultancy.eu/news/2043/deloitte-launches-certified-quality-mark-for-ai-and-robotics.

Council of Europe (2018), 'Algorithms and Human Rights: Study on the Human Rights Dimensions of Automated Data Processing Techniques and Possible Regulatory Implications', accessed 12 August 2019 at https://rm.coe.int/algorithms-and-human-rights-en-rev/16807956b5.

Csernatoni, R. (2018), 'Constructing the EU's High-Tech Borders: FRONTEX and Dual-Use Drones for Border Management', *European Security*, 27(2), 175–200.

De Genova, N. P. (2002), 'Migrant "Illegality" and Deportability in Everyday Life', *Annual Review of Anthropology*, 31, 419–47.

De Leon, J. (2015), *The Land of Open Graves: Living and Dying on the Migrant Trail*, University of California Press.

Edwards, A. (2012), 'A Numbers Game: Counting Refugees and International Burden-Sharing', public lecture delivered at the University of Tasmania Law School, 19 December, accessed on 23 July 2019 at www.refworld.org/docid/512c75de2.html.

European Commission (2014), 'Technology Study in Smart Borders', accessed on 10 September 2021 at https://ec.europa.eu/home-affairs/sites/homeaffairs/files/what-we-do/policies/borders-and-visas/smart-borders/docs/smart_borders_executive_summary_en.pdf.

European Commission (2018), 'Communication on Artificial Intelligence for Europe' (COM (2018) 237, 25 April 2018), accessed on 23 July 2019 at https://ec.europa.eu/digital-single-market/en/news/communication-artificial-intelligence-europe.

Evans, T. & Wilson, P. (1992), 'Regime Theory and the English School of International Relations: A Comparison', *Millennium: Journal of International Studies*, 21(3), 329–351.

Farraj, A. (2011), 'Refugees and the Biometric Future: The Impact of Biometrics on Refugees and Asylum Seekers', *Columbia Human Rights Law Review*, 42(1), 891–942.

Franklin, U. (1990), *The Real World of Technology*. Toronto, ON: House of Anansi Press.

Frenkel, S. (2018), 'Microsoft Employees Protest Work With ICE, as Tech Industry Mobilizes over Immigration', accessed on 10 September 2021 at https://www.nytimes.com/2018/06/19/technology/tech-companies-immigration-border.html.

Gillespie, Tarleton (2016), 'Algorithm', in Ben Peters (ed.), *Digital Keywords: A Vocabulary of Information Society and Culture*, Princeton, NJ: Princeton University Press, pp.18–30.

Glaser, A. (2017), 'ICE Wants to Use Predictive Policing Technology for its "Extreme Vetting" Program', *Slate*, 8 August. Accessed on 10 September 2021 at https://slate.com/technology/2017/08/ice-wants-to-use-predictive-policing-tech-for-extreme-vetting.html.

Government of Canada (2019), 'Algorithmic Impact Assessment', accessed 23 July 2019 at https://www.canada.ca/en/government/system/digital-government/digital-government-innovations/responsible-use-ai/algorithmic-impact-assessment.html.

Government of India, (2018), 'National Strategy for Artificial Intelligence #AIforAll', accessed 23 July 2019 at http://niti.gov.in/writereaddata/files/document_publication/NationalStrategy-for-AI-Discussion-Paper.pdf.

Hao, K. (2019), 'Why AI is a Threat to Democracy – And What We Can Do to Stop It', accessed on 23 July 2019 at www.technologyreview.com/s/613010/why-ai-is-a-threat-to-democracyand-what-we-can-do-to-stop-it/.

Hemmadi, M. (2019), 'Controversial Data-Mining Firm Palantir Signs Million-Dollar Deal with Defence Department', accessed on 10 September 2021 at https://thelogic.co/news/exclusive/controversial-data-mining-firm-palantir-signs-million-dollar-deal-with-defence-department/.

Howden, D., Fotiadis, A., & Loewenstein, A. (2019), 'Once migrants on the Mediterranean were saved by naval patrols. Now they have to watch as drones fly over', accessed on 10 September 2021 at https://www.theguardian.com/world/2019/aug/04/drones-replace-patrol-ships-mediterranean-fears-more-migrant-deaths-eu.

Inda, Jonathan Xavier (2004), *Targeting Immigrants: Government, Technology, and Ethics*, Oxford: Blackwell.

Johns, F. (2017), 'Data, Detection, and the Redistribution of the Sensible in International Law', *American Journal of International Law*, 111(1), 57–103.

Kaurin, D. (2020), 'Space and Imagination: Rethinking Refugees', accessed on 10 September at https://www.unhcr.org/innovation/wp-content/uploads/2020/04/Space-and-imagination-rethinking-refugees-digital-access_WEB042020.pdf.

Keady-Tabbal, N. & Mann, I. (2020), 'Tents at Sea: How Greek Officials Use Rescue Equipment for Illegal Deportations', accessed on 10 September 2021 at https://www.justsecurity.org/70309/tents-at -sea-how-greek-officials-use-rescue-equipment-for-illegal-deportations/.

Kenyan Wall Street (2018), 'Kenya Govt Unveils 11 Member Blockchain & AI Taskforce Headed by Bitange Ndemo', accessed on 23 July 2019 at https://kenyanwallstreet.com/kenya-govt-unveils-11 -member-blockchain-ai-taskforce-headed-by- bitange-ndemo/.

Koliska, Micheal & Diakopoulos, Nicholas, (2018), 'Disclose, Decode and Demystify: An Empirical Guide to Algorithmic Transparency', in Scott A Eldridge II and Bob Franklin (eds), *The Routledge Handbook of Developments in Digital Journalism Studies*, New York: Routledge, pp.559–560.

Laperruque, J. (2018), 'ICE Backs Down on "Extreme Vetting" Automated Social Media Scanning', Project on Government Oversight, accessed 23 July 2019 at www.pogo.org/blog/2018/05/ice-backs -down-on-extreme-vetting-automated-social-media- scanning.html.

Larson, J., Mattu, S., Kirchner, L., & Angwin, J. (2016), 'How We Analyzed the COMPAS Recidivism Algorithm', accessed 23 July 2019 at www.propublica.org/article/how-we-analyzed-the-compas -recidivism-algorithm.

Lewis, Sam & Mok, Opalyn. 2020. Malaysia enforces lockdown compliance with drones. *Privacy International*, 25 March.

Ling, J. (2019), 'Palantir's Big Push into Canada', accessed on 10 September 2021 at https://www .opencanada.org/features/palantirs-big-push-into-canada/.

Macklin, A. (2005), 'Disappearing Refugees', *Columbia Human Rights Law Review*, 36, 101–61.

Magnet, Shoshana (2011), *When Biometrics Fail: Gender, Race, and the Technology of Identity*, Durham, NC: Duke University Press.

Martin, M. (2018), 'Ethical Implications and Accountability of Algorithms', *Journal of Business Ethics*, 160, 835–850.

McFarland, D.A. & McFarland. R.H. (2015), 'Big Data and the Danger of Being Precisely Inaccurate', *Big Data and Society*, 2(2), 1–4.

McGregor, L., Murray, D., & Ng, V. (2019), 'International Human Rights as a Framework for Algorithmic Accountability', *International and Comparative Law Quarterly*, 68(2), 309–343.

Middle East Monitor (2019), EU using Israel drones to track migrant boats in the Med', accessed on 10 September 2021 at https://www.middleeastmonitor.com/20190819-eu-using-israel-drones-to-track -migrant-boats-in-the-med/.

Mijente (2018), 'Who's Behind ICE?', accessed on 10 September 2021 at https://mijente.net/notechforice/ .

Moerel, L. & Storm, M. (2018), 'Law and Autonomous Systems Series: Automated Decisions Based on Profiling – Information, Explanation or Justification? That is the Question!', University of Oxford Faculty of Law Blog, 27 April 2018, accessed 23 July 2019 at www.law.ox.ac.uk/business-law-blog/ blog/2018/04/law-and-autonomous-systems-series-automated-decisions-based-profiling.

Molnar, P (2019), 'Technology on the Margins: AI and Migration Management from a Human Rights Perspective', *Cambridge International Law Journal*, 8(2), 305–330.

Molnar, P. (2020), 'The Human Rights Impacts of Migration Control Technologies', accessed on 10 September 2021 at https://edri.org/the-human-rights-impacts-of-migration-control-technologies/.

Molnar, P. & Gill, L. (2018), 'Bots at the Gate: A Human Rights Analysis of Automated Decision-Making in Canada's Immigration and Refugee System', accessed 23 July 2019 at https://citizenlab.ca/wp -content/uploads/2018/09/IHRP-Automated-Systems-Report-Web-V2.pdf.

Molnar, P & Naranjo, D. (2020), 'Surveillance Won't Stop the Coronavirus', accessed on 10 September 2021 at https://www.nytimes.com/2020/04/15/opinion/coronavirus-surveillance-privacy-rights.html.

Murphy, H. (2017), 'Why Stanford Researchers Tried to Create a "Gaydar" Machine', *The New York Times*, 9 October, accessed 12 August 2019 at www.nytimes.com/2017/10/09/science/stanford-sexual -orientation-study.html.

Nature Editorial Team (2017), 'Data on Movements of Refugees and Migrants are Flawed', accessed 23 July 2019 at www.nature.com/news/data- on-movements-of-refugees-and-migrants-are-flawed-1.21568.

Oberhaus, D. (2018), 'ICE Modified Its "Risk Assessment" Software so It Automatically Recommends Detention', accessed 23 July 2019 at https://motherboard.vice.com/en_us/article/evk3kw/ice-modified -its-risk-assessment-software-so-it-automatically- recommends-detention.

OHCHR (2011) 'The Right to Privacy in the Digital Age', accessed on 23 July 2019 at https://www
.ohchr.org/en/issues/digitalage/pages/digitalageindex.aspx.

Parrish, W. (2019), 'The U.S. Border Patrol and an Israeli Military Contractor are Putting a Native
American Reservation under "Persistent Surveillance"', accessed on 10 September 2021 at https://
theintercept.com/2019/08/25/border-patrol-israel-elbit-surveillance/.

Pasquale, Frank (2015), *The Black Box Society: The Secret Algorithms that Control Money and
Information*, Cambridge, MA: Harvard University Press.

Picheta, R.(2018), 'Passengers to Face AI Lie Detector Tests at EU Airports', broadcasted on CNN 3
November, accessed 23 July 2019 at https://edition.cnn.com/travel/article/ai-lie-detector-eu-airports
-scli-intl/index.html.

Pillay, N. (2013), 'Opening Remarks by Ms. Navi Pillay, United Nations High Commissioner for Human
Rights to the Side-Event at the 24th Session of the UN Human Rights Council: How to Safeguard the
Right to Privacy in the Digital Age?', United Nations Office of the High Commissioner for Human
Rights, 20 September 2013, accessed 23 July 2019 at https://newsarchive.ohchr.org/EN/NewsEvents/
Pages/DisplayNews.aspx?NewsID=13758&LangID=E.

Pillay, N. (2014), 'Opening Remarks by Ms. Navi Pillay United Nations High Commissioner for Human
Rights to the Expert Seminar: The Right to Privacy in the Digital Age', United Nations Office of the
High Commissioner for Human Rights, 24 February, accessed 23 July 2019 at https://newsarchive
.ohchr.org/EN/NewsEvents/Pages/DisplayNews.aspx?NewsID=13758&LangID=E.

Privacy International (2017), 'Submission of Evidence to the House of Lords Select Committee on
Artificial Intelligence', accessed on 12 August 2019 at https://privacyinternational.org/advocacy
-briefing/664/submission-evidence-house-lords-select-committee- artificial-intelligence.

Privacy International (2019), Challenging the Drivers of Surveillance, accessed on 10 September 2021
at https://privacyinternational.org/report/3225/challenging-drivers-surveillance-eu-access-documents
-requests.

Raustiala, Kal & Slaughter, Anne-Marie (2002), 'International Law, International Relations and
Compliance', in Walter Carlsnaes, Thomas Risse & Beth A Simmons (eds), *Handbook of International
Relations*, Thousand Oaks, CA: Sage, pp. 538–539.

Rayfuse, Rosemary (2018), 'Public International Law and the Regulation of Emerging Technologies', in
Roger Brownsword, Eloise Scotford & Karen Yeung (eds), *The Oxford Handbook of Law, Regulation
and Technology*, Oxford: Oxford University Press, pp. 503–512.

Root, B. (2018), 'US Immigration Officials Pull Plug on High-Tech "Extreme Vetting"', accessed on 23
July 2019 at www.hrw.org/news/2018/05/18/us-immigration-officials-pull-plug-high-tech-extreme
-vetting.

Russell, M. & Zamfir, I. (2018), 'Digital Technology in Elections: Efficiency versus Credibility?'
European Parliamentary Research Service Briefing, PE 625.178, September.

Satzewich, V. (2015), *Points of Entry: How Canada's Immigration Officers Decide Who Gets In*,
Vancouver, BC: UBC Press.

Scheel, S. & Ustek-Spilda, F. (2018), 'Why Big Data Cannot Fix Migration Statistics', *Refugees Deeply*,
accessed on 10 September 2021 at https://www.newsdeeply.com/refugees/community/2018/06/05/
why-big-data-cannot-fix-migration-statistics.

Silverman, S. & Molnar, P. (2016), 'Everyday Injustices: Barriers to Access to Justice for Immigration
Detainees in Canada', *Refugee Survey Quarterly*, 35(1), 109–127.

SITA (2019), 'Smart technology for border security', accessed on 23 July 2019 at https://www.sita.aero/
resources/type/infographics/smart-technology-for-border-security.

Staton, B. (2016), 'Eye Spy: Biometric Aid System Trials in Jordan', accessed 23 July 2019 at www
.irinnews.org/analysis/2016/05/18/eye-spy-biometric-aid-system-trials-jordan.

Thomas, R. (2005), 'Biometrics, International Migrants and Human Rights', *European Journal of
Migration and Law*, 7(4), 377–411.

Tufekci, Z. (2015), 'Algorithmic Harms Beyond Facebook and Google: Emergent Challenges of
Computational Agency', *Colorado Technology Law Journal*, 13(203), 216–217.

UN Human Rights Committee (2004), 'General Comment No 31 on the Nature of the Legal Obligation
Imposed on States Parties to the Covenant', UN Doc CCPR/C/21/Rev.1/Add.13, paras 3–8.

UN Human Rights Council (2011), 'Report of the Special Representative of the Secretary-General on the
Issue of Human Rights and Transnational Corporations and Other Business Enterprises, John Ruggie,

on Guiding Principles on Business and Human Rights: Implementing the United Nations "Protect, Respect and Remedy" Framework', UN Doc A/HRC/17/31, principles 1–10.

UNGA Res 68/167 (2014), 'The Right to Privacy in the Digital Age', 21 January.

UNGA Third Committee (71st Session) (2016), 'The Right to Privacy in the Digital Age', UN Doc A/C.3/71/L.39/Rev.1, 16 November.

UNHRC Res 20/8 (2012), 'The Promotion, Protection and Enjoyment of Human Rights on the Internet', 5 July, para 1.

UNHRC Res A/HRC/RES/28/16 (2015), 'The Right to Privacy in the Digital Age', 26 March.

United Nations Human Rights Office of the High Commissioner (2011), 'Guiding Principles on Businesses and Human Rights: Implementing the United Nations "Protect, Respect and Remedy" Framework', 13–16, accessed 23 July 2019 at http://www.ohchr.org/Documents/Publications/Guiding PrinciplesBusinessHR_EN.pdf.

Université de Montréal (2017), 'Montreal Declaration for a Responsible Development of AI', accessed 23 July 2019 at www.montrealdeclaration- responsibleai.com/the-declaration.

Verity, A. & Whipkey, K. (2015), 'Guidance For Incorporating Big Data into Humanitarian Operations, *UN-OCHA*', accessed on 17 March 2019 at http://digitalhumanitarians.com/sites/default/files/resource -field_media/IncorporatingBigDataintoHumanitarianOps-2015.pdf.

WFP (2019), 'Palantir and WFP partner to help transform global humanitarian delivery', accessed on 10 September at https://www.wfp.org/news/palantir-and-wfp-partner-help-transform-global -humanitarian-delivery.

Wickins, J. (2007), 'The Ethics of Biometrics: The Risk of Social Exclusion from the Widespread Use of Electronic Identification', *Science and Engineering Ethics*, 13(1), 45–54.

Wilson, Michele (2017), 'Algorithms (and the) Everyday', *Information, Communication and Society*, 20, 137–147.

Zureik, E. & Hindle, K. (2004), 'Governance, security and technology: the case of biometrics', *Studies in Political Economy*, 37(1), 113–137.

11. Drones and border control: an examination of state and non-state actor use of UAVs along borders

Rey Koslowski[1]

INTRODUCTION

Unmanned Aerial Vehicles (UAVs) are increasingly being used for border control as well as search and rescue of migrants and refugees. Drones initially developed for the US military patrol US borders with Mexico and Canada, watching for drug smugglers and unauthorized border crossers. European military drones have searched for migrants in the Mediterranean; and Frontex, the European Border and Coast Guard Agency, has tested drones and begun using them for border surveillance missions. China and other countries have deployed drones along their borders and coastlines as well. Moreover, various nongovernmental organizations (NGOs) have launched their own drones along international borders for a variety of purposes.

States have long used manned aircraft to surveil those who illegally cross or attempt to cross their territorial boundaries and one may view the use of drones as a simple extension of such border control practices. As opposed to manned surveillance aircraft, however, drones invoke concerns associated with their increasing use in military conflicts for killing people in other states while not risking a pilot. Moreover, with increasing use of computer technology to fly drones, control their movements and gather information, drones are now flying robots (Singer 2009, 32–35) rather than devices similar to the remotely controlled radioplanes used by the US military in WWII for target practice or the small radio-controlled scale model planes flown by hobbyists for decades. Border surveillance drones thereby raise all of the ethical and political issues associated with robotics (Singer 2009, 123–35, 382–428), particularly autonomy from human decision-making and privacy, especially with the rapidly expanding information capabilities that ever more powerful and smaller computers, improved sensors, and artificial intelligence entail.

When neighboring states are at war, the drones flown along borders have military missions in theaters of combat operations, those who operate these drones are subject to the laws of armed conflict and the political implications are similar to those of other uses of technology in war, i.e., whether a new technology helps one state impose a political settlement upon another. When neighboring states are at peace, the context within which drones operate is much more complex. Drones are being flown in airspace that is not a military theater but rather governed by rules of civil aviation intended to facilitate commercial air transport and general aviation with private aircraft, as well as enable commercial drone use (for everything from crop-dusting to wedding picture taking), small-UAV flying by hobbyists and drone use by law enforcement officials and first responders for solving crimes, apprehending those suspected of breaking the law and search and rescue.

This complexity of the peacetime use of drone technology along borders can be illuminated through reflection upon the following hypothetical example: A surveillance drone flies along a country's boundary. The meaning of this flight and the political implications it may have depend on who is flying the drone and what is done with the information collected. If that drone is being flown by that country's border control authorities and the imagery collected is used to direct officers to intercept unauthorized border crossers, it means one thing; if the drone is being flown by an NGO to find people in distress and rescue them, it means something quite different. But not all NGOs have the same agenda. An NGO with an anti-immigration agenda may use a drone to capture images of people crossing borders and post them on the internet to demonstrate that borders are 'out of control' and pressure policymakers to implement more restrictive immigration policies. If the border surveillance drone is being flown by a different kind of NGO, i.e., a transnational criminal organization, the camera may be aimed not at unauthorized border crossers but at the border guards. While a drone flying along a border can have a variety of political implications depending on who is operating it, most analysis of drones and borders has focused on drones operated by state border control authorities (see, e.g., Milivojevic 2015; Marin 2016; Jumbert 2016).

This chapter examines government use of drones for border control along boundaries between states that are at peace, but it considers this state use of drone technology within the broader context of drone use by non-state actors in order to fully explore the policy implications and political ramifications of drones along borders. I conduct this examination in five steps: first, I review various perspectives on using drones for border control generated by differing theoretical approaches and normative stances; second, I describe the use of drones along borders by a growing number of states; third, I examine the proliferation of an increasing variety of hobbyist and commercial drones; fourth, I consider NGO policy positions on drones and increasing NGO use of hobbyist and commercial drones that has led to a commercialization of border surveillance in formal, informal and illicit economies; fifth, I explore emerging developments that may pose future challenges with major political and ethical implications.[2]

Perspectives on Drones and Border Control

Most academic scholarship about UAVs (or 'drones') and borders is to be found in the fields of engineering, science and technology, as a quick Google Scholar search will demonstrate. This examination is oriented toward scholars in migration studies, public policy analysis and security studies. Except for brief references to drone use along the US southern border (Cornelius 2005) and in the Mediterranean Sea (Tazziolli 2016; Albahari 2018), drones (or 'UAVs') have not been discussed in relation to borders in articles published in leading migration studies journals[3] nor in any articles published in leading public policy journals.[4] With respect to scholarship in the fields of foreign policy and security studies, research devoted to the subject of drones has primarily focused on armed military drones in wars and counterterrorism operations (e.g., Singer 2009; Kaag and Kreps 2014) rather than on government use of unarmed drones for purposes such as disaster and humanitarian response, peacekeeping, nuclear safety and border control. Most academic works addressing non-military drones do so within larger arguments about the 'securitization of migration,' border militarization and humanitarian uses (Sandvik and Lohne 2014; Emery 2016; Custers 2016; Sandvik and Jumbert 2016), and surveillance and privacy (Završnik, 2015), with only a few chapters in some of these volumes that specifically deal with the use of drones for border security (Milivojevic 2015; Marin 2016; Jumbert 2016).

The phrase 'securitization of migration' first emerged as Barry Buzan, Ole Waever and their colleagues broadened the range of security analysis beyond such traditional topics as military capabilities, diplomacy and political events, to topics such as ethnicity, national identity and migration (Waever et al. 1993). Buzan and his colleagues highlighted the political usage of the linkage between international migration and security in the discourse of policymakers, who increasingly depicted migration as a security issue, thereby 'securitizing migration.' Leonard (2010) and Marin (2016) argue that the increasing use of drones by border authorities provides another example of how migration is being securitized, namely through unprecedented appli- cations of military technologies.

Similarly, governments' use of drones along borders has been depicted as a manifestation of border militarization, which involves the transference of military values and hardware into domestic life, particularly law enforcement (Wilson 2015; Milivojevic 2015; Jumbert 2016). For example, the US Department of Homeland Security's Bureau of Customs and Border Protection (CBP) and its Border Patrol subunit are heavily influenced by the military. Roughly a third of CBP staff previously served in the military, the CBP has long used equipment originally designed for the military (Dunn 2013) and the National Guard has on occasion been called up to support Border Patrol operations. European border protection has arguably also become more militarized (Anderson 2014) as evidenced by cooperation between border guards and state military forces, even to the extent that armed forces have become directly involved in policing (Bertin and Fontanari 2011).

In contrast, Meier (2015) depicted drones more positively in an examination of the impact of the information revolution on humanitarian aid and intervention, much as Chow (2012), who praised drones for mitigating many of the hazards associated with humanitarian missions and making it easier to provide assistance. Meier went on to found UAViators, a humanitarian UAV network with over 3,300 members in over 120 countries that promotes 'the safe, coor- dinated and effective use of UAVs for data collection and cargo delivery in a wide range of humanitarian and development settings.'[5]

The use of drones along borders for rescue missions and humanitarian assistance illumi- nates the dual use nature of drone technology that, in turn, opens another line of critique. For example, Milivojevic notes that drones could potentially be more threatening in border security contexts than in military conflicts due to the differing degrees of scrutiny. 'The human cost, clearly visible in the drone strikes debate is carefully hidden in border policing narratives or replaced with the notion of combating transnational crime and/or 'rescuing' migrants in distress' (Milivojevic 2015, 94). Milivojevic's assessment coincides with Marin's argument that humanitarian drone missions legitimize the expansion of EU and EU member state sur- veillance capabilities (Marin 2016) as well as Sandvik and Lohne's contention that drones not only show a process of technological borrowing but also 'the transfer of social, cultural and political practices' (Sandvik and Lohne 2014, 150). Emery (2016) warns that drones may undermine humanitarian objectives by legitimizing the drone industry's military ventures, cre- ating the wrong impression about how programs are operating, and introducing the possibility that domestic security drones could be armed.

Policymakers, policy analysts, government agencies and practitioners have paid far more attention than academics to drones' deployment along borders (e.g., Haddal and Gertler 2010; Fleming et al. 2015; GAO 2017), but much of that analysis is harnessed to arguments either advocating more border drone use (e.g., by members of the Unmanned Systems Caucus of the U.S. House of Representatives) or pointing to questions of cost-effectiveness (DHS 2014; Barry 2013) and/or privacy and civil liberties (Hayes and Vermeulen 2012; Stepanovich 2012). Moreover, contrasting academic depictions of non-military drones largely reflect practitioners' opinions, as exemplified by a survey of humanitarian aid workers produced by the Swiss Foundation for Mine Action (FSD), which found that 60 percent of 194 respondents considered drones to have the potential for a positive impact in disaster response operations, while 22 percent viewed drone use following natural disasters negatively. With respect to the use of drones in conflict zones, 40 percent stated that drones should never be used by humanitarian organizations, while 41 percent said they would consider using drones even during armed conflicts (FSD n.d.). The disjuncture between critique and advocacy for drones as humanitarian tools demonstrates the broad range of perspectives informing academic analysis of drone use for border control.

THE EVOLUTION OF BORDER DRONES

Several capabilities and features of drones used by the US military and the Central Intelligence Agency (CIA) have made them attractive as tools of border control. Drones can operate for long periods (over 20 hours) without landing and watch for and follow border crossers longer than piloted vehicles. Drones can operate in areas where it might be difficult to use other border protection techniques, particularly in maritime contexts, and they reduce the risks that border guards face from environmental hazards and armed smugglers.

The use of drones for border surveillance began with the counter drug-smuggling Operation Alliance, when U.S. Marines piloted UAVs along the US–Mexican border in Texas for three weeks in February 1990. Although the purpose was drug trafficking interdiction, the UAVs were also credited for apprehending more than 300 unauthorized border crossers (Zamichow 1990). During this operation, the US Border Patrol considered expanding UAV operations to intercept drugs elsewhere along the border, but then opted against it.

It was not until the September 11, 2001 attacks on the World Trade Center and the Pentagon that members of the US Congress and certain NGOs called for using UAVs to intercept unauthorized border crossers. In Senate and House hearings held shortly after the establishment of the Department of Homeland Security (DHS) in March 2003, several members of Congress called for using UAVs along the border, but the DHS was slow to respond. After an NGO, American Border Patrol,[6] received extensive media coverage of a successful April 25, 2003 test of its border surveillance drone[7] Secretary of Homeland Security, Tom Ridge testified in a May 20, 2003 Congressional hearing, saying, 'It is our goal to have a pilot up by the end of the year using a UAV along some of our land borders.'[8] Accordingly, the DHS tested the General Atomics MQ-9 Reaper (which is usually called the Predator B when it is used for border security) from October 29 to November 12, 2003.

US Customs and Border Protection (CBP), a DHS division, began employing Predator B drones along the US–Mexican and US–Canadian border in 2004. As of 2016, the program had nine operational drones, two of which are stationed along the US–Canadian border. CBP drones are used to patrol the border, conduct surveillance for investigations, conduct disaster damage assessments, and respond to officer safety scenarios. Beginning in March 2013, CBP implemented an automated 'change detection' strategy whereby drones take video of a stretch of the border then repeat 24 hours later. These videos are sent for automated 'processing, exploitation, and dissemination' by the Office of Intelligence and Investigative Liaison, whose analysts detect any changes in the imagery that may indicate a border crossing, or risk of future incursion or of no unauthorized border crossing activity, which, in turn, helps Border Patrol sector managers determine where to deploy agents (Fisher 2015).

A December 2014 DHS Inspector General's report, however, criticized the CBP drone program for poor planning and mismanagement, which limited actual flight time of the entire nine drone fleet to 5,110 hours in 2013 at an estimated cost of $12,255 per hour in fiscal year 2013. The Inspector General concluded that the UAV program was not very cost-effective in assisting Border Patrol agents to apprehend unauthorized border crossers and recommended investing additional resources in alternatives (DHS 2014). A subsequent DHS Office of Inspector General report found that CBP did not secure images and video collected by its Predator drones because CBP was unaware it was required to conduct a privacy threshold analysis for the Intelligence, Surveillance and Reconnaissance (ISR) systems used, thereby, allowing system deployment without CBP Privacy Office oversight. Hence, DHS cannot determine whether data collected by ISR systems on CBP surveillance drones contained data requiring safeguards per privacy laws, regulations, and DHS policy, even as the DHS Office of Inspector General also found a variety of data security deficiencies (DHS 2018).

CBP's ultimate goal is to have the capacity to respond to any event along the border with drone surveillance in three hours or less, but CBP is far from achieving this capability along the 2,000 mile border with Mexico, let alone adding the roughly 5,500 mile border with Canada, given that the entire CBP Predator drone fleet operates for only about 5,000 hours per year. A more cost-effective means for achieving this operational goal may be found in the Small Unmanned Aircraft Systems (sUAS) that CBP began testing in 2017. The Puma, Raven and InstantEye Quadcopter are sUAS systems developed for tactical military surveillance and operation by ground troops that weigh less than 55 pounds, are small enough to be transported in a sport utility vehicle (CBP 2017) and cost a small fraction of the Predator B. Since these drones can only fly for a few hours, they are not used for prolonged surveillance but rather targeted investigations, whereby images and video collected are transmitted to the person controlling the drone, who can then decide if additional prolonged surveillance is required and/or Border Patrol Agents dispatched. According to a CBP spokesperson, these sUAS were flown for a total of 176 hours between October 2018 and April 2019 and responsible for 474 apprehensions of individuals at the border (Ghaffary 2019).

After many years of research, demonstrations and pilot projects, Frontex and the European Maritime Safety Agency began using drones for border surveillance in 2018. Beginning in 2012 Frontex hosted a series of workshops and industry days with private sector contractors to discuss how Remotely Piloted Aircraft Systems (RPAS) could be used for border surveillance and then develop plans for how to test these systems in pilot projects. Then in October 2016 the European Union upgraded Frontex to the European Border and Coast Guard Agency and tasked it to work together with the European Fisheries Control Agency and the European Maritime Safety Agency and enable these agencies to 'launch joint surveillance operations, for instance by jointly operating Remotely Piloted Aircraft Systems (drones) in the Mediterranean Sea' (European Commission 2015). In 2016, the European Maritime Safety Agency issued a call to tender 'Contracts for Remotely Piloted Aircraft System (RPAS) services in support of the execution of Coast Guard functions' for a total of 67 million euros.[9] By September 2019, the European Maritime Safety Agency provided drone services that were used by Portugal,

Spain, Denmark, Greece, Croatia, Italy and Iceland for various operational goals as well as drone services that supported the European Border and Coast Guard Agency with surveillance along the Portuguese coast.[10] In September 2018, Frontex also began testing surveillance drones in Portugal, Greece and Italy. The Israeli-built Heron medium-altitude, long endurance UAVs were equipped with thermal cameras and radars to detect vessels suspected of criminal activities, such as drug and weapon smuggling, and to support search and rescue operations (Frontex 2018a). In October 2019, Frontex issued a tender for maritime drone surveillance services to be delivered in 'Greece and/or in Italy and/or in Malta.' The two-year 50 million Euro service contract will be for 'a complete service providing all the necessary technical and human resources' and 'include reliable close to real time live data streaming and data sharing capacity' (Frontex 2019).

Drones are envisioned to play a significant role in EUROSUR, the European border surveillance system, which was established by the European Commission in 2008 to help EU member states detect cross-border movements and provide a common technical framework for improved operational communication and cooperation, as well as foster the use of cutting-edge technologies for border surveillance. Operated by FRONTEX, EUROSUR uses satellite imagery, ship reporting systems and pre-frontier intelligence feeds to improve situational awareness and reaction capability of member states in their efforts to stop unauthorized migration, combat crime and prevent loss of lives at sea. Information collected by drones contributes to Multipurpose Aerial Surveillance (MAS), which allows planes and drones monitoring the external borders to feed live video and other information directly to the Frontex headquarters and affected EU countries. The Frontex Situational Centre (FSC) operates 24/7 and offers EUROSUR fusion services to Member States, including satellite imagery and Multipurpose Aerial Surveillance (Frontex 2018b). EUROSUR works through having each EU member state establish a National Coordination Center to coordinate the border surveillance activities on national level and exchange information through EUROSUR. National Coordination Centers collect data on border activities, including unauthorized border crossings and crimes, process that data to create a national situational picture then share the relevant information with other member states and FRONTEX. These national data feeds plus information from other sources enables FRONTEX to create a European situational picture and a common pre-frontier intelligence picture of the areas beyond the Schengen Area and EU borders.

Some EU member state border guard agencies also began testing and deploying easily transportable, short-flight Small Unmanned Aircraft Systems. For example, the Polish Border Guard purchased twelve sUAS for 7 million Zloty (1.6 million Euro) in 2015 (Sabak 2015), well before US Customs and Border Protection's first sUAS purchase in 2018. Similarly, the Estonian border guard purchased nine ELIX-XL surveillance drones from ELI Military Simulations for 500,000 Euros in 2018 (ERR News 2018). In 2019, the Lithuanian Border Guard also purchased two drones to patrol the external EU border with Russia and Belarus (LRT 2019). In 2019, the Finnish Border Guard began testing drones for maritime surveillance in the Baltic Sea (Finnish Border Guard 2019).

The US and EU member states have been joined by China, Australia, the UK (and most likely other states) in deploying drones for border control. In 2015, China began deploying drones along its border as part of an integrated border surveillance system in Xinjiang, Tibet, Yunnan and other regions. System designer Mao Weichen of the Southwestern Institute of Technology and Physics explained, 'Our system has been adopted by border defense units … to curb illegal border crossings and drug trafficking …. Compared with traditional border monitoring networks that mainly depend on video surveillance, our system has a wider coverage and more deterrence thanks to the use of drones and acoustic weapons' (quoted in Acosta 2015). The Australian Air Force is flying seven Triton high-altitude long-endurance drones for maritime surveillance and in October 2018, Australian Home Minister Peter Dutton proposed to supplement these capabilities with major purchases of UAVs and other new technologies worth hundreds of millions of dollars while issuing a request for information from aircraft manufacturers as to how to completely overhaul Australia's maritime surveillance capabilities by 2024 (Australian Department of Home Affairs 2018). Airbus, Northrop Grumman and Lockheed Martin among others submitted responses but so far funding for the initiative has failed to materialize (Riordan 2019). In December 2019, the United Kingdom began using drones to patrol the English Channel (BBC 2019).

THE PROLIFERATION OF HOBBYIST AND COMMERCIAL DRONES

Academic and policy analysis has largely focused on the repurposing of military reconnaissance drones, like the US-built Predator and Triton and the Israeli-built Heron, to border surveillance, but the rapid proliferation of hobbyist and commercial drones, particularly those used by photographers for producing aerial videos of real estate, sporting events and weddings, has transformed the scope and nature of border surveillance. Whereas the Predator's deployment along US borders and Frontex plans to deploy similar military surveillance drones have provided evidence for arguments about the militarization of borders, the rapid exponential growth in the number of small commercial drones that have been produced is precipitating the commercialization of border surveillance in formal, informal and illicit economies.

The French company Parrot initiated the consumer drone market at the 2010 Consumer Electronics Show in Las Vegas with the introduction of the AR Drone, a video camera equipped quadcopter that can be controlled by a smartphone. Defining personal consumer drones as internet-connected, camera-equipped UAVs that weigh less than 2 kilograms and can typically fly for up to one hour and 500 meters high, Gartner Research (2017) estimated that two million were sold worldwide in 2016 and projected 2.8 million to be sold in 2017.

Commercial drones, which are usually larger, carry higher payloads, have longer flight times and are designed for specific purposes such as mapping and industrial inspection, have been deployed by companies and utilities in the agriculture, oil and gas, power generation and transportation sectors. The UK, France, Germany and Spain were quick to begin approving commercial operators. France aggressively opened its airspace to commercial drones, with more than 1,200 commercial operators by May 2015 (West 2015). In contrast, it took until August 2016 for the US Federal Aviation Administration (FAA) to issue a rule for operating drones in domestic airspace. In a matter of months, however, more than 620,000 hobbyist drones and 44,000 commercial drones were registered with the FAA in 2016 (FAA 2017, 31–32). It is the use of these much less expensive drones along borders by state and non-state actors that may be the most significant development moving forward.

NON-STATE ACTORS, DRONES, AND THE COMMERCIALIZATION OF BORDER SURVEILLANCE

NGOs have played important roles in the evolution of border drones as advocates against and for government use of drones for border control as well as drone operators themselves. As the costs of acquiring consumer and commercial drones dropped, NGOs operating in individual states, international non-governmental organizations (INGOs) operating in several states, and NGOs from the 'dark side,' such as transnational criminal organizations and terrorist organizations whose activities span borders, are increasingly flying drones along borders for a variety of purposes. Border surveillance is increasingly being commercialized as the percentage of drone surveillance conducted shifts from governments using large expensive drones to non-state actors using smaller drones that anyone can purchase on the open market at prices individuals and NGOs can easily afford.

In terms of advocacy, the US-based anti-war organization Code Pink argues against using drones for border surveillance because they infringe on privacy rights, are wasteful government spending and expand the military–industrial complex into domestic life (Benjamin 2012). Like Code Pink, Statewatch and the Heinrich Böll Foundation are among the European NGOs opposing surveillance drones, and EUROSUR more broadly.[11] They, as well as other anti-drone NGOs,[12] have expressed concerns that drones will lead to intrusive aerial surveillance that will, especially when taken alongside the collection of biometrics and other personal data, threaten privacy and data protection (Hayes and Vermeulen 2012, 9).

In contrast, the NGO, American Border Patrol, has not only advocated that drones be used for border security, but, as discussed above, the publicity surrounding American Border Patrol UAV flights in April 2003 prompted the US government to deploy its own drones along the border. Founded by Glenn Spencer, then a 65-year-old retiree with a background in systems engineering and operations research, American Border Patrol is a non-profit organization that monitors the border, initially with webcams to show live video of unauthorized border crossings on the internet and then with its UAV, a hobbyist remote-controlled model airplane carrying a video camera (American Border Patrol n.d.). After the FAA forced the organization to discontinue its UAV flights in 2005, it began flying manned aircraft. Glenn Spencer's primary objective was to use relatively inexpensive technology to show how people were crossing the border despite the efforts of Border Patrol agents as well as to demonstrate how the border could be monitored more cost-effectively than the systems that CBP was deploying, such as the ill-fated Secure Border Initiative network (SBI*net*), also known as the 'virtual fence' (American Border Patrol n.d.). The 82-year-old Spencer now uses a consumer drone on his ranch on the Mexican border in Southeastern Arizona to spot unauthorized border crossers and then call the local Border Patrol station to report his sightings (Taylor 2017).

In contrast, another NGO, Migrant Offshore Aid Station (MOAS) has used drones to rescue asylum seekers and migrants (Tazziolli 2016, 5). As MOAS founder Christopher Catrambone explained: 'The use of drones has been instrumental to MOAS' successful humanitarian efforts [in] providing real-time daylight and infrared video, widening the view of the crew onboard and enabling them to locate migrants in distress even well beyond the horizon' (quoted in Ball 2016). MOAS claims to have rescued over 40,000 people during search and rescue missions conducted between August 2014 and August 2017 in the Central Mediterranean and Aegean (MOAS 2020). Media organizations have also used drones to capture video and photographs of Syrian asylum seekers landing on Greek islands (Degnbol n.d.), attempting to cross into

Hungary (Hannah 2015), and walking towards northwestern Turkey in order to cross into Europe (McNabb 2020), thereby better showing citizens the circumstances asylum seekers face at and near borders.

NGOs of a different variety, drug cartels, are also using drones for surveillance along the US–Mexican border. Rodrigo Nieto-Gomes, an expert on criminal organizations' technology research and development efforts, points out that cartels use drones 'to try to identify positions of Border Patrol Agents and, therefore, inform smugglers of their positions to allow for the trafficking of drugs through the desert' (quoted by Adams 2017). Cartels have long employed 'an army of civilian lookouts who might receive $100 a month just to keep their eyes open and make a phone call if they notice an uptick in border inspections' (Keefe 2012). Known as 'falcons,' these lookouts position themselves in buildings with views of ports of entry or on hilltops with views across the border into the US. Given that the Mexican government enacted laws in 2015 to promote commercial drone use[13] and the price of small drones that can take and transmit video has rapidly dropped below the cost of employing lookouts, the lookouts 'are in the process of being replaced with a fleet of drones that fly along the U.S.–Mexico border, giving comprehensive real-time intelligence to smugglers on the location and movement of border patrol and other law enforcement officers and vulnerabilities in our border security infrastructure' (Balido 2012).[14] CBP spokesperson Jennifer Gabris indicated that the likelihood is high that many of the small drones seen by CBP along the US Mexican border are used for 'counter-surveillance in conjunction with narcotics smuggling ... because of the areas in which they are operating' (quoted in Nardi 2018). In an effort to reduce drug smuggler counter-surveillance and use of drones to fly drugs across the border, CBP signed a $1.2 million contract to pilot technology that stops drug-smugglers' drones mid-flight by alerting CBP to any drones entering protected airspace and then commandeering the drones and setting them on the ground (Boyd 2019a).

EMERGING DEVELOPMENTS

Perhaps the most important recent developments with the most far-reaching consequences involve the confluence of advances in drone technology, sensor technology, 5G mobile networks, artificial intelligence and facial recognition technology. Advances in computing power, batteries and materials science are making it possible to build drones that can fly longer, are easier to operate and can carry more weight while at the same time becoming much less expensive. Sensors with ever-greater resolution enhance drone surveillance capabilities, e.g., a drone carrying the ARGUS-IS 1.8 gigapixel video surveillance system can see objects only 6 inches wide and track every moving object within 36 square miles.[15]

With 5G's 1 millisecond latency, the time it takes for data from a device to be uploaded and reach its target, and data transmission up to 100 times faster than 4G mobile networks, 5G will dramatically improve functionality of mobile phone-controlled consumer drones and be able to transmit high-definition video in real time (Paul 2019). While artificial intelligence has enabled drones to fly autonomously, autonomous drones can only navigate through drone races at comparatively low speeds and, so far, cannot match human-piloted drones (Swearingen 2019). Given that the human eye and brain process data with a latency of 10 milliseconds, drones communicating on a 5G network with 1 millisecond latency will enable autonomous drones to execute more complicated maneuvers and tasks at higher speeds while avoiding crashing with obstacles as well as each other in a swarm. Prototype testing of such autonomous small unmanned aircraft systems for border security missions has been conducted by the DHS Science and Technology Directorate (Boyd 2019b) and the EU-funded research project ROBORDER seeks to develop 'a fully-functional autonomous border surveillance system with unmanned mobile robots including aerial, water surface, underwater and ground vehicles (UAV1, USV, UUV and UGV), capable of functioning both as standalone and in swarms, and incorporate multimodal sensors as part of an interoperable network.'[16]

Finally, artificial intelligence combined with large databases of facial imagery has improved the accuracy of facial recognition systems and reduced the cost of acquisition of facial recognition capabilities for law enforcement and border control authorities. Many border control authorities currently use facial recognition technology at ports of entry and some are looking toward using drones to collect images for facial recognition analysis. For example, the 2017 CBP solicitation for small, unmanned aircraft systems requested that the drones be '[a]ble to track multiple targets persistently,' and have the capability for '[i]dentification of humans via facial recognition or other biometric at range' (quoted in Feeney 2017). If border control officers could use drones to collect photos or video of unauthorized border crossers' faces, officers may soon also be able to identify who they are encountering at the border. Accessible by mobile phone or laptop, the Clearview facial recognition app draws on three billion publicly-available images scraped from Facebook, YouTube and millions of other websites, to enable officers of over 600 law enforcement agencies (including DHS) to run a photo or video frame of an unknown suspect and receive a positive match 75 percent of the time to an image in the database, often with address and other biographical information attached and at a cost of as little as $2,000 per month. If Clearview is willing to sell its services to NGOs, those NGOs may likewise be able to acquire similar personal data of the border guards whose facial imagery they manage to capture using drones.

When combined, these advances in drone technology, sensor technology, 5G mobile networks, artificial intelligence and facial recognition technology may further empower the states that choose to use them to better surveil and control their borders. Alternatively, non-state actors, particularly those that are more inclined to adopt new technologies, may use these technologies in very different ways along the border to much greater effect than states. In either case, these technologies are evolving so quickly, with the consequences of their use together not fully analyzed, that states, NGOs and the private sector may not fully consider the implications of the confluence of these new technologies for privacy, data protection and human rights.

NOTES

1. This chapter draws from Rey Koslowski and Marcus Schulzke, 'Drones along Borders: Border Security UAVs in the United States and the European Union,' *International Studies Perspectives*, Vol. 19, No. 4 (November 2018), pp. 305–324. I am very grateful to Marcus Schulzke and Oxford University Press for their cooperation in making this chapter possible. Please note, all webpages referenced in the endnotes were accessed on 10 September 2021.
2. Analysis of the use of drones for assisting humanitarian response in displacement settings is beyond the scope of this chapter, but a fuller discussion of some of the policy and ethical issues of drone use in humanitarian response can be found in Koslowski and Schulzke 2018, pp. 308, 312, 315–317, 319–320.

3.   *International Migration Review, International Migration, Journal of Ethnic and Migration Studies, Georgetown Immigration Law Journal, Asian and Pacific Migration Journal, Journal of International Migration and Integration, Global Networks, Migration Studies, Journal on Migration and Human Security, Comparative Migration Studies, International Journal of Migration and Border Studies.*
4.   *Journal of Policy Analysis and Management, Journal of Public Policy, Policy Studies Journal, Science and Public Policy, Policy and Politics, Policy and Society, Policy Science, European Journal of Public Policy, International Journal of Public Policy.*
5.   See: http://uaviators.org
6.   American Border Patrol is a 501 c (3) non-profit corporation that describes itself as 'the only non-governmental organization (NGO) that monitors the border on a regular basis—mostly by air', http://americanborderpatrol.com/About.html.
7.   Including CNN, MSMBC and Reuters. Video clips and a list of media coverage can be found at: http://americanborderpatrol.com/ADMINISTRATION/History.html
8.   'How is America Safer?' A Progress Report on the Department of Homeland Security, Hearing before the Select Committee on Homeland Security House of Representatives, May 20 and 22, 2003, 108-6, pp. 83–84.
9.   Tender no. EMSA/OP/12/2016, see: https://ted.europa.eu/udl?uri=TED:NOTICE:276544-2016: TEXT:EN:HTML.
10. Answer given by Ms Bulc on behalf of the European Commission, Question reference: E-002454/2019, Parliamentary questions, 19 September 2019.
11. See Statewatch reports, for example, Hayes, Jones and Töpfer 2014 as well as the preface written by Barbara Unmüßig, President of the Heinrich-Böll-Stiftung and Ska Keller, Member of the European Parliament, to Hayes and Vermeulen 2012.
12. See, e.g., Article 36, http://www.article36.org/; The International Committee for Robot Arms Control, http://icrac.net/.
13. Regula La SCT El Uso De Aeronaves No Tripulades (Drones) Secretaria de Comunicaciones y Transportes. Comunicado 190. 29/04/2015 in Carlos R. Soltero, 'An Introduction to Mexican Drone Regulations,' February 4, 2016, at: http://www.mcginnislaw.com/images/uploads/news/ Introduction_to_Mexican_Drone_Regulations__Exhibit.pdf.
14. For an example of video of the US–Mexican border taken by a hobbyist in Tijuana, see 'USA– Mexico Border Wall Drone Video (San Diego–Tijuana) Using Litchi App' at: https://www.youtube .com/watch?v=729dt8NEBKg.
15. See video posted on 'Autonomous Real-Time Ground Ubiquitous Surveillance Imaging System (ARGUS-IS),' BAE Systems at: https://www.baesystems.com/en-us/product/autonomous -realtime-ground-ubiquitous-surveillance-imaging-system-argusis.
16. See 'Aims and Objectives,' at: https://roborder.eu/the-project/aims-objectives/

REFERENCES

Acosta, Arvin L. (2015), 'Border Authorities Install More Robust Surveillance System,' *Yibada*, 6 November, accessed 10 November 2019 at: http://en.yibada.com/articles/82049/20151106/border -protection-china-border-patrol-illegal-border-crossing-drug-trafficking-surveillance-system-drones -uav-border-crimes.htm.
Adams, Kimberly (2017) 'Intimate images in the digital age,' *Marketplace Tech*, 24 January, accessed 10 November 2019 at: https://www.marketplace.org/shows/marketplace-tech/012417-mtech.
Albahari, Maurizio (2018) 'From Right to Permission: Asylum, Mediterranean Migrations, and Europe's War on Smuggling,' *Journal on Migration and Human Security* **6** (2), 121–130.
American Border Patrol (n.d.) American Border Patrol Facebook page, accessed 1 April 2020 at: https:// www.facebook.com/Seidarm/.
Anderson, Ruben (2014), *Illegality, Inc: Clandestine Migration and the Business of Bordering Europe*, Oakland, CA: University of California Press.

Australian Department of Home Affairs (2018), 'Securing Australia's borders for the future' media release, 29 October, accessed 10 November 2019 at: https://minister.homeaffairs.gov.au/peterdutton/Pages/Securing-Australia%27s-borders-for-the-future.aspx.

Balido, Nelson (2016), 'Mexican Cartels Patrol Border with Drones—and U.S. Has No Response.' *Fox News*, 19 February, accessed 12 October 2019 at: http://www.foxnews.com/opinion/2016/02/19/nelson-balido-mexican-cartels-patrol-border-with-drones-and-us-has-no-response.html.

Ball, Mike (2016), 'Schiebel CAMCOPTER UAS Supports Migrant Offshore Aid Station Mission,' *Unmanned Systems News*, 21 November.

Barry, Tom (2013), 'Drones Over the Homeland: How Politics, Money and Lack of Oversight Have Sparked Drone Proliferation, and What We Can Do,' Center for International Policy, 23 April, accessed 12 October 2019 at: https://www.ciponline.org/research/html/drones-over-the-homeland.

BBC (2019), Drones monitor south coast of England for migrant boats, BBC 5 December 2019, accessed 20 January at: https://www.bbc.com/news/uk-england-kent-50673241.

Benjamin, Medea (2012), *Drone Warfare: Killing by Remote Control*. New York, NY: OR Books.

Bertin, Francesca, and Elena Fontanari (2011), 'Militarizing the Mediterranean: Enforcing Europe's Borders has Meant Abandoning Some of Its Principles,' *Internationale Politik Journal of German Council on Foreign Relations*, **12** (July/August), 22–26.

Boyd, Aaron (2019a), 'CBP to Deploy Anti-Drone Bubbles along U.S.–Mexico Border,' *Nextgov* 27 September, accessed 20 January 2020 at: https://www.nextgov.com/emerging-tech/2019/09/cbp-deploy-anti-drone-bubbles-along-us-mexico-border/160218/.

Boyd, Aaron (2019b), 'The experimental project reached the fourth and final stage of development: testing in real-world environments,' *Nextgov*, 29 August, accessed 20 January 2020 at: https://www.nextgov.com/emerging-tech/2019/08/cbp-test-autonomous-drones-use-border/159542/.

CBP (2017), 'CBP to Test the Operational Use of Small Unmanned Aircraft Systems in 3 U.S. Border Patrol Sectors,' Customs and Border Protection Press Release, September 14.

Chow, Jack C. (2012) 'Predators for Peace,' *Foreign Policy*, April 27, accessed April 1 2021 at: http://foreignpolicy.com/2012/04/27/predators-for-peace/.

Cornelius, Wayne A. (2005) 'Controlling "Unwanted" Immigration: Lessons from the United States, 1993–2004,' *Journal of Ethnic and Migration Studies* **31**(4), 775–794.

Custers, Bart (ed.) (2016), *The Future of Drone Use*, Berlin: Springer.

Degnbol, Rasmus (n.d.), Europe's New Borders, photography, accessed 1 April 2020 at: http://rasmusdegnbol.com/portfolio-item/europes-new-borders/.

DHS (2014), 'U.S. Customs and Border Protection's Unmanned Aircraft System Program Does Not Achieve Intended Results or Recognize All Costs of Operations,' Office of the Inspector General, Department of Homeland Security, OIG-15-17, December 24.

DHS (2018), 'CBP Has Not Ensured Safeguards for Data Collected Using Unmanned Aircraft Systems,' Office of the Inspector General, Department of Homeland Security, OIG-18-79, September 21.

Dunn, David Hastings (2013), 'Drones: Disembodied Aerial Warfare and the Unarticulated Threat,' *International Affairs* **89** (5), 1237–1246.

Emery, John R. (2016), 'The Possibilities and Pitfalls of Humanitarian Drones,' *Ethics and International Affairs* **30** (2), 153–165.

ERR News (2018), 'Border Guard gets surveillance drones,' ERR News (English-language service of Estonian Public Broadcasting), 12 January, accessed 20 October 2019 at: https://news.err.ee/653720/border-guard-gets-surveillance-drones.

European Commission (2015), 'A European Border and Coast Guard to protect Europe's External Borders,' press release, Strasbourg, 15 December.

FAA (2017) *FAA Aerospace Forecast Fiscal Years 2017–2037*, Washington, DC: Federal Aviation Administration.

Feeney, Matthew (2017), 'Border Patrol Seeking Facial Recognition Drones,' Cato at Liberty blog, 10 April, accessed 20 October 2019 at: https://www.cato.org/blog/border-patrol-seeking-facial-recognition-drones.

Finnish Border Guard (2019), 'In collaboration with other authorities, the Finnish Border Guard has launched a project named Valvonta 2, under which it is looking into deploying unmanned aircraft in maritime surveillance duties', *News From the Border Guard*, 14 October.

Fisher (2015), Testimony of Border Patrol Chief Michael J. Fisher at 'Securing the Border: Understanding Threats and Strategies for the Northern Border,' U.S. Senate Committee on Homeland Security and Governmental Affairs, 22 April.

Fleming, Matthew H., Samuel J. Brannen, Andrew G. Mosher, Bryan Altmire, Andrew Metrick, Meredith Boyle, and Richard Say (2015), 'Unmanned Systems in Homeland Security,' Homeland Security Studies and Analysis Institute, January.

Frontex (2018a), 'Frontex begins testing unmanned aircraft for border surveillance', News Release, 27 September, accessed 20 October 2019 at: https://frontex.europa.eu/media-centre/news-release/frontex -begins-testing-unmanned-aircraft-for-border-surveillance-zSQ26A.

Frontex (2018b), 'Frontex marks two years as the European Border and Coast Guard Agency,' News Release, 6 October, accessed 20 January 2020 at: https://frontex.europa.eu/media-centre/news -release/frontex-marks-two-years-as-the-european-border-and-coast-guard-agency-ECWley.

Frontex (2019), 'Remotely Piloted Aircraft Systems (RPAS) for Medium Altitude Long Endurance Maritime Aerial Surveillance,' 2019/S 202-490010, contract notice, 7 October.

FSD (n.d.), The Swiss Foundation for Mine Action (FSD) Drones in Humanitarian Action: A Guide to the Use of Airborne Systems in Humanitarian Crises, accessed 20 October 2019 at: http://drones.fsd .ch/wp-content/uploads/2016/11/Drones-in-Humanitarian-Action.pdf.

GAO (2017), Border Security: Additional Actions Needed to Strengthen Collection of Unmanned Aerial Systems and Aerostats Data Government Accountability Office. GAO-17-152. Feb 16, 2017.

Gartner Research (2017), 'Gartner Says Almost 3 Million Personal and Commercial Drones Will Be Shipped in 2017,' Press Release, 9 February.

Ghaffary, Shirin (2019). 'The "Smarter" Wall: How Drones, Sensors, And AI Are Patrolling The Border,' *Vox*, 16 May, accessed 20 January 2020 at: https://www.vox.com/recode/2019/5/16/18511583/smart -border-wall-drones-sensors-ai.

Haddal, Chad C. and Jeremiah Gertler (2010), 'Homeland Security: Unmanned Aerial Vehicles and Border Surveillance,' CRS Report for Congress, Congressional Research Service, 8 July.

Hayes, Ben and Mathias Vermeulen (2012), *Borderline: The EU's New Border Surveillance Initiatives*, Berlin: Die grüne politische Stiftung.

Hayes, Ben, Chris Jones and Eric Töpfer (2014) 'Eurodrones Inc.' February, accessed 20 October 2019 at: http://statewatch.org/observatories_files/drones/eu/eurodrones.htm.

Jumbert, Maria Gabrielsen (2016), 'Creating the EU Drone: Control, Sorting and Search and Rescue at Sea,' in Kristin Bergtora Sandvik & Maria Gabrielsen Jumbert (eds), *The Good Drone*, London: Routledge, pp. 89–108.

Kaag, John, and Sarah Kreps (2014) *Drone Warfare*, Malden, MA: Polity Press.

Keefe, Patrick Radden (2012), 'Cocaine Incorporated,' *New York Times*, 15 June.

Koslowski, Rey and Marcus Schulzke (2018) 'Drones along Borders: Border Security UAVs in the United States and the European Union', *International Studies Perspectives*, **19** (4), 305–324.

Leonard, Sarah (2010), 'EU border security and migration into the European Union: FRONTEX and securitisation through practices,' *European Security*, **19** (2), 231–254.

LRT (2019) 'Lithuanian border guards buy two drones to fight cigarette smuggling', Lithuanian National Radio and Television (LRT), 8 February, accessed 20 January 2020 at: https://www.lrt.lt/en/news-in -english/19/1084268/lithuanian-border-guards-buy-two-drones-to-fight-cigarette-smuggling.

Marin, Luise (2016), 'The Humanitarian Drone and the Borders: Unveiling the Rationales Underlying the Deployment of Drones in Border Surveillance,' in Bart Custers (ed.), *The Future of Drone Use*, Berlin: Springer, pp. 115–132.

McNabb, Harry (2020) 'Drones in Journalism: Drone Footage Shows the Plight of Syrian Refugees,' *Dronelife*, 28 February, accessed March 15 2020 at: https://dronelife.com/2020/02/28/drones-in -journalism-drone-footage-shows-the-plight-of-syrian-refugees/.

Meier, Patrick (2015) *Digital Humanitarians: How Big Data Is Changing the Face of Humanitarian Response*, New York, NY: CRC Press.

Milivojevic, Sanja (2015), 'Re-bordering the Peripheral Global North and Global South: Game of Drones, Immobilising Mobile Bodies and Decentring Perspectives on Drones in Border Policing', in Aleš Završnik (ed.), *Drones and Unmanned Aerial Systems: Legal and Social Implications for Security and Surveillance*, Berlin: Springer, pp. 83–100.

MOAS (2020), Migrant Offshore Station, 'MOAS Mission,' accessed 16 January 2020 at: https://www .moas.eu/mission/.

Nardi, William (2018), 'Drone Activity Soaring at the US–Mexico Border,' *Washington Examiner*, 6 July.

Paul, George (2019), 'Verizon wants to become the first carrier to use its 5G network to connect 1 million drone flights,' *Business Insider*, 30 December accessed 16 January 2020 at: https://www .businessinsider.com/verizon-highlights-5g-drone-opportunity-2019-12.

Riordan, Primrose (2019), 'Hi-tech drones plan for borders in slow motion,' *The Australian*, 13 May.

Sabak, Juliusz (2015), 'Polish Border Guard to be Operating UAV's soon,' Defence24.com, 19 February, accessed 15 October 2019 at: https://www.defence24.com/polish-border-guard-to-be-operating-uavs -soon.

Sandvik, Kristin Bergtora, Kjersti Lohne (2014), 'The Rise of the Humanitarian Drone: Giving Content to an Emerging Concept,' *Millennium* **43** (1), 145–164.

Sandvik, Kristin Bergtora and Maria Gabrielsen Jumbert (eds) (2016), *The Good Drone*, Farnham: Ashgate.

Singer, P.W. (2009), *Wired for War: The Robotics Revolution and Conflict in the 21st Century*, New York, NY: Penguin Press.

Stepanovich, Amie (2012), Association Litigation Counsel, Electronic Privacy Information Center, Testimony at Hearing on 'Using Unmanned Aerial Systems Within the Homeland: Security Game Changer?' before the Subcommittee on Oversight, Investigations, and Management of the U.S. House of Representatives, Committee on Homeland Security, 19 July.

Swearingen, Jake (2019), 'A.I. Is Flying Drones (Very, Very Slowly),' *New York Times*, 26 March.

Taylor, Ramon (2017), 'Man Aims to Secure US–Mexico Border with Drone,' Voice of America (VOA), March 7.

Tazziolli, Martina (2016), 'Border displacements: Challenging the politics of rescue between Mare Nostrum and Triton,' *Migration Studies* **4** (1), 1–19.

Waever, Ole, Barry Buzan, Morten Kelstrup and Pierre Lemaitre, (1993), *Identity, Migration and the New Security Agenda in Europe*, New York, NY: St. Martin's Press.

West, Gretchen (2015), 'The Sky's the Limit – If the FAA Will Get Out of the Way,' *Foreign Affairs*, **94** (3), 90–97.

Wilson, Dean (2015), 'Border Militarization, Technology and Crime,' in Sharon Pickering and Julie Ham (eds), *The Routledge Handbook on Crime and International Migration*, New York: Routledge, pp. 141–154.

Zamichow, Nora (1990), 'Marine Drones Used in Drug War,' *Los Angeles Times*, 8 March.

Završnik, Aleš (ed.) (2015) *Drones and Unmanned Aerial Systems: Legal and Social Implications for Security and Surveillance*, Berlin: Springer.

PART III

INTEGRATION, REINTEGRATION AND MIGRANTS' (DIGITAL) (VIRTUAL) (TRANSNATIONAL) IDENTITIES

12. Migrant inclusion 4.0: the role of mobile tech

Céline Bauloz[1]

INTRODUCTION

In today's digital societies, migrants are as likely as any other human being to embrace new technologies. Internet usage has kept on increasing worldwide at an average rate of 10 per cent per year since 2005 and was estimated at 53.6 per cent at the end of 2019 (ITU, 2019, p. 1). While geographic disparities remain, especially for the least developed countries, these differences are attenuated when it comes to mobile subscriptions due to lower costs and better availability compared to fixed connections (ibid., pp. 2 and 5).[2] Mobile technologies, such as cellphones/smartphones, tablets and laptops, have arguably opened up more universal opportunities for connectivity. According to the International Telecommunication Union, '[t]he relatively small difference between developed and developing countries also shows that connectivity is a priority among people in countries at all levels of development' (ibid., p. 5).

This near-universal aspiration to connectivity starkly contrasts with the (over-)mediatization and negative public reactions raised by photos of migrants carrying their smartphones when arriving on European shores in 2015. As noted by Leurs and Ponzanesi, '[t]he appearance of digitally connected refugees was perceived as incongruent with Eurocentric ideas of sad and poor refugees fleeing from war and atrocities' (2018, p. 6; see also Leurs, 2017, p. 680). Far from luxury goods, research has evidenced the importance of mobile devices, especially smartphones, during migrants' journeys. Qualified as a 'lifeline' for migrants, they provide migrants with a wealth of information at hand before, during and after their migration journeys (Alencar et al., 2018).

While mobile technology is omnipresent during the entire migration process, this chapter examines the role it plays for migrants, including refugees, in settling in their new environment. Building on the vast literature on migrants' appropriation of mobile technology and, more broadly, of information and communication technologies (ICTs), it aims to take stock of the potentials and limitations of smartphones for migrants' inclusion in receiving societies. Smartphone usage is arguably not totally distinct between 'connected migrants' and other individuals (Diminescu, 2008). However, a systematic literature review on migrants' motivations for adopting and using ICT found 'adjustment and integration in the host country as the most important motivations for immigrants' use of ICTs' (Bhakta Acharya, 2016, p. 46).

In order to understand the value of smartphones for migrants in receiving societies, the next section of this chapter explores the interrelationships between inclusion and appropriation of mobile technology as exercises of migrants' agency (section 2). While a comprehensive analysis of these two notions and their connections would go beyond the scope of this chapter, this preliminary overview will set the grounds for better understanding migrants' digital practices for their own inclusion and the obstacles they may face (section 3). By reviewing research on the topic, this chapter will highlight a bias existing in the literature towards the so-called 'Global North' (Leurs and Smets, 2018, p. 11). Albeit not peculiar to these issues (IOM, 2019,

p. 139), the lack of research focusing on the 'Global South' will, among others, give the occasion to draw some recommendations for further research in conclusion (section 4).

THE RELATIONSHIP BETWEEN INCLUSION AND APPROPRIATION OF MOBILE TECH: MANIFESTATIONS OF MIGRANTS' AGENCY

While migrants' inclusion has always been an important part of the migration phenomenon, there exists no universally recognized definition. Inclusion is often defined in policy terms, considering policy approaches followed by some countries to frame the relationship between migrants and receiving communities according to their respective values.[3] The most prevalent policy models of inclusion have been – somehow consecutively – those of assimilation, multiculturalism and integration (Bauloz, Vathi and Acosta, 2019). Their common objective has been to preserve social cohesion, that is, the bonds tying a community together through trust and common social norms (Demireva, 2017). However, their approach has differed as to what social cohesion entails and the means to preserve it. Considering diversity brought by migrants as a risk for national identity and social cohesion, assimilation requires the highest degree of adaptation by migrants who have to fully embrace the national identity and values of their receiving society to the detriment of their original ones. By contrast, multiculturalism values diversity and migrants' diverse cultural identities (Castles et al., 2014, p. 270). Standing in between assimilation and multiculturalism, integration is generally accepted as a two-way process of mutual adaptation between migrants and the societies in which they live.

These inclusion policies arguably reflect the different attitudes of receiving countries towards migrants, including their openness to migration and migrants. Each society approaches inclusion differently, depending on its historical, economic, sociocultural and political contexts. The approach can even be different at the local level where inclusion occurs in practice. As spaces of inclusion, an increasing number of cities have for instance taken up an interculturalist approach to inclusion, considering diversity as an advantage and promoting a culture of mutual understanding between migrants and the local population (Zapata-Barrero, 2017). While one's irregular migration status is undoubtedly a first major obstacle to inclusion, some cities have also relied on pragmatic solutions to counter exclusion of irregular migrants, including through provision of ID cards to all residents to ensure their access to various services (Medina, 2015).

These external factors play an important role in influencing the degree of migrants' inclusion in a given destination country, whether in supporting or constraining their inclusion process. However, they represent only one side of the coin as they are unable to account for inclusion as a process that is inherent and personal to migrants' experience. Far from being externally imposed upon migrants, inclusion entails a psychosociological process of adaptation when settling in a new country with a new culture, customs, social values or language (see Berry, 1997).

Similar to the agency exercised during migration (Turton, 2003; McAuliffe et al., 2017; Triandafyllidou, 2018), migrants in receiving societies are often agents of their own inclusion: through more or less intentional and conscious decisions, they adopt certain behaviours and strategies to familiarize with their new environment, interact with different communities, and be incorporated into the different societal areas of the receiving country.[4] In addition to external/structural factors such as inclusion policies and legal frameworks, their capacity to exercise agency and their ensuing degree of inclusion are influenced by individual factors, such as age, gender, level of education, language skills, familial situation or any other social networks (Castles et al., 2002, p. 126). The inclusion process thus differs among migrants, their family members, and even among different 'groups' of migrants, such as refugees or high-/low-skilled migrant workers (see e.g. ibid., p. 119).

From this perspective, the use of mobile technology by migrants, including refugees, can be understood as an illustration of their agency, a means for them to foster their inclusion (see Shah et al., 2019; Kaufmann, 2018, p. 883). In fact, the mere use of ICT by any individual is already in and of itself an exercise of human agency (Díaz Andrade and Doolin, 2019, p. 148):

> Individuals do not engage with ICT as passive subjects without a sense of purpose. They exercise their judgement through critical consciousness in order to assess how and to what extent the function-alities of ICT afford possibilities for action that can help them address their needs and achieve their goals in the specific relational and temporal conditions in which they live ....

Smartphones represent a particularly interesting medium for one to exercise agency. The wide variety of applications they give access to allows a 'personalization par excellence' (Horst and Hjorth, 2013, p. 94), as users can configure their device according to their needs and objectives (Kaufmann, 2018, pp. 883–884). As Jung underlines, 'users decide what a smartphone is for themselves, rather than just adopting a given product' (2014, p. 300). This has in fact been the underlying motto of Apple's marketing strategy with all products starting with the letter 'i' not only to refer to Internet, but also to 'individual', alongside to 'instruct, inform, inspire'.[5] As 'user-empowering' ICT (ibid., p. 301), it is thus unsurprising that researchers now rely on smartphones to develop new research methods for understanding migration journeys from migrants' perspectives, including by examining which are the top ten applications featured on refugees' home screens for fostering their inclusion process (AbuJarour et al., 2019a).

However, mobile technology is not only a means for inclusion but also an object of inclusion, which requires migrants to have access to such technology and be digitally liter-ate (Collin, 2012; see also Codagnone and Kluzer, 2011, p. 16). As an object of inclusion, the access, use and appropriation of mobile technology is thus contingent on structural and individual factors similarly to the process of inclusion. Among the structural factors, the receiving country and its degree of digitalization play an important role: the more the country is digitalized, the more digitally integrated migrants will expected to be. As underlined by Collin, 'the successful integration of migrants requires that their technological integration is as important as the social, political, and economic integration traditionally reported in scientific literature' (Collin, 2012, p. 66; see also Díaz Andrade and Doolin, 2019, p. 157). With the ever-increasing omnipresence of technology, however, the appropriation of mobile technology has also been evidenced in spaces usually considered to be 'less digitalized', such as refugee camps (Lewis and Thacker, 2016, p. 4). In the Za'atari refugee camp in Jordan, for instance, mobile phones were found to be the most popular type of ICT to use the Internet among the 174 young Syrians surveyed in the study (Fisher and Yafi, 2018, p. 5).

While e-inclusion is important to avoid migrants being digitally excluded in the receiving society (see textbox below), digital inequalities and exclusion may also be prompted by indi-vidual factors. Most notably, varying degrees of digital literacy among migrants due to their previous use of technology in their country of origin, level of education, age, gender, language proficiency or culture ultimately impacts on their appropriation of mobile technologies (see

Fisher, 2018, pp. 84–85; Alam and Imran, 2015, pp. 347 and 355 referring to a 'knowledge divide'). Even when coming from a 'digitalized' country, migrants may have to 're-appropriate … their smartphones based on localized digital infrastructures' as their prior usage may not be adapted to the digital information landscape of the receiving country (Kaufmann, 2018, p. 889). Age has also been evidenced as a factor negatively correlated with one's appropriation of ICT, as migrant youth have higher capacities in mastering new technologies (Gifford and Wilding, 2013, p. 560; Hebbani et al., 2015, pp. 43–44; Khvorostianov et al., 2011). Hence, young migrants have found to play 'intrinsic roles in helping others through information and technology', such as their parents, older family members and siblings (Fisher and Yafi, 2018, p. 3). Migrants' perceptions of mobile technology can also influence their appropriation. Migrants – especially refugees – may fear surveillance from their country of origin, as well as from some receiving countries that, for instance, may use mobile phone data to ascertain the credibility of their stories during refugee status determination processes (Gelb and Krishnan, 2018, p. 10; Gillespie et al., 2016, p. 19). Some migrants have thus reverted to self-censorship on social media platforms (Leurs, 2017, pp. 691–692).

Hence, the relationship between their inclusion process and their appropriation of mobile technology is closely intertwined. Migrants' appropriation of mobile technology will depend on their digital inclusion. By contrast, if digitally excluded, migrants run the risk of experiencing information poverty that will undermine their inclusion process (Gelb and Krishnan, 2018, p. 9; Alam and Imran, 2015, p. 347).[6] In other words, digital literacy – understood here as appropriation of mobile technology – 'could be argued to be a key indicator of successful integration and social inclusion' (Gifford and Wilding, 2013, p. 561).

BOX 12.1 BEYOND MIGRANTS' AGENCY: TOP-DOWN APPROACHES TO MIGRANTS' (DIGITAL) INCLUSION

The potential of mobile technology to foster migrants' inclusion does not only depend on migrants' agency, but also structural factors. These include the policies and measures States adopt and implement with respect to both technology and migrant inclusion. Such top-down interventions can have the ultimate effect of supporting or undermining to a certain or greater extent the potential of mobile technology to the benefit of migrants' inclusion.

Since the turn of the 21st century and rapid technological developments, a priority for States, especially developed countries, has notably been to ensure the e-inclusion of all residents, including migrants. Building on the 2006 Riga Declaration of the European Union, Collin explains that:

> e-inclusion's main purpose is to enable everyone to participate in the information society and to reduce as much as possible the digital divide within a population …; on the other hand, they use ICTs to remedy other forms of exclusion and improve different social, political, and economic aspects …. (Collin, 2012, p. 73)

Tackling any existing digital divide – from access and affordability, to age, bandwidth, education, gender, location or useful usage (UN DESA, 2018, p. 34) – has become even more essential with increasing moves towards 'digital first/only' provision of services. Indeed, if e-governments provide a great potential for strengthening resilience through access to information and services, they also entail the risk of further isolating those with insufficient digital access and/or literacy (ibid.).

However, e-inclusion also extends beyond e-governments (Žajdlea Hrustek et al., 2016, p. 439). It more broadly concerns the whole tech community, especially the private sector, which is generally at the forefront of new technological developments such as artificial intelligence (AI) systems increasingly integrated into smartphones and applications. Ensuring the design of inclusive technology by the private sector may be particularly challenging, as attested by the existence of biased algorithms used to make decisions on job interviews or the granting of parole, as well as in voice interfaces (Knight, 2017a and b).

On the other hand, capitalizing on digital technology as a tool to remedy other forms of exclusion entails that such technology is then not used for creating new exclusions. This may be the risk with digital surveillance targeting migrants' Internet activities, including on social media, ultimately 'keep[ing] [migrants] from feeling like full members of society' (Shahshahani, 2018).

THE POTENTIAL OF MOBILE TECHNOLOGY IN MIGRANTS' INCLUSION

By providing direct and instant access to web browsers, geolocation platforms, and an endless number of applications offering services as diverse as online banking, news, gaming, social media, instant messaging or streaming, smartphones are used daily for a vast array of purposes, including entertainment, leisure, business or online shopping. However, as particular types of users, the use of mobile devices takes on additional and special values for migrants settling in a new society and community (see Jung, 2014; and Gillespie et al., 2016, p. 43). As examined in this section, the appropriation and usage migrants have made of their smartphones to foster their inclusion process can broadly be considered as meeting two main objectives: those of maintaining their existing social networks and creating new connections, as well as fostering their inclusion into the different societal areas of the receiving country.

Sustaining Existing and Creating New Social Networks

Unsurprisingly, migrants report that the primary use of their smartphones is to keep in touch with their family and friends who have remained in their country of origin or with the diaspora in other receiving countries (Baldassar et al., 2016; Kabbar and Crump, 2006). Through such appropriation of smartphones, migrants have found a means to further maintain and strengthen their social networks, referred to in sociology as bonding social capital (see most notably Bourdieu, 1980; Putnam, 1993; and Portes, 1998).[7]

Compared to the traditional calling and SMS functions of cell phones, smartphones give access to a broad range of applications providing the opportunity to connect with family and friends for free or at relatively low costs through a Wi-fi or mobile network connection. Among the applications offering services of instant messaging and (video) calls, existing research points to WhatsApp, Skype, Line, Facebook/Instagram messaging and Viber as the most popular applications among migrants (Kaufmann, 2018, p. 892; Shah et al., 2019, p. 2). Social media platforms such as Facebook and Instagram also give migrants the possibility to remotely follow the life of their relatives and friends by checking their posts or photos and interact with them through likes and comments (Komito, 2011). These connections are of course contingent on access of family members to ICT and connectivity issues, especially in some countries of origin where ICT may still be expensive and/or network coverage limited or unreliable.

In 2004, prior to the smartphone boom initiated by Apple with its first iPhone in 2007,[8] Vertovec already considered mobile phones as 'the social glue of migrant transnationalism' (2004). Referring to the 'multiple ties and interactions linking people and institutions across the borders of nation-states' (Vertovec, 1999, p. 447), transnationalism questions traditional understandings of migrants' belonging based on attachment to Nation-States (Castles, 2002, p. 1157). Migrants' transnational ties accordingly frame their identity beyond States' boundaries.

While migrants' transnational lives have sometimes been seen at odds with expectations of migrants as 'settlers', especially in assimilationist approaches to inclusion (Bauloz et al., 2019), existing evidence does not suggest that migrants' use of ICT, such as smartphones, for bonding purposes would create segregation negatively impacting on social cohesion (although it is not excluded all together; Diminescu et al., 2009, p. 195; Codagnone and Kluzer, 2011, p. 39). In today's world where the transnational connectivity offered by the digital space is more generally used by all (McAuliffe et al., 2019, p. 149), research rather converges on the importance of these transnational links for fostering migrants' inclusion in the receiving society.

Connections with family and friends represent a source of emotional support for migrants that positively impacts their mental health. Settling in a new country and society may be a source of stress and anxiety for migrants, which can be relieved to a certain extent through support, comfort and encouragement from family and friends (Shah et al., 2019, p. 2; Chib and Aricat, 2016, pp. 7–8). For instance, a qualitative study on the impact of digital technology on the mental health of 290 refugees in the United States found that ICT, including smartphones, 'helped [refugees] decrease the burden of separation from their families and fulfilled some of their emotional and mental needs by making them feel at home when they were away from their homes' (Shah et al., 2019, p. 6). For refugees especially, these contacts can also relieve the stress caused by worries about the fate of family and friends who have remained in the country of origin and may be facing dangerous situations, including armed conflicts (Kaufmann, 2018). However, these digital communications can be a 'double-edge sword' as they may act for migrants as a reminder of the distance between them and their loved ones, intensifying their feeling of loneliness in their new society (Chib and Aricat, 2016, p. 10). The situation of those left in the country of origin may create more emotional distress and feeling of helplessness for refugees (Shah et al., 2019, pp. 6–7). Migrants may also hide the hardships they are facing in the receiving country from their relatives in order to avoid worrying their relatives about the difficult or precarious conditions they may be living in. In such situations, they have proved to make a strategic use of smartphones by avoiding video calls on Viber, for instance, so as not to disclose their living environment (ibid., pp. 7–8).

That said, when translating into support and comfort, interactions with family and friends through instant messaging, texting, (video) calls and social media can contribute to migrants' well-being in receiving societies, setting the ground for conditions conducive to their inclusion. The appropriation of smartphones for such bonding purposes also has the potential to improve social abilities, including for facilitating connections with the local population (Witteborn, 2015; Codagnone and Kluzer, 2011, p. 39).

This 'bridging' social capital is important for migrants' social inclusion as social networks can support their inclusion into different societal areas, such as employment, and create some sense of inter-cultural understanding. Research indicates that social networks in the receiving countries are mostly created with co-ethnic migrants, rather than with the local population, for reasons of trust and language (Bacinshoga and Johnston, 2013; Popivanov and Kovacheva, 2019). Studies on migrants' use of ICT, more generally, have found that migrants were more at ease to communicate with the local population via digital means, such as emails for instance, as they allow them the time to structure their thoughts and use online dictionaries if needed (Díaz Andrade and Doolin, 2019, pp. 158–159). Smartphone applications have also been created for fostering migrants' community engagement and social cohesion, through both online and offline interactions between migrants' and the local population (Froneberg et al., 2015). However, social media sites appear to have played the greatest role for migrants' social inclusion among the diverse existing digital technologies available on smartphones. These have somehow acted as a source of multicultural exchange through posts, photos and story-telling, where migrants can learn the culture of the receiving community by following local 'friends' and share their experiences with others, including concerning their own culture and identity (Alencar and Tsagkroni, 2019, pp. 190–191; AbuJarour and Krasnova, 2018, p. 2; Leurs, 2017, pp. 691–693). Social media practices, however, come with risks of exposure to intolerance, hate speech, discrimination and misinformation that may ultimately negatively impact on migrants' inclusion and, more broadly, on social cohesion (Alencar and Tsagkroni, 2019, p. 188).

As a result, smartphones stimulate 'a (new) sense of belonging' within the receiving community (Borkert et al., 2018, p. 3), creating transnational identities beyond States' boundaries. Through such appropriation of smartphones, migrants have even been found to be 'pioneers and innovators in the adoption of new technologies' (Gonzalez et al., 2009, p. 43). For instance, a study of transnational communication among Iranian migrant workers in Australia highlighted that ICTs restrictions in their country of origin led these migrants and their families to circumvent censorship of certain applications by using new ones or relying on proxy servers (Farshbaf Shaker, 2018).

Inclusion in the Different Societal Areas of the Receiving Country

Migrants' incorporation in the different societal areas of the receiving country lies at the heart of their inclusion process (Faist, 2018, p. 4). As any other residents, migrants need to be incorporated in the education, health, housing or employment systems of the receiving country that, all together, participate to their inclusion process. In increasingly digitalized societies, smartphones have a great potential for facilitating information and access to such services. Geolocation services offered by smartphone applications such as Google Maps are particularly useful for migrants (Kaufmann, 2018, p. 890). They are often the only way for foreigners, including migrants and tourists, to navigate new surroundings as they have come to virtually replace physical maps. They also assist migrants in locating different services such as those provided by the authorities or civil society organizations. Smartphones and other ICT tools may also be the only way of accessing information and services given increasing moves towards digital-only governmental services (Pinsky and Steinhauser, 2018; Gifford and Wilding, 2013, p. 561).

However, a certain command of the local language is often a prerequisite to make use of such e-government services because, predominantly addressed to residents in general, they are usually not accessible in migrants' languages (AbuJarour et al., 2017, p. 3269). More generally, language has been found important for migrants in navigating their new environment and increasing the likelihood of finding employment and their self-reported health outcomes (Chiswick, 2016; Aoki and Santiago, 2018). Language barriers are often identified by migrants among the first challenges they face upon arrival in a new country and as one of the first areas on which they most critically need information (Borkert et al., 2018, p. 6).

While some national or local authorities support language acquisition through (free) language courses (Bauloz et al., 2019, p. 174; Wiesbrock, 2011, p. 52), language, translation and vocabulary applications have been used by migrants as a tool for language acquisition (Kaufmann, 2018, p. 891). These smartphone applications, such as Google Translate, offer the advantage of immediate translations in everyday life situations (Bradley et al., 2017, p. 6). As time is often considered by migrants as one of the main issues for learning a language (Huddleston and Dag Tjaden, 2012), the mobile nature of this technology also enables them to maximize their time, such as during commuting on public transportation (Kaufmann, 2018, p. 891; Benton, 2014, p. 8). However, these applications have been found to rely on passive approaches to language learning, to the detriment of speaking and pronunciation (Sofkova Hashemi et al., 2017, p. 5; Bradley et al., 2017, p. 7). Migrants' preference thus often lies in online videos and audio recordings, such as language tutorials on YouTube where the national language is taught by co-ethnic migrants in their mother tongue (AbuJarour and Krasnova, 2018, pp. 3 and 7).

While the impact of these applications on migrants' language skills would require further research, Kaufmann notes that '[t]he skill needed to navigate "bureaucratic space[s]" … has not necessarily become easier in the age of smartphones' (2018, p. 892). For instance, migrants keep on relying on their local networks for translating official letters they receive and which they have been unable to translate through digital means (ibid.).

Moreover, beyond their language skills, the use of applications and other digital platforms for information-seeking requires migrants to be first aware of their existence amid the myriad of smartphone applications now available (Kaufmann, 2018, p. 889). Failing so, migrants usually tend to turn to their peer networks in the receiving country to acquire and share information, including through the use of mobile phones (Bacinshoga and Johnston, 2013).

The idea of 'one-stop shop' digital platforms consolidating all relevant information needed by migrants with a single click underlay the proliferation of applications in the follow-up of the large arrival of migrants in Europe in 2015. Developed by the tech sector or/with governments and non-governmental or civil society organizations, these digital tools were set up for supporting newly arrived migrants' orientation and inclusion in receiving countries, especially for admission, housing, education and job opportunities. Despite all good intentions, these applications have, however, quickly shown their limits, spreading a 'digital litter' that has been considered unproductive for migrants' inclusion, if not harmful (Benton, 2019). The duplication of such initiatives combined with the lack of proper follow-up curation and cuts in funding have resulted in outdated and misinformation shared on dormant applications, which often re-direct users to broken links. Migrants can thus lose time 'in an internet rabbit hole', without being able to access the information needed or ultimately making decisions based on outdated and/or incorrect information (ibid.). It is thus not surprising that migrants rather turn to applications not specifically designed for them to get the information they need, including social media platforms, such as Facebook (Gelb and Krishnan, 2018, p. 10; Kaufmann, 2018, pp. 889–890).

These digital 'failures' should, however, not obscure the value of smartphones for migrants to access various information and services in receiving countries. Research underlines the increasing importance mobile technology has taken in migrants' information practices, although so far it has not been conclusive as to the impacts of these applications on migrants' inclusion given its intangible nature (Benton, 2014, p. 5). Today, access to education is facilitated by various digital solutions supported on computer or mobile technology, such as free digital learning, including massive open online courses (MOOCs) (Colucci et al., 2017, p. 20; Lewis and Thacker, 2016, p. 15). The health sector has also been particularly active in digitalizing health services for diverse populations. In addition to accessibility, such digital healthcare and advocacy initiatives tailored to migrants offer the advantage of lower costs and anonymity in an area that may be particularly sensitive for migrants depending upon their cultural and health background (Gillespie et al., 2016, p. 37; Fisher, 2018, p. 95). Migrants' labour market inclusion has also been targeted through digital platforms listing employment opportunities, while the tech sector is now increasingly capitalizing on improving migrants' tech skills, including in mobile technology, for sustainable employment opportunities in receiving countries (see the textbox below).

BOX 12.2 REFUGEEKS? IMPROVING REFUGEES' TECH SKILLS FOR THEIR ECONOMIC INCLUSION

Access to employment is an important facet of migrants' economic inclusion, especially with employment rates usually lower than those of non-migrants (see for instance Eurostat, 2019). Among migrants, refugees and those who have migrated for family reunification are usually less likely to find an employment than other cohorts (Lens, Marx and Vuji, 2018).

Nevertheless, history also accounts for significant contributions made by migrants to their receiving country thanks to their entrepreneurship, especially in the tech sector (McAuliffe et al., 2019, p. 177). In the United States, for example, over half of the tech companies were founded by migrants or their children (Salinas, 2018).

The innovative entrepreneurship migrants can bring along into receiving societies has arguably not escaped the attention of tech companies. An increasing number of tech courses are now offered to migrants, including refugees, especially in digital coding. While migrants consider ICT skills as important to improve their likelihood of being employed (Diminescu et al., 2009), these tech courses more generally correspond to labour needs of developed countries of destination where the tech sector keeps on growing (Mason, 2018, p. 3). In addition to English as a working language, research underlines that, in theory, four main reasons make refugees a 'natural fit' for digital careers: 'high income levels, relatively high social status, few barriers to entry, and portability of skills' (ibid., p. 4).

Moreover, as for any other workers in digitalized societies, training refugees and other migrants for tech jobs arguably improves the sustainability of their employment in the context of the future of work (see Hepp et al., 2009, p. 7). As noted by Benton and Patuzzi, 'Workers in occupations that are likely to be augmented rather than replaced by technology will count amongst the winners in the future world of work' (2018, p. 5). This will arguably be a double-win situation if migrants are to design tomorrow's digital technology tailored to migrants' inclusion needs.

CONCLUSION

With ever-increasing penetration rates and future innovations in the tech sector, the potential of mobile technologies in the lives of human beings, including migrants, is accordingly only at its inception. As empowering-ICT tools, they have already taken a considerable place in migrants' journeys, including during their inclusion process. However, if 'smartphones hold an untapped potential for integration processes' (Kaufmann, 2018, p. 882), it is equally important to acknowledge the 'limits of app-ology' (Horn, 2015).

As an object of inclusion in and of itself, migrants' appropriation of mobile technology remains contingent on their digital inclusion and literacy. As noted nearly 20 years ago, 'Not knowing how to and being unable to access technologies designed to communicate digital information, such as the Internet, may, in the future, be equivalent to not knowing how to read and write today' (Mitchell, 2002). While this assertion could not be more accurate today with regard to the role played by smartphones, how to foster migrants' digital inclusion and literacy remains somewhat under-researched. Indeed, existing research predominantly focuses on destination countries of the so-called 'Global North', while issues related to digital inclusion and literacy arguably already concern countries of origin and extend as well as to countries of destination in the 'Global South'. The ever-increasing pervasiveness of smartphones thus calls for extending the geographical focus of research to better understand the factors and barriers impacting on migrants' appropriation of mobile technology as an object of inclusion.

As a means of inclusion, migrants' appropriation of mobile technology appears to be central for reframing a novel sense of belonging in receiving societies through mutually reinforcing bonding and bridging digital practices, although it will never totally replace the value of offline social interactions (Alencar and Tsagkroni, 2019). Nevertheless, migrants are nowadays not only emotionally attached to their smartphones, but also technically dependent on them (Kaufmann, 2018, p. 893). While this may not fundamentally differ from other human beings who are today transitioning into genuine 'homo digitalis' (Montag and Diefenbach, 2018), it comes with accrued risks of dis/misinformation, discrimination, hate speech and fear of surveillance for migrants who may already be in a more vulnerable situation. These risks need to be factored in any new digital initiative developed to support migrants' inclusion in order to improve migrants' perceptions. Failing so, they are likely to negatively impact on migrants' appropriation of smartphones and, ultimately, on their potential for inclusion.

If the creation of digital one-stop shops can be useful for centralizing information and services for migrants, lessons can still be drawn from the failed digital initiatives prompted by the 'migrant crisis' in Europe in 2015. These digital tools were set up in a quick and ad hoc manner as support measures for migrants who had just arrived in Europe. They thus lacked a longer-term and more sustainable approach to digital support for migrants' inclusion. Beyond issues of funding, they have shed light on the fact that, without raising migrants' awareness, any existing dedicated applications or digital platforms will not be of much value. National and, especially, local authorities accordingly have a greater role to play in providing such information (see Kaufmann, 2018, p. 889). Digital tools also need to be tailored to migrants' needs, not only in terms of language accessibility, but also more generally from an intercultural user experience (AbuJarour et al., 2019b, p. 884). Finally, what ultimately matters is not the quantity of applications offered to migrants, but the quality and reliability of the information they provide (Benton, 2014, p. 9; Benton, 2019).

NOTES

1. The opinions, comments and analyses expressed in this chapter are those of the author and do not necessarily represent the views of any of the organizations or institutions with which the author is affiliated.
2. See however Pew Research Center (2019) finding that there still exists a mobile divide in the 11 emerging economies surveyed in their research, especially due to issues of connectivity, costs and security.
3. It is noteworthy that inclusion policies have so far been mostly adopted by developed States from the so-called 'Global North'. This does not imply that, as a psychosociological process, inclusion does not occur in practice in other countries, but simply that these countries have not set a nation-wide strategy for migrants' inclusion (IOM, 2019, p. 170) due, for instance, to other more pressing socioeconomic challenges (Gagnon and Khoudour-Castéras, 2012).
4. On human agency more generally, see Bandura, 2001.
5. See www.businessinsider.com/what-i-means-iphone-2016-9?r=US&IR=T (accessed on 14 February 2020).
6. As noted by Lloyd et al. (2013, p. 125): 'The idea that limited access to information and associated information skills restricts the capacity of individuals to fully participate in society and to make informed decisions has been seen as an underpinning driver for information poverty. Over time, this reduced capacity can affect ability to extend social networks, to gain employment, maintain health and to improve educationally, thus creating a cycle of alienation, continued marginalisation and disenfranchisement in this sector of the community.'
7. For more discussion on social capital, see most notably Bourdieu, 1980 and Putnam, 1993.
8. The first smartphone was, however, the Simon Personal Communicator commercialized by IBM in 1994 (Smith, 2018).

REFERENCES

AbuJarour, S., H. Krasnova, A. Díaz Andrade, S. Olbrich, C.-W. Tan, C. Urquhart and M. Wiesche (2017), 'Empowering refugees with technology: Best practices and research agenda', Panel, Twenty-fifth European Conference on Information Systems (ECIS), Guimarães, Portugal.

AbuJarour, S. and H. Krasnova (2018), 'E-learning as a means of social inclusion: The case of Syrian Refugees in Germany', Twenty-fourth Americas Conference on Information Systems, New Orleans, United States of America.

AbuJarour, S., C. Bergert, J. Gundlach, A. Köster and H. Krasnova (2019a), '"Your home screen is worth a thousand words": Investigating the prevalence of smartphone apps among refugees in Germany', Twenty-fifth Americas Conference on Information Systems, Cancun, Mexico.

AbuJarour, S., M. Wiesche, A. Díaz Andrade, J. Fedorowicz, H. Krasnova, S. Olbrich, C.-W. Tan, C. Urquhart and V. Venkatesh (2019b), 'ICT-enabled refugee integration: A research agenda', *Communications of the Association for Information Systems*, 44, 874–891.

Alam, K. and S. Imran (2015), 'The digital divide and social inclusion among refugee migrants: A case in regional Australia', *Information Technology & People*, 28(2), 344–365.

Alencar, A., K. Kondova and W. Ribbens (2018), 'The smartphone as a lifeline: An exploration of refugees' use of mobile communication technologies during their flight', *Media, Culture & Society*, 41(6), 828–844.

Alencar, A. and V. Tsagkroni (2019), 'Prospects of refugee integration in the Netherlands: Social Capital, information practices and digital media', *Media and Communications*, 7(2), 184–194,

Aoki, Y. and L. Santiago (2018), 'Speak better, do better? Education and health of migrants in the UK', *Labour Economics*, 52, 1–17.

Bacinshoga, K.B. and K.A. Johnston (2013), 'Impact of mobile phones on integration: The case of refugees in South Africa', *The Journal of Community Informatics*, 9(4). https://doi.org/10.15353/joci .v9i4.3142.

Baldassar, L., M. Nedelcu, L. Merla and R. Wilding (2016), 'ICT-based co-presence in transnational families and communities: Challenging the premise of face-to-face proximity in sustaining relationships', *Global Networks*, 16(2), 133–144.

Bandura, A. (2001), 'Social cognitive theory: An agentic perspective', *Annual Review of Psychology*, 52, 1–26.

Bauloz, C., Z. Vathi and D. Acosta (2019), 'Migration, inclusion and social cohesion: Challenges, recent developments and opportunities', in M. McAuliffe and B. Khadria (eds), *World Migration Report 2020*, Geneva, International Organization for Migration, pp. 167–188.

Benton, M. (2014), 'Smart inclusive cities: How new apps, big data, and collaborative technologies are transforming immigrant integration', Transatlantic Council on Migration, Migration Policy Institute.

Benton, M. (2019), 'Digital litter: The downside of using technology to help refugees', Migration Policy Institute, Migration Information Source, 20 June, accessed 9 February 2020 at www.migrationpolicy .org/article/digital-litter-downside-using-technology-help-refugees.

Benton, M. and L. Patuzzi (2018), 'Jobs in 2028: How will changing labour markets affect immigrant integration in Europe', Robert Bosch Stiftung and Migration Policy Institute Europe, Integration Futures Working Group,

Berry, J. (1997), 'Immigration, acculturation, and adaptation', *Applied Psychology: An International Review/Psychologie appliquée: Revue internationale*, 46(1), 5–34.

Bhakta Acharya, B. (2016), 'A systematic literature review on immigrants' motivation for ICT adoption and use', *International Journal of E-Adoption*, 8(2), 34–55.

Borkert, M., K.E. Fisher and E. Yafi (2018), 'The best, the worst, and the hardest to find: How people, mobiles, and social media connect migrants in(to) Europe', *Social Media + Society*, 1–11.

Bourdieu, P. (1980), 'Le capital social', *Actes de la Recherche en sciences sociales*, 31, 2–3.

Bradley, L., N. Berbyuk Lindström and S. Sofkova Hashemi (2017), 'Integration and language learning of newly arrived migrants using mobile technology', *Journal of Interactive Media in Education*, 1(3), 1–9.

Castles, S. (2002), 'Migration and community formation under conditions of globalization', *The International Migration Review*, 36(4), 1143–1168.

Castles, S., H. de Haas and M.J. Miller (2014), *The Age of Migration: International Population Movements in the Modern World*, Fifth edition, London: Palgrave Macmillan.

Castles, S., M. Korac, E. Vasta and S. Vertoved (2002), 'Integration: Mapping the field', Report for a project carried out by the University of Oxford Centre for Migration and Policy Research and Refugee Studies Centre, Home Office Online Report 28/03.

Chiswick, B.R. (2016), '"Tongue tide": The economics of language offers – Important lessons for how Europe can best integrate migrants', Institute for the Study of Labour (IZA), IZA Policy Paper No. 113, July.

Codagnone, C. and S. Kluzer (2011), 'ICT for the Social and Economic Integration of Migrants into Europe', Joint Research Centre (JRC), JRC Scientific and Technical Report, European Commission, Publications Office of the European Commission, Luxembourg.

Collin, S. (2012), 'ICTs and migration: The mapping of an emerging area of research', *The International Journal of Technology, Knowledge, and Society*, 8(2), 65–77.

Colucci, E., H. Smidt, A. Devaux, C. Vrasidas, M. Safarjalani and J. Castaño Muñoz (2017), 'Free digital learning opportunities for migrants and refugees: An analysis of current initiatives and recommendations for their further use', European Commission, Joint Research Centre (JRC), JRC Science for Policy Report, Luxembourg: Publications Office of the European Union.

Demireva, N. (2017), 'Immigration, diversity and social cohesion', Briefing, Oxford: The Migration Observatory, University of Oxford.

Díaz Andrade, A. and B. Doolin (2019), 'Temporal enactment of resettled refugees' ICT-mediated information practices', *Information Systems Journal*, 29, 145–174.

Diminescu, D. (2008), 'The connected migrant: An epistemological manifesto', *Social Science Information*, 47(4), 565–579.

Diminescu, D., A. Hepp, S. Welling, I. Maya-Jariego and S. Yates (2009), 'ICT supply and demand in immigrant and ethnic minority communities in France, Germany, Spain and the United Kingdom', Joint Research Centre (JRC) Technical Paper, European Commission, Office for Official Publications of the European Communities.

European Union (2006), Ministerial Declaration, Riga, Latvia, 11 June, accessed 23 March 2020 at https://ec.europa.eu/information_society/activities/ict_psp/documents/declaration_riga.pdf.

Eurostat (2019), 'Migrant integration statistics – labour market indicators', May, accessed 10 February 2020 at https://ec.europa.eu/eurostat/statistics-explained/index.php/Migrant_integration_statistics_%E2%80%93_labour_market_indicators.

Faist, T. (2018), 'A primer on social integration: Participation and social cohesion in the Global Compacts', Universität Bielefeld, Fakultät für Sociologie, Centre on Migration, Citizenship and Development (COMCAD), COMCAD Working Paper 161.

Farshbaf Shaker, S. (2018), 'A study of transnational communication among Iranian migrant women in Australia', *Journal of Immigrant & Refugee Studies*, 16(3), 293–312.

Fisher, K.E. (2018), 'Information worlds of refugees', in C.F. Maitland, *Digital Lifeline? ICTs for Refugees and Displaced Persons*, Cambridge, MA: MIT Press, pp. 79–112.

Fisher, K.E. and E. Yafi (2018), 'Syrian youth in Za'atari refugee camp as ICT wayfarers: An explanatory study using LEGO and storytelling', COMPASS' 2018, Menlo Park and San Jose, California, 20–22 June.

Froneberg, E., J. Wiebe, M.K. Phillips, I. Mannino, P. Podieg and A. Ertel (2015), 'Creating community: social media, a tool for integration', Brussels: Vesalius College.

Gagnon, J. and D. Khoudour-Castéras (2012), 'South–South migration in West Africa: Addressing the challenges of immigrant integration', OECD Development Centre, Working Paper No. 312.

Gelb, S. and A. Krishnan (2018), 'Technology, migration and the 2030 Agenda for Sustainable Development', Briefing note, Overseas Development Institute, Swiss Agency for Development and Cooperation, Federal Department of Foreign Affairs.

Gifford, S.M. and R. Wilding (2013), 'Digital escapes? ICTs, settlement and belonging among Karen youth in Melbourne, Australia', *Journal of Refugee Studies*, 26(4), 558–575.

Gillespie, M., L. Ampofo, M. Cheesman, B. Faith, E. Iliadou, A. Issa, S. Osseiran and D. Skleparis (2016), 'Mapping refugee media journeys: Smartphones and social media networks', Research Report, The Open University and France Médias Monde, 13 May.

Gonzalez, V., L.A. Castro and M.D. Rodriguez (2009), 'Technology and connections: Mexican immigrants in U.S.', *IEEE Technology and Society Magazine*, 28(2), 42–48.

Hebbani, A. and K. Van Vuuren (2015), 'Exploring media platforms to serve the needs of the South Sudanese former refugee community in Southeast Queensland', *Journal of Immigrant & Refugee Studies*, 13, 40–57.

Hepp, A., S. Welling and B. Aksen (2009), 'ICT for integration, social inclusion and economic participation of immigrants and ethnic minorities: Case studies from Germany', European Commission, Joint Research Centre (JRC), JRC Technical Notes, Luxembourg: Office for Official Publications of the European Communities.

Horn, H. (2015), 'Coding a way out of the refugee crisis: Apps for migrants to Europe are everywhere. But how much can they really help?', *The Atlantic*, 30 October, accessed 15 February at https://www.theatlantic.com/international/archive/2015/10/apps-refugee-crisis-coding/413377/.

Horst, H. and L. Hjorth (2013), 'Engaging practices: Doing personalized media', in S. Price, C. Jewitt and B. Brown (eds), *The Sage Handbook of Digital Technology Research*, London: Sage, pp. 87–101.

Huddleston, T. and J. Dag Tjaden (2012), *Immigrant Citizens Survey: How Immigrants Experience Integration in 15 European Cities*, Brussels: King Baudouin foundation and Migration Policy Group.

International Organization for Migration (IOM) (2019), *World Migration Report 2020*, Geneva: IOM.

International Telecommunication Union (2019), *Measuring Digital Development: Facts and Figures 2019*, Geneva: ITU Publications.

Jung, Y. (2014), 'What a smartphone is to me: Understanding user values in using smartphones', *Information Systems Journal*, 24, 299–321.

Kabbar, E.F. and B. Crump (2006), 'The factors that influence adoption of ICTs by recent refugee immigrants to New Zealand', *Informing Science Journal*, 9, 111–121.

Kaufmann, K. (2018), 'Navigating a new life: Syrian refugees and their smartphones in Vienna', *Information, Communication & Society*, 21(6), 882–898.

Khvorostianov, N., N. Elias and G. Nimrod (2011), '"Without it I am nothing": The internet in the lives of older immigrants', *New Media & Society*, 1–17.

Knight, W. (2017a), 'Biased algorithms are everywhere, and no one seems to care', *MIT Technology Review*, 12 July, accessed 24 March 2020 at www.technologyreview.com/s/608248/biased-algorithms-are-everywhere-and-no-one-seems-to-care/.

Knight, W. (2017b), 'AI programs are learning to exclude some African–American voices', *MIT Technology Review*, 16 August, accessed 24 March 2020 at www.technologyreview.com/s/608619/ai-programs-are-learning-to-exclude-some-african-american-voices/.

Komito, L. (2011), 'Social media and migration: Virtual community 2.0', *Journal of the American Society for Information and Technology*, 62(6), 1075–1086.

Lens, D., I. Marx and S. Vuji (2018), 'Does migration motive matter for migrants' employment outcomes? The case of Belgium', Institute of Labour Economics, Discussion Paper Series, IZA DP No. 11906.

Leurs, K. (2017), 'Communication rights from the margins: Politicising young refugees' smartphone pocket archives', *The International Communication Gazette*, 7(6–7), 674–698.

Leurs, K. and K. Smets (2018), 'Five questions for digital migration studies: Learning from digital connectivity and forced migration in(to) Europe', *Social Media + Society*, 1–16.

Leurs, K. and S. Ponzanesi (2018), 'Connected migrants: Encapsulation and cosmopolitanization', *Popular Communication: The International Journal of Media and Culture*, 16(1), 4–20.

Lewis, K. and S. Thacker (2016), 'ICT and the education of refugees: A stocktaking of innovative approaches in the MENA region', World Bank Education, Technology & Innovation, SBET-ICT Technical Paper Series, No. 17.

Lloyd, A. M.A. Kennan, K. Thompson and A. Qayyum (2013), 'Connecting with new information landscapes: Information literacy practices of refugees', *Journal of Documentation*, 69(1), 121–144.

Mason, B. (2018), 'Tech jobs for refugees: Assessing the potential of coding schools for refugee integration in Germany', Robert Bosch Stiftung, Migration Policy Institute Europe, Integration Futures Working Group.

McAuliffe, M., A. Kitimbo, A.M. Gossens and A. Ahsan Ullah (2017), 'Understanding migration journeys from migrants' perspectives', in M. McAuliffe and M. Ruhs (eds), *World Migration Report 2018*, Geneva: International Organization for Migration, pp. 171–189.

McAuliffe, M., A. Kitimbo and B. Khadria (2019), 'Reflections on migrants' contribution in an era of increasing disruption and disinformation', in M. McAuliffe and B. Khadria (eds), *World Migration Report 2020*, Geneva: International Organization for Migration, pp. 161–183.

Medina, D.A. (2015), 'Undocumented migrants in New York get ID cards to open bank accounts', *The Guardian*, 12 January.

Mitchell, M.M. (2002), 'Exploring the future of the digital divide through ethnographic futures research', *First Monday*, 7(11), accessed 15 February 2020 at https://journals.uic.edu/ojs/index.php/fm/article/view/1004/925.

Montag, C. and S. Diefenbach (2018), 'Towards homo digitalis: Important research issues for psychology and the neurosciences at the dawn of the Internet of Things and the digital society', *Sustainability*, 10 (2), 415. https://doi.org/10.3390/su10020415.

Pew Research Center (2019), *Mobile Divides in Emerging Economies*, Washington, DC: Pew Research Center.

Pinsky, O. and R. Steinhauser (2018), Online government services could change your life. But only if you have access to the internet, World Economic Forum Agenda, 16 July, accessed 9 February 2020 at www.weforum.org/agenda/2018/07/government-digital-equal-access-internet-mobile/.

Popivanov, B. an S. Kovacheva, 'Patterns of social integration strategies: Mobilising "strong" and "weak" ties of the new European migrants', *Social Inclusion*, 7(4), 28–38.

Portes, A. (1998), 'Social capital: Its origins and applications in modern sociology', *Annual Review of Sociology*, 24, 1–24.

Putnam, R.D. (1993), *Making Democracy Work: Civic Traditions in Modern Italy*, Princeton, NJ: Princeton University Press.

Salinas, S. (2018), 'More than half of the top American tech companies were founded by immigrants or the children of immigrants', CNBC, 30 May, accessed 10 February 2020 at www.cnbc.com/2018/05/30/us-tech-companies-founded-by-immigrants-or-the-children-of-immigrants.html.

Shah, S.F.A., J.M. Hess and J.R. Goodkind (2019), 'Family separation and the impact of digital technology on the mental health of refugee families in the United States: Qualitative Study', *Journal of Medical Internet Research*, 21(9). doi: 10.2196/14171.

Shahshahani, A. (2018), 'Government spying on immigrants in America is now fair game. What next?', *The Guardian*, 12 February, accessed 24 March 2020 at www.theguardian.com/commentisfree/2018/feb/12/government-spying-immigrants-america.

Smith, R. (2018), 'IBM created the world's first smartphone 25 years ago', World Economic Forum Agenda, 13 March, accessed 8 February 2020 at www.weforum.org/agenda/2018/03/remembering-first-smartphone-simon-ibm/.

Sofkova Hashemi, S., N. Berbyuk Lindström, L. Bartram and L. Bradley (2017), 'Investigating mobile technology resources for integration: The technology–pedagogy–language–culture (TPLC) model', in *Proceedings of mLearn*, Larnaca, Cyprus, 30 October to 1 November.

Triandafyllidou, A. (2018), 'The migration archipelago: Social navigation and migrant agency', *International Migration*, 57(1), 5–19.

United Nations Department of Economic and Social Affairs (UN DESA) (2018), *United Nations E-Government Survey 2018: Gearing E-Government to Support Transformation towards Sustainable and Resilient Societies*, New York: United Nations.

Vertovec, S. (1999), 'Conceiving and researching transnationalism', *Ethnic and Racial Studies*, 22(2), 445–462.

Vertovec, S. (2004), 'Cheap calls: The social glue of migrant transnationalism', *Global Networks*, 4, 219–224.

Wiesbrock, A. (2011), 'The integration of immigrants in Sweden: A model for the European Union?', *International Migration*, 40(4), 48–66.

Witteborn, S. (2015), 'Becoming (im)perceptible: Forced migrants and virtual practice', *Journal of Refugee Studies*, 28(3), 350–367.

Žajdlea Hrustek, N., A. Prosser and V. Dušak, (2016), 'A multidimensional model of e-inclusion and its implications', *Proceedings of the Central and Eastern European eDem and eGov Days 2016*, Austrian Computer Society, 435–446.

Zapata-Barrero, R. (2017), 'Interculturalism in the post-multicultural debate: A defence', *Comparative Migration Studies*, 5(1). DOI:10.1186/s40878-017-0057-z.

13. Online technology for promoting the inclusion of refugees into higher education: a systematic review of current approaches and developments

Franziska Reinhardt, Olga Zlatkin-Troitschanskaia, Roland Happ and Sarah Nell-Müller

INTRODUCTION: HIGHER EDUCATION AS A FUNDAMENTAL HUMAN RIGHT

As a human right (United Nations, 1948, Art. 26), education (primary, secondary, and also post-secondary) plays an important role in the integration of migrants, including refugees, in a host country (UNHCR, 2019b;[1] Green, Chesla, Beyene, & Kools, 2018). While primary and secondary education can rely on established infrastructures and programs, the provision of tertiary education for refugees is not commonly guaranteed (Dryden-Peterson & Giles, 2010; Yildiz, 2019). Although access to tertiary education paves the way for future employment through the acquisition of professional skills and personal development, and increasingly leads to social integration in the host country (Cin & Doğan, 2020; UNHCR, 2020), it can also contribute to overcoming the shortage of urgently needed skilled professionals in industrial nations (Streitwieser & Miller-Idriss, 2017).

Integration in higher education also has a beneficial impact on the post-conflict phase in countries of origin. As education supports critical thinking and may turn students into future role models and peace builders, it can protect against possible radicalization and bring stability to crisis regions (UNHCR, 2019a). Educated people can return to their country and use their ideas and skills to reshape the society in the post-conflict phase (Handler, 2018).

Considering that only approximately 16 percent of refugees live in a developed country and that most refugees are in countries immediately adjacent to the conflict regions (UNHCR, 2018), alternative educational formats that are distinct from the local education system are gaining increasing importance. Online education is considered a promising approach for the integration of disadvantaged learner groups in policy and practice, but there is still an increased need for research on its effectiveness and efficiency (Unangst & DeWit, 2020; Colucci et al., 2017).

This chapter presents existing online educational approaches and discusses critically how they can be used to integrate refugees into higher education. First, technologies to make higher education more accessible are described; and, second, the current trend of open education and efforts to promote the educational integration of refugees are explained in more detail. Finally, possible future technologies and their suitability for the integration of refugee students into the educational system are discussed.

ONLINE TECHNOLOGY AS AN EMERGING TOOL IN EDUCATION FOR DIFFERENT NEEDS

In contrast to the efforts of traditional universities in terms of increasing internationalization, most higher education systems in fact still give low priority to the integration of refugees (Crosier & Kocanova, 2019; Baker et al., 2018). For instance, the recent Syrian conflict and the associated refugee influx with massive migration movements made it necessary to provide access to tertiary education for a large number of potential students with a refugee background. Though the Syrian refugees were considered to be highly educated, most of them had to interrupt their studies due to the ongoing conflict since 2011 (De Wit & Altbach, 2016; El-Ghali & Ghosn, 2019; Cerna, 2019) and thus aimed to continue their studies in their current host countries. Prospective students with a refugee background face specific institutional and personal barriers when entering higher education, such as language barriers, procedural barriers such as the rigid bureaucracy in the enrollment process, and financial costs of studying (Lambrechts, 2020). At British universities, for example, students with a refugee background used to be considered as international students (Bowen, 2014). Students with this status face enormously high tuition fees for a degree course that far exceed those of regular students.

Open access education 'for everyone' has been established as a promising approach both for universities to deal with the growing number of potential students and for the students themselves who, due to high costs or institutional hurdles, were unable to begin their studies at traditional universities. Thanks to the advances in online technology, these efforts have become more intense, as it is now easier to make educational resources freely available and to promote collaborative sharing across the globe (Knox, 2013).

The advantages of digital learning include the flexible availability through independence of time and place. Furthermore, it is characterized by cost efficiency, as an unlimited number of students can register digitally for a course, while participation in traditional university courses is usually limited to a certain number of students. The requirements include a functioning technological infrastructure such as a stable internet connection and a device to register for and participate in the online course and to study using video-based material. However, the described characteristics of digital learning do not apply to all offers. Within the wide range of digital education courses, there are just as many courses (especially MOOCs) with restricted access, a limited number of participants or no optional course schedule (Liyanagunawardena et al., 2019).

To provide access to higher education to refugees located in camps or in urban areas, various support measures have been established, such as scholarships for refugee students to study at regular universities in their current (e.g., DAFI program for Kenya) or a new host country (e.g., scholarships from York University) (UNHCR, 2019b). While scholarships allow only a limited number of refugee students to study at traditional universities, the recent measures include digital learning opportunities. Various approaches were developed (Crea, 2016) that differ in their level of online utilization: the provision of learning programs through blended learning in traditional universities, fully online learning platforms, bridging programs for higher education, or training of specific skills (El-Ghali & Ghosn, 2019). Another differentiating feature of the various measures is whether they are camp-based programs or host community programs (Russell & Weaver, 2019).

ICT & Mobile Devices and Required Skills for Refugees

The ubiquity of information and communication technology (ICT) in today's society is also evident in educational settings. Especially in higher education, the availability and use of technological devices is indispensable and is increasing rapidly. ICT influences the educational environment as well as how teaching or learning takes place. Computers and digital presentations have replaced the blackboard as a classic teaching tool, and learning takes place using computers or smartphones in a collaborative exchange with people from all over the world. Furthermore, e-learning and online learning have become commonplace in recent years. New technologies like virtual reality are predicted to be of great importance in the educational context (Coughlan et al., 2019). So far, they are not that frequently used by students, with the focus instead being on online multimedia, social media, and mobile technologies (Coughlan et al., 2019). ICT supports formal learning, implementing it within the traditional educational environment, but also has a great influence on informal learning by using community platforms like YouTube and other social media (El-Ghali & Ghosn, 2019).

Mobile technology, i.e., the use of devices such as smartphones or tablets, also constitutes a form of ICT. Mobile technology, in combination with additional technologies such as mobile learning applications, mobile games or quizzes, and videos is increasingly used in higher education (El-Ghali & Ghosn, 2019), particularly to enable access to education and to support vulnerable groups (Moser-Mercer, 2016; AbuJarour et al., 2019).

Digital technologies are of particular importance for refugees (Twight, 2018). Despite the enormous financial hardship they often face, many refugees still have a smartphone (AbuJarour & Krasnova, 2018) to communicate with their families and to store their personal information. This device therefore provides a sense of stability to refugees who have to deal with great insecurity and instability far away from home (Mendoza Pérez & Morgade Salgado, 2020). Smartphones are also perceived rather positively in the learning context of refugees, as they offer more flexibility than learning spaces with stationary computers (Burkardt, Krause & Velarde, 2019). Mobile technologies enable refugees to become active participants in the learning process and can strengthen their motivation (Fichten et al., 2020). Nevertheless, it should be noted that the use of smartphones for educational purposes requires more digital skills and a stable internet connection than private use. But studies have shown promising results in terms of language learning by migrants using the smartphone (Kukulska-Hulme et al., 2017).

Refugees who have higher levels of education, are of a younger age, and have a good English proficiency use ICT most frequently and in many different ways (O'Mara & Harris, 2016). Localization also has an impact on ICT use. While refugees in urban environments come into contact with a wider range of ICTs, refugees in rural and/or enclosed camps have more limited access to ICT (Taftaf & Williams, 2019, Reinhardt et al., 2018). As ICT skills, or digital literacy, vary widely between refugees, fostering those skills is a popular tool for the education of adult refugees. Increasing their ICT skills not only enables refugees to follow the course content and pass the course, it also has additional effects as refugees use the skills, for example, for future job perspectives (Burkardt, Krause & Velarde, 2019).

EDUCATIONAL EQUITY THROUGH DIGITAL INCLUSION

When broadening educational settings through developing ICT skills and using mobile devices for educational purposes, various online Open Access services have been developed to contribute to educational inclusion, in particular the integration of refugees.

Open Educational Resources (OERs)

Open Educational Resources (OERs) are defined as teaching, learning and research materials that are freely available to educators and learners. OERs are mostly based on existing course materials. OERs can have many formats, but are usually written materials sorted by topic that can be freely accessed, modified, and used by students and teachers (Coughlan et al., 2019; McGreal, 2020). OERs are often used in higher education, for example, as preparation or course materials (Knox, 2013). Their advantage lies in the lower cost compared with expensive textbooks or licenses for scientific literature. Furthermore, through modification and exchange between users, OERs may achieve a higher relevance than traditional resources. The constant control-and-feedback loop in the learning cycle can also guarantee quality.

Students who may not have access to traditional education can use this approach. The possibility for students to choose and use materials freely and to be free from possible institutional structures enables a shift from traditional learning to self-directed learning (Knox, 2013), which requires a high level of responsibility, motivation and higher cognitive resources (Lee, 2017) but is not necessarily associated with obtaining a diploma.

Students with a non-traditional background, however, need more specific support in the learning context, for instance, due to other personal obligations (e.g., family or work). Through self-directed learning, the feeling of social belonging may be less pronounced than in traditional higher education courses. A de-personalized learning environment is a decisive reason for the termination of online learning programs by refugee students (Reinhardt et al., forthcoming). Moreover, OERs are mostly based on materials from established, Western-culture-oriented institutions such as the 'OpenCourseWare' of MIT. The self-determined involvement with unknown academic concepts in a foreign language may lead less to positive experiences of cultural adaptation and rather to exclusion and isolation (Kanu, 2008, Kizilcec et al., 2013), making OERs less suitable for the integration of refugees. OERs have been designed to promote inclusion in higher education (UNESCO, 2012); in fact, they are often used by learners who already have a formal degree (Coughlan et al., 2019). Overall, there is a lack of substantiated research on how students use OERs in real contexts, and on whether OERs actually contribute to targeted educational inclusion.

Massive Open Online Courses (MOOCs)

MOOCs are video-based lectures with additional course material such as standardized tests or assignments (Baturay, 2015). Completed courses, often including an assignment, can be certified by the respective platform or institution (Höfler & Kopp, 2018). In addition to MOOCs, learning content is processed in face-to-face study or discussion groups and wikis, and it is recommended to use social media platforms and forums as additional means of communication with other online learners (Anderson, 2013; Conole, 2014).

Through MOOCs, a large number of students can access and use course content, attend and complete courses, most of which are offered by established universities (Ferguson, 2019). MOOCs in the early stages of their implementation in 2008 (Yuan & Powell, 2013) were considered an approach to meet the growing demands of higher education and especially to provide simplified access for a large number of people, especially in less developed regions (Weinhardt & Sitzmann, 2019; Lambert, 2020; Lee, 2017.) Initially, MOOCs could be accessed for free and from anywhere (Engle, Mankoff, & Carbrey, 2015).

Practical experience reveals, however, that providing high-quality courses was expensive and time-consuming, as the didactic requirements in MOOCs were greater than the mere reproduction of the traditional classroom setting (Ferguson, 2019). Moreover, the initial philanthropic efforts were also replaced by economically oriented goals (Yuan & Powell, 2013). The previously free access to MOOCs was restricted and user fees were implemented (Lambert, 2020). This development led to a move away from the original goal of widening participation, and rather to the exclusion of disadvantaged learners. Today's typical MOOC users are white, male, have a high educational level and come from Western countries (Rolfe, 2015).

Refugee students are considered part of the disadvantaged groups facing insurmountable barriers to accessing basic education (Shah & Santandreu Calonge, 2019; Reinhardt et al., 2018). Moser-Mercer (2014) describes three major types of hurdles: technological, cultural and linguistic barriers. First, technological accessibility makes a globally equal distribution of online learning opportunities impossible. Especially in rural areas, in refugee camps or in restricted conditions in the host countries of refugees, no easy access is guaranteed. In addition to the lack of technical equipment and learning space, a necessary internet connection is not always guaranteed (AbuJarour & Krasnova, 2018; Reinhardt et al., 2018). Second, the level of digital literacy required for the participation in MOOCs is often insufficient among refugees (Mason & Buchmann, 2016; Hatakka, Thapa, & Sæbø, 2019). (iii) Language problems occur, as most courses are offered in English (Weinhardt & Sitzmann, 2019). Thus, course content is more difficult to understand (Russell & Weaver, 2019) and the active participation in MOOCs-related discussion boards as an important principle of conversational learning is restricted (Ferguson, 2019).

Particularly the aforementioned fact that most courses correspond to Western academic culture can be a significant obstacle to the successful participation of MOOCs, especially for refugees. MOOCs are based on Western pedagogies and promote Western-culture-oriented concepts and values (Kizilcec, Saltarelli, Reich, & Cohen, 2017; Russell & Weaver, 2019). Refugees, however, often have great problems in their acculturation. Due to their past experiences, they show strong forms of acculturative stress when integrating into the new culture and society. This may also occur in educational settings, when refugees are confronted with unknown educational concepts and principles (Kizilcec, Saltarelli, Reich, & Cohen, 2017; Russell & Weaver, 2019).

Open Education Programs

Open Education Programs are another form of inclusion of refugees into higher education. Similar to open access education programs e-learning or blended learning approaches are offered in refugee camps and teach skills that go beyond post-secondary education. In most cases, the programs are covered by non-governmental organizations and UN agencies

(Burkardt, Krause & Velarde, 2019). They are less aimed at ensuring that a large number of people can participate in the course, but rather are usually designed specifically for the specific refugee camp context. This causes with further demanding circumstances, e.g., setting up workspaces in camps or realizing transportation and food during the courses (Burkardt, Krause & Velarde, 2019; Russell & Weaver, 2019).

In contrast to local programs, the goal of Open Education Programs is to provide access to educational content for refugees regardless of their current status and residence. They are therefore equally suitable for resettled refugees in host communities or refugees in camps (Castaño-Muñoz et al., 2018). The educational content is prepared online and usually offered in the form of MOOCs. It is usually possible to take an individual course or a number of courses in a specific subject area. In addition, a number of support strategies are offered, such as partnerships with business partners or mentoring and tutoring programs (Lambert, 2020).

There are various programs with different objectives.[2] A distinction can be made by considering the involvement of traditional higher education. For example, several programs offer preparatory courses for university studies (REIs2), others offer an integrated program with a combined curriculum of online and university studies (Kiron), and still others are comprehensive online programs that do not necessarily involve the attainment of a degree (Coursera for Refugees). An example of an outstanding program is the one from Borderless Higher Education for Refugees (BHER). An association of several international higher education institutions (from Canada, Kenya) provide onsite and online educational offers in refugee camps. These are accredited tertiary education offers that lead to the acquisition of credit points and to the graduation certificates. An important aspect of the program is the education of teachers, who then take over teaching in refugee camps. Furthermore the program has been extended to the point where university degrees such as a Bachelor's degree can be obtained (Boškić et al., 2018; Giles, 2018). Hatakka, Thapa, & Sæbø (2019) describe an example of a specific ICT-knowledge program and its impact on the participating refugee students. The skills learned (self organization and ICT skills) enabled the participants to open a micro business, although this was not intended as a learning objective of the course. Another example of an open education program is Kiron Open Higher Education (Kiron). Kiron provides refugees with access to higher education through an online study platform with several study tracks: Computer Science, Business and Economics, Mechanical Engineering, Social Work and Political Science. The credits earned by completing courses and modules are comparable to the standards of the European Credit Transfer and Accumulation System (ECTS). Certain courses are based on MOOCs, which are embedded in the predefined modules via external MOOC providers (Coursera, Udacity, Edix, Sailor). The study tracks are structured according to a predetermined curriculum, aligned to international educational standards and are comparable to traditional university study programs. If students can provide the necessary documents, applications and language certificates, an assisted transition from the online study phase to an offline formal education is possible. During the study program, Kiron cooperates with higher education institutions to support the transfer to a partner university to obtain an academic degree. Furthermore, the Kiron system offers a variety of online support programs such as professional language courses, counseling or mentoring programs. In addition, Kiron offers refugees a collaboration platform for all participants via the student forum (Kiron, 2020; Zlatkin-Troitschanskaia et al., 2018).

The advantage of online education programs over MOOCs lies in the possibility of obtaining a graduation certificate. The goal of online education programs is the possible recognition of their certificates in formal higher education. However, most of these promising ambitions are not met so far (El-Ghali & Ghosn, 2019). One reason is that such programs are provided by not-state-run actors what makes formal certification more difficult, further the quality and level of access is difficult to determine, and there is little information on their long-term impact (Pherali & Abu Moghli, 2019). Moreover, the question of financing and the long-term implementation of these programs is still unresolved and can usually only be realized through extensive donations (Lambert, 2020).

Open Access Education as the Remedy or rather an Extension of the Digital Divide?

In addition to the aforementioned difficulties in formal recognition of the online education degree, research has shown that purely online education programs were not very effective yet. Often, less than 10 percent of participants successfully completed the online education program (Gomez-Zermeno & Aleman de la Garza, 2016; Xing et al., 2016). The number of dropouts is significantly higher than within conventional educational programs (Xiong et al., 2015). The completion rates for refugee students only indicate higher completion rates, but still to a small extent (Deribo et al., 2019).

A blended learning approach is increasingly recommended for refugees (Russell & Weaver, 2019; Coughlan et al., 2019; Castaño-Muñoz et al., 2018), because the entry conditions are low threshold in contrast to higher education institutions, where the lack of documentation or educational certificates usually makes access difficult or completely impossible. Dropout rates are lower compared to the high dropout rates of MOOCs (Deribo et al., 2019; see also Reinhardt et. al. forthcoming). MOOCs are more likely to be used by those learners who are better educated (Greene et al., 2015) and have better access to education (Gomez-Zermeno & Aleman de la Garza, 2016; Reinhardt et al., 2018).

Overall, this implies that online resources may expand the educational inclusion only of advantaged learners rather than integrate disadvantaged groups such as refugees. Behind these promising developments in higher education, there are specific risks for a digital divide among refugees, which is reflected in the differences in access, usage and reception of such opportunities between learner groups (Rohs & Ganz, 2015). Not everyone, and especially not refugees, has equal access to digital services, or digital media (access gap) (AbuJarour et al., 2019), not every refugee student can use the available information to the same extent (usage gap, Mason & Buchmann, 2016), and not every refugee student has the ability to take advantage of the available services (reception gap) (Hatakka, Thapa & Sæbø, 2019). MOOCs and OERs alone, therefore, cannot solve the specific educational problems of refugees in disadvantaged situations. Nevertheless, MOOCs or OERs embedded in an established educational program present a promising approach (Russell & Weaver, 2019).

FUTURE DIRECTIONS IN HIGHER EDUCATION AND THE POTENTIAL IMPACT ON DISADVANTAGED LEARNER GROUPS

The specific challenge of the integration of refugee students is an example of the changing higher education landscape due to transforming social conditions and global changes (Makrakis & Kostoulas-Makrakis, 2020). The Open Education efforts form a solid basis for achieving the goal of education for all. However, in terms of online education programs, there is a risk that these efforts may increase rather than reduce the digital divide. The particularly strong heterogeneity of refugee students in terms of educational experiences and language skills (Zlatkin-Troitschanskaia et al., 2018) presents particular challenges, and requires new approaches. Due to widely diverging knowledge, there can be no generally valid recommendation for the integration of refugees into higher education; instead, the individual needs of each student must be adequately addressed. This can be most effectively achieved by using ICT and the principle of technology enhanced learning.[3]

In addition, there are new developments such as adaptive learning environments, artificial intelligence (AI), machine learning or learning analytics that may have a significant impact on future higher education, as the new technology is already available and can be used in educational practice. The Open University is already using adaptive learning environments[4] to improve the quality of teaching and learning and – in terms of research – to gain a better understanding of the complex relationships in education (Rienties & Jones, 2019).

In terms of integration into higher education, these developments could have a significant positive impact as they could address the heterogeneity of disadvantaged students such as refugees (Coughlan et al., 2019). Various levels of achievement, different languages and language skills can be taken into consideration. An increase in motivation and learning can be achieved more effectively with an individualized solution than with a one-size-fits-all approach. As refugees belong to the high-risk groups with regard to dropout (Deribo et al., 2019), an early intervention strategy is required. Data analytics could also form the basis for possible practice recommendations, as the data allow conclusions to be drawn about the learning process of refugee students in a study course.

Technologies for speech recognition (speech to text and vice versa) are another potential use of adaptive learning systems for refugee students. This would address the low language proficiency of refugee students and possibly facilitate the work in courses (via automatically translating texts and audio). This approach is already used to make it easier for disabled students to follow the course content, and these technologies are also already widely used in smart systems such as Amazon's Alexa (Coughlan et al., 2019). These examples demonstrate that the new technologies can be used in education as a means of integrating disadvantaged learning groups.

The open and free availability of the required technologies might be the main obstacle, since there is an enormous economic interest in the development of these adaptive systems, and less motivation to achieve philanthropic goals such as the integration of refugees. Furthermore, the use of new ICTs in higher education teaching requires a rethinking of the way knowledge is imparted in higher education. The use of, for instance, mobile devices as a tool should be fully justified today, learning materials should be fully available in electronic form, software and hardware should be open to continuous development (Fichten et al., 2020; El-Ghali & Ghosn, 2019). To learn to use adaptive algorithms also requires new technological skills that need to be acquired by students (French & Poole, 2020).

Especially with regard to the reported low digital literacy of many refugee students und uncertainties in the handling of ICTs (AbuJarour et al., 2017), the trend should be towards enabling highly complex processes and procedures to compensate for the lack of knowledge and skills. It should be ensured that new technologies do not serve to increase the digital divide, but rather to support marginalized groups and to identify at-risk students early on (Niemi, 2020). The prototypical refugee student of the past does not have to be the university student of the future (Black et al., 2019).

CONCLUSION

The UNHCR has as a declared goal to increase enrollment into college to up to 15 percent for eligible refugee students by 2030. This goal is to be achieved by promoting technical and vocational training, scholarships and targeted programs in the host countries (UNHCR, 2019b). Though online education is not explicitly mentioned in the list, it is a suitable way to reach many people in different places, promoting a global network of students in higher education. This approach offers more than just the development of students' individual skills as it offers them a unique opportunity to learn from each other, to connect and to solve societal issues (Black et al., 2019).

Online education, however, is only one of the ways in which the use of technology in higher education may shape the future. Through individual adjustment, adaptive learning methods represent a possibility for the integration of marginalized groups into higher education. Especially vulnerable learner groups do not have the possibility to afford certain learning tools. It is therefore a concern to offer high-quality education to everyone, and adaptive learning systems can make an important contribution to this objective (Niemi, 2020). Using new ICTs in higher education requires a framework that provides clear definitions and provisions with regard to ethical components, legal conditions, security and human rights (French & Poole, 2020).

In 2020, the global pandemic called for a radical shift in the way traditional university courses are offered. Online education was considered to be an appropriate solution to maintain the necessary educational provision of universities in the context of the pandemic restrictions. On the one hand, the widespread usage of online education in the context of an academically accredited degree program could lead to a rethinking of the acceptance of online education. Disadvantaged groups such as refugee students, who were previously denied access to university courses but who have taken own responsibility for online studies are expected to benefit from these special circumstances. At the same time, the current situation can promote social inclusion and social equity in particular. To achieve this, access to online courses should not only be open exclusively to enrolled students; opportunities must be created for easier access in these exceptional times. At the same time, the mandatory shift to online education creates an infrastructure that can be used even after the end of the worldwide restrictions. This will also have positive effects for disadvantaged students, since possible effort and costs would probably have been less spent on integration efforts of disadvantaged groups. And finally, the exceptional current situation also leads to a special situation for educational research. In this way, large representative samples can currently be used to examine challenges and special needs in more detail to identify multi-factorial interrelations that also allow conclusions to be drawn about special challenges faced by vulnerable learners.

NOTES

1. The UN Refugee Agency will almost double the budget in 2020 for all educational activities compared with the previous year and invest about USD 42 million in programs related to education (UNHCR, 2020).
2. A more detailed overview of individual initiatives can be found in Streitweiser et al. (2019) or on the platforms of specific projects such as InZone (InZone, 2020).

3.   Technology enhanced learning (TEL) comprises all innovations and technical resources in the education sector, which are used as learning, instruction and support systems (Wang & Hannafin, 2005).
4.   Two major topics in the use of adaptive systems are (i) the paradigms of personalization of the learning process and (ii) the data analytics that are derived by the technology to immediately use and utilize learning process data (Rienties & Jones, 2019).

REFERENCES

AbuJarour, S. and H. Krasnova (2018), 'E-Learning as a means of social inclusion: The case of Syrian refugees in Germany', *Proceedings of the Americas Conference on Information Systems*.

AbuJarour, S. A., M. Wiesche, A. Díaz Andrade, J. Fedorowicz, H. Krasnova, S. Olbrich, C. Tan, C. Urquhart and V. Venkatesh (2019), 'ICT-enabled refugee integration: a research agenda', *Communications of the Association for Information Systems*, **44**, 874–890.

Anderson, T. (2013), 'Promise and/or peril: MOOCs and open and distance education', *Commonwealth of Learning*, **3**, 1–9.

Baker, S., G. Ramsay, E. Irwin and L. Miles (2018), '"Hot", "Cold" and "Warm" supports: Towards theorising where refugee students go for assistance at university', *Teaching in Higher Education*, **23** (1), 1–16.

Baturay, M. H. (2015), 'An overview of the world of MOOCs', *Procedia-Social and Behavioral Sciences*, **174**, 427–433.

Black, D., C. Bissessar and M. Boolaky (2019), 'Online education as an opportunity equalizer: The changing canvas of online education', *Interchange*, **50** (3), 423–443.

Boškić, N., Sork, T. J., Irwin, R., Nashon, S., Nicol, C., Meyer, K., & Hu, S. (2018). Using technology to provide higher education for refugees. In E. Jean-Francois (ed.), *Transnational Perspectives on Innovation in Teaching and Learning Technologies* (pp. 285–304). Leiden: Brill Sense.

Bowen, A. (2014), 'Life, learning and university: An inquiry into refugee participation in UK higher education', accessed 1 January 2020 at http://eprints.uwe.ac.uk/22944/1/ALBowen_PhD%20thesis_final.pdf.

Burkardt, A. D., N. Krause and M. C. R. Velarde (2019), 'Critical success factors for the implementation and adoption of e-learning for junior health care workers in Dadaab refugee camp Kenya', *Human Resources for Health*, **17** (1), art. 98.

Castaño-Muñoz, J., E. Colucci and H. Smidt (2018), 'Free digital learning for inclusion of migrants and refugees in Europe: A qualitative analysis of three types of learning purposes', *The International Review of Research in Open and Distributed Learning*, **19** (2), 1–21.

Cerna, L. (2019), 'Refugee education: Integration models and practices in OECD countries', *OECD Education Working Papers*, 203, Paris: OECD Publishing.

Cin, F. M. and N. Doğan (2020), 'Navigating university spaces as refugees: Syrian students' pathways of access to and through higher education in Turkey', *International Journal of Inclusive Education*, **2**, 298–312.

Colucci, E., H. Smidt, A. Devaux, C. Vrasidas, M. Sarfar Jalani and J. C. Munoz (2017), *Free Digital Learning Opportunities for Migrants and Refugees: An Analysis of Current Initiatives and Recommendations for Their Further Use*, Luxembourg: Publications Office of the European Union.

Conole, G. (2014), 'A new classification schema for MOOCs', *The International Journal for Innovation and Quality in Learning*, **2** (3), 65–77.

Coughlan, T., K. Lister, J. Seale, E. Scanlon and M. Weller (2019), 'Accessible inclusive learning: foundations', in R. Ferguson, A. Jones and E. Scanlon (eds), *Educational Visions: Lessons from 40 years of innovation*, London: Ubiquity Press, pp. 51–73.

Crea, T. M. (2016), 'Refugee higher education: Contextual challenges and implications for program design, delivery, and accompaniment', *International Journal of Educational Development*, **46**, 12–22.

Crosier, D. and D. Kocanova (2019), *Integrating Asylum Seekers and Refugees into Higher Education in Europe: National Policies and Measures*, Luxembourg: Publications Office of the European Union.

De Wit, H. and P. Altbach (2016), 'The Syrian refugee crisis and higher education', *International Higher Education*, **84**, 9–10.

Deribo, T., F. Reinhardt, O. Zlatkin-Troitschanskaia, R. Happ and S. Nell-Müller (2019), 'Refugees as a high risk group of dropping out of Online Education', paper presented at the Annual Meeting of the American Education Research Association (AERA), April 2019 Toronto/Canada.

Dryden-Peterson, S. and W. Giles (2010), 'Higher education for refugees', *Refuge: Canada's Journal on Refugees*, **27**(2), 3–9.

El-Ghali, H. A. and E. Ghosn (2019), 'Towards connected Learning in Lebanon', accessed 1 January 2020 at http://www.aub.edu.lb/ifi.

Engle, D., C. Mankoff and J. Carbrey (2015), 'Coursera's introductory human physiology course: Factors that characterize successful completion of a MOOC', *The International Review of Research in Open and Distributed Learning*, **16** (2), 46–68.

Ferguson, R. (2019), 'Teaching and learning at scale: Futures', in R. Ferguson, A. Jones and E. Scanlon (eds), *Educational Visions: Lessons from 40 years of innovation*, London: Ubiquity Press, pp. 33–50.

Fichten, C., D. Olenik-Shemesh, J. Asuncion, M. Jorgensen and C. Colwell (2020), 'Higher education, information and communication technologies and students with disabilities: An overview of the current situation', in J. Seale (ed.), *Improving Accessible Digital Practices in Higher Education*, Cham: Palgrave Pivot, pp. 21–44.

French, L. and M. Poole (2020), 'New competencies for media and communication in an AI era', in *Humanistic Futures of Learning: Perspectives from UNESCO Chairs and UNITWIN Networks*, Paris: UNESCO, pp. 136–140.

Giles, W. (2018), 'The borderless higher education for refugees project: Enabling refugee and local Kenyan students in Dadaab to transition to university education', *Journal for Education in Emergencies*, **4** (1), 164–184.

Gomez-Zermeno, M. G. and L. Aleman de la Garza (2016), 'Research analysis on MOOC course dropout and retention rates', *Turkish Online Journal of Distance Education – TOJDE*, **17** (2), 1–14.

Green, E., K. Chesla, Y. Beyene and S. Kools (2018), 'Ecological factors that impact adjustment processes and development of Ugandan adolescent immigrant females', *Journal of Child and Family Studies*, **27** (1), 34–46.

Greene, J. A., C. A. Oswald and J. Pomerantz (2015), 'Predictors of retention and achievement in a massive open online course', *American Educational Research Journal*, **52**, 925–955.

Handler, H. (2018), 'Economic links between education and migration: An overview', *Flash Paper Series, Policy Crossover Center Vienna-Europe*, **4**, 1–17.

Hatakka, M., D. Thapa and Ø. Sæbø (2019), 'Understanding the role of ICT and study circles in enabling economic opportunities: Lessons learned from an educational project in Kenya', *Information Systems Journal*, **30** (4), 1–35.

Höfler, E. and M. Kopp (2018), 'MOOCs und mobile learning', in C. de Witt and C. Gloerfeld (eds), *Handbuch Mobile Learning*, Wiesbaden: Springer, pp. 543–564.

InZone (2020), *Who we are*, accessed 01 January 2020 at https://www.unige.ch/inzone/who-we-are/.

Kanu, Y. (2008), 'Educational needs and barriers for African refugee students in Manitoba', *Canadian Journal of Education*, **31** (4), 915–940.

Kiron (2019), accessed 1 January 2020 at https://kiron.ngo/en.

Kizilcec, R. F., A. J. Saltarelli, J. Reich and G. L. Cohen (2017), 'Closing global achievement gaps in MOOCs', *Science*, **355** (6322), 251–252.

Knox, J. (2013), 'Five critiques of the open educational resources movement', *Teaching in Higher Education*, **18** (8), 821–832.

Kukulska-Hulme, A., M. Gaved, A. Jones, L. Norris and A. Peasgood (2017), 'Mobile language learning experiences for migrants beyond the classroom', in *Council of Europe Symposium*, Berlin: De Gruyter Mouton, pp. 219–224.

Lambert, S. R. (2020), 'Do MOOCs contribute to student equity and social inclusion? A systematic review 2014–18', *Computers & Education*, **145**, 103693.

Lambrechts, A. A. (2020), 'The super-disadvantaged in higher education: Barriers to access for refugee background students in England', *Higher Education*, **80** (4), 1–20.

Lee, K. (2017), 'Rethinking the accessibility of online higher education: A historical review', *The Internet and Higher Education*, **33**, 15–23.

Liyanagunawardena, T. R., K. Lundqvist, R. Mitchell, S. Warburton and S.A. Williams (2019), 'A MOOC taxonomy based on classification schemes of MOOCs', *European Journal of Open, Distance and E-learning*, **22** (1), 85–103.

Makrakis, V. and N. Kostoulas-Makrakis (2020), 'The quest for meaningful learning through ICTs', *Humanistic Futures of Learning: Perspectives from UNESCO Chairs and UNITWIN networks*, Paris: UNESCO, accessed 30 August 2021 via https://www.academia.edu/44072359/The_quest_for_meaningful_learning_through_ICTs.

Mason, B. and D. Buchmann (2016), 'ICT4Refugees: A report on the emerging landscape of digital responses to the refugee crisis', accessed 1 January 2020 at https://regasus.de/online/datastore?epk=74D5roYc&file=image_8_en.

McGreal, R. (2020), 'Open educational resources and global online learning', *Humanistic Futures of Learning: Perspectives from UNESCO Chairs and UNITWIN Networks*, Paris: UNESCO, pp. 122–126.

Mendoza Pérez, K. and M. Morgade Salgado (2020), 'Mobility and the mobile: A study of adolescent migrants and their use of the mobile phone', *Mobile Media & Communication*, **8** (1), 104–123.

Moser-Mercer, B. (2014), 'MOOCs in fragile contexts', in U. Cress and C. D. Kloos (eds), *Proceedings of the European MOOCs Stakeholders Summit*, Lausanne: EMOOCS, pp. 114–121.

Moser-Mercer, B. (2016), 'Participatory innovation – mobile and connected learning as a driving force in Higher Education in Emergencies', accessed 1 January 2020 at https://www.unige.ch/inzone/files/5214/8517/5103/2016ThinkPiecediscussion.pdf.

Niemi, H. (2020), 'Artificial intelligence for the common good in educational ecosystems', in *Humanistic Futures of Learning: Perspectives from UNESCO Chairs and UNITWIN Networks*, Paris: UNESCO, pp. 148–151.

O'Mara, B. and A. Harris (2016), 'Intercultural crossings in a digital age: ICT pathways with migrant and refugee-background youth', *Race Ethnicity and Education*, **19** (3), 639–658.

Pherali, T. and M. Abu Moghli (2019), 'Higher Education in the context of mass displacement: Towards sustainable solutions for refugees', *Journal of Refugee Studies*, https://discovery.ucl.ac.uk/id/eprint/10087985/.

Reinhardt, F., O. Zlatkin-Troitschanskaia, R. Happ, S. Nell-Müller and J. Fischer (2020), 'Refugee students' dropout factors for online Higher Education', *in review*.

Rienties, B. and A. Jones (2019), 'Evidence-based learning: Futures', in R. Ferguson, A. Jones and E. Scanlon (eds), *Educational Visions: Lessons from 40 years of innovation*, London: Ubiquity Press. pp. 109–125. DOI: https://doi.org/10.5334/bcg.g. License: CC-BY 4.0.

Rohs, M. and M. Ganz (2015), 'MOOCs and the claim of education for all: A disillusion by empirical data', *International Review of Research in Open and Distributed Learning*, **16** (6), 1–19.

Rolfe, V. (2015), 'A systematic review of the socio-ethical aspects of Massive Online Open Courses', *European Journal of Open, Distance and E-learning*, **18** (1), 52–71.

Russell, C. and N. Weaver (2019), *Language, Teaching, and Pedagogy for Refugee Education: Innovations in Higher Education Teaching and Learning*, vol. 15, Bingley: Emerald Publishing Limited.

Shah, M. A. and D. Santandreu Calonge (2019), 'Frugal MOOCs: an adaptable contextualized approach to MOOC designs for refugees', *International Review of Research in Open and Distributed Learning*, **20** (5), 1–19.

Streitwieser, B. and C. Miller-Idriss (2017), 'Higher Education's response to the European refugee crisis: Challenges, strategies and opportunities', in H. de Wit, J. Gracel-Àvila, E. Jones and N. Jooste (eds), *The Globalization of Internationalization*, London: Routledge, pp. 53–63.

Streitwieser, B., B. Loo, M. Ohorodnik and J. Jeong (2019), 'Access for refugees into higher education: A review of interventions in North America and Europe', *Journal of Studies in International Education*, **23** (4), 473–496.

Taftaf, R. and C. Williams (2019), 'Supporting refugee distance education: A review of the literature', *American Journal of Distance Education*, **34** (38), 1–14.

Twight, M. A. (2018), 'The mediation of hope: Digital technologies and affective affordances within Iraqi refugee households in Jordan', *Social Media + Society*, **4** (1), 2056305118764426.

UN High Commissioner for Refugees (UNHCR) (2018), 'Stepping up: refugee education in crisis', accessed 01 January 2020 at https://www.unhcr.org/5d08d7ee7.pdf.

UN High Commissioner for Refugees (UNHCR) (2019a), 'Global Trends, Forced Displacement', accessed 01 January 2020 at https://www.unhcr.org/steppingup/.

UN High Commissioner for Refugees (UNHCR) (2019b), 'Refugee education 2030: A strategy for refugee inclusion', accessed 01 January 2020 at https://www.unhcr.org/publications/education/5d651da88d7/education-2030-strategy-refugee-education.html.

UN High Commissioner for Refugees (UNHCR) (2020), 'Global Appeal 2020–2021', accessed 1 January 2020 at http://reporting.unhcr.org/sites/default/files/ga2020/pdf/Global_Appeal_2020_full _lowres.pdf.

Unangst, L. and H. de Wit (2020), 'Non-profit organizations, collaborations, and displaced student support in Canada and the USA: A comparative case study', *Higher Education Policy*, **2**, 223–242.

United Nations (1948), 'Universal declaration of human rights', *UN General Assembly*, **302** (2).

United Nations Educational, Scientific and Cultural Organisation (UNESCO) (2012), 'Paris OER Declaration', accessed 1 September at http://www.unesco.org/new/fileadmin/MULTIMEDIA/HQ/CI/WPFD2009/English_De claration.html.

Wang, F. and M. J. Hannafin (2005), 'Design-based research and technology-enhanced learning environments', *Educational Technology Research and Development*, **53** (4), 5–23.

Weinhardt, J. M. and T. Sitzmann (2019), 'Revolutionizing training and education? Three questions regarding massive open online courses (MOOCs)', *Human Resource Management Review*, **29** (2), 218–225.

Xing, W., X. Chen. J. Stein and M. Marcinkowski (2016), 'Temporal predication of dropouts in MOOCs: Reaching the low hanging fruit through stacking generalization', *Computers in Human Behaviour*, **58**, 119–129.

Xiong, Y., H. Li, M. L. Kornhaber, H. K. Suen, B. Pursel and D. D. Goins (2015), 'Examining the relations among student motivation, engagement, and retention in a MOOC: A structural equation modeling approach', *Global Education Review*, **2**, 23–33.

Yildiz, A. G. (ed.) (2019), *Integration of Refugee Students in European Higher Education Comparative Country Cases*. Izmir: Yaşar Üniversitesi Publications.

Yuan, Li and S. Powell (2013), *MOOCs and open education: Implications for higher education white paper*, CETIS, University of Bolton.

Zlatkin-Troitschanskaia, O., R. Happ, S. Nell-Müller, T. Deribo, F. Reinhardt and M. Toepper (2018), 'Successful integration of refugee students in Higher Education: Insights from entry diagnostics in an online study program', *Global Education Review*, **5** (4), 158–181.

14. Using ICTs to be here and not 'here': African migrants and religious transnationalism

Henrietta Nyamnjoh[1]

INTRODUCTION

With the advent of globalization, proliferation and appropriation of Information and Communication Technologies (ICTs) has led to the redundancy of assimilation in favour of migrants seeking to maintain a life of double engagement. This enables a foothold in both the host society and the home country (Grillo & Mazzucato 2008), as well as within multiple societies or social fields. Since its conception by Schiller et al. (1992), the concept of transnationalism has mainly focused on the study of migrants' abilities to navigate and stay connected in the host and home countries. While acknowledging that transnationalism between the host and home countries is predominantly the case, it does not, however, preclude the fact that migrants are increasingly engaged in transnational social fields beyond home. Additionally, a new rise in migrants' activities and changing dynamics in the development of the internet and social media shows that processes of transnationalism have moved beyond the home and host countries and are now enabled to a degree not previously possible (see also Kissau & Hunger, 2010: 246). Although ICTs by themselves cannot create a community, they enable the communication that is essential between migrants and transnational social fields and are pacesetters of transnational activities. According to Schiller, 'transnational social fields are networks of networks that link individuals directly or indirectly to institutions located in more than one nation-state' (2010: 112; see also Levitt, 2007: 107). These transnational social fields of migrants, she argues, rightly so, can 'contribute to, be shaped by or contest the local or transnational reach of various states' military, economic and cultural powers' (ibid.: 112).

This chapter explores how Cameroonian migrants in Cape Town, South Africa, owing to their appropriation of ICTs, have sought to pursue a religious transnational life and to navigate multiple transnational social fields either physically or virtually. Transnational engagement is by no means universal among migrants, but it is very common that a large majority of them will at least practise some form of transnationalism. The chapter goes beyond the often-one-dimensional notion of transnational practices that focuses exclusively on the host and home countries to explore migrants' religious transnational practices that focus away from home to other social fields. As pointed out by Schiller (2010: 111), too often, scholars of transnational migration or diaspora have often bound their unit of study along the lines of national or ethnic identities. Making a case against national or ethnic identities, Morales and Jorba (2010: 290) reiterate that 'transnational practices are certainly not restricted to links with homeland actors and social fields.' Hence, I seek not to reproduce a normative concept of religious transnationalism that evokes home and host countries as the locus of activities. Rather, I propose a broader approach able to embrace the ambiguity and multiplicity of experiences and social fields that pertain to the discourse of recalibrating 'home'. This shift is informed by arguments by Levitt and Schiller (2004: 4) that central to the project of transnational migration studies and to the scholarship on other transnational phenomena is a reformulation of the concept of society:

> Our analytical lens must necessarily broaden and deepen because migrants are often embedded in multi-layered, multi-sited transnational social fields, encompassing those who move and those who stay behind. As a result, basic assumptions about social institutions such as the family, citizenship, and nation-states need to be revisited.

Against this backdrop, this chapter looks at two major transnational practices that migrants' lives revolve around. It explores the transnational religious and economic practices among Cameroonian migrants in Cape Town, South Africa. The focus on these two practices is premised on the gospel of holistic health and material wealth that has a far greater appeal to migrants and the marginalized than the spiritual narcissism of mainline churches (Hunt, 2000). A transnational religious concept seeks to understand religious transnationalism for those who move and for those who stay put.

As regards transnational economic activities, I examine the religious concept of wealth, which is pivotal to prosperity gospel, and how it informs migrants' mobility to various churches to seek prophecies and blessings for the success of their business. The focus on spiritual health and material wealth underscores how migrants' lifeworld centres around improving their standard of living – a continuous life of economic activities to make ends meet (Nyamnjoh, 2020). I argue that migrants' lifeworld, especially for the marginalised, is a seesaw where they are constantly trying to make ends meet economically but equally seeking to make sense of the challenges they face by going closer to God to find answers for the shortcomings, failures and unsuccessful ventures of their migrant trajectories.

For these migrant entrepreneurs, faith and prosperity gospel are the driving forces that underpin the quest for wealth. This is because, as born-again Christians, poverty is not a virtue, since it denies all that Christ won through his death – that is, prosperity for born-again believers (Matthew, 1987). According to faith and prosperity gospel, neither Christ nor his disciples were poor. Consequently, hard-working believers – through their faith in God – are destined to reap the proper fruits of their labour and therefore provide the model lifestyle for all true believers (Price, 1984: 12). Health and wealth, Hunt (2000: 334) indicates, can be demanded and enjoyed immediately through the 'currency' of faith. It is this currency of faith and expectation of God's intervention that take the focus away from home to seek holistic health and wealth beyond home.

Studies in transnational research are informed by multi-sitedness. However, this is hindered by the inability of researchers to be present simultaneously for data collection. Hence, this research is informed by ethnographic methods that draw on interviews from migrants about their transnational accounts and practices. Also, non-participant observations of how migrants conduct their businesses – listening to phone conversations on how goods are ordered, and making travel arrangements and arrangements with transporters were followed closely. This method follows suggestions by Mazzucato (2010) that propose a simultaneous matched-sample methodology (SMS), used in a study of two-way flows, in this case between Cape Town and Lagos. Mazzucato's analysis sheds light on the asymmetric two-way flows between different regions that migrants carry out their transnational practices. This methodological tool covers transnational flows across localities. This chapter draws on data that were collected between 2016 and 2019 regarding migrants' transnational religious and economic activities. Interviews were conducted with over thirty immigrants, more than two hundred hours of informal conversations at their business locations/homes and following migrants to various Pentecostal Churches (PCs). There was a gender disparity concerning interviews on economic activities wherein more men were interviewed than women. Contrastingly, more women than men participated in interviews on religious activities. This could be attributed to the gender roles that men and women assume; where men are traditionally seen as the 'bread winners' of the family and women tend to take on the role of (pious) care giver (that may include praying for their families). Critical discourse analysis (CDA) was also important in understanding migrants' perspectives of holistic healing. Among the questions posed are, what meaning do migrants attribute to the sermons? Why do migrants have to travel far to seek intervention from other preachers in Nigeria, for instance? How different is the healing in Nigeria from that of Cape Town? Significantly, CDA was helpful towards answering the overarching questions: What informs migrants' religious transnational lives? How does Pentecostal religion position itself in the age of globalization? To what extent are migrants' economic and religious activities intertwined and how do they make sense of these activities and mitigate the challenges that emanates from them? What meaning do these lives have? What can be done to give them meaning?

In the first section of this chapter, I look at how migrants' religious and economic migration activities are in sync with the concept of transnationalism and how the practices are attainable owing to the domestication of ICTs. In the next section, I examine the actual practices of religious transnationalism for those who are mobile and how they navigate multiple states to make sense of their lived realities and find solutions to their challenges. I go further to look at virtual religious transnational practices for those unable to move because of lack of valid documentation to navigate nation-state borders.

MIGRANTS' ECONOMIC AND RELIGIOUS TRANSNATIONALISM

Schiller et al. (1995) have drawn on transnational studies to focus on various areas that have explored transnational social spaces, fields and formations, and have used the concept to connote everyday practices of migrants engaged in various activities. Transnationalism refers to processes that transcend international borders and therefore appear to describe vividly more abstract phenomena (Faist, 2010). Vertovec (2004) conceived transnational practices from the holistic point of view, as transformation in the 'socio-cultural', 'political' and 'economic' domains. This study speaks to that of Vertovec as it draws on two aspects discussed by Vertovec – cultural (religion) and economic. Perhaps, the work of Dahinden (2010) helps us to distinguish between more sedentary and nomadic forms of cross-border movements and ties. She looks at 'transnationalism through mobility', which is characterized by circulation and the perpendicular movement across borders where the focus is not necessarily between home and host country but focused on migrants' mobilities and the connections that result thereof. Such forms of mobilities (economic or religious) speak to the time–space compression that migrants can navigate corporeally or virtually. Dahinden's analysis therefore exemplifies how we could think through transnational mobilities contemporarily as well as incorporate migrants' transnational everyday practices and processes (mobile traders and those seeking holistic healing). In this regard, I draw on Sheller and Urry's (2006) 'new mobilities paradigm' that posits geographical mobility as a ubiquitous phenomenon of general societal importance to make sense of migrants' movements and to understand the proclivity to particular geographic spaces. Similarly, Morales and Jorba (2010: 277), although focusing on migrants' association, explore the multidimensionality of transnational practices and fields. They also give insights into the multiple forms of transnational exchanges that take place among individual migrants and the specific area to which transnational engagement is commonly oriented. Of interest to this chapter, therefore, is evidence of sustained transnational practices rather than sporadic ones. As indicated elsewhere (Nyamnjoh, 2019), Christianity, especially Pentecostalism, has not been left behind in the movement towards a global and transnational practice because religions have always crossed borders (Levitt, 2003). Whereas globalization has the concept of nation states, it has equally led to fluidities among which is the rise of the global phenomenon of Pentecostalism. These fluidities have become the hallmark for understanding religion and identities in a globalizing context and the experiences of those seeking healing as the spectrum of lived religion (Casanova, 2011; Gornik, 2011). Underlying these practices is access to communication technologies. How have ICTs made it possible for migrants to be transnational in multiple fields? Charismatic Pentecostal Churches (CPCs) in Africa, prior to the coming of the internet began with print media and audio recordings of audio cassettes and radio broadcast. The internet led to the expansion from radio to television broadcast, then mass production of video tapes to DVDs, and of late online streaming and even online churches that enables worshippers to log in via Facebook (see also de Witte, 2010), and recently Zoom. The appropriation of mass media achieves its mission of reaching out to adherents beyond the borders and eventually leading some to travel and physically participate in the service and meet the man of God. Hence the use of ICTs, especially, visuals and imagery as Coleman (2010) indicates, offer a combination of proximity and distance, the optic and the haptic that simultaneously evokes a sense of the wider world and an individual experience.

Focusing on Ghana, Marleen de Witte (2010) equally echoes how a transnationally circulating format of televangelism became paradigmatic for charismatic churches' media production. ICTs and social media have indeed offered PCs the possibilities to create their own identities and fashion their messages to the needs of the masses (Brown, 2011: 6–7). Arguing further, de Witte notes, 'through the local production, national and global circulation, the local consumption of Charismatic–Pentecostal television broadcasts and video tapes, a Christian Pentecostalized public crystallizes, that is thoroughly transnational' (2010: 88).

For the migrants in Cape Town whose transnational Pentecostalism is virtual or corporeal, holistic healing is received as the Holy Spirit travels easily via the mass mediated images, carried by transnational television networks and mediated by faith. The corollary has been a flexibility and adaptability of ICTs by PCs relevant to the local context as well as modifiable to ensure that the discourse reaches out to health- and wealth-seeking migrants on a global scale, effectively shaping globalization from below and above (Ribeiro, 2009; Vasquez, 2008). This underscores the fact that, in the twenty-first century, as Peggy Levitt (2007) posits, transnational activities have engendered practices where many people belong to several societies and cultures at once, and they will use religion to do so. Religion and religious practices are not simply reacting to global processes but generating global interconnectedness (Hüwelmeier & Krause, 2010).

As Levitt and Schiller (2004, see also Hüwelmeier and Krause, 2010; Vertovec, 2009; Levitt et al., 2003) intimate, such transnational 'ways of being and belonging' require relinquishing the geographically fixed approach to identity and community while recognizing the continuing significance of borders and state policies in mitigating and controlling movements. Consequently, transnational actors have steadily maintained religious or economic activities through the agency of individual actors in the different public/social spheres. The search for spiritual, physical and economic healing is both corporeal and virtual as will be illustrated later. This also undergirds the fact that appropriation of ICTs is done by migrants and evangelicals who use a constellation of ICTs to attain healing and to reach out to the faithful and a global audience respectively. While ICTs and social media have been credited in the globalization of religion, Thomas Csordas cautions about the importance to 'identify what travels across geographical and cultural space' (2007: 260), by assessing the portability of religious practices and the transposability of the religious messages. This call is quite significant as it underscores not only the transposability of the messages but shows the far-reaching effects of the messages on those who tune in globally.

Therefore, approaching migrants' religious practices from a transnational perspective takes into consideration the networks of individuals and institutions that emerge and how they operate. Significantly, it reveals the many layers and sites that make up the social fields that migrants occupy beyond the home and host country nexus (Levitt, 2007). When faced with inexplicable illness, physical or psycho-social, Pentecostalism seems to be the pit stop for answers. Similarly, when their economic gains are marginal or when business is not productive, the ripple effect is psycho-social and economic insecurities that cause migrants to seek Divine intervention to make sense of the challenges before them. Prayers for prosperity and business success or reliance on trans-local religious networks are often the immediate quick fixes and in serious cases migrants seek to travel in search of prophecies on their fortunes and destiny. From this perspective, Rijk van Dijk (2010: 102) explores the extent to which migrants' religious groups can be regarded as comforting niches, circles of security or coping mechanisms in a context of migration. According to Van Dijk (ibid.), contemporary Pentecostalism presents an ideology that fosters an entrepreneurial spirit of taking on challenges as a way forward for their modern believers by creating a context where private initiative is highly valued. While this is a similar case among migrant churches in Cape Town, when such entrepreneurial spirit is met with failure, migrants are forced to look to other churches, and at times, virtually, for solutions.

VIRTUAL/ONLINE TRANSNATIONALISM

For most interlocutors who are documented, religious or economic transnationalism is commonplace. This, nevertheless, is interspersed with virtual or online transnational activities such as following prayers online or attending church services virtually by watching TV channels of the desired church or Man of God that they feel is powerfully anointed. However, undocumented migrants do not enjoy such flexibility and are contented with virtual/online religious or economic transnational activities. Owing to the lack of a Permanent Residence (PR) permit, the mobility of undocumented migrants or asylum seekers/refugees has been severely hampered and they are thus reduced to being 'mobile while staying put' and to participate virtually in services and prayers across borders (see also Levitt, 2007: 111; Ammerman, 2007). For this group of migrants, the internet, Pentecostal TV channels and social media, especially WhatsApp and Facebook, have been domesticated to produce a similar effect for them, as for those who travel. Migrants' actions dovetail Thomas Lloyd's (cited in Porterfield, 2005: 162) findings that the Internet is not simply a convenience for believers; it allows people to reveal diseases and obsessions that they might not otherwise reveal or seek prayer for in face-to-face gatherings like church, hence blurs the lines between private and public lives. At the same time, the speed and virtual reality of Internet communication affect the way people think about divine communication.

Zora, for instance, is an asylum seeker in Cape Town who has a temporary refugee status that she needs to renew every three months. This makes it impossible for her to travel out of the country, hence she can only travel domestically. This, however, has not deterred her from being virtually transnational. According to Zora, she feels there is a python in her stomach that moves to her foot causing foot rot and a swollen right leg that she is constantly seeking healing for from different churches nationally, unsuccessfully. She resorted to virtual healing by following the healing church services of Prophet T. B. Joshua of Synagogue Church of All Nations and Apostle John Chi of Mercy Ministries International, on the TV from her home, and would place the affected area on the screen. Zora narrates:

> When the movement of the python is too much inside me, I turn to Mercy TV channel, I pull my TV closer to the bed, I lie on my side and lean my stomach against the TV. When the pastor is firing prayers, I can feel the python's movements as if it is struggling to come out. The thing will be disturbed and it will run inside my skin, physically I would feel like *vruu-vruu-vruu*, all over, all over, it would be going down on my toes, I would feel my toe, like this, *kekekekeke* again. Then after sometimes, it stays quiet and the pains also reduce. This is the same feeling I experience when I am in the presence of a TV channel that is full of anointing. Even at night I sleep like this (*with her stomach against the TV*), I lean my stomach against the TV, and it would disturb the python.[2]

For Zora, staying tuned to a Pentecostal TV channel all night long, being virtual and following the instructions of the Man of God on TV, enhances her Pentecostal transnational activities. Besides putting the affected area against the TV, she is also contented with online/TV prayers by the various men of God to whom she subscribes. Switching between God's prophets on various TV channels are sources of healing that Zora acknowledges to have benefited from. She avows that in this virtual context God is working through the TV screens. This also indicates God's involvement in her everyday life and the centrality of ICTs in her spiritual life that momentarily give her the ability to transpose and free herself from the encumbered body despite the constrains on her transnationality. Attesting to the success of getting healing through televised services, she intimates:

> Yeah, the TV helps a lot, I do get the anointing. Me, I can tell you what channels that have got anointing and channels which don't have. Immediately, you put Mercy TV, this one, when I'm coming from far, it's like my head, its turning, it's like I can fall, just the channel, the same thing with TB Joshua's, those channels have anointing. For instance, I had a terrible cough [that caused me to cough out some slimy stuff, which] was the okra of this python. I went to all the hospitals and they can't treat it, they gave me *Bactrim* that didn't work. Because I didn't have John chi (*Mercy TV*) on my TV. I went to my sister's house just to pray with John chi over the TV … I put my hand on the TV as the pastor instructed, as he was praying, I felt that spirit, I was delivered from that and the cough finish.[3]

Like Zora, Alice is a refugee hence cannot travel. But her TV is permanently tuned to Emmanuel TV of Prophet TB Joshua. She follows the service and fasts with the church when the Man of God calls for a fast. Leaving her TV on 24/7 helps her to connect her faith with those who experience similar challenges as her and she can receive healing through the broadcast. On days – especially Thursdays, when they re-broadcast Sunday service – that she is too tired to follow the service on TV she connects via YouTube on her phone and lies in bed. She usually does this on Thursdays, when they re-broadcast Sunday service. She explains,

> I try to close from work by seven pm on Thursdays in order to follow the service. It has a calming effect especially when I see how God is using Prophet TB Joshua mightily to heal many people. I believe in the power of healing, God is anointing him, and through those that are healed I also receive my healing. I connect my faith to theirs.[4]

For Alice, the distance is not a hindrance, but what counts is one's faith in order to get healing. Consequently, it is faith that propels her to believe in the healing powers of the tele-service and that she will be delivered from ancestral curses and spirit spouses that have intercourse with her at night in her dreams (see also Gifford, 2014), which she attests to having received healing. Social media platforms such as YouTube in addition to the TV channels of mega PCs constitute integral tools that enable Alice and Zora to live transnational virtual lives. According to Alice, leaving her TV on or following the service on YouTube helps her to live continuously in the spirit with God in order not to backslide into alcohol addiction, night clubbing and having multiple partners. She has also been able to pray for her finances because each time she earns some considerable amount of money there is a crisis at home that warrants that she sends it home. To this effect, she listened attentively to a prophecy that was given by someone on TV regarding his finances and acknowledged that it speaks to her as well: 'God is showing me something about your finances. Whatever the devil has stolen from you, God is going to multiply it for you.' According to Alice, all that she seeks healing from is an indication of spiritual emptiness that affects her destiny and inhibits her from getting married.

The 'trans-virtual' participation at religious services and prayers of Zora and Alice are common practice among migrant communities unable to move but facilitated by the global traffic of Pentecostal ideas owing to the appropriation of the various forms of ICTs – digital, social media and visual. As espoused by Vasquez and Marquardt (2003: 93), the constellations of these technologies 'obliterate distance' and 'dispense physical presence'. Through them, migrants can create transnational ways of religious belonging (Levitt & Schiller, 2004). To Zora and Alice, these communication technologies cease to be just a regular tool, but they acknowledge them as 'ritual objects that connect the spirit and human worlds' and 'contribute to the mediation process by increasing the potency of the ritual practices that bring spirits into presence' (Hüwelmeier & Krause 2010: 5). This is because the act of watching/looking constitutes a powerful practice of belief (Morgan, 1999).

I would like to underscore the fact that, while trans-virtual religious participation is not limited to migrants only, it is important to recognise that the demand for such forms of worship is higher among migrants who are attracted to PCs because most have been relegated to the margins of society to deal not only with challenges of the margins but, at times, are equally stripped of their dignity. Turning to the church not only helps them to make sense of their challenges but restores parts of their dignity and humanity.

Although Zora has not found a permanent relief from her illness, she has come to accept the intermittent healings that occur and sees this onset or persistence of suffering as part of religious life, while also celebrating relief from suffering as a sign of the power and meaning of her faith (Porterfield, 2005). Nevertheless, she also questions the supposedly spiritual powers of some of the big Prophets that she went to for crusade hoping to get healing but left without receiving healing despite what she paid (R5000, approximately 320 euros), to be given diplomatic status (it entails a seat at the front row, the pastor will lay hands on them and a one-on-one meeting with the pastor). She notes, unsatisfactorily, you come and stand in a long queue in front of the church, he still does same thing [*lay hands*].[5] I didn't fall down, I didn't feel dizzy. I was praying to even pretend as if I want to fall, so that the pastor will take his time to deliver me, you know. Nothing did happen to me. He doesn't have that time to say to you come and, like, sit in the office, to speak to you one-on-one. No, it doesn't happen like that. I was not happy because even if you wanted to explain to him your problem, he didn't even look at me.[6]

Judging from Zora's insights, requesting Christians to pay for a diplomatic service or status appears to be a façade to get money from migrants knowing that they would be enticed by the idea of sitting in front of the church and would have privileged access to the pastor, which was not the case.

Conversely, despite Zora's mobility in search of healing, she is also quite critical of some of the pastors who lack ethics regarding the way they prophesise to people in church. She notes, 'it's not everything you see about someone that you say in front of everybody in church.' Attending a crusade, Zora was prophesised to by a junior prophet as follows:

the spirit, which is in me, even if I befriend a man, the man will break down, the man will not become rich. The type of spirit that I have in me, is the type of spirit that even if I get married or I befriend the pastor, if he has a church, the church will break down. That is what he did say in front of the church, how can you be prophesying like that to someone? I decided I will never go to their crusade again.[7]

Zora echoes the sentiments of most of my interlocutors who like to have a one-on-one session with the pastor rather than publicize their problems to all in church.

Similarly, while informants are willing to pay to meet the pastors or buy healing products such as holy water, anointing oil, stickers and handkerchiefs, some are not convinced about the price tag of the goods and are quite critical of the pastor's self-enrichment. This notwithstanding, when migrants' business is failing, they are prepared to do what it takes to get prophesies from the pastors.

MIGRANTS' ECONOMIC ACTIVITIES: FROM TRANS-LOCAL TO TRANSNATIONAL

Jonas used to be a successful businessman who quit his job and used his savings to invest in a business that failed after two-and-a-half years and ate through all his savings. Dejected and contemplating suicide, he mustered courage to travel to Lagos in search of prophecy that would turn his life around. He went to Prophet T. B. Joshua of SCOAN to seek healing for financial prosperity and the restoration of stability in his business life. In response to my question on whether he saw his business as a failure, he said:

At that stage yes, it was a failure, for me, it was a failure. Life was not worth living anymore. I was owing about 3 months' house rent and the matter had to go to court. I was sent a paper to vacate the house and I didn't have food, I didn't have petrol money, I didn't have airtime, but I had to move. At that stage of life my option was to go to Nigeria because I believed that my dream could come true knowing very well that I was not going to have money in Nigeria, but I needed a divine intervention. So, I decided that if it's God will, I will go to Nigeria, maybe the prophet will be able to locate me and maybe prophesies that maybe my destiny still lies in Cameroon or the film industry, but I need to know. I had to sell one of my cars to go to Nigeria because I didn't have any money. So, I went to Nigeria and spent about 5 days. I managed to go into the church just once because of the population and it happened on the last Sunday prior to my departure the next day. We prayed with Prophet T. B. Joshua, it was a mass prayer, it's not like I had the opportunity to see him in person, no. So, I had the mass prayer, the next day I travelled back but still believing that the fact I've been there my prayer had been answered. So, when I came back it was still not easy. At the point I contemplated suicide and went to the beach to drown. You know it was so painful when I thought about my beautiful boys [...] But deep within, I trusted that my journey to Nigeria was not in vain. I went back to the film industry but was waiting to be called for an assignment. A week later I was called for work. I had about 6/7 days of commercial we were doing. [...] After that job, we had to shoot a movie for 6 months that I had to go out of the country. By the time I came back, I never remembered that I had problems before in my whole life. I believed God intervened at the right moment.[8]

Many scholars (van Dijk, 1997; Kalu, 2008; Mensah, 2008; Adogame, 2013) have shown how migrants veered to these churches in search of prophecy that forms part of healing to restore their economic success in whatever they do and to avoid sickness and economic downturn. It is perhaps in this respect that, upon his return to Cape Town and confronted by financial insecurity and unemployment, Jonas decided to sell his car and use the money to travel to Nigeria despite his financial challenges. His travelling to Lagos could also be steeped in the Pentecostal belief that Born-Again Christians have a right to enjoy prosperity by God's Grace.

Jonas' story underscores the inextricable links between migration, mobility and transnationalism on the one hand, and the fusing of fundamentalism with elements of a materialistic culture so that Christian scriptures are given a unique and unorthodox interpretation that privileges health, wealth and happiness, and refutes suffering as part of Christian life. For instance, Christians like Jonas, in search of prosperity, draw on the scriptures where they derive a sense of hope that their demands will be fulfilled by God – 'my God will supply every need of yours according to his riches of glory in Christ Jesus' (Philippians 4:19). The Gospel of Prosperity, which views wealth and material success as signs of God's love, is both preached and believed (Sabar, 2019; Gifford, 2004). Hence, journeys in search of pastors are not only viewed as the last hope to make meaningful change in their lives but are reinforced by these teachings that support the notion that migrants can reverse the downturn of their businesses through prayers and prophecies. This also begs the question whether such prosperity manifests in migrants' lives, or should they be contented with the view of the rich in the church in anticipation of God's miracle in their own lives? Although believers are made to oscillate between the dream of individual, ethically sound prosperity and the reminder of the difficulty to attain prosperity, as Meyer (2007) posits, the search for material wealth and health still leads migrants/believers to acknowledge the 'power' of God, as reduced to a spiritual force that can be tapped by various formulas in order to appropriate physical, psycho-social and material benefits (Hunt, 1998). What this has often created, it would appear, are worshippers who have reneged their right to critically think of solutions to their challenges than believe wholesale in the power of deliverance and healing.

While I have highlighted cases of migrants' transnational activities to seek health and economic wellbeing, it does not preclude the fact that pastors do not equally engage in transnational healing crusades to meet communities that have been following them virtually. These journeys are followed by the formation of a spiritual mentor–mentee relationship between the visiting pastors and Christians that continues over transnational social fields and spaces held by ICTs. This means that, inasmuch as migrants attend a Pentecostal church in Cape Town, they also maintain a transnational relationship with a pastor. In his absence, followers stay closely in touch with him via WhatsApp calls or they follow prayers/services that are streamed live on Facebook.

CONCLUSION

I have, in this chapter, attempted to show how the different religious manifestations occupy a particularly significant place in the lives of the migrants/believers, for it is through their faith and in their religious-based conviction that they actively seek material wealth and health. Challenging the transnational practices that have too often been linked to the home and host countries, I show how migrants' religious and economic transnational practices could be away from home and link migrants to different social fields and spaces. This is made possible through the highly mediatized era that is rapidly changing the landscape of Pentecostal Christianity, thanks to the time–space compression of globalization. This has resulted in Charismatic Pentecostalism being inherently transnational, having domesticated ICTs as tools (audio-visual and social media) that facilitate its mobility as well as the mobility of converts who subscribe to the teachings. The global relevance of transnational Pentecostalism is not only because of the scale of its success, but as Stephen Hunt (2000: 331, see also Nyamnjoh, 2019) indicates, it is 'because of its distinctive teaching related to divinely-blessed "health and wealth" which has enjoyed considerable acceptance in different parts of the world', especially among migrants. The data provided above support the fact that the concepts of wealth, health and deliverance help to provide consolation and support to victims of life's vicissitudes.

The study also offers explanations about the causality of helplessness and failures (Stoker, 2007: 168). Pentecostal soteriology of deliverance and healing, as such, fulfils a combination of spiritual and practical needs, and engages in a variety of activities within the psycho-social and economic sphere. These practices (religiosity and economy) form the essence of migrants' lived realities in their migration journeys, which have become part and parcel of their life world. For migrants, the journey is often begun virtually through the consumption of performances that are circulated through audio-visual technologies of how miracles are performed. TV, internet and social media provide new possibilities for engagement, communication, representation and imagination (Kissau & Hunger, 2010), for those who can move and those unable to. According to Derrida (2001: 63), 'there is no need any more to believe, one can see. But seeing is always organized by a technical structure that supposes the appeal to faith' and engenders a transnational practice. Religious transnationalism exists therefore to respond to migrants' existential needs of here and now, while making sense of life's challenges. This means that physical or virtual mobility in search of holistic healing is a search for a Christianity that is authentic to them.

NOTES

1. The research project this article draws from was funded by Templeton administered by Nagel Institute. Grant ID: 2016-SS150.
2. Interview with Zora, Cape Town: 29/06/2017. Words in italics are my addition for clarity.
3. Words in italics are my addition.
4. Interview with Alice, Cape Town: 31/08/2018.
5. Author's addition in italics.
6. Interview with Zora, Cape Town: 29/06/2017.
7. See footnote 9.
8. Interview with Jonas, Cape Town: 01/05/2017.

REFERENCES

Adogame, Afe (2013), *The African Christian Diaspora: New Currents and Emerging Trends in World Christianity*, London: Bloomsbury Academic.

Ammerman, Nancy. T (eds) (2007), *Everyday Religion: Observing Modern Religious Lives*, Oxford: Oxford University Press.

Brown, G. Candy (2011), *Global Pentecostal and Charismatic Healing*, Oxford: Oxford University Press.

Casanova, Jose (2011), 'Religion, the new millennium, and globalization', *Sociology of Religion*, **62** (4), 415–441.

Coleman, Simon (2010), 'Constructing the globe: a charismatic sublime', in G. Hüwelmeier, and Krause, K. (eds), *Travelling Spirits: Migrants, Markets and Mobilities*, New York: Routledge, pp. 186–202.

Csordas, Thomas (2007), 'Introduction: modalities of transnational transcendence', *Anthropological Theory*, **7** (3), 259–272.

Dahinden, Janine (2010), 'The dynamics of migrants' transnational formations: between mobility and locality', in Bauböck Rainer and Thomas Faist (eds), *Diaspora and Transnationalism: Concepts, Theories and Methods*, Amsterdam: Amsterdam University Press, pp. 51–71.

De Witte, Marleen (2010), 'Religious media, mobile spirits: publicity and secrecy in African pentecostalism and traditional religion', in G. Hüwelmeier and K. Krause (eds), *Travelling Spirits: Migrants, Markets and Mobilities*, New York: Routledge, pp. 83–100.

Derrida, Jacques (2001), 'Above all, no journalists! in religion and media', in H. de Vries and S. Weber (eds), *Religion and Media*, Stanford, CA: Stanford University Press, pp. 56–93.

Faist, Thomas (2010), 'Diaspora and transnationalism: what kind of dance partners?', in R. Bauböck and T. Faist (eds), *Diaspora and Transnationalism: Concepts, Theories and Methods*, Amsterdam: Amsterdam University Press, pp. 9–34.

Gifford, Paul (2004), *Ghana's New Christianity: Pentecostalism in a Globalizing African Economy*, London: Hurst & Company.

Gornik, Mark (2011), *Word Made Global: Stories of African Christianity in New York City*. Grand Rapids, MI: Wm. B. Eerdmanns Publishing Co.

Grillo, Ralph and Mazzucato, Valentina (2008), 'Africa Europe: a double engagement', *Journal of Ethnic and Migration Studies*, **34** (2), 175–198.

Hunt, Stephen (1998), 'Magical moments: an intellectualist approach to the NeoPentecostal faith ministries', *Religion*, **28** (3), 271–280.

Hunt, Stephen (2000), 'Winning ways: globalisation and the impact of the health and wealth gospel', *Journal of Contemporary Religion*, **15** (3), pp. 331–347.

Hüwelmeier, Gertrud and Krause, Kristine (2010), 'Introduction', in (eds), *Travelling Spirits: Migrants, Markets and Mobilities*, New York: Routledge, pp. 1–16.

Kissau, Kathrin and Hunger, Uwe (2010), 'The internet as a means of studying transnationalism and diaspora', in R. Bauböck and T. Faist (eds), *Diaspora and Transnationalism: Concepts, Theories and Methods*, Amsterdam: Amsterdam University Press, pp. 245–266.

Levitt, Peggy (2003), 'You know, Abraham was really the first immigrant: religion and transnational migration', *International Migration Review*, **37** (3), 847–873.

Levitt, Peggy (2007), 'Redefining the boundaries of belonging: the transnationalization of religious life', in Ammerman, T. Nancy (ed.), *Everyday Religion: Observing Modern Religious Lives*, Oxford: Oxford University Press, pp. 103–120.

Levitt, Peggy and Schiller, N. Glick (2004), 'Conceptualizing simultaneity: a transnational social field perspective on society', *International Migration Review*, **38** (3), 1002–1039.

Matthew, Stephen (1987), *Money Matters*, School of the Word Study Series: Harvestime.

Mazzucato, Valentina (2010), 'Operationalising transnational migrant networks through a simultaneous matched sample methodology', in Bauböck, R. and Faist, T. (eds), *Diaspora and Transnationalism: Concepts, Theories and Methods*, Amsterdam: Amsterdam University Press, pp. 205–226.

Mensah, Joseph (2008), Ghanaian immigrants use religion to affirm their identity. At https://yorkspace.library.yorku.ca/xmlui/handle/10315/29099; URI: http://hdl.handle.net/10315/29099 (accessed on 24 August 2021).

Morales, Laura and Jorba, Laia (2010), 'Transnational links and practices of migrants' organisations in Spain', in Bauböck, R. and Faist, T. (eds), *Diaspora and Transnationalism: Concepts, Theories and Methods*, Amsterdam: Amsterdam University Press, pp. 267–293.

Morgan, David (1999), *Visual Piety: A History and Theory of Popular Religious Images*, Berkeley, CA: University of California Press.

Nyamnjoh, Henrietta (2019), 'Globalized healing and evangelism: the quest for health and healing among Cameroonian migrants in Cape Town (South Africa)', in Bieler, Andrea, Isolde, Karl, HyeRan, Kim-Cragg, and Nord, Ilona (eds), *Religion and Migration: Negotiating Hospitality, Agency and Vulnerability*, Leipzig: Evangelische Verlagsanstalt, pp. 183–199.

Nyamnjoh, Henrietta (2020), 'Entrepreneurialism and innovation among Cameroonian street vendors in Cape Town', *African Identities*, **18** (3), 295–312.

Ogbu, Kalu (2008), *African Pentecostalism: An Introduction*, Oxford: Oxford University Press.

Porterfield, Amanda (2005), *Healing in the History of Christianity*, New York: Oxford University Press.

Price, Fred (1984), *High Finance: God's Financial Plan*, Tulsa, OK: Harrison House.

Ribeiro, L. Gustav (2009), 'Non-hegemonic globalizations: alternative transnational processes and agents', *Anthropological Theory*, **9** (3), 297–329.

Sabar, Galia (2019), 'Re-thinking the study of religion: lessons from field studies of religions in Africa and the African Diaspora', in Karen Lauterbach and Mike Vähäkangas (eds), *Faith in African Lived Christianity*, Brill: Rodopi, pp. 80–108.

Schiller, Nina Glick (2010), 'A global perspective on transnational migration: theorising migration without methodological nationalism', in Bauböck Rainer and Thomas Faist (eds), *Diaspora and Transnationalism: Concepts, Theories and Methods*, Amsterdam: Amsterdam University Press, pp. 109–129.

Schiller, Nina Glick, Basch, Linda and Blanc-Szanton, Cristina (1992), 'Transnationalism: a new analytic framework for understanding migration', *Annals of the New York Academy of Sciences*, 1992, 1–24.

Sheller, Mimi and Urry, John (2006), 'The new mobilities paradigm', *Environment and Planning A*, **38**, 207–226.

Stoker, Wessel (2007), 'Is God violent?: on violence and religion: philosophical reflections', *Currents of Encounter*, **31**, pp. 147–166.

Van Dijk, Rijk (1997), 'From camp to encompassment: discourses of transsubjectivity in the Ghanaian Pentecostal diaspora', *Journal of Religion in Africa*, **27** (2), pp. 135–159.

Van Dijk, Rijk (2010), 'Social catapulting and the spirit of Entrepreneurialism: migrants' private initiative, and the Pentecostal Ethic in Botswana', in Hüwelmeier, G. and Krause, K. (eds), *Travelling Spirits: Migrants, Markets and Mobilities*, New York: Routledge, pp. 101–117.

Vasquez, Manuel (2008), 'Studying religion in motion: a networks approach', *Method and Theory in the Study of Religion*, **20**, 151–184.

Vasquez, Manuel and Marquardt, F. Marie (2003), *Globalizing the Sacred: Religion across the Americas*. New Brunswick, NJ: Rutgers University Press.

15. ICTs and transnational householding: the double burden of polymedia connectivity for international 'study mothers'

Yang Wang and Sun Sun Lim

INTRODUCTION

Over the past several decades, the growing accessibility of information and communication technologies (ICTs), especially the mobile phone and the internet, have emancipated people from temporal and spatial constraints, and brought about unprecedented flexibilities in social communication (Fortunati, 2002; Turkle, 2011). For transnationally split households, ICTs assume particularly crucial roles as they constitute the only viable way to keep affective family bonds alive after physical separation (Paragas, 2009; Vertovec, 2004; Wilding, 2006). ICT-mediated communication, which enables information, emotions and care to transcend national boundaries, allows distant family members to stay updated on one another's emotional well-being and provide instrumental help anytime and anywhere (Parreñas, 2005; Uy-Tioco, 2007; Madianou & Miller, 2011; Baldassar, 2016).

Despite hallowed expectations around ICTs sustaining long-distance intimacy, transnational communication via ICTs is not without burdens. Indeed, they may also herald detrimental implications for family relationships. In particular, the 'polymedia' (Madianou & Miller, 2012) environment imposes an obligation of constant availability for mediated communication, and thus brings about emotional burdens and guilt of family members who are unable or unwilling to maintain virtual co-presence (e.g., Baldassar, 2016; Nedelcu & Wyss, 2016; Peng & Wong, 2013; Wilding, 2006). This constant connectivity afforded by ICTs also gives rise to digital surveillance of one another's whereabouts and daily life routines between transnational family members (e.g, Cabanes & Acedera, 2012; Chib, Malik, Aricat, & Kadir, 2014; Hannaford, 2015; Madianou, 2016). Moreover, ICTs can also introduce new dimensions of inequalities between transnational family members in terms of differential accessibility to technological infrastructures, quality of communication digital skills and so forth (e.g., Benítez, 2012; Cabalquinto, 2018; Horst, 2006; Lim, 2016; Parreñas, 2005).

In view of the equivocal implications of ICT-mediated communication for transnational households, this chapter seeks to provide a comprehensive insight into the role of ICTs in shaping life experiences of transnational families and their members. The analysis draws on both literature review and empirical evidence. Specifically, we bring together and review previous research on ICTs, mediated communication and long-distance intimacy, with special focus on various efforts to stay in perpetual contact with remote family members and the emergence of burdens, tensions as well as inequalities during transnational communication. The empirical case study presented in the chapter is derived from a two-year ethnographic research on ICT domestication by a group of Chinese migrant mothers in Singapore.

ICTs AS THE 'SOCIAL GLUE OF TRANSNATIONALISM': VIRTUAL CO-PRESENCE AND POLYMEDIA AFFORDANCES

After transnational relocation, mediated communication through ICTs becomes the only viable option for physically split households to keep alive affective family bonds (Horst, 2006; Paragas, 2009; Wilding, 2006). By virtue of the constant connectivity bestowed by ICTs, international migrants and their left-behind family members can remain involved in the mundane experiences of each other's everyday lives and perform familial responsibilities from afar on a daily basis (Parreñas, 2005; Wilding, 2006; Uy-Tioco, 2007; Madianou & Miller, 2011; Baldassar, 2016). Therefore, ICTs are widely venerated as the 'social glue of transnationalism' (Vertovec, 2004) that helps to preserve a coherent sense of familyhood characterized by collective memories, goals and welfare (Bryceson & Vuorela, 2002; Bacigalupe & Lambe, 2011).

According to previous studies, mediated communication strategies deployed by international migrants and their remote loved ones are largely determined by the technologies available in the context of certain eras (Wilding, 2006). There has been a transition from delayed and non-interactive connections such as cassette tapes and telegrams, to sporadic and expensive interactions through landline telephones, and all the way to synchronous and flexible communication via the internet and mobile devices. With this evolving panoply of technological tools, virtual co-presence via ICTs has become the habitual lifestyle of many contemporary transnational households (Madianou & Miller, 2012; Madianou, 2014; Wilding, 2006). With advanced ICTs, especially videocall software such as Skype, migrants and their left-behind families are able to 'live stream' each other's daily routines in a shared mediated space where information and emotions are reciprocated in a rich and continual manner as if they still live together (Wilding, 2006; King-O'Riain, 2015; Longhurst, 2015).

Research on ICTs and transnational communication has delved extensively into the ways in which transnational family members incorporate a variety of ICTs into their daily routines to reconstitute family intimacies. A major strand of current literature on ICTs and transnational householding delves into the renegotiation of parenthood, especially motherhood, in transnational families with children staying behind in their home countries. For migrant mothers, geographical relocation is always accompanied by the deviation from traditional gender ideologies of being major caregivers and nurturers for their children, which not only brings about emotional pain and a sense of guilt but also causes tensions in their relationships with children and other family members back home (Parreñas, 2005; Fresnoza-Flot, 2009; Chib *et al.*, 2014).

Instead of forsaking parental responsibilities after physical separation, migrant mothers often seek to reconstitute or even strengthen their gender identity as 'ideal mothers' via virtual involvement in diverse facets of their children's daily routines on a daily or even hourly basis (Uy-Tioco, 2007; Madianou, 2012; Peng & Wong, 2013). They rely heavily on a variety of ICTs to carry out regular conversations with their left-behind children so as to stay updated on their physical and emotional well-being. These mediated mothering practices often go into extremely detailed aspects of everyday life such as waking them up in the morning and saying goodnight before going to bed, checking their dressing and appearance for school, reminding them to have meals and take medicine on time, as well as providing comfort when they are depressed (Parreñas, 2005; Uy-Tioco, 2007; Madianou & Miller, 2011; Madianou, 2012; Peng & Wong, 2013; Chib *et al.*, 2014).

Apart from ritual and emotional interactions, migrant mothers also provide instrumental and practical assistance in their children's educational and professional lives. Specifically, mothers with school-age children tend to take advantage of the simultaneous communications enabled by advanced ICTs to provide real-time guidance in their homework (Parreñas, 2005; Madianou, 2012; Chib *et al.*, 2014). Mothers with adult children, in the same vein, often share work experiences and provide practical advice on professional development to their children (Peng & Wong, 2013). In the event that their children get into trouble or encounter unfair treatment in their home country, migrant mothers can also advocate for their children and safeguard their rights via ICTs (Peng & Wong, 2013).

In a similar vein, transnationally separated couples are also found to employ a variety of ICTs to keep abreast of each other's daily activities and reproduce conjugal intimacies across vast geographical distances. Specifically, they can 'hang out' and enjoy daily conversations with each other in mediated spaces in which sharing mundane bittersweet occurrences in everyday life and expressing love to compensate for the missing 'face-to-face' contact (Aguila, 2009; Neustaedter & Greenberg, 2012; King-O'Riain, 2015). Moreover, transnational relocation of one partner usually requires couples to reconstitute long-standing familial obligations and routines. Simultaneous communications via ICTs allow real-time coordination around family matters between distant husbands and wives, and thus facilitate the smooth functioning of family life despite the physical separation (Cabanes & Acedera, 2012; Kang, 2012; Peng & Wong, 2013).

With the wide prevalence of smart ICT devices and proliferation of digital applications, transnational households are increasingly enveloped by an environment of 'polymedia' (Madianou & Miller, 2012) wherein ICTs of different functionalities coexist and combine to offer integrated multi-faceted structure of affordances. Living a polymedia life, transnational families can strategically mobilize a constellation of ICTs to meet different communication needs and appropriately manage intimate relationships (Madianou, 2014; Madianou & Miller, 2012; Wilding, 2006).

For instance, webcam software such as Skype and Facetime are particularly suitable for intensive 'deep conversations' where distant family members participate in each other's everyday routines and provide real-time emotional support in a quasi-face-to-face manner (Cabalquinto, 2017; Francisco, 2013; King-O'Riain, 2015; Longhurst, 2015). By keeping webcams on, they know that their loved ones are 'right there' for them, ready to talk, listen and respond, even though they engage in different tasks respectively without exchanging a word for hours (Francisco, 2013; King-O'Riain, 2015; Longhurst, 2015). Meanwhile, text-based communication such as SMS, instant messaging and emails, are used as complements for visual and audial interactions to maintain continual greetings, updates and coordination (Peng & Wong, 2013; Thomas & Lim, 2011; Uy-Tioco, 2007; Vancea & Olivera, 2013). In addition, many transnational families also stay in 'ambient co-presence' on social network sites like Facebook and Twitter to get a sense of one another's quotidian bittersweet moments even without direct interaction (Madianou, 2016).

ICTs WITHOUT GUARANTEE OF POSITIVE IMPLICATIONS: BURDENS, TENSIONS AND INEQUALITIES

Despite hallowed expectations around ICTs strengthening long-distance intimacy, some transnational households acknowledge the inherent technological deficiencies and have concerns about their potential negative implications for family functioning and the well-being of family members. In particular, the constant connectivity afforded by ICTs means far less than the immediate company of loved ones, especially when mediated interactions are reduced to mere rituals without intensive emotional investment (Madianou & Miller, 2011; Cabanes & Acedera, 2012; Vancea & Olivera, 2013). Even when transnational family members manage to provide visual companionship to each other and engage in real-time emotional exchanges with webcam applications, they still fail to enjoy full intimacy in the mediated space since they cannot physically touch, hug or kiss each other (Madianou, 2012; King-O'Riain, 2015). Sometimes it is precisely the simulated togetherness that reminds them of actual geographical distances lying between them, and thus engenders accentuated feelings of guilt, anxiety and loneliness on both sides (Parreñas, 2005; Wilding, 2006; Uy-Tioco, 2007; King-O'Riain, 2015; Baldassar, 2016).

Moreover, while the polymedia environment grants migrants unprecedented opportunities for constant connection with remote loved ones, it also brings about emotional burdens of constant availability for mediated interactions (Horst, 2006; Longhurst, 2015; Su, 2015). Typically, after long periods of physical separation, some members of transnational families may have less motivation to share life stories with each other (Thomas & Lim, 2011; Cabanes & Acedera, 2012; Peng & Wong, 2013). In this context, over-enthusiasm and high expectations of reciprocity by certain family members in transnational communications often become burdens for their remote loved ones (Wilding, 2006; Baldassar, 2016; Nedelcu & Wyss, 2016). For example, children of transnational families often lament that their parents constantly request mediated communication and intervene excessively in their daily routines (Madianou & Miller, 2011; Longhurst, 2013; Nedelcu & Wyss, 2016; Pham & Lim, 2016). In a similar vein, migrant mothers were also burdened by continuous family responsibilities as mother, wife, daughter, sister etc. due to the increasing availability of long-distance communication (Rakow & Navarro, 1993; Thomas & Lim, 2011; Baldassar, 2016). Virtual co-presence, instead of facilitating the renegotiation of household power relations and labour division, often renders migrants and their family members to be virtually 'thrust back' into their family lives and hold on to their previous family roles (Uy-Tioco, 2007; Cabanes & Acedera, 2012; Madianou, 2012).

The constant connectivity bestowed by the affordances of polymedia can also descend into digital surveillance when transnational family members utilize various ICTs to monitor daily behaviours of left-behind children or spouses. Specifically, some migrant mothers tend to closely monitor the whereabouts and media use of their left-behind children, even to the extent of scrutinizing every corner of their rooms through webcam, checking their emails and having access to their passwords to online platforms, which children regard as an invasion of privacy and denial of autonomy (Madianou, 2012, 2016; Francisco, 2013; Chib *et al.*, 2014; Yoon, 2016). As for separated couples, ICTs also served as watchful eyes for interrogation into their dressing styles, physical movements, financial arrangements, social interactions and so forth, which tend to exacerbate estrangement and tensions between them instead of generating emotional closeness (Cabanes & Acedera, 2012; Hannaford, 2015).

Economic burdens of international migrants can also become aggravated due to the increasing convenience of transnational communication via ICTs. Especially with the prevalence of the mobile phone, migrants nowadays remain constantly accessible to their left-behind children and other relatives in the homeland who keep making financial and material demands such as asking for additional remittances or expensive gifts (Barber, 2008; Peng & Wong, 2013; Tazanu, 2015). In this sense, the possibility of real-time communication actually creates more burdens and anxieties to the already stressful life of migrants instead of providing emotional comforts and support (Barber, 2008; Peng & Wong, 2013; Tazanu, 2015).

Apart from emotional and economic burdens, ICTs are also found to create new inequalities within transnational households. Living in different geographical and socio-cultural contexts, transnational family members often have differential access to and experiences of using ICTs, which brings about various inequalities in mediated family communication as well as household power hierarchies (Benítez, 2012; Parreñas, 2005; Plüss, 2013). Specifically, previous studies have scrutinized different accessibility to ICT infrastructures between migrants and their left-behind families, usually with the former enjoying superior connectivity and higher quality of mediated communication than the latter (e.g., Cabalquinto, 2018; Cheong & Mitchell, 2016; Madianou, 2014; Parreñas, 2005). Gaps in digital skills and competency are also widely noted within transnational households where women and elder family members tend to be less technologically savvy and consequently marginalized in family communication (e.g., Kang, 2012; Pham & Lim, 2016). In addition, the increasingly convenient coordination of remittances also contributes to establishing and reinforcing authority and power hierarchies in the household (e.g., Madianou, 2012, 2014; Mckay, 2007; Parreñas, 2005). These inequalities, either clearly recognized or remaining unconscious to transnational families, can trigger self-deprecation and sense of insecurity among disadvantaged family members, which in turn, undermine family cohesion over time.

TRANSNATIONAL HOUSEHOLDING OF CHINESE 'STUDY MOTHERS': A CASE STUDY

To better illustrate the ambivalent implications of ICTs for transnational householding, we will present a case study on ICT use and transnational family communication of Chinese 'study mothers' (*peidu mama*) who accompany their school-going children to pursue education in Singapore (Huang & Yeoh, 2005, 2011; Wang & Lim, 2018). Study mothers originate from the burgeoning family arrangement of 'education migration' among middle- and upper-middle-class households across East Asia. In typical households undertaking education migration, mothers uproot and resettle along with their children who receive education in more developed, typically English-speaking countries, while the fathers remain in the home country to continue working and provide financial support (Huang & Yeoh, 2005; Lee, 2010; Waters, 2002). This trend is widely witnessed among middle- and upper-middle-class households across Asia, including 'astronaut families' from Hong Kong and Taiwan (e.g., Chee, 2003; Waters, 2002), '*kirogi* families' from South Korea (e.g., Jeong et al., 2014; Lee, 2010), and '*peidu* families' from Mainland China (e.g, Huang & Yeoh, 2005, 2011; Wang & Lim, 2018).

For *peidu* families, Singapore is one of the most popular destinations due to its cultural proximity to Chinese society, bilingual education system (English with Malay/Mandarin/ Tamil), incentive schemes for foreign students and their care-givers, as well as their relative affordability (Huang & Yeoh 2005, 2011). Chinese *peidu mama* started to flock to Singapore in rising numbers after the Singapore government issued a special type of long-term social visit pass to 'Mother or Grandmother of a child or grandchild studying in Singapore on a Student's Pass' (website of Singapore 'Immigration & Checkpoint Authority'). This gendered immigration policy classifies the study mothers as dependants but also nurturers of their children, and their legal status terminates once their children quit the overseas education or enter college.

Therefore, unlike business or skilled immigrants who enter the destination country under their own steam as citizens or potential citizens, study mothers mostly remain as 'transient sojourners' (Huang & Yeoh, 2005) who are marginalized in the host society and prepare to return to their homeland upon their children finishing studies abroad.

Compared to their husbands who retain their existing jobs and social relationships, study mothers tend to pay a higher price for education migration as they disrupt their lives and sacrifice their own career aspirations, social lives and conjugal intimacy to care for their children (Chee, 2003; Huang & Yeoh, 2005; Jeong et al., 2014). Becoming de facto 'single mothers' after transnational relocation, they are faced with broadened parenting obligations and domestic workloads, while at the same time being expected to continue fulfilling family responsibilities and maintaining affective bonds with left-behind family members. ICTs, which enable the flow of information, emotions and care to straddle geographical boundaries, thus play a critical role in these women's everyday negotiation of transnational family relationships (Wang & Lim, 2018).

Empirical data presented in the case study was collected from 40 study mothers through ethnographic methods including participant observation, semi-structured interviews and media diary (see also Wang, 2020; Wang & Lim, 2018). Analysis of qualitative data reveals that the Chinese migrant mothers studied were living a polymedia lifestyle where, on the one hand, they benefit from technological affordances of various ICTs to maintain long-distance intimacy and seek emotional supports, while on the other hand, also experienced multifarious emotional burdens and digital asymmetries originating from transnational communication.

The Bright Side of ICTs: Affective Streaming, Transnational Coordination and Emotional Compartmentalization

In the face of physical separation of households, the study mothers in this case study relied heavily on mediated communication to 'stream' their quotidian life experiences to their remote family members and properly coordinate multi-sited family life. Multifarious ICT devices and applications, each providing different technological affordances and deployed for different communication needs, were woven seamlessly into the fabrics of participants' transnational lives and helped them to negotiate family intimacies from afar.

For most study mothers in the current study, constant 'streaming' of quotidian life experiences to each other had become the default state of their transnational life (Wilding, 2006; King-O'Riain, 2015; Longhurst, 2015). Specifically, they reported continual or at least regular conversations via ICTs with their families back home, especially their left-behind husbands, to stay updated with each other's physical and emotional well-being. Instead of merely routinized greetings or discussions of important family events, these conversations mostly went deeply into mundane activities or feelings in their everyday lives such as what they ate for dinner, interesting stories they had heard, the weather and the like. The virtual presence of absent family members allowed information and emotions to flow smoothly across geographical borders, and thus reproduce and renegotiate family relationships on a continual basis.

Video calls were identified by many study mothers as their favourite approach of mediated communication with their left-behind family members. The visual affordances of webcams allowed participants and their remote loved ones to actually 'see' each other and engage in collective activities despite vast geographical distances (Wilding, 2006; King-O'Riain, 2015; Longhurst, 2015). During the observations, many participants were found to leave the webcam on for an extended period of time while engaging in domestic chores or other activities at home. Instead of being fully concentrated on the mediated conversations with their husbands or other close family members, they engaged in sports, replied to messages, performed skin care routines, undertook domestic chores and so forth along with discreet chatting, as if they were still living together and talking to each other from time to time. For both parties, this 'half-hearted' visual communication created the perceived reality of tangible involvement in each other's quotidian bittersweet encounters and a reassurance of togetherness without disrupting their regular life rhythms (Wilding, 2006; King-O'Riain, 2015; Longhurst, 2015).

Prolonged video calls were also supplemented by other mediated communication approaches, including voice and text messages, photos, hyperlinks, emojis and so on, to stay in continual or at least regular contact between study mothers and family members, especially when spatial and temporal conditions did not permit face-to-face or visualization interactions. A typical scenario is the tradition of 'food show' practised by many participants wherein they and their far-away family members, usually their husbands, took photos of every meal they had and exchanged these photos in a real-time manner (Wilding, 2006; King-O'Riain, 2015; Longhurst, 2015). Photos of daily clothing, natural landscapes, public events and so on were also frequently reciprocated via IM or SNS platforms, which allowed them to keep abreast of each other's mundane life experiences even without direct mediated interactions. Besides photos, these mothers were also found to share interesting or useful hyperlinks, videos, stories, riddles and so on with their remote loved ones. For transnational families like them, more important than the specific photos or fragmentary information exchanged was the continual existence of the interaction itself.

Apart from the aimless 'small talk' demonstrated above, transnational households of study mothers also employed ICTs to coordinate family activities and provide instrumental help for each other across vast geographical distances (Cabanes & Acedera, 2012; Kang, 2012; Peng & Wong, 2013). Owing to the synchronized interactions enabled by ICTs, the majority of domestic affairs, such as decision making, online purchase, schedule confirmation and so on, can all be negotiated and solved collectively between transnational family members in a real-time and precise manner. During the participant observation, such transnational coordination process often happened naturally and smoothly, without distracting other daily tasks of both sides. For example, when a participant coordinated with her husband to buy airline tickets back home, they virtually 'sat together' to solve a series of problems including selecting the optimal air route, entering passenger information online, re-sending the verification code, downloading a digital itinerary, etc. The entire coordination process happened on WeChat through a combination of video call and text messages. Neither of them sensed any inconvenience despite the vast physical distances between them.

Unlike migrants from less-well-off backgrounds, such as refugees and foreign domestic workers who could not always access or afford ICTs (e.g., Cabalquinto, 2018; Parreñas, 2005; Wilding, 2012), the study mothers in this study hailed from relatively well-off middle-class families and were unconstrained by such issues. Instead, these migrant mothers were located in an integrated environment of 'polymedia' where the social, cultural and emotional considerations of choosing between multiple ICTs are more salient than prosaic issues of cost and access (Madianou & Miller, 2012). In their transnational communication routines, these migrant mothers created idiosyncratic personal repertoires of ICT use in which each ICT device or platform is attached with unique symbolic meanings, and multiple available ICTs are strategically deployed in an alternate manner to manage various relationships. In particular, they were found to compartmentalize their mediated communication according to cultural inclinations, potential audiences and technological affordances of different ICTs (see also Wang & Lim, 2021).

In the context of mediated family communication, study mothers demonstrated strong capabilities in identifying implicit emotional cues of different ICTs and selecting the most appropriate communication approaches in different circumstances. For instance, visual and voice calls, which are known to facilitate 'focused interactions' (Burchell, 2015), were commonly used by these mothers when they were eager to express strong feelings and affection to intimate family members (see also Longhurst, 2013, 2015; King-O'Riain, 2015). By contrast, IM and SNS platforms, which often convey 'frequent and short interactions' (Licoppe, 2004) and 'ambient co-presence' (Madianou, 2016), were likely to be chosen when they wanted to exchange pragmatic and fragmentary information without devoting intense emotions (see also Licoppe, 2004; Köhl and Götzenbrucker, 2014; Longhurst, 2015).

Apart from choosing among multiple ICTs, the participants also made the most of the polymedia affordances to adjust mediated communication routines in terms of content, tones, gestures, form of expression and so on. For example, when a participant shared photos taken at her and her child's volunteer activity, she was found to share dozens of unadulterated photos with her husband via WeChat on an ongoing basis, while posting only three elaborately embellished photos in her family chat group comprising more than twenty extended family members. Such compartmentalized photo sharing routines allowed her to negotiate the most appropriate emotional distances with different family members. With her husband, who was her most intimate family member, she could freely 'spam' him massive amounts of 'useless' information without concerns about timing and quality of sharing. With relatively less-intimate relatives, she chose to project herself as a well-groomed and considerate person who was willing to disclose her life experiences yet not inundate her friends with too many personal issues.

When ICTs Backfire: Emotional Labours of Constant Connectivity and Digital Asymmetries

While ICTs and the polymedia lifestyle bestow unprecedented opportunities to sustain long-distance intimacy, they also introduce new burdens and dilemmas (e.g., Horst, 2006; Parreñas, 2005). In the case of Chinese study mothers, physical distance with left-behind family members did not exempt them from fulfilling family obligations, but rather exacerbated their workloads of transnational coordination and multi-cultural navigation. Therefore, many participants had experienced circumstances where transnational communication with remote family members translated into burdensome undertakings.

For these migrant mothers, the possibility of constant connectivity via ICTs actually imposes an obligation of constant participation in or at least availability for mediated interactions, and poses threats to family intimacy whenever an expected interaction does not happen or fails to gain sufficient concentration (see also Burchell, 2015; Licoppe, 2004; Longhurst, 2015; Su, 2015). During participant observation, many participants engaged seamlessly in family decision making and daily small chats with their left-behind family members, especially their husbands, parents and siblings, on an hourly basis with a combination of video calls, voice and text messages, photos, hyperlinks and so on. Specifically, they were found to coordinate various mundane affairs in real time, such as buying new clothes, house renovation and itinerary planning, and provide greetings, congratulations and emotional support whenever something delightful or sad happened to their relatives. Such virtual companionship constantly reminded study mothers of their family roles as wives, (grand)daughters, sisters, nieces for their faraway loved ones, and brought about stresses when they were perceived as 'incompetent' in fulfilling related family obligations.

For instance, physical absence could no longer exempt these migrant mothers from the obligation of emotionally supporting their family members due to the wide availability of real-time mediated communication. It was not rare to see them suspend work tasks or domestic chores to send congratulatory messages and digital 'red packets' (a function of WeChat for small amounts of monetary exchange) to relatives who announced good news such as weddings, children's entrance into good high schools and moving into new homes. Failure in promptly responding to such joyful events would relegate the participants to be regarded as indifferent and ungrateful by the entire family. Similar stresses also emerge when the participants were constantly asked by left-behind family members to help purchase commodities from abroad. Compared with asking for remittances or gifts, which brings economic burdens to international migrants (e.g., Barber, 2008; Peng & Wong, 2013; Tazanu, 2015), their new role as 'purchasing agent' tended to trigger more emotional strains and tensions to family relationships, especially when they failed to find the requested items or to buy them at the expected low price.

The constant connectivity bestowed by ICTs also gave rise to the emotional labour of constant performativity (see also Wang & Lim, 2020, 2021). In particular, many study mothers reported feeling burdened by the perpetual pressure of appropriate self-presentation in transnational communication. Uprooted from the homeland to venture into unfamiliar foreign terrain, these mothers often feel obliged to display a strong and optimistic image in front of their left-behind kin, so as to gain respect from them and alleviate their concerns and anxiety. As a result, they made efforts to present the brightest and happiest side of their everyday lives during mediated communication, usually to the extent of hiding all the negative emotions and concertedly tuning their modes of expression. For example, in the aforementioned case of compartmentalized photo sharing, the participant embellished photos before sharing in family chat group to present remote relatives a well-groomed and happy transnational life, even though she was indeed tired of such 'face work'. Although similar emotional labours existed long before the prevalence of advanced ICTs, the increasingly convenient and synchronized mediated communication have made it a day-to-day emotional undertaking for contemporary transnational households.

Another common challenge faced by study mothers concerns their efforts to properly choose among and alternate between multiple ICT platforms for mediated family communication. As previously mentioned, in the integrated environment of polymedia affordances, different ICT platforms have different cultural inclinations and potential audiences, thus engendering different norms and expectations of mediated interactions. In this context, some study mothers can be trapped by 'context collapse' (Davis & Jurgenson, 2014; Marwick & Boyd, 2011) when they brought seemingly inappropriate contents or expressions from one mediated context to another (see also Wang & Lim, 2021). A typical example of context collapse was the 'misuse' of English during family communication. Several participants complained about their unpleasant experiences of being misunderstood by their left-behind kin as 'show-off' when she unconsciously mixed English words in their mediated conversations on WeChat or other China-based ICT platforms where Chinese was perceived as the 'appropriate language'.

Apart from emotional labours derived from the polymedia and perpetual connectivity lifestyle, ICTs also created new dimensions of inequalities within transnational households. Unlike many other transnational families suffering from unequal access to ICT infrastructures and quality of mediated communication (e.g., Cabalquinto, 2018; Cheong & Mitchell, 2016; Madianou, 2014; Parreñas, 2005), the study mothers and their remote family members were beleaguered by more invisible and nuanced 'digital asymmetries' (Lim, 2016) characterized by gaps in routines, emotional experiences as well as outcomes of ICT use (Wang & Lim, 2020). In particular, digital asymmetries could emerge when a study mother resorted to asking her adolescent child to set up video calls with remote family members ('competency asymmetry'), while she waited a long time to receive a perfunctory greeting from her left-behind husband ('expectation asymmetry'), and when she unconsciously arranged every video call according to the work schedule of her husband instead of her own preferences and needs ('autonomy asymmetry'). While these asymmetrical mediated routines sometimes remained invisible even to the participants themselves, they could indeed trigger frustration and disappointment in certain family members, and in turn, damage family intimacy in the long run.

CONCLUSION

In this chapter, we present extant literature and empirical data on both benefits and burdens of transnational communication, which help to understand not only the role of ICTs in maintaining long-distant relationships but also potential emotional labours and asymmetries that might emerge during mediated interactions.

The analysis reveals that ICTs, as well as the polymedia lifestyle they forge, can contribute to either mobility or immobility of transnational family members. On the one hand, ICT-mediated communication facilitates information, emotions and care to transcend geographical boundaries in real time, thus allowing international migrants to bring their home along with them wherever they go (Horst, 2006; Paragas, 2009; Wilding, 2006). On the other hand, the same virtual co-presence that grants migrants greater mobility also anchors them to their previous family roles and obligations, which renders them emotionally immobile albeit physically apart (Uy-Tioco, 2007; Cabanes & Acedera, 2012; Madianou, 2012).

As far as the Chinese study mothers are concerned, despite physical separation with left-behind family members, they remained intimately involved in their quotidian bittersweet experiences and could dutifully fulfil family responsibilities as wives, daughters, nieces etc. as if they never left. As 'transient sojourners' (Huang & Yeoh, 2005), these migrant mothers tended to devote more time and energy to maintaining intimate ties back home than acculturate to and assimilate into the host society (Wang & Lim, 2020). Therefore, they appreciated the polymedia affordance of ICTs, which allows real-time coordination of family matters and emotional exchanges with their remote loved ones.

However, the same polymedia connectivity can backfire when it imposes obligations over these mothers yet fails to provide practical support for their pressing needs. Due to the convenience of long-distance communication, these women are expected to participate actively in family activities, respond promptly to requests of left-behind family members, and provide regular updates of their own life experiences. As de facto 'single mothers' after relocation, they took up additional parental tasks and domestic labours, which could hardly be alleviated by virtual support from their families. In this sense, ICTs actually conspire with geographical distance to create a double burden for study mothers in which physical separation deprived them of hands-on familial support yet these ICTs constantly conveyed the obligations of transnational householding and reminded these women of their familial responsibilities.

REFERENCES

Aguila, A. P. N. (2009), 'Living long-distance relationships through computer-mediated communication', *Social Science Diliman*, **5** (1–2), 83–106.

Bacigalupe, G. and S. Lambe (2011), 'Virtualizing intimacy: Information communication technologies and transnational families in therapy', *Family Process*, **50** (1), 12–26.

Baldassar, L. (2016), 'De-demonizing distance in mobile family lives: Co-presence, care circulation and polymedia as vibrant matter', *Global Networks*, **16** (2), 145–163.

Barber, P. G. (2008), 'Cell phones, complicity, and class politics in the Philippine labor diaspora', *Focaal*, **51**, 28–42.

Benítez, J. L. (2012), 'Salvadoran transnational families: ICT and communication practices in the network society', *Journal of Ethnic and Migration Studies*, **38** (9), 1439–1449.

Bryceson, D. and U. Vuorela (2002), 'Transnational families in the twenty-first century', in D. Bryceson and U. Vuorela (eds), *The transnational family: New European frontiers and global networks*, Oxford and New York: Berg, pp. 3–30.

Burchell, K. (2015), 'Tasking the everyday: Where mobile and online communication take time', *Mobile Media & Communication*, **3** (1), 36–52.

Cabalquinto, E. C. B. (2018), '"We're not only here but we're there in spirit": Asymmetrical mobile intimacy and the transnational Filipino family', *Mobile Media & Communication*, **6** (1), 37–52.

Cabanes, J. V. A. and K. A. F. Acedera (2012), 'Of mobile phones and mother-fathers: Calls, text messages, and conjugal power relations in mother-away Filipino families', *New Media & Society*, **14** (6), 916–930.

Chee, M. W. L. (2003), 'Migrating for the children: Taiwanese American women in transnational families', in N. Piper and M. Roces (eds), *Wife or worker?: Asian women and migration*, New York: Rowman & Littlefield, pp. 137–156.

Cheong, K. and A. Mitchell (2016), 'Helping the helpers: Understanding family storytelling by domestic helpers in Singapore', in S. S. Lim (ed.), *Mobile communication and the family: Asian experiences in technology domestication*, Dordrecht: Springer, pp. 51–69.

Chib, A., S. Malik, R. G. Aricat and S. Z. Kadir (2014), 'Migrant mothering and mobile phones: Negotiations of transnational identity', *Mobile Media & Communication*, **2** (1), 73–93.

Davis, J. L. and N. Jurgenson (2014), 'Context collapse: Theorizing context collusions and collisions', *Information Communication and Society*, **17** (4), 476–485.

Fortunati, L. (2002), 'The mobile phone: Towards new categories and social relations', *Information, Communication & Society*, **5** (4), 513–528.

Francisco, V. (2013), '"The internet is magic": Technology, intimacy and transnational families', *Critical Sociology*, **41**, 173–190.

Fresnoza-Flot, A. (2009), 'Migration status and transnational mothering: The case of Filipino migrants in France', *Global Networks*, **9** (2), 252–270.

Hannaford, D. (2015), 'Technologies of the spouse: Intimate surveillance in Senegalese transnational marriages', *Global Networks*, **15** (1), 43–59.

Horst, H. A. (2006), 'The blessings and burdens of communication: Cell phones in Jamaican transnational social fields', *Global Networks*, **6** (2), 143–159.

Huang, S. and B. S. A. Yeoh (2005), 'Transnational families and their children's education: China's "study mothers" in Singapore', *Global Networks*, **5** (4), 379–400.

Huang, S. and B. S. A. Yeoh (2011), 'Navigating the terrains of transnational education: Children of Chinese "study mothers" in Singapore', *Geoforum*, **42** (3), 394–403.

Jeong, Y.-J., H.-K. You and Y. I. Kwon (2014), 'One family in two countries: Mothers in Korean transnational families', *Ethnic and Racial Studies*, **37** (9), 1546–1564.

Kang, T. (2012), 'Gendered media, changing intimacy: Internet-mediated transnational communication in the family sphere', *Media, Culture & Society*, **34** (2), 146–161.

King-O'Riain, R. C. (2015), 'Emotional streaming and transconnectivity: Skype and emotion practices in transnational families in Ireland', *Global Networks*, **15** (2), 256–273.

Köhl, M. M. and G. Götzenbrucker (2014), 'Networked technologies as emotional resources? Exploring emerging emotional cultures on social network sites such as Facebook and Hi5: A trans-cultural study', *Media Culture and Society,* **36** (4), 508–525.

Lee, H. (2010), '"I am a kirogi mother": Education exodus and life transformation among Korean transnational women', *Journal of Language, Identity & Education*, **9** (4), 250–264.

Licoppe, C. (2004), '"Connected" presence: The emergence of a new repertoire for managing social relationships in a changing communication technoscape', *Environment and Planning D: Society and Space*, **22** (1), 135–156.

Lim, S. S. (2016), 'Asymmetries in Asian families' domestication of mobile communication', in S. S. Lim (ed.), *Mobile communication and the family: Asian experiences in technology domestication*, Dordrecht: Springer, pp. 1–9.

Longhurst, R. (2013), 'Using skype to mother: Bodies, emotions, visuality, and screens', *Environment and Planning D: Society and Space*, **31** (4), 664–679.

Longhurst, R. (2015), 'Mothering, digital media and emotional geographies in Hamilton, Aotearoa New Zealand', *Social & Cultural Geography*, **17** (1), 1–20.

Madianou, M. (2012), 'Migration and the accentuated ambivalence of transnational motherhood: New media in Filipino migrant families', *Global Networks*, **12** (3), 277–295.

Madianou, M. (2014), 'Smartphones as polymedia', *Journal of Computer-Mediated Communication*, **19** (3), 667–680.

Madianou, M. (2016), 'Ambient co-presence: Transnational family practices in polymedia environments', *Global Networks*, **16** (2), 183–201.

Madianou, M. and D. Miller (2011), 'Mobile phone parenting: Reconfiguring relationships between Filipina migrant mothers and their left-behind children', *New Media & Society*, **13** (3), 457–470.

Madianou, M. and D. Miller (2012), *Migration and new media: Transnational families and polymedia*, London: Routledge.

Marwick, A. E. and D. Boyd (2011), 'I tweet honestly, I tweet passionately: Twitter users, context collapse, and the imagined audience', *New Media and Society*, **13** (1), 114–133.

Mckay, D. (2007), 'Sending dollars shows feeling': Emotions and economies in Filipino migration', *Mobilities*, **2** (2), 175–194.

Nedelcu, M. and M. Wyss (2016), '"Doing family" through ICT-mediated ordinary co-presence: Transnational communication practices of Romanian migrants in Switzerland', *Global Networks*, **16** (2), 202–218.

Neustaedter, C. and S. Greenberg (2012), 'Intimacy in long-distance relationships over video chat', in *Proceedings of the CHI (2012),* Austin, Texas, USA, pp. 753–762.

Paragas, F. (2009), 'Migrant workers and mobile phones: Technological, temporal, and spatial simultaneity', in R. Ling and S. Campbell (eds), *The reconstruction of space and time: Mobile communication practices*, New Brunswick, NJ: Transaction Publishers, pp. 39–65.

Parreñas, R. (2005), 'Long distance intimacy: Gender and intergenerational relations in transnational families', *Global Networks*, **5** (4), 317–336.

Peng, Y. and O. M. H. Wong (2013), 'Diversified transnational mothering via telecommunication: Intensive, collaborative, and passive', *Gender & Society*, **27** (4), 491–513.

Pham, B. and S. S. Lim (2016), 'Empowering interactions, sustaining ties: Vietnamese migrant students' communication with left-behind family and friends', in S. S. Lim (ed.), *Mobile communication and the family: Asian experiences in technology domestication,* Dordrecht: Springer, pp. 109–126.

Plüss, C. (2013), 'Chinese migrants in New York: Explaining inequalities with transnational positions and capital conversions in transnational spaces', *International Sociology*, **28** (1), 12–28.

Rakow, L. F. and V. Navarro (1993), 'Remote mothering and the parallel shift: Women meet the cellular telephone', *Critical Studies in Mass Communication*, **10** (2), 144–157.

Su, H. (2015), 'Constant connection as the media condition of love: Where bonds become bondage', *Media, Culture & Society*, **38** (2), 232–247.

Tazanu, P. M. (2015), 'On the liveness of mobile phone mediation: Youth expectations of remittances and narratives of discontent in the Cameroonian transnational family', *Mobile Media & Communication*, **3** (1), 20–35.

Thomas, M. and S.S. Lim (2011), 'On maids and mobile phones: ICT use by female migrant workers in Singapore and its policy implications', in J. E. Katz (ed.), *Mobile communication: Dimensions of social policy*, New Brunswick: Transaction Publishers, pp. 175–190.

Turkle, S. (2011), *Alone together: Why we expect more from technology and less from each other*, New York: Basic Books.

Uy-Tioco, C. (2007), 'Overseas Filipino workers and text messaging: Reinventing transnational mothering', *Continuum*, **21** (2), 253–265.

Vancea, M. and Olivera, N. (2013), 'E-migrant women in Catalonia: Mobile phone use and maintenance of family relationships', *Gender, Technology and Development*, **17** (2), 179–203.

Vertovec, S. (2004), 'Cheap calls: The social glue of migrant transnationalism', *Global Networks*, **4** (2), 219–224.

Wang, Y. (2020), 'Parent–child role reversal in ICT domestication: media brokering activities and emotional labours of Chinese "study mothers" in Singapore', *Journal of Children and Media*, **14** (3), 267–284.

Wang, Y. and S. S. Lim (2018), 'Mediating intimacies through mobile communication: Chinese migrant mothers' digital "bridge of magpies",' in R. Andreassen, M. N. Petersen, K. Harrison and T. Raun (eds), *Mediated intimacies: Connectivities, relationalities and proximities*, London; New York: Routledge, pp. 159–178.

Wang, Y. and S. S. Lim (2020), 'Digital asymmetries in transnational communication: Expectation, autonomy and gender positioning in the household', *Journal of Computer-Mediated Communication*, **25** (6), 365–381.

Wang, Y. and S. S. Lim (2021), 'Nomadic life archiving across platforms: Hyperlinked storage and compartmentalized sharing', *New Media & Society*. **23** (4), 796–815.

Waters, J. L. (2002), 'Flexible families? 'Astronaut' households and the experiences of lone mothers in Vancouver, British Columbia', *Social & Cultural Geography*, **3** (2), 117–134.

Wilding, R. (2006), '"Virtual" intimacies? Families communicating across transnational contexts', *Global Networks*, **6** (2), 125–142.

Wilding, R. (2012), 'Mediating culture in transnational spaces: An example of young people from refugee backgrounds', *Journal of Media & Cultural Studies*, **26** (3), 501–511.

Yoon, K. (2016), 'The cultural appropriation of smartphones in Korean transnational families', in S. S. Lim (ed.), *Mobile communication and the family: Asian experiences in technology domestication*, Dordrecht: Springer, pp. 93–108.

16. In support of return and reintegration? A roadmap for a responsible use of technology

Nassim Majidi, Camille Kasavan and G. Harindranath

INTRODUCTION

As migration has drawn increased attention globally, shifting political contexts, and a hardening of both migration rhetoric and policies have led to returns becoming the option of choice for most destination country governments (DRC et al. 2019). Return modalities exist across a spectrum, from voluntary to involuntary, and forced; it represents a particular stage in an 'ongoing migration cycle of spatial mobility' (Ruben et al. 2009, p. 911). Return policies and political contexts play a key role in determining the modality of return migration, and, conversely, these modalities determine specific support and reintegration policies available or needed upon return (Battistella 2018). This global emphasis on returns has led more recently to a renewed examination of the reintegration process – that is, of what happens in the long term after return, and what sustainable reintegration might look like (see Box 16.1).

Recognising this shift, in 2017 IOM revised its definition and adopted an integrated approach to reintegration, noting that: 'Reintegration can be considered sustainable when returnees have reached levels of economic self-sufficiency, social stability within their communities, and psychosocial well-being that allow them to cope with (re)migration drivers. Having achieved sustainable reintegration, returnees are able to make further migration decisions a matter of choice, rather than necessity' (IOM 2017, p. 3). In other words, successful reintegration is multi-dimensional, beyond the economic dimension; multi-levelled, beyond the individual returnee; and is not a linear or uni-directional process. Reintegration becomes about the ability to make a well-informed choice to stay, to leave or to return, in dignity.

BOX 16.1 GLOBAL DISCUSSIONS ON REINTEGRATION: OBJECTIVE 21 OF THE GCM

The Global Compact for Safe, Orderly, and Regular Migration (GCM), adopted by over one hundred states in 2018 partly reflects global discussions and renewed interest in return and reintegration. Objective 21 of the compact calls on the need to 'co-operate in facilitating safe and dignified return and readmission, as well as sustainable reintegration.'

Specifically, subsection h. of Objective 21 puts forth the right and responsibility of states to support and facilitate long-term reintegration after return from a holistic perspective, taking into account the social, psychosocial and economic dimensions needed in order to move towards a sustainable and inclusive return.

The academic and grey literature on return and reintegration has grown in parallel to these expanding global discussions (Cassarino 2004; DRC et al. 2019). However, studies focusing solely on the use and impact of information and communications technologies (ICTs) on the twin processes of return and reintegration remain scarce. This gap in the literature remains even as technology has been increasingly used to manage returns by states as well as by returnees themselves in preparing and embarking on their return and reintegration journeys.

While migration is often seen from an immigration and host country perspective, this chapter reverses this lens by discussing the linkages between technology and return migration. We discuss both state and migrant perspectives on return in order to examine the potential of technology to support dignified returns and sustainable reintegration processes. This chapter seeks to consolidate the available knowledge on return, reintegration and technology in order to provide an overview of the technological innovations being used in existing or planned programming, the use of ICT by returnees as well as an analysis of the potential consequences of use. The chapter will serve as an introduction to theory and recent developments concerning the relationship between technology, return and reintegration.

This chapter is divided into two sections:

- First, an assessment of the migrant perspective, examining returnees' use of technology to support their own return and reintegration, in particular the ways in which technology influences the decision-making processes around pre-return planning, return and post-return outcomes. This section also includes a discussion around the role of technology in circular migration, which requires a different kind of return and reintegration.
- Secondly, an examination of the use of technology in formal return management on the part of destination countries or migration management agencies, in particular the use of technology to manage forced returns, and the human rights implications of this usage.

This division will allow for an examination of how technology has been adapted or used to maintain various return modalities and will encompass both positive and negative aspects of technology in return and reintegration, analysing perspectives and practices of returnees and those linked to institutional migration management structures.

THE INFLUENCE OF TECHNOLOGY ON DECISION MAKING, PREPAREDNESS FOR RETURN AND SUSTAINABLE REINTEGRATION

The discussion on migrants' use of technology to support return and reintegration requires a preliminary discussion on the importance of information and preparedness. In 'Theorising Return Migration', Cassarino (2004) highlights the importance of return preparedness for returning migrants to 'become actors of change and development at home' (p. 271) and also breaks down what preparedness entails: it relates to both the *willingness* to return (whether returns are voluntary or involuntary and possibly forced) and the *readiness* to return. Both components require resource mobilisation, including adequate support and resources in relation to conditions at home. Information is a central component of both of these processes, and hence of return preparation – people can make a choice to return if they have adequate information; they can also prepare for return if they have this information *and* the communication tools to plan transnationally (pre-return) for their reintegration. ICT plays an increasingly important role in influencing and structuring returnee preparedness (Figure 16.1). One of the global lessons learned on returns centres on the need to enhance preparedness through information sharing, and improving returnees' awareness of their rights (DRC et al. 2019).

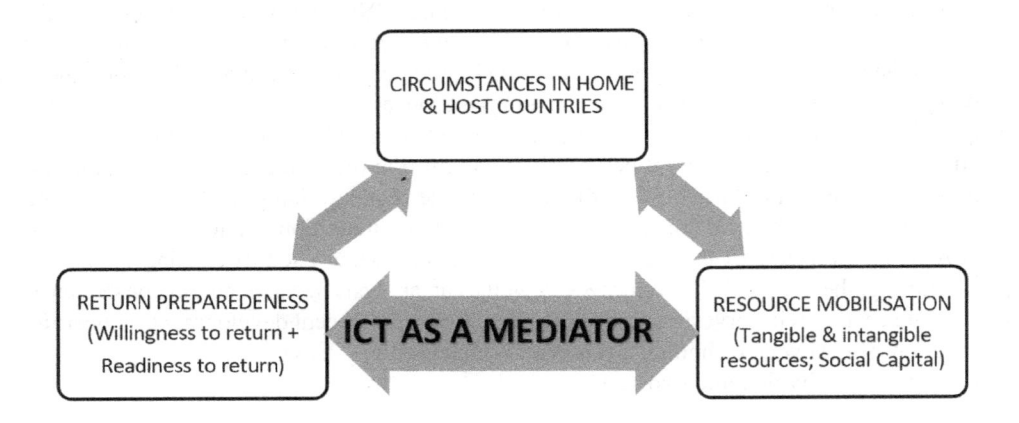

Source: Based on Cassarino, 2004, p. 271.

Figure 16.1   Return migration and the role of ICT

While broader contextual factors in both the origin and host countries necessarily play a significant role in return decision making, pathways to return and sustainable reintegration are being paved through the extensive use of ICTs. The common challenges of return and reintegration include, among others, lack of documentation, limited awareness of existing services for returnees, and difficulties in validating education and skill credentials (Ruiz Soto et al. 2019). Mapping out technology's potential to address these challenges can provide a roadmap to support return and reintegration in a way that is migrant centred. For instance, blockchain technology has the potential to function as an electronic wallet for identification documents, where migrant and returnee biodata cannot be lost and would remain protected. In terms of awareness and access to existing services, apps such as IOM's MigApp[1] have been developed to enhance migrants' awareness of their rights. Apps have also been adapted or developed to address gaps in services, such as the effort to provide refugees with accommodation through Airbnb's #weaccept,[2] a testament to the potential of technology to address migrants' need for housing. Difficulties validating education or skills credentials are being addressed in destination countries through a 'qualifications passport' scheme to address barriers to employment, piloted in a refugee camp in Zambia by UNESCO (UNESCO 2019).

While some of these technological interventions have not to date been aimed at the return context, these pilots can potentially serve as models for future reintegration programming.

ICTs also play a significant role in facilitating returnee preparedness through sustained social engagement with cross-border systems of ethnic and family relationships, and within networks of migrants and non-migrants (Cassarino 2004). ICTs are often instrumental in the mobilisation of intangible resources and social capital. They allow for the micro-management of affairs back home, assisting the return process and reintegration (Hunter 2015; Hamel 2009; Panagakos & Horst 2006), and may allow dual national migrants to return to their home country and leverage technological and other skills acquired as a 'bridge' between two cultures or work forces (Wang 2016). The rapid development of both transport links and ICTs have altered the previously one-way process of international migration towards a more circular set of practices characterised by seasonal and/or circular migration involving frequent, and often strategic, journeys back and forth between home and host countries for work (Wang 2017).

Thus, ICTs are increasingly playing a critical role in the migrants' decision-making process relating to return, preparations for return, and their reintegration process. These three components are discussed in this section with a review of the available evidence and through supporting cases.

Does Technology Support or Impede the Decision to Return?

Current trends in return migration and technology, from a transnational network perspective, highlight the significant shift that has occurred for migrants with the advent of cheaper and more accessible communication systems. This access may have facilitated the ability to stay in touch with one's family while on the move, but it has also raised family expectations, notably with regards to migrants' financial contributions during their time away. While previously the cycle of letters written, sent, received and replied to provided migrants with ample time to find opportunities for revenue generation, there is now an expectation of instant answers, notably surrounding the frequency and amount of remittances that migrants send back or will bring back upon return (Hunter 2015).

This has added pressures on migrant return decision-making processes: through ICTs, the financial expectations of the family at the point of origin is made clearer and, as shown in the case of West African migrants to France, can be a deterrent to return (Hunter 2015). Returning without sufficient (or any) financial contributions to the household's or the community's development is seen as a failed migration. The new communications bond between migrants at destination and their families at origin is experienced by the migrant as a source of pressure. This shift from the uprooted migrant to the connected migrant (Diminescu 2008) has now brought to the fore the self-perception as a failed migrant who cannot deliver on his or her family's expectations, and who will hence decide not to return. Drawing on Cassarino's concept of return preparedness, Hunter (2015) reports the ambivalence shown by older West African migrants in France towards ICT. Mobile phones, for instance, can add pressure on men abroad to be financially supportive of their families at home. This mobile phone use may impede return in two ways: by amplifying the remittance burden (pressure from home) and by facilitating networks of solidarity with other migrants in the host country. Migrants in hostels in France, according to Hunter's ethnographic study, find reassurance in the new networks they build, turning themselves away from an eventual return towards a system of solidarity among fellow migrants. Instead of using technology to bring them closer to their return, access to information and communication through mobile phones reinforce links in the country of immigration, impeding or discouraging return. The age dimension is particularly important in Hunter's work – the technological shift that happened in the early 2000s was experienced as an abrupt change for migrants who were not well versed in using mobile phone technologies. Instead of viewing this technology as a positive force to reinforce bonds with home communities, they perceived and lived it as a detrimental and invasive one.

This ambivalence towards ICTs can also be seen in other contexts. For example, Madianou highlights the case of Filipina migrants' use of ICT, where ICTs are shown to play a central role in decision making in relation to both migration and return (Madianou 2012). ICTs here become a way to justify staying in migration and not returning indefinitely, helping to perpetuate a trend of temporary return and cyclical migration, and allowing women (and men) to keep one foot in their families 'from a distance' and one foot out. In other words, instead of contributing to a closeness of ties that would make women migrants want to return, ICTs have empowered Filipina migrants to more easily live their family life at a distance: connected enough to be present virtually, but also helping them realise the fact that they want to preserve the independence and autonomy acquired abroad. While decisions to return to the Philippines, get married and have children are often made at the end of a contract period, that return is then followed by renewed labour migration to sustain the expanded family, facilitated by cheap communications flows enabling 'motherhood at a distance'. Thus, for Filipina migrants, return becomes a dual process – either return home to expand the family and devote oneself to one's child at a critical juncture in their childhood and upbringing; or return to the country of work to earn money and fulfil the need for financial support that is also critical to a child's upbringing. Technology offers Filipina migrants the ability to bridge this gap, allowing them to speak to and see their children online, therefore removing the need to go back to check on them. While in the past Filipina migrants' investments may have been diverted, mothers can today, through mobile technology and video systems, check that their children are properly clothed, have received their remittance, and are benefiting from their hard work. In this situation ICT provides a system of 'remote control' of children's wellbeing and of the impact of their hard work and remittances, which does not then necessitate a physical return to check on the family.

While transnational migrants may therefore adopt new ICTs to suit their communications and networking needs in relation to migration planning, transit and integration (see Hamel 2009; Alonso and Oiarzabal 2010; Kozachenko 2013; Andrade and Doolin 2015), the impact on return is either largely unknown or problematic, as seen in the above cases. It cannot be assumed that technology – whether through the Internet, mobile phones or other ICTs – will result in migrants feeling more empowered to make their return decision. Indeed, the link between technology and migration can represent a paradox (Nie et al. 2002): it lowers the threshold for the initial migration decision (Dekker and Engbersen 2012), but increases the threshold for a decision to return, by empowering migrants *not* to return.

However, state funding and institutional efforts are going towards the creation of virtual networks intended to facilitate migrants' return. Will these virtual networks have a different impact on return decisions? It is most often a state or institutional assumption that online platforms can help steer migrants towards return. For example, the Integrated Approach regarding Information on Return and Reintegration in Countries of Origin (IRRICO) was set up to provide potential returnees with information about conditions and prospects in 20 countries of origin. Country sheets were drafted to provide information on the economic situation but were also as much a communication tool between IOM country offices as a source of information for potential returnees. However, there is evidence that migrants rarely consult online platforms as they often rely on 'face-to-face and personal communications in the course of decision making processes' (Sanchez et al., 2018 p. 44). More recent attempts at sharing information have used technology through the release of videos from returnees about their experiences of return, made available by IOM through a dedicated website. Videos are seen as a tool to provide first-hand accounts of both disappointment with migration and happiness in the return process. In West Africa, a similar initiative by IOM was rolled out under the title of 'Migrants as Messengers', for migrants to put forward their own accounts of return migration though social media. However, these narratives remain facilitated by agencies and institutions, largely state funded, and may not have the level of credibility and trust required to help migrants to prepare for return. Such use of technology is also most often limited to the spectrum of IOM's assisted voluntary return and reintegration (AVRR) programme.

On the other hand, use of social media that is not supported or driven by institutional support, but rather by migrants themselves, has gained more traction. Teleconferencing and visual conferencing – such as through WhatsApp – or group calls such as through Viber, can open new channels for supporting returns. These tools, which are encrypted, visual and real time, have a stronger potential to become a space that is used to advise each other – across borders – on the viability to return (for instance, see Frouws et al. 2016).

Technology as a Reintegration Tool?

For those who have crossed the threshold of return decision making, the ability to imagine positively their reintegration in their origin country will make them more likely to return sustainably. But if they believe this process will be difficult – if they are not ready or lack resources – they will not be able to return and sustainably reintegrate. Sustainable return migration has also been characterised as a process of 'mixed embeddedness' where embeddedness 'refers to the ways how individuals find and define their position in society, feel a sense of belonging and possibilities for participation in society' (Ruben et al. 2009, p. 910). Here, sustainable reintegration is seen as multi-dimensional involving economic embeddedness (relating to material conditions for sustainable livelihoods), social network embeddedness (relating to information and access to social networks) and psychosocial embeddedness (relating to a sense of wellbeing and safety linked to the ability to construct one's identity). While each of these dimensions is multi-faceted and dependent on migrants' characteristics, conditions in host and home societies and assistance received in the home country (Ruben et al. 2009), ICTs can play a powerful role in supporting such mixed embeddedness as a route to reintegration. This section details some recent efforts led by civil society organisations in this direction.

Supporting the belief that technology can play a role in imagining return outcomes, in 2016 in Nairobi, *Techfugees*[3] set up a local chapter in response to the announcement by the Government of Kenya of the closure of Dadaab refugee camp and the potential involuntary return of half a million Somali refugees. The initiative brought together the technology sector, with representatives of Safaricom – a leading mobile operator in Kenya – and international players such as GSMA, research organisations, and non-governmental organizations, to discuss how they could best support each other and collaborate to support Somali refugees' return and reintegration process. Service mapping – in particular, for basic services such as health and education – and the provision of mobile apps for family tracing as well as remittances have been identified as two key needs in the eventuality of a return process. There have also been discussions around the use of blockchain technology to support cross-border payments for refugees returning to their ancestral lands. Recognising that migrants have their own assets that need to travel with them (whether land titles or savings), and that migrants need to show creditworthiness and a financial history, blockchain has been introduced by companies such as *BitPesa*, as a traceable, interoperable, secure, immutable, and mobile means to support people on the move. However, the potential of such technology for reintegration remains to be tested (Barabas & Zuckerman 2016).

BOX 16.2 THE DIGITAL GENDER DIVIDE IN THE CONTEXT OF REINTEGRATION

While a digital gender divide persists for both migrant and non-migrant women, including in sending and receiving countries, digital resources and technologies can also be leveraged into initiatives to allow women to 'leapfrog' towards enhanced livelihood opportunities (OECD 2018).

Returnee women may also benefit from such initiatives. Suri and Jack (2016) highlight, for instance, the potential for mobile money technology to influence migration and livelihood possibilities for women (including potentially returnees), reducing their need to depend on scattered occupations and allow them to run their own businesses and moves towards strengthened reintegration within their community.

In Mexico, a social enterprise called *HolaCode*[4] has tapped into technology to connect returned Mexicans and migrants in Mexico to jobs in the tech industry, thereby addressing one of the most important needs upon return: employment. Schmidtke and Chuayffet (2018) note that between 2009 and 2014, approximately one million people returned to Mexico, some voluntarily and others involuntarily, revealing the importance of investing in their integration back into Mexican society at all levels, whether economic, social or psychosocial. HolaCode aims to support reintegration through technology education, access to employment and financial inclusion. Returnees undergo software development training programmes as a way to obtain employment and to facilitate their reintegration into society. HolaCode, which operates in Mexico City and Tijuana, provides an example of an initiative that can be scaled and potentially transferred to other return contexts.

Similarly, in a study of 11 countries and 69 communities of return in West and Central Africa, a Samuel Hall series of reports (2018a, b, c) reveals the potential for the technology industry to integrate returnee youth in need of labour market integration, offering a better match to returnees' aspirations than more traditional sectors such as the agricultural sector. Specifically, the Senegal report shows that the training most adapted to both the demand and the supply side of labour markets in return communities requires the use of ICT training to ensure that youth can offer the skills expected of them in the labour market. The Nigeria report recognises the link between technology and the strengthening of labour markets in areas of return. It emphasises the importance of mobilising a range of stakeholders alongside the technology industry, such as research institutes, microfinance banks and other state-owned lenders who could support a more sustainable reintegration of returnees.

Finally, the case of Afghanistan – which has seen over 1 million returns in 2016 alone, both voluntary and forced, from Iran, Pakistan and Europe (Majidi 2017) – also highlights the opportunities for better linkages between technology and reintegration. Mobile money transaction schemes such as M-Paisa in Afghanistan can bring potential benefits to institutions supporting reintegration and to returnees themselves. This scheme was based on the Kenyan example of M-Pesa, which has transformed mobile banking in Kenya. The Afghan scheme currently focuses on peer-to-peer money transfer, airtime purchases, and salary payments for government employees. The system, a collaboration with the First MicroFinance Bank of Afghanistan, provides an opportunity for private sector engagement in reintegration assistance. In a context such as Afghanistan, mobile money can address security and access concerns. Investment in technology-oriented solutions could not only replicate the practice that migrants already engage in but provide new ways to support their reintegration process. With changes in operational modalities due to COVID-19, it might also no longer be safe for returnees to be handed cash grants physically in person. Piloting new modalities that provide an alternative to the physical exchange of money, through digital wallet schemes, can provide a safer, and more traceable option.

As shown here, the use of ICTs can have a significant role in supporting the effective and sustainable return and reintegration of migrants. However, such opportunities remain to be defined and require the orchestration of efforts by a range of civil society and state-level actors to address possible barriers in information and trust that may hinder returnees' access to and use of digital financial services. While many reintegration support initiatives have implicitly targeted economic embeddedness of returnees, initiatives to support social and psychological embeddedness are far fewer if not non-existent. This represents an opportunity for intervention by civil society organisations and local actors working closely with migrants and returnees.

Although this section has focused on the use of ICTs by migrants, civil society organisations and international agencies, governments are increasingly using ICTs for formal return management – particularly in the management of forced returns – which often puts vulnerable migrant populations at increased risk. The next section of this chapter will examine this flipside of technology, looking at how technology has been used as a tool for managing forced returns, potentially hindering possibilities for safe, dignified, and sustainable return and reintegration.

THE USE OF TECHNOLOGY IN FORMAL RETURN MANAGEMENT AND GOVERNANCE: A TOOL NOT WITHOUT CONSEQUENCES

Objective 21 (subsection c) of the Global Compact for Safe, Orderly and Regular Migration includes explicit discussion of the use of biometrics and digital technologies for identification of a state's own nationals 'for safe and dignified return and readmission in cases of persons that do not have the legal right to stay on another State's territory'. While the GCM recognises deportation processes and highlights these technologies as a means to ensure that this is done in ways that are legal and organized, the question of rights to privacy and safety remains unresolved – at the state level these technologies have also been used in development with private sector actors, in ways that may violate these rights.

Private–Public Partnerships and the Use of Technology in Managing Forced Returns

Forced returns – largely through the mechanism of deportation, understood as the physical removal of someone against their will from the territory of one state to that of another (Schuster & Majidi 2013) – are one formal element of migration management. Used to return migrants – including failed asylum seekers – who have not managed to obtain legal status within a destination country, the administration and organisation of forced returns have in some contexts been increasingly assisted by technological innovations developed privately and shared with governments.

The most striking publicly reported instances of this have been in the United States, where Immigration and Customs Enforcement (ICE), the federal office in charge of migration management (including forced return), have reportedly relied on the expanded use of artificial intelligence – most notably facial recognition – and increased monitoring of social media and online activities in order to identify potential deportees and streamline mechanisms of forced return (Molnar 2019b; Rivlin-Nadler 2019; Simon 2020). In widely publicised instances in Utah, Vermont and Washington in the United States, ICE officials used facial recognition technology to mine millions of state drivers' licences in order to identify undocumented migrants (Mak 2019; Edmondson 2019). In early 2020, advocacy groups investigating increases in raids and detention of undocumented migrants in Maryland found that unrestricted ability to run facial recognition searches on drivers' licences in the state had detected 275,000 licences belonging to undocumented migrants (Harwell & Cox 2020). While deportation statistics related to facial recognition technology are not publicly available, anecdotal evidence suggests that there is a link between this unfettered use of facial recognition technology, detentions, and ensuing deportations (Johnson 2020; Harwell & Cox 2020).

A landmark report published jointly in 2019 by the National Immigration Project, Immigrant Defense Project and the non-profit organisation Mijente, examined in detail the linkages and relationships between the tech industry, specific technologies and deportations carried out by ICE. The report explicitly names Amazon and data giant Palantir as two key entities who have provided ICE with key support in collecting, storing and managing data related to potential deportees, and further highlights the central importance of data and the support of these tech companies in driving deportation raids (Mijente 2019).

This technological support includes linking local police and municipal databases to national ICE databases, thus undermining the ability of urban areas to maintain their roles as 'sanctuary cities'. ICE's case management software, developed by Palantir, allows immigration agents to scour local, state and federal databases across the US to create profiles of immigrants and their networks based on both private and publicly available information. As both ICE and local law enforcement agencies use the same Palantir-created systems, local usage by law enforcement agents feeds further data for ICE as they conduct immigration raids (Mijente 2019).

BOX 16.3 USING TECHNOLOGY TO BETTER SUPPORT HUMAN RIGHTS IN RETURN?

ICTs can be used in return and post-return monitoring in order to support the international protection of returning migrants and to ensure transparency, accountability and human rights protection within returns processes (Pirjola, 2015). While the use of technology in return has largely been used at the state level in order to better organise and manage these returns (with sometimes significant human rights consequences), technology platforms developed at the civil society level may provide an avenue for better support of human rights for returning migrants in the future.

Existing web and mobile platforms such as eyeWitness to Atrocities* app and Ushahidi** rely on crowdsourcing to document and report human rights violations and share these with actors who may be able to act on them. Initially developed to monitor election violence and other conflict-related human rights violations, there may be space for linking these platforms to local actors, community members and returnees themselves seeking to ensure their rights and safety are respected, although no publicly documented cases of these technologies being used in this manner yet exist.

As highlighted in this chapter, the promise of technology almost always comes with potential for abuse. Proportionality, accountability and ethical principles ought to serve as key criteria when using technologies in this context (see Dijstelbloem 2017; Molnar 2019b).

Contracts between tech giants such as Palantir and Amazon and migration management agencies such as ICE play a significant role in how migrant data is managed, and a transition from national data centres towards cloud services – hosted in the US primarily by Amazon Web Services – is key to immigration enforcement mechanisms, including managing and streamlining data which facilitates deportation (Mijente 2019).

These public–private contractual partnerships between tech companies and the public sector are not limited solely to the United States. While data protection laws in Europe (both within and beyond the European Union) are more robust than those in the United States, several European countries including Norway (Brekke & Staver 2019), Germany (BDI 2017) and Denmark (Crouch & Kingsley 2016; Bilefsky 2016) nonetheless maintain laws allowing immigration officials to access and keep data from asylum seekers' smartphones, which may then be used as evidence in deportation cases when asylum seekers are rejected. This is even though data taken off cell phones may not accurately reflect asylum seekers' journeys, and this data mining may be in violation of existing privacy and data protection norms and regulations (UNHCR 2018). These laws are reportedly implemented in conjunction with private sector companies specialising in software used to extract smartphone data (Meaker 2018).

The Use of Algorithms and Biometrics in Returns and Its Human Rights Consequences

The increased use of technology in ways described above has implications for migrants and their ability to exercise their rights. Maitland (2018) raises the spectre of the 'digital refugee' (or indeed, migrant) who is co-constructed through her own ICT usage (digital self-construction) as well as through her interactions with humanitarian organisations and state agencies that leave behind digital traces. These traces then form the basis for official digital 'selves' of the migrants, which may or may not reinforce the individual digital selves. Vulnerable migrants are left with no protection in situations where these two digital selves clash. For example, the reliance on social media and phone data to confirm or verify an asylum seeker's journey, potentially leading to rejection and deportation, may not only be in violation of European data privacy laws, but can also lead to grave errors in cases where deportation or return decisions are dependent on mobile phone or computer derived algorithms. Molnar (2019b) has examined some of the tangible human rights consequences in cases where technology fails, including one drastic instance where over 7,000 foreign students in the UK were wrongfully deported after an algorithm mistakenly concluded that they had cheated on their language tests.

Beyond the personal devastation faced by those wrongly deported, the legal consequences of algorithmic decisions (and mistakes) on forced return are complicated: 'If you want to challenge an algorithmic decision like this in a court of law, is it the designer, the coder, the immigration officer or the algorithm itself who is liable?' (Molnar 2019b, p. 2). Algorithmic decision making, thus, occupies a complex legal space that lacks clear guidelines relating to individual rights and interests as well as mechanisms of redress (Molnar 2018).

This increased collection of data that is essential to algorithmic decision making comes with inherent protection and privacy risks (see Box 16.4). Biometric identification, increasingly used by state-level actors and humanitarian agencies in migration contexts, has also been shown to fail including both iris recognition and fingerprint technologies (Kingston 2018). As the use of algorithms – and technological innovations such as biometrics – in managing forced returns is being implemented before global or national regulatory legal frameworks have been established, the likely result is that those who will suffer the worst consequences of errors or technological inflexibility will be asylum seekers and other migrants who are unprepared to navigate or manage the impact that technology has on their ability to stay in a country, and therefore on their potential return, forced or otherwise. Furthermore, biometrics are open to corruption and misuse by states and rogue actors, and algorithms are easily biased against marginalised groups (Molnar 2018). The use of technologies in many countries to monitor the movement of people as well as track and trace disease during the COVID-19 pandemic has created even more opportunities for targeting migrants and other vulnerable groups (Molnar 2020). Their extensive use in surveillance raises significant ethical concerns and has implications for the rights and well-being of migrants (UNESCO 2013) who are increasingly being criminalised through technology (Singler 2018).

As the digitalisation of the migrant gathers pace, it is incumbent upon agencies such as UNHCR and IOM to find tools and methods for redress as well as the protection of their digital identities (Maitland 2018). Indeed, there is a growing need to revisit the values that underpin the use of technology in migration contexts where the needs and aspirations of migrants are rarely taken into account (Harindranath 2019).

BOX 16.4 DATA PROTECTION AND PRIVACY ISSUES FOR RETURNEES

Migrants whose data has been entered in a national system – for instance asylum seekers, or those seeking to regularise their status – by default find themselves needing to provide significant amounts of data to national authorities before they know the results of their application. These data, which may be used for protection and support purposes, may also be used for the opposite.

In the United States, for instance, undocumented young adults qualifying for protection from deportation under the Deferred Action for Childhood Arrivals (DACA) programme[5] must submit significant amounts of personal information in order to receive this protection, including home addresses and identifying information. While these data are nominally protected from disclosure to immigration authorities who manage deportation, in practice this protection is dependent on national policies and not legally binding, leaving DACA participants and their families in precarious positions of dependence on national and political goodwill towards this database.

Returnee and potential returnee databases, however, are also a key tool for providing effective return and post return support. Globally, those who are returned formally – whether through forced or voluntary measures – provide significant amounts of personal data to the organisations facilitating or imposing their return. These data can be crucial to ensuring post-return support and monitoring the effectiveness of reintegration programming, and strict data protection laws are in place, in particular for EU countries and entities receiving EU funding who must comply with General Data Protection Regulation (GDPR) laws.

CONCLUSION

This chapter has reviewed the limited available research on technology, return and reintegration, and highlighted a specific knowledge gap and roadmap for both researchers and practitioners to address. Given the type of challenges that exist surrounding decision making relating to return, circular migration and reintegration planning, and given the centrality of information to migrants' preparedness for return, ICTs may ultimately be able to fill a significant protection gap. Whether by addressing challenges related to housing, education, health, or by profiling returnees' skills and potential on the job market, or simply by reuniting families and creating transnational networks of solidarity, technology can support returnees, their families and their communities in many ways.

Our analysis highlights both the promise and limits of technology in the migration context and return in particular. It specifically draws attention to the potential for human rights abuses (as well as possibilities for leveraging technology to better support human rights in future), and state-centric use of technology that disregards the imperative of seeing returnees as legal persons (Benhabib 2004), i.e., people who are entitled to the full spectrum of rights, even in contexts where return may be involuntary and regardless of their migration status. While technology can help returnees in their return planning and reintegration, they can also cause harm when used for surveillance, identification and exclusion by states. These tendencies need to be carefully monitored to ensure that the rights of migrants and returnees are fully respected.

Thus, technology solutions to address the challenges faced by returnees need to put migrants and sustainable reintegration processes at the centre. The substantial digital divide between Sub-Saharan Africa, Asia and Europe, which has not been examined within this chapter, also needs to be urgently addressed in this regard, as do gender divides. While there are more migrants and returnees in East Africa, for instance, than in all European countries combined, most technological innovations on behalf of refugees and other migrants have been undertaken in Europe. A recent study found that over 100 technological innovations have been built for refugees worldwide since 2015 and almost all of these focus on Europe and the Syrian crisis (Samuel Hall 2018b). This gap in supportive technological innovation can be addressed if investments are made to support migrants in their decision-making process, and preparedness for return, and should they choose to return, for their sustainable reintegration.

Very little attention has also been given within the migration and technology literature to southern contexts hosting migrants and returnees. Much of the focus has been on the use of mobile technology for migration into Europe and the US, with little research conducted to assess the technology uses and needs of returnees in other contexts. Further research is therefore needed to investigate the potential of technology to support a human rights-based approach to return and reintegration that recognises the social, economic, psychological and political consequences of technology use for migrant lives.

NOTES

1. At https://www.iom.int/migapp (all webpages in the endnotes were accessed on 5 September 2021).
2. At https://www.airbnb.com/weaccept.
3. At https://techfugees.com/.
4. At https://holacode.com/.
5. Established in 2012, the Deferred Action for Childhood Arrivals (DACA) programme provides deferred action from deportation for two years for undocumented migrants who were brought to the United States as children, providing a path to obtaining work permits and towards regularisation.

REFERENCES

Alonso, A. & P.J. Oiarzabal (2010), *Diasporas in the New Media Age: Identity, Politics, and Community*. Reno, Nevada: The University of Nevada Press.

Andrade, A.D. & B. Doolin (2015), 'Temporal enactment of resettled refugees' ICT-mediated information practices', *Information Systems Journal*, 29, 145–174.

Barabas, C. & E. Zuckerman (2016), 'Can Bitcoin be used for good?' *The Atlantic*, 7 April. Accessed 20 April 2020 at https://www.theatlantic.com/technology/archive/2016/04/bitcoin-hype/477141/.

Battistella, G. (2018), 'Return Migration: A Conceptual and Policy Framework', Center for Migration Studies Essay Series, published on 8 March. Accessed 20 April 2020 at https://cmsny.org/publications/2018smsc-smc-return-migration/.

Benhabib, S. (2004), *The Rights of Others: Aliens, Residents, and Citizens*. Cambridge: Cambridge University Press.

Bilefsky, D. (2016), 'Danish Law requires asylum seekers to hand over valuables', *The New York Times*, 26 January.

Brekke, J. & A. Staver (2019), 'Social media screening: Norway's asylum system', *Forced Migration Review* (FMR 61). Accessed 5 September 2021 at: https://www.fmreview.org/sites/fmr/files/FMRdownloads/en/ethics/brekke-balkestaver.pdf.

Cassarino, J. (2004), 'Theorising return migration: The conceptual approach to return migrants revisited', *International Journal on Multicultural Societies*, 6 (2), 253–279.

Crouch, D. & P. Kingsley (2016), 'Danish parliament approves plan to seize assets from refugees', *The Guardian*, 26 January. Accessed 20 April 2020 at https://www.theguardian.com/world/2016/jan/26/danish-parliament-approves-plan-to-seize-assets-from-refugees.

Dekker, R. & G. Engbersen (2012), 'How social media facilitate migrant networks and facilitate migration', International Migration Institute (IMI) Working Paper published online 20 November 2011. Accessed 20 April 2020 at https://www.migrationinstitute.org/publications/wp-64-12.

Dijstelbloem, H. (2017), 'Migration tracking is a mess', *Nature*, 543 (7643), pp. 32–34.

Diminescu, D. (2008), 'The Connected Migrant: An epistemological manifesto', *Social Science Information*, 47 (4), pp. 565–579.

DRC/NRC/IRC/ReDSS/DSP/ADSP/Samuel Hall (2019), 'Unprepared for (Re)integration: Lessons Learned from Afghanistan, Somalia, and Syria on Refugee Returns to Urban Areas', European Research Council. Accessed 20 April 2020 at https://www.samuelhall.org/publications/redss-unprepared-for-reintegration-lessons-learned-from-afghanistan-somalia-and-syria-on-refugee-returns-to-urban-areas.

Edmondson, C. (2019), 'ICE used facial recognition to mine state driver's license databases', *The New York Times*, 7 July. Accessed 20 April 2020 at https://www.nytimes.com/2019/07/07/us/politics/ice-drivers-licenses-facial-recognition.html.

Frouws, B., M. Phillips, A. Hassan & M. Twigt (2016), 'Getting to Europe the Whatsapp way: The use of ICT in contemporary mixed migration flows to Europe', *Regional Mixed Migration Secretariat Briefing Paper*. Accessed 20 April 2020 at https://ssrn.com/abstract=2862592.

Hamel, J.-Y. (2009), 'Information and Communication Technologies and Migration', *Human Development Research Paper* 2009/39. UNDP.

Harindranath, G. (2019), 'Why migrant technology research needs "values" at its core', *Migration for Development and Equality (MIDEQ)*, 31 August. Accessed 20 April 2020 at https://www.mideq.org/en/blog/migrant-technology-research-values/.

Hunter, A. (2015), 'Empowering or impeding return migration? ICT, mobile phones, and older migrants' communications with home', *Global Networks*, 15 (4), 485–502.

IOM (2017), *Towards an Integrated Approach to Reintegration in the Context of Return*, International Organisation for Migration, Geneva. Accessed 20 April 2020 at https://www.iom.int/sites/default/files/our_work/DMM/AVRR/Towards-an-Integrated-Approach-to-Reintegration.pdf.

Johnson, K. (2020), 'ICE's warrantless facial recognition searches trigger Maryland bill', *VentureBeat*, 27 February. Accessed 30 June 2020 at https://venturebeat.com/2020/02/27/ice-warrantless-facial-recognition-searches-trigger-maryland-bill/.

Kingston, L.N. (2018), 'Biometric identification, displacement and protection gaps', in C.F. Maitland (ed.), *Digital Lifeline? ICTs for Refugees and Displaced Persons.* Cambridge, MA: The MIT Press, pp. 35–53.

Kozachenko, I. (2013), 'Horizon Scanning Report: ICT and Migration', *Working Papers of the Communities and Culture Network+*, 2. ISSN 2052-7268.

Madianou, M. (2012), 'Migration and the accentuated ambivalence of motherhood: the role of ICTs in Filipino Transnational Families'. *Global Networks*, 12 (3), 277–295

Maitland, C.F. (2018), 'The ICTs and displacement research agenda and practical matters', in C.F. Maitland (ed.), *Digital Lifeline? ICTs for Refugees and Displaced Persons.* Cambridge, MA: The MIT Press, pp. 239–258.

Majidi, N. (2017), 'Home sweet home! Repatriation, reintegration, and land allocation in Afghanistan', *Revue des mondes musulmans et de la Méditerranée*, 133, pp. 207–225.

Mak, A. (2019), 'ICE's reckless use of facial recognition tech', *Slate*, 10 July. Accessed 20 April 2020 at https://slate.com/technology/2019/07/ice-is-using-facial-recognition-technology-at-dmvs.html.

Meaker, M. (2018), 'Europe is using smartphone data as a weapon to deport refugees', *Wired*, 28 July 2018. Accessed 20 April 2020 at https://www.wired.co.uk/article/europe-immigration-refugees -smartphone-metadata-deportations.

Mijente, 2019. *Who's Behind ICE? The Tech Companies and Data Fueling Deportations.* Accessed 5 September 2021 at https://mijente.net/wp-content/uploads/2018/10/WHO%E2%80%99S-BEHIND -ICE_-The-Tech-and-Data-Companies-Fueling-Deportations-_v1.pdf.

Molnar, P. (2018), 'The contested technologies that manage migration', *Center for International Governance Innovation*, 14 December 2018. Accessed 20 April 2020 at https://www.cigionline.org/ articles/contested-technologies-manage-migration.

Molnar, P (2019a), 'Emerging voices: immigration, iris-scanning, and iBorderCTRL – the human rights impacts of technological experiments in migration', *OpinioJuris*, 19 August. Accessed 20 April 2020 at http://opiniojuris.org/2019/08/19/emerging-voices-immigration-iris-scanning-and-iborderctrl-the -human-rights-impacts-of-technological-experiments-in-migration/.

Molnar, P. (2019b), 'New technologies in migration: human rights impacts', *Forced Migration Review* (FMR 61:7–9).

Molnar, P (2020), 'Technology, migration, and illness in the times of COVID-19', *EDRI*, 15 April 2020. Accessed 30 June 2020 at https://edri.org/technology-migration-and-illness-in-the-times-of-covid-19/

Nie, N. L. Erbring & S. Hillygus (2002), 'Internet use, interpersonal relations and sociability: Findings from a detailed time diary study', in B. Wellman & C. Haythornwaite, *The Internet in Everyday Life*, ed. Oxford: Blackwell Publishers, pp. 215–243.

OECD (2018), *Bridging the Digital Gender Divide: Include, Upskill, Innovate*, accessed 20 April 2020 at http://www.oecd.org/internet/bridging-the-digital-gender-divide.pdf.

Panagakos, A. & H. Horst (2006), 'Return to Cyberia: Technology and the social worlds of transnational migrants', *Global Networks*, 6 (2), 109–124. Accessed 5 September 2021 at https://www.dhi.ac.uk/ san/waysofbeing/data/data-crone-panagakos-2006.pdf.

Pirjola, J. (2015), 'Flights of shame or dignified return? Return flights and post-return monitoring', *European Journal of Migration and Law*, 17, 305–328.

Rivlin-Nadler, M. (2019), 'How ICE uses social media to surveil and arrest immigrants', *The Intercept*, 22 December. Accessed 5 September 2021 at https://theintercept.com/2019/12/22/ice-social-media -surveillance/.

Ruiz Soto, A., L. Rodrigo Dominguez-Villegas & R. Capps (2019), *Sustainable Reintegration: Strategies to Support Migrants Returning to Mexico and Central America.* Washington D.C.: Migration Policy Institute (MPI).

Ruben, R. M. van Houte & T. Davids (2009), 'What determines the embeddedness of forced-return migrants? Rethinking the role of pre- and post-return assistance', IMR, 43(4), 908–937.

Samuel Hall (2018a), 'Cartographie et profil socio-économique des communautés de retour au Senegal', Report. Accessed 20 April 2020 at https://www.samuelhall.org/publications/cartographie-et-profil -socio-conomique-des-communauts-de-retour-au-sngal.

Samuel Hall (2018b), 'Innovating mobile solutions for refugees in East Africa', Report. Accessed 20 April 2020 at https://static1.squarespace.com/static/5cfe2c8927234e0001688343/t/5d1f1b83bfecef0 001fb6621/1562319758732/Innovating_mobile_solutions_report_2018.pdf.

Samuel Hall (2018c), 'Mapping and socioeconomic profiling of communities of return in Nigeria', Report. Accessed 20 April 2020 at https://static1.squarespace.com/static/5cfe2c8927234e0001688343/t/5d1dcb3e83dfbe00019e3543/1562233672396/IOM+-+Nigeria+-+Full+Report.pdf.

Sanchez, G., R. Hoxhaj, S. Nardin, A. Geddes, L. Achilli & R.L. Kalantaryan (2018), 'A study of the communication channels used by migrants and asylum seekers in Italy, with a particular focus on online and social media', European University Institute Migration Policy Centre, with funding from the European Commission. Accessed 5 September 2021 at https://cadmus.eui.eu/handle/1814/61086.

Schmidtke, R. & R. Chuayffet (2018), 'Tapping into Tech: How one organization is connecting returned Mexicans and migrants in Mexico to jobs in the tech industry', *Wilson Center Mexico Institute*, 4 October. Accessed 1 July 2020 at https://www.wilsoncenter.org/article/tapping-tech-how-one -organization-connecting-returned-mexicans-and-migrants-mexico-to-jobs.

Schuster, L. & N. Majidi (2013), 'What happens post-deportation? The experience of deported Afghans', Migration Studies, 1 (2), 221–240. Accessed 5 September 2021 at https://academic.oup.com/migration/article-abstract/1/2/221/990706.

Simon, M (2020), 'Investing in immigrant surveillance: Palantir and the #NoTechforICE Campaign', *Forbes*, 15 January. Accessed 20 April 2020 at https://www.forbes.com/sites/morgansimon/2020/01/15/investing-in-immigrant-surveillance-palantir-and-the-notechforice-campaign/#26f3cdddb770.

Singler, S (2018), 'The role of technology in the criminalisation of migration', Border Criminologies Blog, 28 November. Accessed 5 September 2021 at https://www.law.ox.ac.uk/research-subject -groups/centre-criminology/centreborder-criminologies/blog/2018/11/role-technology.

Suri, T. & W. Jack (2016), 'The long-run poverty and gender impacts of mobile money', *Science*, 354 (6317), 1288–1292.

UNESCO (2013), *Ethical and Societal Challenges of the Information Society: Background document distributed for the session on Ethics of the Information Society*, 28 May. Accessed 20 April 2020 at http://unesdoc.unesco.org/images/0022/002209/220998e.pdf.

UNESCO (2019), 'UNESCO qualifications passport for refugees and vulnerable migrants', Paris. Accessed 20 April 2020 at https://en.unesco.org/themes/education-emergencies/qualifications -passport.

UNHCR (2018), 'Asylum procedure: cell phone search has no benefits'. Accessed 20 April 2020 at https://www.unhcr.org/blogs/asylum-procedure-cell-phone-search-no-benefits/.

Wang, L.K (2016), 'The benefits of in-betweenness: Return migration of second generation Chinese American professionals to China', *Journal of Ethnic and Migration Studies*, 42 (12), 1941–1958.

Wang, L.K (2017), '"Leftover Women" and "Kings of the Candy Shop": Gendering Chinese American ancestral homeland migration to China', *American Behavioral Scientist*, 61(10), 1172–1191.

PART IV

CONNECTIVITY AND MIGRATION: TRENDS AND IMPACTS

17. Technology for engaging and empowering migrant workers

Angela Kintominas, Laurie Berg and Bassina Farbenblum

INTRODUCTION

Mobile phones and digital technologies are now a ubiquitous feature of migrant workers' everyday lives. Technology plays an indispensable role for migrants navigating the migration experience including maintaining their family life transnationally and overcoming social isolation. It is also clear that technology can play a role in helping migrants find decent work, gain information on their rights as workers, crowdsource knowledge from peers, improve access to services and legal remedies, and build broader networks of solidarity. In this context, the chapter explores the potential to leverage the growth in smartphone penetration and digital connectivity to better engage and empower migrant workers. It explores four core and overlapping domains where digital technology interventions seek to improve the engagement and empowerment of migrants: providing migrant workers with accurate and up-to-date information on their legal rights and market conditions; empowering migrant workers to rate and review their employers and recruiters; promoting solidarity and collective organizing; and finally, supporting migrants' access to justice. The focus is primarily upon initiatives driven by civil society and community actors such as migrant worker organizations, legal advocates and trade unions as part of their efforts to engage and empower hard-to-reach and isolated migrant workers. However, businesses and governments face similar sets of concerns when deploying technology in this sphere.[1]

What emerges from a review of these tools and initiatives is that digital technology has enormous potential to support the engagement of migrant workers more rapidly, cheaply and at greater scale. But despite tendencies towards 'tech-utopianism' (Toyama 2015), technological interventions are far from a panacea to complex barriers to empowerment and engagement, and can rarely transform the structural drivers of migrant workers' precarity. Technocratic enhancements to migrant workers' ability to record their hours and wages owed are of little use where court processes remain discriminatory, corrupt, slow and prohibitively costly. More information about legal rights and vast numbers of reviews on employers and recruiters may not change the fact that migrant workers have few other options but to accept exploitative work. Digital interventions likewise introduce a suite of new risks. This includes heightened risks to migrant workers' privacy and data security, legal risks for migrants and organizations relating to defamation and libel, and challenges in ensuring financial sustainability is values-aligned. It is impossible to eliminate all risks involved in collecting data from migrant workers, and so organizations face a series of ongoing trade-offs between desirable ends. For example, collecting and/or storing less personal data may minimize potential risks of harm to migrant workers but reduces an organization's capacity to follow up over time and respond to each individual's problems.

Ultimately, digital initiatives are most likely to be successful in engaging and empowering migrant workers where they are well integrated into an organization's broader (and, typically, offline) programs with well-conceived theories of change. This involves understanding the broader political, economic and other forces structuring migrant workers' precarity and a clear understanding of how technology will expand or strengthen an organization's core activities and functions, as well as recognizing limits and risks.

METHOD

This chapter is informed by qualitative empirical research and a desk-based review of the emergent literature on technology-based initiatives for migrant worker engagement.[2] Literature included legal materials, academic literature, and grey literature such as media reporting, public reports, policy briefs and website content. A mapping exercise was conducted primarily in 2018 to identify and categorize emerging digital tools and technology-based initiatives. Where feasible, the authors downloaded the relevant app, made an inventory of its features and identified the developer or affiliated organisation.

This literature review and mapping exercise guided the authors' approach to a series of preliminary conversations (conducted via Skype/Zoom) with a range of individuals with expertise and experience in the design, use, funding, regulation or research of technology for migrant worker engagement and empowerment. These exchanges were followed by more detailed interviews and email exchanges. Participants were selected via snowball sampling until data saturation was achieved. Interviewees included migrant organizations and legal advocates (n=5), trade unions and other migrant worker organisations (n=9), a multi-stakeholder initiative with business and worker organizations (n=1), government agencies and regulators (n=3), researchers/consultants (n=8) and donors/investors (n=9).[3] Interviews were semi-structured and participants were asked to explain the functions and mechanisms of the tools they had developed, funded or implemented (if relevant), reflect on their challenges in implementation, funding and managing risks including safety and data privacy, as well as for general reflections on the state of the emerging field. Interviews were recorded (with consent) to enable the authors to engage in multiple rounds of thematic analysis. In addition, a two-day global convening of experts was hosted in February 2018, comprising plenary panels, small group discussions and anonymous reflections.

This is a fast-moving field with new digital tools being launched, tested and retired at rapid pace. This chapter therefore does not seek to provide an exhaustive catalogue of all digital technology initiatives for migrant worker engagement and instead seeks to thematically capture key trends and general challenges and risks.

Technological tools typically transgress geographical boundaries as they are funded, developed and rolled-out across a wide range of jurisdictions. This necessitates a global view to examine emerging and cross-cutting trends. At the same time, the contexts in which these tools are implemented vary substantially. Even within the Global North, access to justice, the standing of trade unions and worker organizations, and the political economy of migration varies enormously. This preliminary mapping of the field therefore raises questions that must be answered by local-level empirical research and user-testing within specific contexts and communities. It is likely that findings about the suitability, safety and success of the same tool could vary substantially in different contexts with different groups of migrants.

Direct engagement with migrant workers is essential to understand the operation and outcomes of digital technology initiatives that seek to empower them. There is a pressing need for future research in this area. Regrettably limited funding and time did not permit the authors to undertake fieldwork with migrant workers directly or to conduct user-testing as part of this study. The authors nevertheless sought to incorporate migrant workers' experiences through interviews with representative organisations, legal service providers, trade unions and other worker organizations.

PROVIDING MIGRANT WORKERS WITH RESPONSIVE AND TAILORED INFORMATION

The profound information asymmetry between workers and their recruiters and employers heightens their vulnerability to exploitation. Democratizing access to information can in some circumstances enable migrant workers to make safer and better decisions, or take action when problems arise.

The theory of change assumes that possession of more information empowers migrant workers as rational and self-reliant agents and ultimately creates a more transparent and competitive marketplace. However, this assumption has been challenged in some contexts. Empirical research has established, for example, that underpaid temporary migrant workers in Australia (such as international students and working holiday visa makers) are generally well aware that they are being paid below the minimum wage but do not feel they can expect to receive that wage given market conditions (Berg and Farbenblum 2017). Findings like this raise broader questions about the power of information-delivery on its own to prevent injustices due to the ways in which precarious employment is produced in a broader social and economic context. Even after being better informed about the risks related to particular recruiters, employers or industries, migrant workers in restricted employment or recruitment markets may have few options other than to accept exploitative and undesirable work. Nevertheless, it is also clear that misinformation and misperceptions, for instance about applicable legal protections or firm policies, can lure migrant workers into exploitative situations or present barriers to seeking assistance.

In this context, countless digital initiatives have emerged seeking to improve the quality and quantity of information available for migrants making decisions about migration and work. Many provide information on migrant workers' legal rights and how to practically navigate legal and bureaucratic processes. Other tools provide information on the job market to set realistic expectations on pay and labour conditions in their country of employment, as well as on reputable versus blacklisted recruiters and employers. More innovative tools seek to keep up with rapidly changing legal rules and regularly update their lists of accredited and blacklisted recruitment agents in order to provide information that would otherwise be difficult or impossible for migrant workers to find (e.g., Golden Dreams). Some tools do this in the form of AI chatbots to provide answers in real-time to a worker's specific questions (e.g., WorkIt). Others create a Q&A area where users can submit questions and receive direct responses (e.g., Shuvayatra). Seeking to fill information gaps in particular migration contexts, some tools focus on providing country- or corridor-specific information (e.g., My Labour Matters). Some go beyond know-your-rights-style information and also include information on financial literacy specifically relevant to migrant workers such as on loan repayments, interest accumulation and foreign exchange (e.g., Shuvayatra, SaverAsia). In order to improve accessibility, there have been efforts to use animations, comics, videos, podcasts and radio dramas (often designed to require limited internet bandwidth), create short online training modules, or offering customisable alerts or newsfeeds (e.g., Smart Domestic Workers, Contratados, Shuvayatra). Some organizations acknowledge the limited reach of their own app or platform and therefore strategically partner with popular commercial and community radio programs to multiply their dissemination pathways (e.g., Shuvayatra). These tools are generally being developed by civil society organizations, trade unions and not-for-profit organizations, although there is similar potential for governmental agencies to adopt digital tools to provide pre-departure information and resources on navigating bureaucratic processes and accessing services.

Digitally enabled information-provision is attractive because it can be more rapidly and cheaply uploaded, updated and transmitted between organizations and hard-to-reach and geographically dispersed migrant workers via their mobile phones. Digital tools and initiatives including online pre-departure training and outreach can often be best placed to reach migrant workers before they have left their country of origin.

Nonetheless, there are major and ongoing resourcing challenges to ensure that information is accurate and up to date. For example, ATUC acknowledges that, even if their information service ATIS became 'fully operational[,] its database capability can never match that of governments' because registration or enrolment in the *ATIS* database is purely voluntary' (Bitonio Jr 2018: 8). Trade unions, worker organizations and legal advocates may therefore lack the ability to provide comprehensive and accurate information in a way that is sustainable over time without access to official government data, which of course may not be forthcoming (cf. the model used by Recruitment Advisor, discussed further below). Even where governments are willing to share data, they may not actually collect sufficient or sufficiently detailed information on non-compliant recruiters and employers. Their listings may not include data on migrant worker complaints or enforcement actions (Athreya (United States Agency for International Development) 2018), or if they do, these data may be rarely updated making them less reliable (Silitonga (Tifa Foundation) 2018). More broadly, information provision practices raise complex questions about whether non-governmental organizations seek to duplicate government functions or strategically refocus their efforts elsewhere (Bitonio Jr 2018: 8).

There are also challenges in determining the content and mode of information delivery most suitable for diverse groups of migrant workers at different stages of migration cycles. Moreover, there are significant risks of harm associated with disseminating out-of-date or inadvertently inaccurate information, for example, representing a recruitment agency as accredited when it in fact has a record of engaging in extreme exploitation or trafficking. Crucially, even the most accurate information is of little utility when it does not reach migrant workers who need it. Emerging qualitative research has painted a sobering picture of migrant workers' awareness of information-delivery initiatives by non-governmental and international organisations (see e.g., Ruget and Usmanalieva 2019). In one study, Filipino migrant workers were more inclined to trust the opinions of their peers on informal job forums and Facebook than government sources (Latonero et al. 2015: 26). Padding out tools with non-related features that will initially attract users and motivate their ongoing engagement (such as social/ community, entertainment, news functionality and content) appears essential. This is in addition to the obvious need for multiple rounds of user-testing, promotion, training and updating to make sure an app or tool works in the way that is intended.

ENABLING MIGRANT WORKERS TO RATE AND REVIEW THEIR RECRUITERS AND EMPLOYERS

Some of the most popular interventions to address 'informational failures' in the migrant worker job market are tools that crowdsource data directly from migrant workers themselves (e.g., Contratados, Recruitment Advisor, Pantau PJTKI, Golden Dreams). Rate and review platforms typically ask migrant workers about their experiences with employers, recruiters or other private intermediaries for the purpose of informing and alerting their migrant worker peers. They are typically structured as a series of closed questions that require answers using ratings (e.g., 1–5), ticking check boxes or selecting binary options (e.g., yes/no). Less commonly rate and review tools also allow open-text answers. The design principles of rate and review tools are generally inspired by their previous proliferation in other contexts, most famously TripAdvisor and Yelp. Rate and review tools for migrant workers are usually only based online (e.g., Contratados), although in initial set-up stages some organizations opted to collect a base level of information through in-person surveys that were then transcribed manually by staff (e.g., Recruitment Advisor).

Some platforms only allow migrant workers to post reviews anonymously, omitting their email address, phone number or other identifying information (e.g., Contratados). Other tools require users to register with an account using an email address or connect with Facebook (e.g., Recruitment Advisor, Golden Dreams). However, in these cases, platforms often still unlink reviews from account details afterwards in an effort to safeguard anonymity if the website is compromised (Recruitment Advisor, n.d.). Platforms then generally consolidate this data into larger datasets that are searchable, publicly available and can be found online if a migrant worker searches the name of the employer or recruiter on the internet.

By crowdsourcing knowledge directly from migrant workers, these tools seek to overcome some of the endemic challenges in ensuring information is accurate and up to date, as outlined above. The creation of a digital or online forum likewise creates a means for migrant workers across dispersed worksites, speaking different languages, or contending with restrictions on their freedom of movement and expression to more easily communicate with one another. The ability to share information anonymously may alleviate cultural reluctance to criticize someone in their community or reduce fears of retribution. Where these tools integrate other functionalities, such as news sections and blog posts, they may also help build solidarity, initiate broader conversations, kick-start organizing or build worker power (Jones and Nuriyati, n.d.).

A significant critical mass of reviews is needed in order for rate and review platforms to be reliable. This is a mammoth task given the sheer number of recruiters and employers of migrant workers this entails. For example, one of the most successful initiatives in this space, Contratados by bi-national migrant workers' rights organization Centro de los Derechos del Migrante, Inc. (CDM), has close to 20,000 recruiters/employers listed in its database. Even where critical mass is reached, the subjectivity of reviews and the personal circumstances of each reviewer raise problems of reliability. For example, one platform found that migrant workers who disclosed significant rights violations nonetheless provided a 4 out of 5 rating of the recruiter whose licence was subsequently revoked and whose director and staff were under investigation for human trafficking offences (Jones and Nuriyati, n.d.: 24). Even negative reviews can understate the severity of a problem or mask more serious issues that a reviewer may decline to report (Rende-Taylor (Issara Institute) 2018).

PROMOTING SOLIDARITY, PEER-TO-PEER CONNECTIONS AND COLLECTIVE ORGANIZING AMONG MIGRANT WORKERS

Worker organizations and trade unions operate in a climate of acute threats to their capacity to engage, empower and organize migrants and other low-wage and vulnerable workers. In the Global North, the standard employer–employee relationship is declining and being replaced by casualized and more precarious forms of work. This fragmentation of work has had consequences downstream as work is outsourced in opaque supply chains to offshore worksites where labour costs (typically migrant) are low. Challenges to worker organizing are most extreme in the Global South, where freedom of expression and collective action is actively suppressed or illegal, and labour law protections are most tenuous. Migrant workers face additional barriers in this difficult climate. Migrant workers are often not seen by unions and worker organizations as part of their traditional core membership base, and they typically work in informal industries without sufficient labour law coverage (such as horticulture and agriculture, meat processing, cleaning, construction, security, labour hire, and domestic and care work). Moreover, migrant workers are often based in rural and isolated locations away from large city and regional centres, they do not speak the same language as other workers or organizers, and they can be unfamiliar with the work of trade unions and other advocates.

Against this backdrop of escalating isolation and fragmentation of work, and the intensified risks that migrant workers face in that context, an extensive body of research has surveyed the ways in which migrants use new media more generally for interpersonal communication, connectedness, support, maintaining transnational family life and trans-local communities (see e.g. Madianou et al. 2013). Similarly, migrant worker advocates (typically outside or on the fringes of the formal labour movement) are now increasingly experimenting with social networking platforms and technology-based tools to engage and organize migrant workers. Some organizations leverage the power of pre-existing mobile message apps and social networking sites to foster connections, share information and exert pressure on brands and employers. Elsewhere, organizations are investing significant resources to build bespoke digital platforms to create communities where migrant workers can more safely engage collectively.

Leveraging the power of social networking sites can be highly attractive to migrant worker organizations, legal advocates and other organizers, especially where funding and resources are limited. Where migrant workers already participate in vibrant and active online communities, it may be highly effective to engage potential future users where they are located, using technology with which they are already familiar. This can increase an organization's capacity for engagement due to high traffic. It can also help organizations to identify grassroots and emerging leaders and organizers. Organizations can likewise help steer pre-existing community conversations and correct misinformation at the point it is shared.

For these reasons, many migrant worker organizations use Facebook (and similar platforms, such as Line, WeChat, WhatsApp and Twitter) as one of their primary means of engaging their constituents (e.g., My Labor Matters, OFW Watch). There are a number of challenges and risks involved, however, that would benefit from consideration before organizations rely on social networking sites for migrant worker organizing. On a practical level, social networking sites often have only rudimentary search functions, which inhibits users' ability to quickly access specific information previously provided by other users (Dehlendorf (Organization United for Respect) 2018; Micah-Jones (Centro de los Derechos del Migrante, Inc.) 2018). Platform hosts may also confront the risk that a social media company could change or remove a functionality being relied upon for migrant worker organizing without prior consultation. In addition, because the social media company owns the platform, in the event of a split within the organization that established the forum, page or group, the external company may ultimately determine which entity maintains ongoing control (Dehlendorf (Organization United for Respect) 2018).

It has also been observed, perhaps most problematically, that social networking sites "occupy a liminal territory between 'open' and 'closed'" status (Burkell et al. 2014: 975). Users often share highly personal information without realizing who else can view it or what unknown third parties can do with their data pursuant to data sharing protocols. A user does not typically own the information they share online, and even if they delete a post it generally remains on the platform's servers. Complex overlays, where apps are built on top of other platforms like Facebook, can produce further uncertainties for data ownership and privacy (Latonero (Data & Society) 2018). Social media sites likewise carry a risk of infiltration by someone using a fake online profile to undermine the migrant workers' action, or obtain information to benefit the employer or recruiter (Bradshaw and Howard 2017; Forestal 2017).

Tailored peer-to-peer platforms bring their own set of benefits, risks and trade-offs. Designing a new peer-to-peer platform carries significant cost and requires multiple rounds of user testing. Moreover, building a new website or service does not guarantee that an audience, let alone the particular target audience, will use it (Civic Patterns, n.d.). On the other hand, bespoke platforms benefit from mediators, gate-keepers, and experts who facilitate the forum, making it less chaotic than unmediated social network spaces (Berfield 2016). These platforms are also designed to overcome risks for migrant workers of surveillance and retaliation by employers and recruiters that may monitor social media, though they cannot alleviate risks of infiltration completely.

There also remains a question as to whether a digital platform can facilitate connections between migrant workers where they do not already exist, or whether migrants need pre-existing connections before a platform has the potential to make that collaboration quicker, more efficient, secure and/or scalable. In either context, organizers seeking to use digital platforms confront the challenge of how to establish trust and engagement. Both existing social media and bespoke digital platforms can significantly enhance organizing or collective action but carry a risk of rendering invisible the work that needs to happen in order to transfer the connection and energy that develops in the online space into offline action and enduring change (Tufekci 2014).

SUPPORTING MIGRANT WORKERS TO ACCESS JUSTICE

When things go wrong, migrant workers face significant challenges in accessing remedies in both formal and informal institutions of justice (Harkins and Åhlberg 2017; Farbenblum et al. 2013). Many of these barriers to justice reflect protracted and structural inequalities with legal and migration systems. For a start, many migrant workers, especially seasonal workers and domestic workers, are explicitly excluded from the scope of domestic labour law such as minimum wage protections (Parreñas and Silvey 2018). Even when not explicitly excluded, the application of statutory protections and entitlements to non-citizens can be unclear. In situations where migrant workers are undocumented, working in breach of their visa requirements, or have a visa that is tied to a particular employer, contacting state labour protectorates or seeking legal remedies carries significant risk that their visa will be cancelled or they will be deported (Farbenblum and Berg 2017). Migrant workers therefore legitimately fear job loss, deportation, other negative migration consequences and employer retaliation if they raise grievances such as wage theft. They are also more likely to have their legal needs unmet due to the unavailability of legal assistance and inaccessibility of justice mechanisms for migrants. This often deters migrant workers from reporting violations and seeking remedy in the first place. Where they do seek help, migrant workers regularly encounter discriminatory and dismissive attitudes from institutional gatekeepers, government agencies and courts. Administrative and legal processes can be expensive, slow and corrupt. There is often limited scope for reform due to under-resourcing and lack of political will. Even in cases where a migrant worker successfully navigates the system and justice is formally 'won', they routinely face evasion of liability by employers and recruiters who disappear, declare bankruptcy or disregard orders.

Digital tools are not a replacement for institutional and structural reform, and their effectiveness will be limited unless they operate in tandem with broader interventions. There are nonetheless some promising ways in which migrant worker and civil society organizations, trade unions and legal service providers (as well as government agencies) have begun to use these tools to help reduce the barriers to engaging with imperfect legal processes. This includes the strategic and well-integrated use of documentation and evidence-gathering tools, tools that help facilitate referrals, complaints and claims-making, as well as tools for managing transnational litigation.

One of the key grievances that migrant workers face is non-payment or underpayment of wages. However, a major barrier to accessing remedies is that many migrant workers are unable to make claims because they lack required evidence. A number of digital tools have been developed to assist migrant workers overcome this obstacle, for example by generating a complete and reliable record of hours worked. Some tools rely on migrant workers to manually enter the hours they work (e.g., DOL Timesheet, Outflank), others use an automatic location-tracking function, or a combination (e.g., Jornaler@/Reporter NICE). Other tools use geo-fencing technology, which allows the user to pre-program when they wish to record their location within a pre-defined geographic boundary, such as their worksite (e.g., FWO's Record My Hours).

Automatic location-tracking functions relieve migrant workers from dedicating time and energy to manually record their hours, which can often be erratic or spread across multiple worksites. Yet, in order to create a complete record of time spent in the workplace, migrant workers need to be motivated to engage with the digital record-keeping tool from the outset, rather than after a problem has arisen. Advocates have observed that migrant workers, when they begin a new job, often prefer to assume that their new employer will comply with labour obligations, and so resist taking the precautionary measures offered by the digital tool (Figueroa (Cornell University) 2018). Even when used consistently, there is a broader risk that over time an automatic location-tracking tool could increase the evidentiary burden on migrant workers across a jurisdiction by creating an expectation they always produce external verification of hours worked (Ticona 2016). Perhaps most critically, the collection and storage of such data carry significant privacy and security risks. These risks are particularly acute for migrant workers, such as the possible severe harms associated with location-sharing data being intercepted by employers for the purposes of surveillance or by immigration officials for the purposes of enforcement or deportation (discussed further below).

Some digital tools take this evidence-gathering a step further and seek to enable migrant workers to calculate their wages owed for the purposes of a formal wage recovery claim (e.g., Jornaler@/Reporter NICE). Accurate evaluation of wage underpayment is typically costly, time-consuming, and requires technical and accounting expertise. Wage calculation tools have the potential to cut across this complexity through their use of algorithms that apply the appropriate statutory or contractual base rate of pay in the migrant worker's industry and jurisdiction to the number of hours worked, as well as additional entitlements such as overtime (e.g., Outflank). These tools can therefore significantly reduce the workload for community legal service providers seeking to meet migrant workers' legal needs by reducing the level of expertise and time that a paralegal or lawyer would need to make complex calculations. However, because wage structures vary widely between and within countries, wage calculation tools may either need to be confined to individual jurisdictions and/or industries, or require significant resources to be replicated for different contexts (Figueroa (Cornell University) 2018).

Several platforms have been developed that seek to connect migrant workers with legal advocates, trade unions and other service providers (e.g., WorkerReport, Jornaler@, ATIS). For example, the Asian Trade Union Council (ATUC) Information System for Migrant Workers (ATIS), launched by ASEAN Trade Union Council in 2018, is an online complaints mechanism and referral system. After registering an account or logging in via Facebook, migrant workers can submit an inquiry or complaint, which is then passed onto national trade union focal points who can provide appropriate information or service, or refer migrant workers onto a relevant support organisation. A risk associated with these tools is that small migrant worker organizations and legal advocates may not have the capacity to meet dramatic increases in their demand for services (Figueroa (Cornell University) 2018; Rojas (The Workers Lab) 2018). If a platform can be downloaded anywhere, but an organization only caters for migrant workers in a particular region, this can leave migrants with unmet expectations of assistance (Figueroa (Cornell University) 2018).

Finally, some organizations have designed digital tools to facilitate transnational litigation that address legal challenges in intercepting deceptive and fraudulent recruitment across jurisdictions. An example in this context is the RADAR database, which supports the Project for Economic, Cultural, and Social Rights' (Proyecto de Derechos Económicos, Sociales y Culturales / ProDESC) litigation on behalf of migrant workers in Mexican and US courts. ProDESC uses the database to capture, securely store and analyse de-identified information provided by migrant workers of labour abuses across complex supply chains. The tool is intended to enable ProDESC to put recruiters, employers and companies on notice of potential labour violations in their supply chains and thereby satisfy an element of liability under US law. Parts of the database are shared with US-based lawyers to assist them in legal actions on behalf of migrant workers in the United States.

CROSS-CUTTING CHALLENGES AND RISKS

Design and Implementation for Uptake

Despite the exponential growth of smartphone ownership and online connectivity, access to digital technology and the quality of access remains unequally distributed depending on age, income, education, gender, IT environment and migration status (Pew Research Centre 2018;

International Labour Organization 2019). Migrant workers face particular barriers in accessing digital technology. This includes the cost of paying for data, inability to use their home phone in their country of employment with related loss of access to data linked to their phone number, isolation in areas with poor connectivity (such as fishing vessels, farms and factories), long working hours, employer surveillance, and limited literacy in their own language or the language of their country of employment.

These inequities of digital access compound the usual challenges to design and uptake of digital technology tools. Generally speaking, very few apps are used more than a handful of times before they are deleted, and tools developed in this space are no exception (Rojas 2017). One organization explained that despite extensive co-design with the migrant worker community, out of 110 migrant workers who were assisted in downloading their app only 10 actually used it for some time (Figueroa (Cornell University) 2018).

Inclusive and iterative design processes with multiple rounds of piloting, user-testing and updating are therefore essential to long-term adoption and impact of a digital tool. Adopting good design principles, such as designing for accessibility, configuring apps in a user's native language and permitting content to be downloaded and accessed offline, are likewise key to address migrant workers' barriers to access and meaningful engagement (Kalbag 2017), as is the possibility of accessing content simply online without the need to download an app or new software.

Migrant Workers' Privacy and Security

Collecting sensitive data from vulnerable populations such as migrant workers involves significant risk. Centralized databases can be hacked or can unintentionally leak data through a security mistake. Government agencies (such as labour inspectorates and immigration departments), employers and recruiters may also seek to access migrant workers' data by subpoenaing it through legal processes. Where data are stored locally on a migrant worker's device, there are risks that third parties, such as their employer, will confiscate or otherwise access their phone to access sensitive data. These risks are especially acute for those in isolated worksites such as live-in migrant domestic workers and migrant workers on fishing vessels and rural farms. Even where data are not surveilled, hacked, leaked or subpoenaed, identities can be accidentally revealed. For example, an insufficient number of reviews about a particular recruiter or employer poses risks that an individual review can be traced back to the migrant worker who wrote it despite their formal anonymity. These sorts of data compromises expose migrant workers to significant risks of harm. This includes acute threats to personal security and safety, for example where automatic location-tracking technology is misused to track their exact whereabouts. Workers who are linked to a negative review they made about an employer or recruiter face risks of job loss, criminal prosecution for defamation and physical retaliation (discussed further below). Where immigration authorities gain access to data revealing unauthorized work, workers likewise face significant risks of detention and deportation.

Generating and extracting sensitive data therefore comes with a heightened responsibility to protect migrants' privacy and security. This is especially the case when security risks may not be visible to migrant workers and when data is collected in countries with weak security and rule of law (Rahman 2018). One way to systematically address data security risks is to formulate a 'theory of harm' establishing a taxonomy of the worst possible harms to migrant workers, along with mitigation strategies (Chambers 2015; Latonero (Data & Society) 2018).

Organizations must obtain migrant workers' informed consent to the use of their data, including informing migrant workers why the data are being collected, what they will be used for, who they will be shared with and the potential risks to them (Rahman (The Engine Room) 2018). Nevertheless, obtaining informed consent from migrant workers does not discharge larger obligations to protect workers' security. It remains essential that workers' data are collected, stored and used responsibly and in accordance with legal data protection frameworks and best practice. Critically, responsibility for ensuring migrant workers' security cannot be outsourced to technology developers who may not understand the relevant risks in this specific context (Rahman (The Engine Room) 2018).

Data security good practice requires collecting and storing the minimum amount of data needed and only holding those data for as long as needed (The Engine Room et al. 2016). Organizations should carry out regular data audits (mapping out what data is held, where and why), risk assessments, and threat modelling to assess potential risks and harms (Rahman 2018). Ultimately, however, it is impossible to eliminate all risks associated with data collection. Organizations therefore face a series of difficult trade-offs between possible benefits to migrant workers arising from collecting certain types of data and the extent to which the risks associated with that data collection can be managed.

Legal Risks

Many of the initiatives outlined in this chapter invite migrant workers to provide information or views about employers and recruiters. This exposes both organizations and migrant workers to potentially serious risks of civil and criminal liability for defamation or libel. These risks are present in public forums (such as websites) as well as notionally private spaces (such as on apps, on Facebook groups and via email). Defamation and libel laws also regularly have global reach, meaning that users and platforms can be liable despite being based in an entirely different jurisdiction, with potential imposition of criminal penalties and imprisonment. These laws can be used to harass organizations through costly vexatious litigation, or may be used to compel organizations to reveal information about the user who made a particular post.

Organizations that crowdsource views from migrant workers should advise migrant workers about the risks involved and explain that statements must not be untrue or misleading, and negative statements should be their honest opinion based on facts. Unless organizations have a policy of not collecting *any* identifying data and also permanently delete users' IP addresses, they may be obligated to advise migrant workers that they are not anonymous and there is a risk the organization could be legally compelled to reveal their IP address or other data. Organizations have taken a range of measures to reduce these risks to migrant workers, and protect against risks to their own organization including eliminating moderation or editorialization of content (Farbenblum et al. 2018: 36–8).

These risks require organizations to carefully consider the jurisdiction in which data will be stored, evaluating data retention requirements in certain jurisdictions (up to seven years) that have resource implications and long-term risks of compulsory acquisition of user data by governments or third parties in litigation.

Financial Sustainability

Complex social problems can rarely be fixed quickly, whether via a "digital" intervention or otherwise. The success of an initiative will likely hinge upon the time and funding available (sometimes over many years) to adequately research the drivers of a problem in a particular context, understand the perspectives of migrants and service providers, and then to design, test and refine that intervention iteratively, building a community of users over time.

As well as securing funding related to the *technological* aspects of an intervention (such as designing and testing software or tools and securing space to safely store new data), organizations need funding to support the essential ongoing offline work that accompanies and underpins successful digital tools – often for at least several years. For example, platforms that direct migrant workers to take action or seek help may result in exponential increases in demand for an organization's services, with a need for new organizational roles and responsibilities. Platforms that provide information on recruiters and employers will need to devote new resources to continually update that information to ensure it is accurate and safe. Where organizations run a Q&A service or supervise an AI chatbot answering migrants' questions, a round-the-clock team of experts needs to be working in the back-end. In order to ensure a critical mass of data on a rate and review platform, organizations may need to conduct extensive offline surveys and then, once a platform is launched, staff need to monitor these posts on an ongoing basis to remove contact details or other identifying information that might endanger migrants and delete spam and predatory content. Where organizations seek to integrate social networking platforms into their organizing, staff are needed for the ongoing close management and curation of those spaces, responding to inquiries via comments and direct messages, deleting inappropriate or dangerous content, and publishing and pushing out content. There is a risk that the innovative technological dimensions to an intervention mask the substantial volume of ongoing human resourcing beneath the surface.

Despite the complexity, resource-intensity and long-term demands of ensuring the success of a tech-based intervention, there is an increasing trend to finance initiatives in the digital innovation space by means of short-term seed funding, competitive seed grants and hackathons. Hackathons and other open competitions to solve a particular challenge generally give a shortlist of ventures support to develop their ideas with the potential prize of further funding. These competitions 'risk contributing to the "pilot and crash" phenomenon, by which new programs keep being introduced but then cannot find the long-term financial support they need' (Benton and Glennie 2016: 18–19). Hackathons foreground the role of technologists who design platforms without a deep understanding of the structural forces driving the problem and the concerns and vulnerabilities of potential migrant worker users (Latonero and Gold 2015). Hackathons have been criticized for favouring 'quick and forceful action with socially similar collaborators' over the 'slow construction of coalition across difference' (Irani 2015). Donors must therefore think carefully about the provision of early-stage grants without sufficient longer-term funding, and consider the significant risk that either the platform will fail or will continue with deep problems.

These constraints and limitations that can be posed by the political economy of philanthropic funding put pressure on migrant organizations, trade unions, legal advocates and other not-for-profits to find alternative pathways towards values-aligned financial sustainability (Farbenblum et al. 2018: 42–3).

CONCLUSION

Digital technologies are changing the landscape of migrant worker engagement and empowerment. New tools are improving migrant workers' access to justice through enhanced evidence-gathering, facilitating referrals, complaints and claim-making, as well as assisting transnational litigation. Other initiatives are addressing information asymmetries using AI chatbots that provide migrant workers with answers in real-time or curating information innovatively in the form of radio dramas, comics and animations. Other tools seek to crowd-source data directly from migrant workers themselves by asking a series of questions to rate and review employers, recruiters, and other private intermediaries. Organizations are likewise using mobile message apps, social networking sites or bespoke digital platforms to create communities where migrant workers can more safely share information, engage collectively and build worker power.

When they are thoughtfully integrated within a migrant worker organization's broader functions and mission, digital technologies bring amplified and sometimes unprecedented opportunities in scaling and sustaining the work they are already doing. However, these tools bring their own set of ethical and practical challenges, risks and trade-offs. Inclusive and iterative design processes with multiple rounds of piloting, user-testing and updating are essential for a digital tool's uptake and impact. Generating and extracting sensitive data from migrant workers comes with a heightened responsibility for ensuring privacy and security. Organizations likewise need to take seriously the risks of civil and criminal liability for defamation or libel when asking migrant workers to provide their views on recruiters and employers and to consider their data storage frameworks. There remain unrealistic expectations about what organizations can do with small amounts of seed-funding and limited ongoing support to develop and sustain initiatives over the longer term.

As the market for these tools continues to evolve and expand rapidly, a key area for future research is closer investigation of how digital interventions operate in specific local contexts. This should include fieldwork with migrant workers themselves and conducting user-testing. As many of these tools are considerably new, longitudinal research can help to illuminate the longer-term outcomes and impacts of these tools for migrant workers and the organizations which use them. In assessing the potential of these interventions it is important that techno-solutionism – a belief that technology can by itself solve complex social and political problems – does not render invisible the myriad of factors aiding any particular intervention's success or distract from the broader struggles for justice that technology is ill-equipped to address.

NOTES

1.   This chapter focuses on digital interventions regarding empowering and engaging migrant workers, excluding, for example, interventions regarding e-recruitment, anti-corruption and fraud prevention measures, mobile money and remittances and blockchain. For a summary of worker-reporting tools by multinational businesses and brands in the supply chain context, see Berg, Farbenblum and Kintominas, 2020. See also Farbenblum, Berg and Kintominas, 2018.
2.   Research was conducted in accordance with HC180181 approved by the Human Research Ethics Office of University of New South Wales, Sydney and funded by the Open Society Foundations.

3. There is some overlap among these categories, for example, some donors have also developed their own app or commissioned or conducted research. The number of interviewees in some instances exceeds the number of organisations interviewed per category, for example, where group interviews were conducted or multiple people from one organisation were interviewed separately.

REFERENCES

Athreya (United States Agency for International Development), B. Interviewed by the authors. (2018).

Benton, M. and A. Glennie (2016), *Digital Humanitarianism: How Tech Entrepreneurs Are Supporting Refugee Integration*, TransAtlantic Council on Migration / Migration Policy Institute, October, p. 35.

Berfield, S. (2016), 'Labor Group Gets IBM's Watson to Help Walmart Workers', *Bloomberg Businessweek*, 5, 14 November.

Berg, L. and B. Farbenblum (2017), *Wage Theft in Australia: Findings of the National Temporary Migrant Work Survey*, Migrant Worker Justice Initiative.

Berg, L., B. Farbenblum and A. Kintominas (2020), 'Addressing Exploitation in Supply Chains: Is technology a game changer for worker voice?', *Anti-Trafficking Review*, **14**, 47–66.

Bitonio Jr, E. R. (2018), *Action Paper on the Use of Digital Technology to Promote Decent Work For Migrant Workers in ASEAN*, ATUC, p. 12.

Bradshaw, S. and P. N. Howard (2017), *Troops, Trolls and Troublemakers: A Global Inventory of Organized Social Media Manipulation*, Working paper 2017.12, University of Oxford, p. 37.

Burkell, J., A. Fortier, L. (Lola) Y. C. Wong and J. L. Simpson (2014), 'Facebook: public space, or private space?', *Information, Communication & Society*, **17** (8), 974–85.

Chambers, L. (2015), 'Utopian and Dystopian Theories of Change: A Template', accessed 10 June 2019 at https://responsibledata.io/2015/03/16/utopian-and-dystopian-theories-of-change-a-template/.

Civic Patterns, *Build It They Wont Come* (n.d.), accessed 13 February 2020 at http://civicpatterns.org/patterns/build-it-they-wont-come/.

Dehlendorf (Organization United for Respect), A. Interviewed by the authors (2018).

Farbenblum, B. and L. Berg (2017), 'Migrant workers' access to remedy for exploitation in Australia: the role of the national Fair Work Ombudsman', *Australian Journal of Human Rights*, **23** (3), 310–31.

Farbenblum, B., L. Berg and A. Kintominas (2018), *Transformative Technology for Migrant Workers: Opportunities, Challenges, and Risks*, Open Society Foundations, accessed 13 June 2019 at https://www.mwji.org/publications.

Farbenblum, B., E. Taylor-Nicholson and S. Paoletti (2013), *Migrant Workers' Access to Justice at Home: Indonesia*, Open Society Foundations, 16 October, accessed 26 February 2020 at https://www.opensocietyfoundations.org/publications/migrant-workers-access-justice-home-indonesia#publications_download.

Figueroa (Cornell University), M. Interviewed by the authors (2018).

Forestal, J. (2017), 'The Architecture of Political Spaces: Trolls, Digital Media, and Deweyan Democracy', *American Political Science Review*, **111** (1), 149–61.

Harkins, B. and M. Åhlberg (2017), *Access to Justice for Migrant Workers in South-East Asia*, Bangkok, Thailand: ILO Regional Office for Asia and the Pacific, accessed 19 May 2020 at https://www.ilo.org/asia/publications/WCMS_565877/lang--en/index.htm.

International Labour Organization (2019), *Mobile Women and Mobile Phones: Women Migrant Workers' Use of Information and Communication Technologies in ASEAN*, accessed 27 February 2020 at https://www.ilo.org/wcmsp5/groups/public/---asia/---ro-bangkok/documents/publication/wcms_732253.pdf.

Irani, L. (2015), 'Hackathons and the Making of Entrepreneurial Citizenship', *Science, Technology, & Human Values*, **40** (5), 799–824.

Jones, K. and D. Nuriyati (n.d.), *Increasing Transparency in International Recruitment: An Evaluation of "PantauPJTKI" (Recruitment Watch)*, Centre for Trust Peace & Social Relations, Coventry University.

Kalbag, L. (2017), *Accessibility for Everyone*, A Book Apart, accessed 27 February 2020 at https://abookapart.com/products/accessibility-for-everyone.

Latonero (Data & Society), M. Interviewed by the authors (2018).

Latonero, M. and Z. Gold (2015), *Data, Human Rights & Human Security: Primer*, Data & Society Research Institute, 22 June, accessed 11 June 2019 at https://datasociety.net/pubs/dhr/Data -HumanRights-primer2015.pdf.

Latonero, M., B. Wex, M. Dank and S. Poucki (2015), *Technology and Labor Trafficking in a Network Society: General Overview, Emerging Innovations and Philippines Case Study*, USC Annenberg Center on Communication Leadership & Policy, University of Southern California, February, accessed 14 February 2020 at https://communicationleadership.usc.edu/files/2015/10/USC_Tech-and -Labor-Trafficking_Feb2015.pdf.

Madianou, M., D. Miller and D. Miller (2013), *Migration and New Media: Transnational Families and Polymedia*, Abingdon and New York: Routledge.

Micah-Jones (Centro de los Derechos del Migrante, Inc.), R. Interviewed by the authors (2018).

Parreñas, R. S. and R. Silvey (2018), 'The Precarity of Migrant Domestic Work', *South Atlantic Quarterly*, **117** (2), 430–38.

Pew Research Centre (2018), 'Demographics of Internet and Home Broadband Usage in the United States', accessed 10 June 2019 at https://www.pewinternet.org/fact-sheet/internet-broadband/.

Rahman, Z. (2018), 'RD 101: Responsible Data Principles', accessed 10 June 2019 at https:// responsibledata.io/2018/01/24/rd-101-responsible-data-principles/.

Rahman (The Engine Room), Z. Interviewed by the authors (2018).

Recruitment Advisor (n.d.), 'Data privacy & security policy', accessed 13 February 2020 at https://www .recruitmentadvisor.org/about/data-privacy-security-policy.

Rende-Taylor (Issara Institute), L. Interviewed by the authors (2018).

Rojas, C. (2017), 'Innovating the Labor Movement: The Workers Lab Library', accessed 14 February 2020 at https://medium.com/the-workers-lab-library/innovating-the-labor-movement-f83ea6a603cf.

Rojas (The Workers Lab), C. Interviewed by the authors (2018).

Ruget, V. and B. Usmanalieva (2019), 'Can smartphones empower labour migrants? The case of Kyrgyzstani migrants in Russia: Central Asian Survey', *Central Asian Survey*, **28** (2), 165–80.

Silitonga (Tifa Foundation), N. F. Interviewed by the authors (2018).

The Engine Room, Benetech and Amnesty International (2016), *DATNAV: How To Navigate Digital Data for Human Rights Research*, June, accessed 10 June 2019 at https://www.theengineroom.org/wp -content/uploads/2016/09/datnav.pdf.

Ticona, J. (2016), 'New Apps like Jornalero Aim to Protect Low-Income Workers. Here's How They Could Backfire.', *Slate*, 21 March, accessed 10 June 2019 at https://slate.com/technology/2016/03/ new-apps-like-jornalero-aim-to-protect-low-income-workers-here-s-how-they-could-backfire.html.

Toyama, K. (2015), *Geek Heresy: Rescuing Social Change from the Cult of Technology*, New York: Hachette UK.

Tufekci, Z. (2014), 'Social Movements and Governments in the Digital Age: Evaluating A Complex Landscape', *Journal of International Affairs*, **68** (1), 1–18.

18. Mobile money and financial inclusion of migrants in sub-Saharan Africa

Adrian Kitimbo[1]

INTRODUCTION

Since the early 2000s, sub-Saharan Africa has experienced a sharp increase in mobile technology use. This has particularly been the case when it comes to mobile phones. The quick growth in the use of mobile phones has been striking, especially because these devices were barely visible in the region as late as the 1990s, even as their use was relatively widespread in regions such as North America and Europe. Within just the last 10 years, the mobile phone penetration rate in sub-Saharan Africa has increased by nearly 10 per cent, jumping from 32 per cent in 2012 to 44 per cent in 2018 (GSMA, 2019). By the end of 2018, the region had an estimated 456 million unique mobile subscribers, with this figure projected to increase to 623 million (or a penetration rate of 50 per cent) by 2025 (GSMA, 2019).

What is also remarkable about the proliferation of mobile phones in sub-Saharan Africa is that unlike Europe, for example, where fixed landlines were widely used prior to adopting mobile phones, landline use was never widespread across the region; most people jumped straight to mobile phones once they became affordable. The high costs of setting up landlines, in addition to poor infrastructure and long distances, made their expansion a lot more difficult and slower across sub-Saharan Africa. The advent and wide use of mobile phones has often been used as an example to illustrate the 'leap-frogging' power of new technology. In this case, most African countries 'leap-frogged' or bypassed landlines (UNCTAD, 2018b).

Mobile phone technology has made it possible to develop and adopt powerful platforms such as mobile money. In just a few years, the mobile phone quickly transitioned from being just a tool for voice and text messaging, to one that can also be used to offer extensive financial services. In a region where many people do not have a traditional bank account, inhibited by poor infrastructure and difficulties accessing financial institutions, mobile money has been key in extending financial services to those whose access to traditional banking institutions is limited. Migrants within and from sub-Saharan Africa have been among the biggest beneficiaries of mobile money platforms.

The growth of mobile technologies in Africa coincided with the significant increase in the continent's internal and international migrants. The region also remains host to among the largest number of displaced populations in the world (IDMC, 2020; UNHCR, 2020). The large number of migrants and displaced populations from and within the region has called for innovative solutions to help meet their various needs, including the need for financial inclusion, which is key to enabling their participation in the formal economy. The Global Compact on Migration (GCM), which was adopted by the United Nations General Assembly in 2018, recognizes the significance of enhancing the financial inclusion of migrants and promoting the "faster, safer and cheaper transfer of remittances." To achieve these goals, the GCM draws from several actions, including objective 20 (e), which stresses the need to "develop innovative technological solutions for remittance transfer, such as mobile payments, digital tools or e-banking" and objective 20 (i), which focuses on providing "access to and develop banking solutions and financial instruments for migrants" (UN, 2018).

Against this background, this chapter provides an overview of how mobile money is used in the context of migration and displacement in sub-Saharan Africa, as well as exploring the ways in which it has fostered the financial inclusion of migrants. The next section defines both mobile money and financial inclusion after providing a brief overview of migration and displacement in the region. These are followed by key examples and analysis of how mobile money enables the financial inclusion of migrants, including the ways in which it has facilitated remittance transfers, its use in displacement contexts and how it has the potential to enable migrants' access to credit as well as support their livelihoods. This chapter then discusses areas for further research and current challenges.

It is important to stress that the chapter does not interrogate or measure the extent to which mobile money has been successful in driving financial inclusion for migrants in sub-Saharan Africa; that is not only beyond its scope, but scientific studies on the subject also remain scant. Rather, this chapter emphasizes the significance of financial inclusion for migrants, drawing from available literature to provide key examples and analysis of how mobile money is providing or has the potential to enable migrants' access to financial services.

CONTEXT AND CONCEPTS

Migration in Africa is characterized by a significant number of intra-regional migrants. The vast majority of international migrants who reside in sub-Saharan Africa (89 per cent) were born within the region (UN DESA, 2019b). While the number of international migrants from Africa to other regions such as Europe has sharply increased since 2000, growing from under 15 million to 19 million in 2019, the number of intra-regional migrants, which stands at 21 million, remains larger (UN DESA, 2019a).

Several factors drive migration from and within sub-Saharan Africa. Some of these include income disparities between origin and destination countries, conflict and violence, unemployment, labour shortages, and environmental disasters (McAuliffe and Khadria, 2020). There are also large numbers of seasonal migrant workers in subregions such as West Africa (McAuliffe and Khadria, 2020). Regional Economic Communities (RECs), such as the Economic Community of West African States (ECOWAS) and the East African Community (EAC), many of which have frameworks to facilitate the free movement of persons within their respective subregions, have played a role in the increase in intra-regional migration within the continent. In ECOWAS, for example, all 15 Member States have ratified the Protocol on free movement and all ECOWAS citizens can visit Member States (visa-free) and have access to a common passport (UNCTAD, 2018a). In EAC, nationals of most Member States can also move within the subregion without visas (UNCTAD, 2018a).

Displacement remains a defining feature of migration dynamics in sub-Saharan Africa. In 2019, the largest number of new internal displacements due to conflict and violence globally were recorded in the region, with an estimated 4.6 million new displacements triggered by conflict, while an additional 3.4 million were due to disasters (IDMC, 2020). The number of new displacements as a result of conflict and violence in sub-Saharan Africa, which have increased during the past decade, accounted for almost 54 per cent of the worldwide total

(IDMC, 2020). The number of refugees in the region has also increased over the last ten years, tripling from 2.2 million to 6.3 million, with countries such as South Sudan, Somalia, the Democratic Republic of the Congo and the Central African Republic producing some of the largest refugee populations (UNHCR, 2020). In addition to being the origin of millions of refugees, several countries in sub-Saharan Africa also host some of the largest numbers of people displaced across borders. Uganda and Ethiopia, for example, were among the top ten host countries globally in 2019, with nearly 1.4 million and over 700,000 refugees respectively (UNHCR, 2020).

What is Mobile Money?

Mobile money refers to financial transactions on a mobile phone or a platform that makes it possible for users to make electronic financial transactions using a cell phone (Aron and Muellbauer, 2019; Jack and Suri, 2011). These transaction services can potentially be used by anyone with a mobile phone, including people who do not have access to traditional banks or the so-called "unbanked" who are not the usual target of commercial banks aiming to make a profit (Aron and Muellbauer, 2019). In other words, a traditional bank account is not required to use mobile money services.

Mobile money encompasses three main services, which include mobile payments, mobile transfers and mobile banking (ITU, 2013). Mobile payments allow users to purchase or sell goods and services via mobile phone, bypassing the need to use cash, while mobile transfers make it possible for users to transfer money to another mobile phone, including across international borders. Before any of these services can be used, users have to install an application (or set up a mobile money account) connected to a SIM card. The account, commonly known as a "mobile wallet", is usually set up with a mobile money services provider. To make transactions, users need to deposit cash in their mobile account and, in exchange, they get electronic money, which is then added to their mobile money account. These deposits are made to an agent, usually of a mobile telecommunications company (ACP and IOM, 2014). The digital money can be withdrawn as cash (from a mobile money agent/outlet), transferred to another user or saved in the mobile wallet for later use. Mobile banking, which makes it possible for users to manage their bank accounts remotely (ITU, 2013), is only available to people who already have traditional bank accounts (ACP and IOM, 2014). This type of banking is most widely used in developed economies, where most people already have pre-existing traditional bank accounts (Aron and Muellbauer, 2019). In developing economies, where financial inclusion through formal banking institutions is much more limited, mobile payments and transfers are more popular than mobile banking.

The use of mobile money has rapidly expanded across sub-Saharan Africa. With an estimated 395.7 million mobile money accounts at the end of 2018, sub-Saharan Africa accounted for nearly half of all mobile money accounts worldwide and the highest penetration rate globally (GSMA, 2019). This number is expected to jump to 500 million by 2020 (GSMA, 2019). Unlike other regions where most adults largely depend on traditional banks to access various financial services, sub-Saharan Africa has more adults using mobile money accounts (21 per cent) (Klapper, Ansar, Hess, Singer, 2019). Despite the widespread use of mobile money platforms, however, access and use of these services remains highly gendered. Men in sub-Saharan Africa have more access to mobile phones and use mobile money more frequently than women (Chamboko et al., 2018).

Kenya's M-Pesa ("M" for mobile and "Pesa" for money in Swahili), which started in 2007, is the most successful mobile money platform in the world. The service, which is owned by Safaricom, the largest cellular provider in Kenya, set the stage for the spread of mobile money platforms across Africa and the world. By the end of 2018, M-Pesa had more than 25 million active users, or around half of Kenya's population (Reuters, 2019). The platform is credited with bringing financial services to the many poor who had been excluded from banking services (Mbiti and Weil, 2016). However, some early studies of M-Pesa contested the conclusion that the platform had brought banking services to the mostly unbanked, showing that its users were, on average, "wealthier, better educated and already banked" (Aker and Mbiti, 2010).

Box 18.1 provides a brief history of mobile money growth in sub-Saharan Africa.

BOX 18.1 BRIEF HISTORY OF MOBILE MONEY IN SUB-SAHARAN AFRICA

With M-Pesa, Kenya is at the forefront of the mobile money revolution: the number of agents across the country increased by 40 per cent in 2013. It is now estimated that 24.8 million subscribers use mobile money services, like M-Pesa, in Kenya (Communication Commission of Kenya, 2013). According to the Pew Research Center's 2013 survey report, the number of Kenyans using mobile wallets to make or receive payments is higher than any of the other 24 countries surveyed: 50 per cent of the Kenyan adult population uses mobile money services (Pew Research Center, 2013).

Mobile money services have spread rapidly in many developing countries. However, only a handful of these initiatives has reached a sustainable scale, in particular GCASH and Smart Money in the Philippines; Wizzit, MTN Mobile Money and FNB in South Africa; MTN Mobile Money in Uganda; Vodacom M-Pesa and Airtel in Tanzania; Celpay Holdings in Zambia and MTN Mobile Money, Orange Money in Côte d'Ivoire.

In South Africa, MTN Mobile Money was launched in 2005 as a joint venture between the country's second-largest network operator MTN and a large commercial bank, Standard Bank.

In Uganda, MTN was the first operator to launch mobile money services in 2009 and remains, by far, the market leader (Intermedia, 2012). By law, each mobile money provider has to partner with a bank. However, users do not need a bank account to use mobile money services.

In Tanzania, Airtel was the first mobile network operator to introduce a phone-to-phone airtime credit transfer service, "Me2U," in 2005 (Intermedia, 2013). Airtel partners with Citigroup and Standard Chartered Bank to provide m-money services, including bill payments, payments for goods and services, phone-to-phone and phone-to-bank money transfers, and mobile wallets. In 2008, Vodacom Tanzania launched the second East African implementation of the Vodafone m-money transfer platform, M-Pesa.

Finally, in Côte d'Ivoire two mobile operators, Orange and MTN, are competing head to head in the mobile money market (CGAP, 2012). Orange Money was launched in 2008 by Orange in partnership with BICICI (BNP Paribas), and MTN Mobile Money was launched in 2009 by MTN in partnership with SGBCI (Société Générale) (GSMA, 2014).

Source: Abridged excerpt from ACP and IOM 2014.

What is Financial Inclusion?

There is no single definition of *financial inclusion*. The term tends to be used slightly differently depending on context. The World Bank defines financial inclusion as ensuring that "individuals and businesses have access to useful and affordable financial products and services that meet their needs – transactions, payments, savings, credit and insurance – delivered in a responsible and sustainable way" (World Bank, 2018). Unlike the World Bank, which includes *businesses* in its definition, others solely focus on *individual* access to financial products and services. Aron and Muellbauer (2019) stress that most definitions of financial inclusion emphasize the need to pull people into the *formal* banking sector and that only recently has *mobile money* been included in definitions of the term. A recent paper by Demirguc-Kunt et al. (2017) stresses that financial inclusion, at a fundamental level, "begins with having a deposit or transaction account at a bank or other financial institution or through a mobile money service provider, which can be used to make and receive payments and to store or save money."

Financial inclusion has garnered widespread attention and gained significant traction in academic and policy discussions in recent years, as it is increasingly linked to many development outcomes. Most notably, the Maya declaration, launched in 2011 and endorsed by dozens of financial institutions from developing and emerging markets, commits them to "concrete financial inclusion targets" (AFI, n.d.). And since the adoption of the Sustainable Development Goals (SDGs) by the United Nations (UN) Member States in 2015, financial inclusion has been touted as a key enabler of these goals (Klapper et al., 2016). Greater financial inclusion of women, for example, has been associated with driving gender equality. Globally, women not only continue to earn less than men, but also own less, which has significant negative implications for many areas of their lives. A relatively recent study in Kenya found that access to the mobile money platform (M-Pesa) not only lifted a significant number of households out of poverty, but that this outcome was more noticeable for female-headed households; the authors further conclude that financial inclusion, rather than additional capital, may be the way out of poverty for women (Suri and Jack, 2016). Meanwhile, access to saving products has been shown to have a positive effect on women's empowerment in countries such as Malawi, Uganda and Ghana (Karlan et al., 2016). And because unequal division of labour persists between the genders and women, too often, lack control over economic resources, poverty levels for women are higher than men, making financial inclusion vital to both reducing poverty and empowering women (Holloway et al., 2017). Research by Siddik (2017) reveals that financial inclusion "increases women's income, purchasing power, living standards and position in the family."

Financial inclusion can also have positive effects at various levels of economic development, including at household, firm and national levels (Kim et al., 2018). At the household level, by making it possible for people to manage financial risks while also investing in their future, financial inclusion has been linked to both a reduction in poverty and inequality (Demirguc-Kunt et al., 2017). For the poor and groups that are disadvantaged, access to inclusive financial systems can enable investment in areas such as education and entrepreneurship, which can help improve their economic standing (Demirguc-Kunt and Klapper, 2013). Sahay et al. (2015) echo this conclusion and stress that for people in developing countries, access to finance can improve their economic conditions. Yet, even as financial inclusion has been shown by multiple studies to have wide-ranging and positive development impacts, work by researchers such as Mader (2018) cautions against fixating on financial inclusion as key to driving economic growth and development, arguing that the evidence base for such claims is thin and that the causal connection remains unclear.

In the context of migration and displacement, financial inclusion is especially important for migrants, who often find themselves with limited access to financial services. In some cases, this is due to their legal status or lack of sufficient documentation to open accounts at traditional banking institutions. For irregular migrants, formal banking institutions/regulated financial services are nearly impossible to access. As McAuliffe (2020, 425) observes in regard to irregular migration and remittances, "those who enter and stay irregularly are likely to find it more difficult to use formal remittance channels and may only have access to informal transfer networks." IFAD and the World Bank (2015) also note that, in many countries, undocumented migrants cannot open bank accounts. The informal channels many migrants often resort to come with risks. Migrant workers without banks, for example, are vulnerable to dodgy employers and are often left exposed to predatory lenders. People in displaced situations, such as refugees, are even more underserved by traditional banking institutions. Refugee camps and settlements are often located in remote areas, which are less connected to financial infrastructure, while the unfamiliarity of refugees with financial services, as well as their legal status, remain significant barriers to their financial inclusion (UNHCR and UNCDF, 2019).

Recent technologies have become important tools in addressing the lack of access to inclusive financial services, especially in developing countries. Technology has become a significant catalyst for financial inclusion with mobile phones, especially, becoming important players in providing banking services to previously unbanked communities, including migrants. There is an abundance of research on how mobile phones have enhanced financial inclusion in sub-Saharan Africa. Recent research in Uganda by Bongomin et al. (2018), for example, shows that there is a strong positive relationship between mobile money usage by the rural poor and their financial inclusion. Navis (2019) argues that mobile money has changed the lives of people in developing countries, overcoming the constraints that have often been imposed by traditional banking systems. Many people in countries such as Tanzania, Gabon, Uganda and Zimbabwe have been brought under the "financial inclusion" umbrella as a result of mobile money accounts. Mothobi, Moshi and Deen-Swarray's recent work on the link between mobile money and financial inclusion in sub-Saharan African countries (2018) suggests that mobile money in these countries is "more likely to be used by individuals who have no access to a bank account" and that these platforms make financial services that would otherwise have not been provided by traditional banks available to the poor. Work by Lashitew et al. (2019), which took an in-depth look at Kenya's M-Pesa, reiterates these findings, showing that mobile money has extended financial services to many unbanked people, partly as a result of a supportive local regulatory system. Llewellyn-Jones (2016) asserts that mobile money has been key to driving financial inclusion in many parts of Africa, where the penetration rate of retail banks remains thin and many people remain unbanked.

While research on how mobile money enhances financial inclusion in sub-Saharan Africa is extensive, very limited research exists on how mobile money is more specifically driving the financial inclusion of migrants, and how these platforms are being used in migration and displacement contexts. Remittances remain the exception, with extensive literature on how mobile money has enhanced the transfer of money. The next section turns to this very aspect and provides an overview of how mobile money has driven down costs, while making it safer and quicker for migrants to send money to and within sub-Saharan Africa.

MOBILE MONEY AND REMITTANCES

An important component of mobile money is the way it has facilitated the transfer of remittances, both domestically and internationally. Mobile money services have not just made the transfer of remittances cheaper but have also enabled faster and safer ways for migrants to send money to their families and friends. As Aker et al. (2020) observe, "existing research shows that technology, in the form of mobile money, is reducing the cost of transferring money between individuals and businesses in sub-Saharan Africa as compared with traditional money transfer systems."

The money migrants send to their families and friends has been shown to have wide-ranging positive impacts on the people, communities and the countries that receive them (McAuliffe, Kitimbo and Khadria, 2019). For households in sub-Saharan Africa, remittances are key to meeting their basic necessities, allowing them to spend on a variety of needs. Migrants have long supported their friends and families in meeting needs such as shelter and food, while also making it possible for them to invest in areas such as education and health (IFAD and World Bank, 2015). When faced with shocks, remittances have helped households to smooth consumption and, over time, have reduced poverty (Blumenstock et al., 2015).

Mobile money has also resulted in an increase in the amount of money migrants send home. A recent study on the impact of introducing mobile money in rural areas in Mozambique found that its introduction drove up the inflow of remittances, which in turn improved the welfare of rural households, since the remittances decreased their vulnerability to shocks (Batista and Vicente, 2019). But the same study acknowledges that the surge in remittances due to the introduction of mobile money also led to divestment in income-generating activities such as agriculture, with households instead opting to invest in the migration of family members to urban areas (Batista and Vicente, 2019). Early studies of M-Pesa usage show that users sent remittances to rural areas more frequently, which resulted in the overall increase in remittances (Morawcyznski, 2009). Mbiti and Weil (2011) reiterate these findings, showing that the adoption of M-Pesa resulted in the increase in the frequency of money transfers. Jack and Suri (2014) also find that M-Pesa not only reduced transaction costs and enhanced domestic money transfers; it led to a reduction of poverty in the medium term. Hennebry et al. (2017) argue that increased access and reduced costs have resulted in more frequent transfer of remittances, including by women remitters. Yet, while mobile money reduces poverty, increases rural consumption and has driven up the frequency of remittance transfers, it has also been shown to apply added pressure on migrant workers to increase remittances (Lee et al., 2018). For communities, the inflow of migrants' remittances means that there is greater demand for products and services. This not only supports local markets and businesses, but also helps to enhance their financial infrastructure (IFAD and World Bank, 2015). Moreover, migrants' remittances have helped to drive job creation in their local communities, both through investment and philanthropy (ibid.).

But the positive impact of remittances is not just limited to households and communities. Remittances are significant contributors to economic growth at the aggregate level. In 2018, an estimated USD 689 billion was sent in remittances globally, with the largest share (USD 529 billion) going to low- and middle-income countries (World Bank, 2019b). The amount sent to low- and middle-income countries was expected to reach USD 551 billion in 2019 (Ratha et al., 2020). Remittances to sub-Saharan Africa alone amounted to USD 46 billion in 2018 and make up a significant share of several countries' GDP (World Bank, 2019b). For Comoros, remittances comprised over 19 per cent of its GDP, while for countries such as The Gambia, Lesotho, Cabo Verde, over 12 per cent of their GDP was from remittances (ibid.). Remittances continue to be a significant source of foreign exchange for recipient countries and over the last 25 years, remittances have exceeded Official Development Aid (ODA) and were projected to surpass Foreign Direct Investment (FDI) to low- and middle-income countries in 2019 (Ratha et al., 2020).

Despite their positive effects, the cost of sending international remittances remains high. The SDGs have a target of reducing remittance costs to 3 per cent by 2030.[2] Yet, for many countries, the costs remain well above this target. The average cost of sending USD 200 to low- and middle-income countries in the first quarter of 2019, for example, stood at nearly 7 per cent (Ratha et al., 2019). For sub-Saharan Africa, the average cost is at a staggering 9 per cent, the highest of any region in the world (Gandhi, 2019; Ratha et al., 2020). Banks and dominant Money Transfer Operators (MTOs) such as MoneyGram and Western Union remain the most expensive avenues for sending international remittances (Ratha et al., 2019). The high costs, coupled with the fact that banks and MTOs are scarce outside of cities, makes it difficult for migrants, especially those in geographically isolated areas, to send money. Even where banks are available, the costs of opening a bank account are often prohibitive, as banks often ask for deposit amounts that many migrants are unable to afford. These factors, as mentioned earlier, force migrants to use informal channels to send money. This may involve middlemen travelling considerable distances from cities to rural areas, for example, which is not just slow but can be fraught with risks; senders and receivers of remittances may be susceptible to crime, as travelling with large amounts of cash potentially makes them targets. Indeed, while some informal channels may come with advantages, such as being easier to access and cheaper to use, they are also generally considered to be at higher risk of theft and loss (Kosse and Vermeulen, 2014). In contrast, mobile money platforms make it possible for migrants to make money transfers at lower costs and without the added risks associated with using informal money transfer channels. Mobile Financial Services (MFS) such as mobile money "require a much simpler and affordable registration process, as well as offer faster and easier transactions compared with formal financial institutes" (Kim et al., 2018, p. 8). Aron and Muellbauer (2019) add that mobile money does not only reduce the cost of sending international remittances; the technology also makes them more traceable, hence making these platforms safer than using informal money transfer channels.

MOBILE MONEY IN DISPLACEMENT CONTEXTS

A growing number of humanitarian organizations working in displacement contexts are increasingly turning to digital cash transfers, using mobile money, as opposed to directly providing cash to refugees or Internally Displaced Persons (IDPs). Cash Based Interventions (CBIs), as they are commonly known, have become popular in recent years. CBIs refers to monetary transfers, either through cash or vouchers, and are used as a way to help meet the needs of various beneficiaries. The shift from more traditional and "tangible" forms of assistance such as shelter and food to CBIs in some contexts is in recognition of the benefits that the latter bring, including their potential to provide self-sufficiency to those who have been displaced for long periods, the improvements in health and financial security that they bring to recipients as well as their ability to enhance a sense of dignity, as beneficiaries are able to purchase goods and services that best meet their needs (GSMA, 2017a).

Mobile money has become particularly ideal for these types of transfers, as the large majority of displaced populations who are among the recipients of CBIs are located in developing countries, where formal banking infrastructure such as ATMs are often limited (GSMA, 2017a). An estimated 84 per cent of refugees are hosted in developing regions, while the overwhelming majority of IDPs are also in developing countries (UNHCR, 2019; IDMC, 2019). Uganda, for example, hosted the third largest number of refugees in the world in 2018 (1.2 million) after Turkey (3.7 million) and Pakistan (1.4 million) (UNHCR, 2019). In settings such as these, mobile money opens up new channels to reach people with cash transfers (Bailey, 2017), and there is increasing recognition that new technologies such as mobile money may be more efficient and reliable than cash distributions (UNHCR, 2012). As Aron and Muellbauer (2019) put it, digital cash transfers eliminate "unreliable paper bureaucracy through instant and transparent digital records, reduces delays and makes cash transfers more cost effective for both the donor and recipient." They further add that "mobile technology can securely provide vital assistance to displaced populations through remote digital cash transfers, perceived as an often more rapid and effective method of providing help" (ibid.).

The use of digital cash transfers in displacements contexts is most visible in East Africa, where they have been implemented in several refugee camps. In 2015, for example, the World Food Program (WFP) introduced electronic cash transfers for refugees in Kenya's Kakuma refugee camp. Dubbed "bamba chakula" (Swahili for "get your own food"), the mobile money transfers, which replaced food rations, have provided refugees with the choice to buy the food that they wish to eat, while making cash transfers safer and faster (Karimi, 2015). The same cash transfers were launched a year later in Kenya's Dadaab, one the largest refugee settlements in the world (WFP, 2016). In Northern Uganda, humanitarian organizations and cellular companies have worked closely in recent years to deliver assistance to refugees. This has especially been the case in Bidi Bidi refugee settlement, which was only established in 2016 but has rapidly expanded in recent years with an influx of refugees from South Sudan. Mobile networks operators such as MTN and Airtel have partnered with organizations including the International Rescue Committee (IRC) and Mercy Corps to use their mobile money platforms to facilitate humanitarian cash transfers, which have not only resulted in enhancing the dignity of refugees by giving them the power to choose, but have also cut logistical costs for humanitarian organizations (GSMA, 2017b).

BOX 18.2 USING MOBILE MONEY FOR CASH TRANSFERS IN NORTHERN UGANDA

The IRC has been working in Bidi Bidi to provide emergency health response since August 2016 and now serve all five zones in the settlement through a mixture of health, protection and sexual and gender-based violence (SGBV) programs. IRC's Economic Recovery and Development department is implementing unconditional humanitarian cash transfers in northern Uganda through mobile money, in partnership with MTN Uganda. In early 2017, IRC's Emergency Response Team visited Uganda to support in-country staff during the crisis and suggested exploring the use of mobile money for cash transfers, given IRC's commitment to having active cash transfer programmes in 75 per cent of their country offices by 2020 (see Box 5), and their proven successes using mobile money as a modality in some of IRC's other operating countries, including Pakistan and Cameroon.

There were also a number of specific advantages of leveraging mobile money as the modality of choice for cash transfers:

- Accountability and compliance: Generating reports gives real-time information on the delivery of cash to beneficiaries.
- Flexibility and freedom: Beneficiaries have the choice and liberty to withdraw cash at any time.
- Improved safety: The replacement of physical cash with digital cash removes the need for IRC staff to move large amounts of cash from banks two hours' drive away, preventing some very real dangers. In recent years, an NGO lost a member of staff caught up in a violent attack whilst delivering cash in Gulu.
- Ease of monitoring and follow-up: Ability to easily follow up with beneficiaries in case of a problem.

ACCESS TO CREDIT AND SUPPORTING ECONOMIC ACTIVITY

Beyond enabling the quicker, cheaper and safer transfer of remittances, as well as aiding electronic cash transfers to those who are displaced, mobile money is also becoming an important instrument to facilitate access to credit. For many people in developing regions, access to credit remains difficult, and barriers to loans and other forms of credit are even more pronounced for migrants, including refugees. Legal status and lack of required documentation are often significant impediments to migrants' ability to secure personal or business loans. But even in cases where migrants have the legal right to get credit, the lack of collateral, often demanded by traditional banks in developing countries, makes getting credit almost impossible. Because of the large asymmetries in information between borrowers and lenders, the latter ask for guarantees that migrants are usually unable to meet. Moreover, for many migrants, their lack of credit history, which would help to bridge the information gap between them and the banks, means that lenders do not have the necessary information to 'trust' migrants' ability to pay back borrowed money.

Mobile Financial Services such as Mobile Money are changing the rules and the ways in which previously unbanked people can access credit (Macmillan, 2016). A growing number of these platforms are making it possible for people in developing countries to access small amounts of credit without the need for collateral or other prohibitive guarantees; this could potentially benefit migrants who cannot meet the more traditional requirements for obtaining credit. This has sometimes been done in collaboration with third party companies, which monitor the financial behaviour of customers and provide credit scores (World Bank, 2019a); this financial history is critical for migrants, many of whom would not have previously had any recorded financial behaviour, and has the potential to be used to getting access to larger amounts of credit or loans, including from formal banking institutions. The potential of mobile money services to enhance credit access for displaced populations, in particular, is emphasized by a recent GSMA report (2017a) on 'Mobile Money, Humanitarian Cash Transfers and Displaced Populations':

The self-sufficiency of long-term displaced populations can be bolstered by access to financial services. Certain financial services, including lending, require a previous financial history. As of December 2015, mobile money providers across sixteen markets had partnered with credit institutions to offer credit services to customers based on previous mobile money account transaction history. Many of these loan recipients would not otherwise have access to credit services. This is an aspirational example of services that could be increasingly utilised by displaced individuals via a mobile money account in the future.

However, it is important to highlight that the ability of migrants to use mobile money to access financial services such as credit depends on countries' regulations. While opening a mobile money account does not require as much documentation as opening a traditional bank account, in many countries a form of identification is a prerequisite before obtaining a SIM card, which is needed before setting up a mobile money account. Migrants without required ID often remain excluded from mobile money services (Langat, 2019).

Increasingly, mobile money platforms are playing an important role in supporting the economic activity of migrants, providing them with financial services that have eased their ability to run their businesses. In countries that are undergoing economic turmoil and where there are limits on cash withdrawals, for example, the ability to quickly transfer cash from one mobile phone to another has been used by various organizations to help drive migrants' economic activity. In Sudan, the International Organization for Migration (IOM) recently partnered with MTN, one of the largest telecommunications companies in Africa, to enable returned migrants to use its mobile money service (MoMo) to "re-establish their livelihoods in the country" (IOM, 2020). Using MoMo, 2,000 returned migrants starting small businesses were given the ability to pay their suppliers using electronic transfers rather than to worry about cash in a country with a liquidity crisis. The ability to quickly make or receive payments through mobile money has also been widely used by refugee entrepreneurs in countries such as Uganda. Research by Betts et al. (2014) has documented how refugees in Uganda have used mobile money transfers to facilitate business transactions and doing so to such a degree that they have attracted the attention of some of the largest telecommunications corporations in the country. In response to the widespread use of mobile money services by refugees, these companies established even more communications infrastructure, including larger radio towers in areas most populated by refugees, in order to further enhance their use of these services (Betts et al., 2014).

CONCLUSION AND CHALLENGES

Migrants, especially those in displacement contexts, remain among the most financially excluded populations. As discussed in this chapter, this reality has far-reaching implications, affecting migrants' ability to support loved ones, start businesses or save for the future. But the lack of inclusive financial services does not just affect individual migrants; it also takes a toll on communities and on the countries in which they live, as they are unable to fully participate and contribute to the formal economy.

This chapter has provided an overview of how mobile money extends financial services to previously unbanked migrants. As earlier discussed, while there is a plethora of research on the subject of *mobile money and financial inclusion*, there is a dearth of literature on how this technology has been in use in the context of migration and displacement. This chapter contributes to this area that remains largely underexplored.

With this said, the following implications for policy and research are offered:

- There is clearly a need for more scientific studies examining the success, or lack thereof, of mobile money platforms in enhancing the financial inclusion of migrants across various services, including the impact on their ability to save, transfer money, access credit, among others.
- Access to mobile phones and mobile money services remains highly gendered, especially in displacement contexts. With men more than twice as likely to own a mobile phone and mobile money account than women, there is a need to understand and address the gender gap that persists.
- Mobile money platforms have undoubtedly had extensive reach, penetrating areas where traditional banks had never existed and providing financial services to people who were previously unbanked or underbanked. However, for some migrants, particularly those without required identification documents, their access to mobile money remains constrained; solutions are needed to further extend the financial services that this technology offers to some of the most vulnerable populations.

NOTES

1. The opinions, comments and analyses expressed in this chapter are those of the author and do not necessarily represent the views of any of the organizations or institutions with which the author is affiliated.
2. SDG target 10.C: "By 2030, reduce to less than 3 per cent the transaction costs of migrant remittances and eliminate remittance corridors with costs higher than 5 per cent." At https://sustainabled evelopment.un.org/sdg10 (accessed 4 January 2020).

REFERENCES

ACP Observatory on Migration and International Organization for Migration (IOM) (2014), 'Mobile money services: "A bank in your pocket" overview and opportunities', accessed 15 December 2019 at https://publications.iom.int/books/mobile-money-services-bank-your-pocket.

Aker, J. and I. Mbiti (2010), 'Mobile Phones and Economic Development in Africa: Working Paper 211', Center for Global Development (CGD), accessed 15 December 2019 at https://www.cgdev.org/publication/mobile-phones-and-economic-development-africa-working-paper-211.

Aker, J., S. Prina and C. J. Welch (2020), 'Migration, Money Transfers and Mobile Money: Evidence from Niger', *AEA Papers and Proceedings*, accessed 25 May 2020 at https://www.aeaweb.org/articles ?id=10.1257/pandp.20201085.

Alliance for Financial Inclusion (AFI) (n.d.), *Maya Declaration*, accessed 15 December 2019 at https://www.afi-global.org/maya-declaration.

Aron, J. and J. Muellbauer (2019), 'The Economics of Mobile Money: harnessing the transformative power of technology to benefit the global poor', Oxford Martin School, accessed 15 December 2019 at https://www.oxfordmartin.ox.ac.uk/downloads/May-19-OMS-Policy-Paper-Mobile-Money-Aron -Muellbauer.pdf.

Bailey, S. (2017), 'Humanitarian cash transfers in the Democratic Republic of Congo', Overseas Development Institute (ODI), accessed 15 December 2019 at https://www.odi.org/sites/odi.org.uk/files/resource-documents/11416.pdf.

Batista, C. and P. C. Vicente (2019), 'Is Mobile Money Changing Rural Africa?' Evidence from a Field Experiment, accessed 05 January 2020 at http://www.catiabatista.org/batista_vicente_mm _experiment.pdf.

Betts, A., L. Bloom., J. Kaplan and N. Omata (2014), 'Refugee Economies Rethinking Popular Assumptions', University of Oxford, accessed 15 December 2019 at https://www.rsc.ox.ac.uk/files/ files-1/refugee-economies-2014.pdf.

Blumenstock, J., N. Eagle and M. Fafchamps (2015), 'Risk sharing and mobile phones: evidence in the aftermath of natural disasters', accessed 15 December 2019 at http://www.jblumenstock.com/files/ papers/jblumenstock_mobilequakes.pdf.

Bongomin, G. O. C., J. M. Ntayi, J. C. Munene and C. A. Malinga (2018), 'Mobile Money and Financial Inclusion in Sub-Saharan Africa: The Moderating Role of Social Networks', *Journal of African Business*, **19**(3): 361–384.

Chamboko, R., H. Soren and V. W. Morne (2018), 'Women and Digital Financial Services in Sub-Saharan Africa : Understanding the Challenges and Harnessing the Opportunities', World Bank Group, accessed 15 December 2019 at http://documents.worldbank.org/curated/en/139191548845091252/ Women-and-Digital-Financial-Services-in-Sub-Saharan-Africa-Understanding-the-Challenges-and -Harnessing-the-Opportunities.

Communications Commission of Kenya (2013), Quarterly Sector Statistics Report: Fourth Quarter of the Financial Year 2012/2013, accessed 21 December 2020 at www.cck.go.ke/resc/downloads/Q4 _201213_STATISTICS_final_25th_oct_2013.pdf.

Demirguc-Kunt, A. and L. Klapper (2013), 'Measuring Financial Inclusion: Explaining Variation in Use of Financial Services across and within Count', Brookings Institution Press, accessed 14 December 2019 at https://www.brookings.edu/wp-content/uploads/2016/07/2013a_klapper.pdf.

Demirguc-Kunt, A., L. Klapper and D. Singer (2017), 'Financial inclusion and inclusive growth', World Bank Group, accessed 14 December 2019 at http://documents.worldbank.org/curated/en/ 403611493134249446/Financial-inclusion-and-inclusive-growth-a-review-of-recent-empirical -evidence.

Gandhi, D (2019), 'Figure of the week: Migration and remittances in sub-Saharan Africa', Brookings Institution, accessed 11 January 2020 at https://www.brookings.edu/blog/africa-in-focus/2019/11/14/ figure-of-the-week-migration-and-remittances-in-sub-saharan-africa/.

GSMA (2014) Mobile money in Côte d'Ivoire: A turnaround story, GSMA, accessed 21 December 2020 at http://www.gsma.com/mobilefordevelopment/wp-content/uploads/2014/05/MMU_Cote_dIvoire _Turnaround_Story.pdf.

GSMA (2017a), Humanitarian Payment Digitisation: Focus On Uganda's Bidi Bidi Refugee Settlement, accessed 21 December 2019 at https://www.gsma.com/mobilefordevelopment/wp-content/uploads/ 2017/11/Humanitarian-Payment-Digitisation.pdf.

GSMA (2017b), 'Landscape Report: Mobile Money, Humanitarian Cash Transfers and Displaced Populations', accessed 21 December 2019 at https://www.gsma.com/mobilefordevelopment/ resources/mobile-money-humanitarian-cash-transfers/.

GSMA (2019), 'The Mobile Economy Sub-Saharan Africa 2019', accessed 11 January 2020 at https:// www.gsma.com/r/mobileeconomy/sub-saharan-africa/.

Hennebry, J., J. Holliday and M. Moniruzzaman (2017), 'At What Cost? Women Migrant Workers, Remittances and Development', United Nations Entity for Gender Equality and the Empowerment of Women (UN Women), accessed 21 December 2019 at https://www.unwomen.org/en/digital-library/ publications/2017/2/women-migrant-workers-remittances-and-development.

Holloway, K., Z. Niazi and R. Rouse (2017), 'Women's Economic Empowerment through Financial Inclusion', Innovation for Poverty Action, accessed 11 January 2020 at https://www.poverty-action .org/sites/default/files/publications/Womens-Economic-Empowerment-Through-Financial-Inclusion -Web.pdf.

InterMedia (2012), Mobile Money in Uganda: Use, Barriers and Opportunities, The Financial Inclusion Tracker Surveys Project, accessed 21 December 2020 at https://www.findevgateway.org/sites/default/ files/publications/files/mfg-en-paper-mobile-money-in-uganda-use-barriers-and-opportunities-oct -2012.pdf.

Internal Displacement Monitoring Centre (IDMC) (2019), *Global Report on Internal Displacement'*, accessed 11 January 2020 at www.internaldisplacement.org/sites/default/files/publications/documents/2019-IDMC-GRID.pdf.

Internal Displacement Monitoring Centre (IDMC) (2020), Global Report on Internal Displacement 2020, accessed 17 May 2020 at https://www.internal-displacement.org/global-report/grid2020/.

International Fund for Agricultural Development (IFAD) and the World Bank (2015), 'The use of remittances and financial inclusion', accessed 11 January 2020 at https://www.ifad.org/documents/38714170/40187309/gpfi.pdf/58ce7a06-7ec0-42e8-82dc-c069227edb79.

International Organization for Migration (IOM) (2020), 'Migrant Returnees Turn to Mobile Money in Sudan', accessed 17 May 2020 at https://www.iom.int/news/migrant-returnees-turn-to-mobile-money-sudan.

Jack, W. and T. Suri (2014), 'Risk Sharing and Transactions Costs: Evidence from Kenya's Mobile Money Revolution', *American Economic Review*, **104** (1): 183–223.

Karimi M (2015), 'In Kenya, WFP Introduces Electronic Cash Transfers for Refugees in Kakuma Camps', 11 January 2020 at https://reliefweb.int/report/kenya/kenya-wfp-introduces-electronic-cash-transfers-refugees-kakuma-camps.

Karlan, D., J. Kendall, R. Mann, R. Pande, T. Suri and J. Zinman (2016), 'Research and impacts of digital financial services', Working Paper 22633, National Bureau of Economic Research, 21 December 2019 at https://www.nber.org/papers/w22633.

Kim, M., H. Zoo, H. Lee and J. Kang (2018), 'Mobile Financial Services, Financial Inclusion, and Development: A Systematic Review of Academic Literature', *Electronic Journal of Information Systems in Developing Countries*, **84** (5): 1–17.

Klapper, L., S. Ansar, J. Hess and D. Singer (2019), 'Mobile money and digital financial inclusion', The World Bank, accessed 21 December 2019 at https://globalfindex.worldbank.org/sites/globalfindex/files/referpdf/FindexNote1_062419.pdf.

Klapper, L., M. El-Zoghbi and J. Hess (2016), 'Achieving the Sustainable Development Goals: The Role of Financial Inclusion', Consultative Group to Assist the Poor (CGAP), accessed 21 November 2019 at https://www.unsgsa.org/files/5614/6118/2625/sdgs_paper_final_003.pdf.

Kosse, A. and R. Vermeulen (2014), 'Migrants' Choice of Remittance Channel: Do General Payment Habits Play A Role?', European Central Bank, accessed 21 November 2019 at https://www.ecb.europa.eu/pub/pdf/scpwps/ecbwp1683.pdf.

Langat, A. (2019), 'In a Kenyan refugee camp, business ideas but little access to credit', The New Humanitarian, accessed 23 November 2019 at https://www.thenewhumanitarian.org/feature/2019/05/01/kenyan-refugee-camp-business-ideas-little-access-credit.

Lashitew, A., R. van Tulder and Y. Liasse (2019), 'Mobile Phones for Financial Inclusion: What Explains the Diffusion of Mobile Money Innovations?' *Research Policy*, **48** (5): 1201–1215.

Lee, J., J. Morduch., S. Ravindran., A. S. Schonchoy and H. Zaman (2018), 'The Many Dimensions of Mobile Money: Evidence from Bangladesh', VoxDev, accessed 24 November 2019 at https://voxdev.org/topic/technology-innovation/many-dimensions-mobile-money-evidence-bangladesh.

Llewellyn-Jones, L. (2016), 'Mobile Money: Part of the African Financial Inclusion Solution?' *Economic Affairs*, **36** (2): 212–216, available at https://onlinelibrary.wiley.com/doi/10.1111/ecaf.12177.

Macmillan, R (2016), 'Digital Financial Services: Regulating for Financial Inclusion an ICT Perspective, The International Telecommunication Union (ITU), accessed 23 November 2019 at https://www.itu.int/en/ITU-D/Conferences/GSR/Documents/GSR2016/Digital_financial_inclusion_GDDFI.pdf.

Mader, P. (2018), 'Contesting Financial Inclusion', *Development and Change*, **49** (2): 461–483, available from https://ideas.repec.org/a/bla/devchg/v49y2018i2p461-483.html.

Mbiti, I. and D. N. Weil (2011), 'Mobile Banking: The Impact of M-Pesa in Kenya', National Bureau of Economic Research, Working Paper 17129, accessed 23 November 2019 at https://www.nber.org/papers/w17129.

McAuliffe, M. and B. Khadria (eds) (2019), World Migration Report 2020, Geneva: International Organization for Migration, accessed 21 December 2020 at https://publications.iom.int/books/world-migration-report-2020.

McAuliffe, M., A. Kitimbo and B. Khadria (2019), 'Reflections on migrant's contributions in an era of increasing disruption and Disinformation', in Marie McAuliffe and Binod Khadria (eds), *World Migration Report 2020*, Geneva: International Organization for Migration, pp. 172–174.

McAuliffe, M. (2020), 'On the Margins: Migrant Smuggling in the Context of Development', in Tanja Bastia and Ronald Skeldon (eds), *Routledge Handbook of Migration and Development*, New York: Routledge, pp. 421–433.

Morawcyznski, O. (2009), 'Exploring the Usage and Impact of "Transformational" Mobile Financial Services: The Case of M-PESA in Kenya', *Journal of Eastern African Studies*, **3** (3): 509–525.

Mothobi, O., G. Moshi and M. Deen-Swarray (2018), 'Mobile Money and Financial Inclusion in Sub-Saharan Africa Countries', CPR South 2018 (Policy Brief), accessed 24 November 2019 at https://ssrn.com/abstract=3275133.

Navis, K. (2019), 'Where is Mobile Money Making the Biggest Difference for Financial Inclusion for Young People?', Center for Global Development (CGD), accessed 14 December 2019 at https://www.cgdev.org/blog/where-mobile-money-making-biggest-difference-financial-inclusion-young-people.

Ndiaye, O. and S. R. Parker (2012), Côte d'Ivoire: A Perfect Time for Mobile Money? CGAP, accessed 18 December 2020 at https://www.cgap.org/blog/cote-divoire-perfect-time-mobile-money.

Pew Research Center (2013), Emerging Nations Embrace Internet, Mobile Technology, accessed 21 December 2020 at https://www.pewresearch.org/global/2014/02/13/emerging-nations-embrace-internet-mobile-technology/.

Ratha, D., S. De., E. J. Kim., S. Plaza., G. Seshan and N. D. Yameogo (2019), Data Release: Remittances to Low- and Middle-Income Countries on Track to Reach $551 Billion in 2019 and $597 Billion by 2021, accessed 4 January 2020 at https://blogs.worldbank.org/peoplemove/data-release-remittances-low-and-middle-income-countries-track-reach-551-billion-2019.

Ratha, D., S. De, E. J. Kim, S. Plaza, G. Seshan, N. D. Yameogo (2020), 'Data Release: Remittances to Low- and Middle-Income Countries on Track to Reach $551 Billion in 2019 and $597 Billion by 2021', accessed 24 November 2019 at https://blogs.worldbank.org/peoplemove/data-release-remittances-low-and-middle-income-countries-track-reach-551-billion-2019.

Reuters (2019), 'M-Pesa helps drive up Kenyans' access to financial services – study', accessed 14 December 2019 at https://uk.reuters.com/article/kenya-banking/m-pesa-helps-drive-up-kenyans-access-to-financial-services-study-idUKL8N21L2HK.

Sahay, R., M. Čihák, P. N'Diaye, A. Barajas, S. Mitra, A. Kyobe, Y. Nian Mooi and S. Reza Yousefi (2015), 'Financial Inclusion: Can it Meet Multiple Macroeconomic Goals? International Monetary Fund (IMF), accessed 24 November 2019 at https://www.researchgate.net/profile/Srobona_Mitra/publication/281821048_Financial_Inclusion_Can_It_Meet_Multiple_Macroeconomic_Goals/links/55f97e7a08aec948c493e404.pdf.

Siddik, N. A. (2017), 'Does Financial Inclusion Promote Women Empowerment? Evidence from Bangladesh', *Applied Economics and Finance*, **4** (4): 169–177.

Suri, T. and W. Jack (2016), 'The long-run poverty and gender impacts of mobile money', *Science*, **354** (6317): 1288–1292.

The International Telecommunication Union (ITU) (2013), 'The Mobile Money Revolution', accessed 24 November 2019 at https://www.itu.int/dms_pub/itu-t/oth/23/01/T23010000200001PDFE.pdf.

United Nations (UN) (2018), *Global Compact for Safe, Orderly and Regular Migration*, accessed 4 January 2020 at https://refugeesmigrants.un.org/sites/default/files/180713_agreed_outcome_global_compact_for_migration.pdf.

United Nations Conference on Trade and Development (2018a), 'Economic development in Africa Report 2018: Migration for structural transformation', accessed 14 December 2019 at https://unctad.org/en/PublicationChapters/edar2018_ch2_en.pdf.

United Nations Conference on Trade and Development (UNCTAD) (2018b), 'Leapfrogging: Look before you leap', accessed 14 December 2019 at https://unctad.org/en/PublicationsLibrary/presspb2018d8_en.pdf.

United Nations Department of Economic and Social Affairs (UN DESA) (2019a), *International Migrant Stock 2019*, United Nations, accessed 14 December 2019 at https://www.un.org/en/development/desa/population/migration/data/estimates2/estimates19.asp.

United Nations Department of Economic and Social Affairs (UN DESA) (2019b), *The Number of International Migrants Reaches 272 Million, Continuing an Upward Trend in All World Regions, Says UN*, accessed 14 December 2019 at https://www.un.org/development/desa/en/news/population/international-migrant-stock-2019.html.

United Nations High Commissioner for Refugees (UNHCR) (2012), *An Introduction to Cash-Based Interventions in UNHCR Operations*, accessed 14 December 2019 at https://www.unhcr.org/en-ie/515a959e9.pdf.

United Nations High Commissioner for Refugees (UNHCR) (2019), *Global Trends: Forced Displacement in 2018*, accessed 14 December 2019 at www.unhcr.org/globaltrends2018/.

United Nations High Commissioner for Refugees (UNHCR) (2020), *Global Trends: Forced Displacement in 2019*, UNHCR, Geneva, accessed 27 June 2020 at https://www.unhcr.org/5ee200e37.pdf.

United Nations High Commissioner for Refugees (UNHCR) (n.d.), 'Financial Inclusion of Forcibly Displaced Persons and Host Communities', accessed 14 December 2019 at http://reporting.unhcr.org/node/20859.

United Nations High Commissioner for Refugees (UNHCR) and the United Nations Capital Development Fund (UNCDF) (2019), Financial Inclusion of Forcibly Displaced Persons and Host Communities, accessed 4 January 2020 at https://www.uncdf.org/unhcr-uncdf-joint-programme-for-financial-inclusion-of-forcibly-displaced-persons-and-host-communities.

World Bank (2018), *Financial Inclusion*, accessed 24 November 2019 at https://www.worldbank.org/en/topic/financialinclusion/overview.

World Bank (2019a), *Disruptive Technologies in the Credit Information Sharing Industry: Developments and Implications*, accessed 4 January 2020 at http://documents.worldbank.org/curated/en/587611557814694439/pdf/Disruptive-Technologies-in-the-Credit-Information-Sharing-Industry-Developments-and-Implications.pdf.

World Bank (2019b), *Migration and Remittances: Recent Developments and Outlook*, accessed 4 January 2020 at https://www.knomad.org/publication/migration-and-development-brief-31.

World Food Programme (WFP) (2016), *Bamba Chakula reaches 73,000 Dadaab Households* (Bamba Chakula Update: Jan-Feb 2016), accessed 4 January 2020 at https://www.wfp.org/publications/bamba-chakula-reaches-73000-dadaab-households-bamba-chakula-update-jan-feb-2016.

19. The gender dimensions of technology in the context of migration and displacement: a critical overview

Ibrahim L. Saïd

INTRODUCTION

The rapid development of information and communication technology[1] (ICT) has changed the way we experience the world (Castells, 2000; Wellman, 2001). The increased access to information, speed of communication and knowledge production, sharing and dissemination have transformed the way humanity lives, works, governs and relates to one another. It has also had a profound impact on people's movement and mobility (Inda & Rosaldo, 2008; McAuliffe et al., 2017). Importantly, the rise of ICT around the world has come to transform the way people experience migration and forced displacement and ultimately altered the conception of migration itself (Madianou, 2014). What was once perceived as a state of permanent uprootedness characterised by a radical detachment from one's community of origin, gave way to the 'connected migrants' (Diminescu, 2008) that maintain links, develop networks, and interact with multiple worlds.

While important, ICTs, however, are not neutral tools (Lohan, 2000; Lohan & Faulkner, 2004) with an overarching, universal effect on international migrants and migration processes. Various demographic, environmental, political, social and economic factors including income, education, age, gender and infrastructure among others influence people's access to and use of ICT, creating as such a digital divide at multiple levels between those who 'have' and those who 'have not' (Bezuidenhout et al., 2017). In particular, gender is increasingly becoming central to discussions on the interplay between migration and ICT. The roles, expectations and power dynamics associated with being a man, woman, boy or girl, significantly affect all aspects of the migration process (IOM, 2019) including international migrants' access and use of ICT, shaping as such their experiences accordingly. While considerable progress has been made to understand the gender dimensions of migration, we have just begun to understand the gender dimension of technology in the context of migration.

This chapter aims to revisit the current debates on the gender dimension of ICT in the context of migration with the aim to chart the direction of future research in the field. More specifically, this chapter aims to describe, analyse and critically appraise, where applicable, the framing of debates around gender and technology in the context of migration in existing literature. Gender here is not approached as a mere variable or the opposition of man versus woman, but as a principle of social organisation, a production of inequality and power relations that is socially constructed (Schoellkopf, 2012). It is thus not limited to women, femininity and womanliness, but also includes men, masculinity and manliness as conceptualised in the interaction between technology and migration. Ultimately, this chapter aims to move beyond the apparent difference between women and men in their access and use of ICT to reflect on the way gender and gender relations are shaped, performed and transformed in the context of migration in the digital society.

The argument presented in this chapter is laid out across three sections. First, the chapter presents a brief review of the literature on migration and technology. Second, it presents a genealogy of the debates on gender and technology before it explores, in the third section, how these debates are reflected in the literature on migration and ICT. The chapter concludes with recommendations for future avenues for research in this field.

MIGRATION AND ICT

The development of new ICT since the early 21st century has come to play an important role in migrants' everyday life, and at all levels of the migration process (Collin & Karsenti, 2012; McAuliffe et al., 2017). Studies found that access to ICT provides an effective medium for social networking and communication, and facilitates international migration (Thulin and Vilhelmson, 2016), migrant movement (mobility) (Zijlstra & Van Liempt, 2017), and integration in destination societies (Codagnone & Kluzer, 2012; Bauloz et al., 2019). It has also come to transform migration networks (Dekker and Engbersen, 2014), and the level and character of interaction within transnational populations constituting as a result what is often described as the 'social glue of migrant transnationalism' (Vertovec, 2004, p. 219). This process is not only restricted to the maintenance of family bonds (Vertovec, 2004), but was also found to strengthen transnational ties (Kok & Rogers, 2017), and empower migrants in the construction of their diasporic spaces, agency, and political, social and economic aspirations (Tona & Whelan, 2009; Brinkerhoff, 2012; Burrell & Anderson, 2008; Nedelcu, 2012).

Since the 2015 'reception crisis' in Europe, ICT has also come to attract considerable attention in the context of irregular migration and forced displacement, in both the academic and policy spheres. Access to ICT, especially smartphones, it has been argued, has become a 'lifeline' for migrants and refugees, as 'important as water and food' (Gillespie et al., 2018, p. 1) and as critical as roads (Gillespie et al., 2018, p. 2). Access to wider information sources through ICT was found to empower migrants and asylum seekers to make informed decisions about routes and migration destinations and help them overcome the various challenges they may encounter in the migration process (Khoury, 2015; Smets, 2018; Wall et al., 2017; Kaufmann, 2018). In light of this, scholars have come to speak not only of 'connected migrants' with access to ICT but also of 'connective migrant routes' emphasising as a result the 'sociality of the routes and information sharing' made possible in part by new ICT (Sánchez-Querubin & Rogers, 2018, p. 2).

This techno-optimist approach to ICT has come to characterise a large number of the literature and policy debates about ICT use by international migrants, especially since 2015 (Awad & Tossell, 2019). This approach tends to view ICT as tools that can continually be improved and can improve the lives and experience of migrates at all levels of migration processes. While important, ICTs, however, are not neutral tools (Lohan, 2000; Lohan and Faulkner, 2004) nor a mere utilitarian resource that international migrants use at their individual discretion to fulfil a function and cover needs (Awad & Tossell, 2019). ICTs are rather a socio-material configuration, a constitutive entanglement between the technical and the social (Wajcman, 2009; MacKenzie & Wajcman, 1999; Lohan & Faulkner, 2004), tied together in a circular relationship. ICTs are concurrently shaped as a result of complex social processes and shape social relations, processes and subjectivities. The way ICTs are produced, adopted and used, and the meaning these technologies acquire in certain contexts are the product of this densely interactive relationship embedded within a larger socio-material assemblage of social, economic, political and cultural practices (Lohan & Faulkner, 2004; Wajcman, 2006, 2009). Thus, contrary to common misconceptions, ICT has the potential not only to empower individuals and transform social relations but also to replicate existing social inequalities and create new forms of stratifications depending on the context in which it unfolds (Sassen, 2002). Various factors such as income, education, age and gender, as well as infrastructure, for instance, affect how ICT are accessed and used and the processes produced as a result of such interaction.

In recent years, gender, in particular, is increasingly becoming central to discussions on migration and ICT. Little analysis, however, is currently available on the state of the literature examining the gender dimension of technology in the context of migration. Examining the interrelation between gender and ICT in the context of migration and forced displacement as it is manifested in the available literature is a central aim of this chapter. In the context of migration and forced displacement, social constructs such as gender play a central role in shaping the dynamics of migration (see Box 19.1 on migration and gender), as well as migrants' access to and use of ICT in the various migration processes.

BOX 19.1 MIGRATION AND GENDER

International migration has a strong gender dimension. Nowadays, women represent almost half (48%) of the 272 million migrants and a half (49%) of the 25.9 million refugees (IOM, 2019). The increasing number of women migrants and the 'feminization' migration that followed have altered migration patterns and processes globally (IOM, 2019). The latest data from the International Labour Organization, for instance, show that women represent almost three-quarters of all migrant domestic workers and just over 80 per cent of migrant domestic workers in high-income level countries. There is also an increasing number of young women and adolescent girls undertaking migration independently (as opposed to being part of a family unit), including via irregular migration and smuggling routes (IOM, 2019). These developments have a significant impact on both men and women, and on the gender dynamics in migration societies, which are imperative to examine and document.

International migration and forced displacement have different consequences for women and girls than for men and boys. Gender influences the risks and threats both men and women experience on their journey as well as their coping and protective mechanisms. The roles, expectations, relations and power dynamics associated with being a man or a woman, or whether one identifies as lesbian, gay, bisexual, transgender, queer, intersex, asexual and/or other identity (LGBTIQA+) significantly affect all aspects of the migration process. Gender roles, relations and inequalities often determine who migrates and to where, why people migrate, how they migrate, the networks they use, opportunities and resources available for them across the migration process (IOM, 2019).

International migration may provide the opportunity to challenge traditional or restrictive gender roles. It can lead to a greater degree of empowerment for both women migrants and those who remain at home. Simultaneously migration may also reproduce patriarchal and gender relations. In certain contexts, there is a strong masculinity aspect to migration, which is often perceived as a rite of passage to adulthood and a step towards manhood (Monsutti, 2007).

Gender influences reasons for migrating, who migrates and to where, how people migrate, the networks they use, opportunities and resources available (IOM, 2019) – including access and use of ICT – prior to departure, during the journey and at destination including relations with the country of origin. Concurrently, migration and ICT both have the potential to reinforce and (re)shape gender dynamics. In this sense, access to ICT, as migration itself, is gendered in essence. The gendered dimension of ICT, however, is not always acknowledged in the mainstream studies on migration and technology. In the cases where it is acknowledged, the analysis in the main tends to overlook the deeper power dynamics entrenched in the relationship between gender and technology and the way it shapes both migration and gender dynamics. This limited analysis, it is argued, is rooted – to a large extent – in the flawed conceptualisation of the interrelation between gender and technology that tends to dominate the discussions in the field and which has found its way uncritically to migration studies. This flawed conceptualisation influences the way ICT is approached in the context of migration, which in turn explains the limitations in the available literature.

THE GENDER DIMENSION OF TECHNOLOGY

In general, studies on ICT and society have centred in the main around technologies as drivers of growth and transformation of economies (Vu et al., 2020). New ICTs are seen to hold a tremendous promise for economic and social development that the international community needs to support by widening access to technology, reducing barriers to it, and providing training that boosts awareness and proficiency of its use. In this spirit, studies on gender and technology have explored how access to and use of technology can empower women and increase their political, social and economic rights. These studies argued that, just as ICT may facilitate key functions of daily life, it may also increase women's access to public space, provide them with new opportunities and strengthen their agency (Sida, 2015). Digital inclusion, it was argued, gives women the possibility to access finance, integrate into national and even regional and global value chains, and provide them with the opportunity to become fully fledged members of the information society.

To fulfil ICT's promise of development, many of the available studies on gender and technology focused on examining the inequalities of access to, and use of, ICT. More specifically, they explored the differences between men and women's access and use of ICT and its impact on gender equality (Hilbert, 2011). In this respect, these studies concluded that a gender digital divide exists, reflected in the lower number of women accessing ICT and possessing the skills and experience needed to use the technology compared to men (Hilbert, 2011). While recognising the potential of ICT to economically empower women and promote gender equality, the persistence of gender discrepancies in access and use of ICT, it is argued, will most likely contribute to widening the gender gap. Men's privileged access and capability to use ICT compared to women, it was found, situate them in a better position to take advantage of the new technology subsequently reinforcing their social, economic and political power in society. Unless this gap is addressed, it is concluded, the gender digital divide is likely to exacerbate existing inequalities and situate women at a significant disadvantage (Hilbert, 2011).

This conclusion comes to constitutes a crucial step towards leveraging ICT's potential to contribute to women's empowerment, especially from a policy point of view, which it comes to inform. Consequently, policies were put in place to enhance communication and media literacy for women with a view to building their capacity to access and use ICT (Hilbert, 2011). While important, the implicit assumption behind such an approach, however, was gendered in essence, affecting as such the way we come to understand the gender technology nexus. Technology in these studies was portrayed as a male domain (Faulkner, 2001). Specifically, men were seen as more interested in technology and more tech-savvy than women, who were ascribed a certain computer anxiety and were seen more likely to be technophobic (Hilbert, 2011). This assumption certainly highlights not only the gender-biased conceptions of technology but also of the mainstream writings on gender and technology in general. These studies trap women and men within the existing gender dynamics while overlooking the power relations that form gender roles and gender inequalities (Lohan & Faulkner, 2004, Wajcman, 2009).

This gendered assumption, still, to a large extent, guides the policies tailored towards addressing the gender digital divide (Liff et al., 2004; see also Sey & Hafkin, 2019). While scaling up women's IT skills and capacity to access and use ICT are indeed important interventions to increase their digital inclusion, they nevertheless are based on reductionist ideas. These ideas assume: first, that gender identities are fixed categories; second, that ICT are neutral in essence; and third, that society's technology determines social change (Faulkner, 2001; Lohan & Faulkner, 2004). This approach produces an overly simplistic gender binary of technophobic women and technophilic men. The construction of women in terms of lack or difference when compared with men, indirectly creates a hierarchy in which men become the standard with respect to which women must be compared (Pujol & Montenegro, 2015). This construction, in essence, undermines the complex power dynamics that shape both ICT and gender norms and that guide the interrelation between the two. It further reduces women's emancipation to the act of (economic) empowerment assuming a linear relationship between the latter and gender equality (Lohan, 2000). Moreover, it tends to underestimate the possibility that ICT may contribute to the entrenchment of social and gender inequalities, thus undermining our understanding of the gender dimensions of ICT.

Feminist scholarship, while acknowledging the gender digital divide, points out that women's limited access to and usage of ICT compared with those of men is not due to their being technophobic. These discrepancies are instead the result of the social construction of gender roles in society, and the longstanding gender-related inequalities it generated as a result (Wajcman, 1991; Cockburn, 1985; Kleif & Faulkner, 2003). These gender power relations, it is argued, have come to be embedded within technology and techno-science and, as such, have come to shape the design, content and development of technology itself. This constructivist approach locates technology within a heterogeneous network of socio-technical assemblage that is constituted from and constitutes a social world (Lohan, 2000; Lohan & Faulkner, 2004). This approach views ICT and gender as mutually constituted in which the specific technological artefacts may be gender shaped and may have gender consequences, a process that can be chartered both in the design and use of technologies (Lohan & Faulkner, 2004). In other words, 'existing power relations in society doesn't only define gender relations and identities, but also determines who benefits from and shapes the design, content, development and use of ICT' (Gurumurthy, 2004, p. 1 in O'Donnell & Sweetman, 2018, p. 217). Within this understanding, mere access to and simple use of ICT, they argued, do not guarantee concrete social outcomes. Alternatively, for ICT to reach its full potential as a force for change, attention should be paid to the effective use of ICT and the inequalities produced by the patriarchal capitalist society (Pujol & Montenegro, 2015).

This approach to gender and technology came to transform the conception of the digital divide. It shifted the focus from the disparity of access and use as related to gender to the apparent difference between women and men in their access to and effective use of ICT (Liff et al., 2004). Access here is not associated with possession of a mobile phone, for instance, or internet access, but with the 'experience of use and whether or not such an experience is of sufficient value to individual users' (Sun, 2016, p. 509). This conception asserts that it is rather 'redundant to have access to and skills in digital technology without effectively making a meaningful improvement for the target community in economic, political or cultural terms' (Sun, 2016, p. 509). It thus urges policymakers, researchers and practitioners to move beyond documenting the statistics on access to ICT to focus more on questions of agency, empowerment and emancipation (O'Donnell & Sweetman, 2018; Liff et al., 2004).

Indeed, while the literature on ICT and women's empowerment is relatively large, it rarely draws a connection between women, economic power, and their voice and influence in society and politics (Hilbert, 2011). Moreover, the way new technologies affect the conceptions of masculinity, femininity or gender relations is also an under-researched field (Gurung, 2018). Therefore, to fully understand the gender dimensions of ICT, feminist scholarship urges the analysis of what having a phone or going online can actually do for the user, the processes it generates and how it can change or reinforce gender dynamics (O'Donnell & Sweetman, 2018; Sassen, 2002). In other words, it urges research on ICT to capture life experiences of the digital divide based on gender in terms of their impact on power relations and gender hierarchies. Examining the gender dimension of ICT from this angle would allow us to better understand the technology gender nexus and reflect on the potential of ICT to promote gender equality and achieve social change. Based on this conceptual reflection, this chapter thus aims to understand the gender dimension of ICT in the process of migration from a multidimensional perspective that moves beyond mere access to incorporate effective use and impact on gender dynamics and gender relations. Ultimately, it aims to reflect on the ways gender has been approached, and the framing of debates around gender and technology in the context of migration.

REFLECTIONS ON THE GENDER DIMENSIONS OF TECHNOLOGY IN THE CONTEXT OF MIGRATION

Although a substantial body of literature has been produced, especially in the last few years on migration and technology, the gendered dimensions of technology in migration have not been adequately examined (Leurs, 2012; Leurs & Ponzanesi, 2013). The available studies examining the gender dimension of technology tend to vary based on the phase of migration in which it was examined, the methods used, and the level and degree of the gender analysis applied, which may range from a focus on simple access and use (Dekker et al., 2018; 2015; Gillespie et al., 2016; Smets, 2018) to more in-depth gender analysis (Witteborn, 2018; 2015). Broadly speaking, most of the literature that performs an 'in-depth' gender analysis on the use of technology in the context of migration, tends to concern the transnational experiences of female (labour) migrants in the country of destination and their relationship with their families in countries of origin (Parreñas, 2005, Tona & Whelan, 2009; Uy-Tioco, 2007), with some few exceptions (Witteborn, 2018). Conversely, a limited number of scholars have examined the gender dimension of technology in pre-migration (Venables, 2008) and during the migration journey. The available studies tend to approach gender from a more basic perspective that does not move beyond examining mere ICT access or use (Gillespie et al., 2016, 2018; Dekker et al., 2015, 2018). Limited studies have examined the differences in experience based on gender, or the social and gender relations produced and reproduced in the context of technology use during migration.

It is worth noting also that the term 'gender', both as an object of study and as a category of analysis, is commonly understood and used in the available literature as something that concerns women, or women in their relationship with men. The gendered experience of men tends to be largely overlooked in these studies despite their constituting a large portion of the research samples of the available studies examining and theorising technology in the context of migration. Overall, the insufficient attention to the gender category in migration and technology studies makes specific conditions and 'experiences of migrant women invisible and occults gender asymmetries that are reproduced at different migratory stages' (Zapata-Barrero & Yalaz, 2018, p. 18). It also oversimplifies the experience of men and serves to obscure a deeper understanding of the gender dimension of technology in the context of migration.

THE (UN)GENDERED EXPERIENCE OF MEN: A PRELIMINARY REFLECTION

Except for the studies examining 'ICT use' among migrants in countries of destination, most studies on gender and technology in the context of international migration, especially those documenting the use of ICT in the pre-migration and during journey process, focus, for the main, on men's experience. While in principle, men's experience is also gendered, reflecting their position in society, most of the available literature, nevertheless, tends to be 'gender blind' in its analysis. It overlooks the gender dimension of technology and instead approaches ICT as a utilitarian, useful tool that migrants and/or potential migrants use in their own discretion to make decisions about migration, migration routes and destinations. This approach to ICT, however, while important, nevertheless tends to simplify migrants' experience and connectivity and bind it to provisional interests or a mere response to arising needs in the migration process. The way the use of ICT among men in the various migration processes shape or reshape masculinity or influence gender relations and power structures in a patriarchal society is hardly touched upon. Also, a more in-depth analysis of the intersection between age, education and gender might be useful to explore in this context. For instance, while abundant studies on gender and technology in the context of migration have acknowledged gender and age as variables that determine access to and use of ICT (Dekker et al., 2018; Gillespie et al., 2016), no study examines how younger men's privileged access to ICT compared to seniors, or moreover, to less technology-literate men, for instance, influences the power relations within a patriarchal structure in the process of migration. Also, little is known about the form of authority ICT proficiency in the context of migration may grant young migrants over decision-making processes and how that might impact gender dynamics, and in particular masculinity. Studies examining the intersection between age and gender or those carrying out intergenerational analysis in the context of digital migration, tend to focus for the main on second- and third-generation migrant women's and girl's experience and the way they negotiate their gender roles in patriarchal families using ICT (Leurs and Ponzanesi, 2013). The gender dimension of such an interaction in the processes of migration, especially in relation to masculinity, is under-studied.

THE GENDER DIMENSION OF TECHNOLOGY IN THE PRE-MIGRATION PHASE

The role of technology in the pre-migration process is a relatively new field that may require further analysis. The available studies, in the main, focus on the role of ICT in informing decision making about migration. These studies found that ICT, such as news, films and advertisements, for instance, are among the most important sources of information for people considering migration (Wood & King, 2013; Hamel, 2009). This type of ICT may create the image of wealth and prosperity of developed countries in developing countries, which can be crucial in terms of making the final decision to migrate, particularly for those who already consider migration as an option (Hamel, 2009). Others have suggested that access to ICT increases people's awareness of the difficulties of the migratory process and supports a more balanced and nuanced understanding of migration (Horst, 2006). Modern ICT also makes it easier to access information and facilitate the development and strengthening of different types of networks. The development of transnational networks, and the increase in people's capacity to interact with family and friendship networks both in countries of origin and destination, enabled people to make a better-informed decision about migration, migration processes, and destination (Dekker & Engbersen, 2014). Harriot Beazley (2019, p. 52), for instance, found that ICT influences intergenerational decision to migrate, sustaining, as a result, the 'intergenerational cycles of female labour migration in the South-East Asia and the Pacific region.'

Gender norms, however, and other socio-economic and structural constraints often limit women's access to ICT, especially in developing countries. Studies found that limited access to the internet, for instance, can limit the ability of migrants to gather information about their future migration opportunities (Gelb and Krishnan, 2018). In this way, access and use of ICT, it has been argued, could have an impact on the selectivity of migration, as it initially attracts those who have access and possess the skills to use technology (Hamel, 2009). This point might explain, to some extent, the limited number of studies that document women's experience in using technology in the pre-migration process. Indeed, studies found that while the gender divide in access to ICT has been significantly reduced in developed countries, it remains an issue in developing countries. Figures estimate that, in low- and middle-income countries, women are on average 10 per cent less likely to own a mobile phone than men and 26 per cent less likely to use mobile internet (GSMA, 2018). The gap is much higher in South Asia and sub-Saharan Africa, which stands at a staggering rate of 70 per cent and 34 per cent, respectively.

Overall, the studies examining gender dimensions of technology in the pre-migration phase either incorporate gender as a variable to measure the difference in access to ICT between men and women, or analyse women's experience of access and use, though the latter tends to be an under-researched area (Venables, 2008). Studies that include gender as a variable in their analysis of ICT use in the context of pre-migration found no statistical gender difference between men and women, both in social media use as a source of information (MMC, 2018) and in terms of ICT use to connect with various transnational networks. Dekker et al. (2015, p. 547) argue that the irrelevance of the digital divide based on gender for non-migrants 'communicating with people in potential destination countries [...] [P]erhaps can be explained in terms of a shifting digital divide from access to equality of use.' They add that 'gender is considered important when it comes to internet access but less relevant when it concerns type and equality of use' (Dekker et al., 2015, p. 547). These are undoubtedly important findings that might demonstrate the positive outcome of global efforts to reduce inequalities in access. This understanding, however, fails to move beyond the mainstream conceptions of the gender digital divide as the inequality of access and use. In addition, it overlooks the complex power dynamics embedded in the interrelations between ICT and gender beyond 'quality of use', which in itself is rather a needs-oriented category that approaches ICT as a neutral tool that might make people's lives better.

More research may be required to better understand the power dynamics and processes embedded in and triggered, respectively, by women's ICT use in the pre-migration phase. A more in-depth analysis is required to understand the impact of access to information on decision making, and on women's empowerment, agency and aspirations. In fact, while there is extensive anecdotal evidence on the relationship between international migration decisions and the use of ICTs, there is generally still very little empirical research (Cummings et al., 2015). This gap needs to be addressed while providing in-depth analysis of the gender dynamics at play both when accessing information and in making an informed decision about migration. There is a further need to examine the characteristics of social networks developed by women compared with men in their search for information and how such dynamics and processes sustain or challenge the existing power and gender dynamic. Often certain ICT engagements in the pre-migration period may enforce gender relations and gender stereotypes; these accounts should be better examined.

Emilie Venables's (2008) study of young Senegalese women, for instance, provides a good example to such a paradox. Venables, in her research, examined how young Senegalese women use internet technologies to communicate with European men, in the hope of forming a relationship that may lead to transnational migration. Rather than assessing the degree to which women are successful in their aspiration, the author considered how online communication enables them to assert their agency and imagine a better future for themselves. She found that, while these women did not find love through the internet, or realise their migratory dream, their online practices allowed them to 'construct a more optimistic future for themselves' (Venables, 2008, p. 488). Needless to say, whereby online practices enabled women to exercise agency, it concurrently trapped them within existing gender dynamics that subordinated them to male dominance and exposed them to risks of exploitation (see Groes-Green, 2014). This paradox certainly highlights the complex relations between women empowerment, agency and emancipation in a patriarchal capitalist society. More research is needed to explore and understand these contradictory dynamics in the context of (pre)migration in general.

THE GENDER DIMENSION OF TECHNOLOGY IN THE MIGRATION JOURNEY

Similar to the studies about the pre-migration process, the field of study examining technology use in the migration journey is fairly new. It received an increased interest following the media circulation of Syrian refugee images using mobile phones in their migration route to Europe in 2015. Studies in this field highlight the importance of ICT, especially mobile phones and their influence on travel methods, on shaping decision making about routes, developing networks, building trust and becoming self-reliant (Zijlstra & Van Liempt, 2017). In essence, ICT has been found to increase migrant mobility and interconnectivity, especially in irregular migration. These studies tend to mainly document men's experience, although it also often incorporates some accounts of women's experiences of ICT use during the journey. Overall, it is hard to ascertain based on existing literature the differences between men and women's experience with ICT or provide any account of a gendered experience of any sort. None of the available studies conducts a thorough gender analysis to examine the gender dimensions of ICT in the migration journey. The literature, in the main, tends to be gender-neutral and provide a utilitarian account of the influence of ICT on the migration journey. These studies document the way ICT assists irregular migrants in their journey and the way it enables them to overcome challenges, without providing accounts on the gender dimension of such an experience (Gillespie et al., 2016).

Needless to say, some of the available literature provides basic anecdotes about ICT use among women and men. ICT4Refugees, for instance, indicate that even among Syrian refugees, for whom smartphone use is almost universal, there are disparities and complexities, pointing to a potential difference in the experience based on gender and age (GSMA, 2017). This plain statement tends to characterise much of the literature on the topic, whereby it acknowledges the potential of disparities but still without performing an in-depth gender analysis to explore the processes it produces (Dekker et al., 2018). Borkert et al. (2018, p. 6) provide a more nuanced account in this regard. Disaggregating their results on the use of ICT during the journey by gender and age group, they found that 'gender was not a significant factor to determine the use of the ICT to seek for help or finding needed information' about route maps or country of destinations. While important, this account is certainly not enough. More detailed studies are needed to better understand the gendered dimension of ICT in the migration journey. The available literature offers little reflection on the way ICT use may shape social and gender relations in the migration journey. Specifically, little is known about the way access and use of ICT affects women's position in society, or whether ICT access and use have actually changed in one way or another women's position during the migration process. As such, it is important to move beyond the question of how refugees use ICT during the journey to explore ICT use within the broad socio-cultural and political conditions in which they unfold (Smets, 2018). While research in the field has examined the way ICT use shapes the migration route and obviously vice versa (Dekker et al., 2018), there is a need to explore the migrant's internal group dynamics, and the way ICT shapes social and gender relations (Wall et al., 2017; Smets, 2018) and the way this shapes the migration experience of both men and women, respectively.

Adopting a similar line of thought Bayramoğlu and Lünenborg (2018) show how the use of ICT in journeys among queer refugees make both migration and queer intimacy possible. Building on the non-media-centric approach to media studies, the authors argue for the need to transform the way researchers view digital media practice from 'an […] "add-on" to everyday life, […] to seek to understand how they are inseparably embedded within socio-cultural relations, the experience of migration, and the affectivity of the body' (Bayramoğlu & Lünenborg, 2018, p. 4). The affective nature of digital media, they argue 'has the potential to foster new intimacies and a sense of belonging, especially in moments when queer refugees feel disadvantaged, immobilized, stuck, or lost.' 'In particular, social media platforms such as Facebook, Twitter, and Instagram need to be understood not only as a tool that effectively connect[s] users with one another and/or circulate[s] information that may be otherwise unobtainable, but also circulate representations of queer intimacy that may be "unrepresentable" in traditional offline media in certain contexts' (Bayramoğlu and Lünenborg 2018, pp. 4–5). More studies along these conceptual lines are required to better understand the migration journey and migrants' dynamics in the digital society. Moreover, considerable attention should be given to examining gender dimensions of ICT in refugee camps, which tends to be under-represented in the literature.

THE GENDER DIMENSION OF TECHNOLOGY IN THE TRANSNATIONAL SPHERE

The use of ICT among migrants in host societies has been touched upon considerably in the literature. ICT facilitates immigrants' economic and social integration (Codagnone & Kluzer, 2011; Garrido et al., 2010), allows migrants to maintain relations with their countries of origin and build diasporic spaces. Contrary to the literature on the pre-migration and journey process, which tends to be gender blind, a large number of the studies on migrants' access and use of ICT in host societies have examined the experience of women including from a gender perspective. Indeed, ICT plays a central role in the life of migrants in destination countries. Access to ICT, for instance, encourages the economic integration of migrant women (Garrido et al., 2010). It enables women to enrol in online training and helps them acquire the necessary information and forms of knowledge needed to develop their skills, which might be vital to find employment. ICT has also allowed the formation of women diaspora online groups and influenced political debates and ultimately political decisions. These online groups may constitute a platform in which relevant gender discourses are presented and constructed in accordance with women's conditions and aspirations. They may concurrently constitute a platform for the negotiation of gender identities, as well as a space that may reinforce gender relations and gender identities (Lee, 2013; Leurs & Ponzanesi, 2013). Saskia Witteborn (2015), for instance, demonstrates how virtual practices provided asylum seekers in Germany, both men and women, the possibility to present themselves as the person they want to become. The subjective feeling of 'lack,' she argues, 'was the main topic that sums up what interviewees expressed repeatedly across research sites, gender and nationality: lack of certainty, health, work and productive legal status. Becoming imperceptible was the effort of positioning oneself as something other than a person lacking fundamental certainties' (Witteborn, 2015, pp. 355–356), which was made possible through ICT.

Other studies have engaged with the notion of affect, transnational motherhood and remote care among transnational families, especially in the context of female labour migrants (Cogo, 2017). These studies show how ICT helps labour migrant women to foster emotional links with their family and friends. Uy-Tioco (2007), for instance, found that the mobile phone enables Filipino migrant workers in the US to remotely assert their role as mothers. They reinforced their love for their children through text messaging and maintained their presence at home despite the geographical distance. In the same vein, Parreñas (2005) observed that, in contrast to migrant fathers, who were inclined to maintain only instrumental communication with their families, migrant mothers sought to foster intimate ties with their children through regular communication via phone calls, text messages or letters. These studies argue that the fact that women are expected to maintain ties with countries of origin facilitated, and/or increased their access to ICT and smartphones. It has also, to some extent, played a crucial role in supporting their decision to migrate, since it allowed them to maintain and foster emotional bonds with their families and perform care duties from a distance (Madianou, 2014). This certainly represents the complexity of technologically mediated sociality, which, while offering the opportunity to transfer family intimacy, also constitutes a challenge and may reinforce existent gender narratives and gender relations. Similarly, studies on left-behind women show that male migration has had a profound positive impact on women and families who remained. While 'major decisions still tended to be made by migrant men through mobile phone communication,' studies found 'a clear increment in women's authority to make decisions' (McAuliffe et al., 2017, p. 185).

These studies indeed demonstrate the potential and limits of ICT use among women within a capitalist patriarchal society. ICT in the context of labour migration allows women to balance between their duties as mothers and carers and as workers and ultimately empowers them economically as a result. It, however, also reinforced gender dynamics and gender roles. Thus, rather than contributing to the emancipation of women, ICT has facilitated their continuous subjugation in the patriarchal society, and as such, it may have even fostered women's double burden. In the event of failure to maintain transnational family ties, women are the ones to blame for the neglect of their children. This may not be even an issue in the case of male migrants. This discrepancy has its roots in the gender-specific distribution of care responsibilities, which found its way to migration and ICT the way gender identities shape ICT practices (Lutz & Palenga-Mollenbeck, 2012).

Studies also underscore the role of emotions and gender in migrants' digital practices (Witteborn, 2011, 2015, 2018; Khoury, 2015). Studies showed that women tend to share more feelings in their online practices and seek and provide more emotional support. Men, on the other hand, tend to use ICT in more 'practical' ways, such as to read the news or look for work. Witteborn (2011) found that, contrary to women, men often tend to conceal their emotions in their virtual interaction with their families in countries of origin. They would be likely to turn off the camera to hide emotions that could distress or disappoint distant loved ones. Whereby women used ICT to share their problem and seek emotional support, men who live in refugee camps in Germany tend to convey a sense of shame of their living conditions and would rather portray their life as a success instead (Witteborn, 2015). This analysis demonstrates the way gender norms of masculinity play out in the use of ICT among men, and its impact on the relations, and transnational dynamics between men and their family and friends in the country of origin. The feeling of shame and precisely the act of shaming of a person who does not adhere to a perceived way of living, as Witteborn (2018) notes, was a common practice applied by men and women to deter refugee women from shared computer rooms. This analysis moves beyond the utilitarian narrative on ICT, highlighting the importance of emotions in constructing the gender experience of ICT. More studies along these lines may allow for a better understanding not only of the gender dimension of ICT but also of migration processes and dynamics, which is much needed.

CONCLUSION

Despite the abundant literature on the role of ICT in the migration process, a limited number of studies focused on examining the gender dimension of ICT in the context of migration.

Most of the available studies, especially in the context of pre-migration process and during the journey are predominantly gender blind. The limited studies that include gender or women's experience approach gender from the narrow perspective of access whereby gender itself is treated as a variable. Gender is predominantly approached as a category that concerns women or women's relationship with men; limited studies have looked at the experience of men from gender or a feminist lens. This shortcoming of the literature largely stems from the reductionist approach to ICT as a neutral tool that determines social change. More in-depth analysis that seeks to understand the way ICT is embedded within socio-cultural relations, and the way it produces social relations is needed to understand better the gender dynamics of ICT in the context of migration.

Studies on the gender dimensions of ICT in the transnational sphere are much more developed than those examining the use of ICT in the pre-migration phase and during the migration journey. Most of these studies tend to focus on women labour migrants, and the way ICT facilitates intimacy and remote care, what it commonly describes as practices of transnational mothering. These studies demonstrate women labour migrants' predicament in the patriarchal capitalist societies in the age of information, whereby ICT empowers women to seek employment, it simultaneously serves as a means to practise their gendered roles as mothers and carers. More studies are needed to examine the way ICT serves to negotiate gender roles and gender relations. Moreover, more studies need to examine men's experience using ICT from a gender perspective. Whereby some studies offer some anecdotes, more work is still required. Approaching ICT as a socio-technical assemblage that is constituted by and constitutes social relations would allow us to examine the complex interrelation between ICT, gender and technology in the context of migration.

NOTE

1. ICT includes the use of hardware such as mobile phones, tablets, laptops and PCs, and software like apps, social media platforms, webpages or digital radio.

BIBLIOGRAPHY

Awad, I. & J. Tossell (2019), 'Is the smartphone always a smart choice? Against the utilitarian view of the "connected migrant,"' *Information, Communication & Society*, 24(4), 1–16.

Bauloz, Celine, Zana Vathi & Diego Acosta (2019), 'Migration, Inclusion and Social Cohesion: Challenges, Recent Development and Opportunities,' in Marie McAuliffe and Biond Khadria (eds), *World Migration Report 2020*, Geneva: IOM, pp. 185–208.

Bayramoğlu, Y. & M. Lünenborg (2018), 'Queer migration and digital affects: Refugees navigating from the Middle East via Turkey to Germany,' *Sexuality & Culture*, 22(4), 1019–1036.

Beazley, Harriot (2019), 'Intergenerational cycle of migrating for work: Young women and girls migrate for work in South East Asia and the Pacific', in IOM, *Supporting Brighter Futures: Young Women and Girls and Labour Migration in South East Asia and the Pacific*, Geneva: IOM, pp. 45–62.

Bezuidenhout, L. M., S. Leonelli, A. H. Kelly & B. Rappert (2017), 'Beyond the digital divide: Towards a situated approach to open data,' *Science and Public Policy*, 44(4), 464–475.

Borkert, M., K. E. Fisher & E. Yafi (2018), 'The best, the worst, and the hardest to find: How people, mobiles, and social media connect migrants in (to) Europe,' *Social Media + Society,* 4(1), 1–11.

Brinkerhoff, J. M. (2012), 'Digital diasporas' challenge to traditional power: The case of TibetBoard,' *Review of International Studies*, 38(1), 77–95.

Burrell, J. & K. Anderson (2008), '"I have great desires to look beyond my world": Trajectories of information and communication technology use among Ghanaians living abroad,' *New Media & Society*, 10(2), 203–224.

Castells, Manuel (2000), *The Rise of the Network Society*, Oxford: Blackwell.

Choudaha, R. and Roy, M. (2015), 'Mobility Patterns and Pathways of Indian Engineers to the U.S.', WES Research & Advisory Services, WENR. https://wenr.wes.org/2015/11/mobility-patterns -pathways-indian-engineers-u-s. accessed on 13 August 2021.

Cockburn, C. (1985), *Machinery of Dominance: Women, Men and Technical Know-how*, London: Pluto.

Codagnone, Cristiano & Stefano Kluzer (2011), *ICT for the Social and Economic Integration of Migrants into Europe*, Publication Office of the European Union.

Cogo, D. (2017), 'Communication, migration and gender: transnational families, activism and ICT use,' *Intercom: Revista Brasileira de Ciências da Comunicação*, 40(1), 177–192.

Collin, Simon & Thierry Karsenti (2012), 'ICT and Migration: A conceptual framework of ICT use by migrants,' Paper presented at the World Conference on Educational Multimedia, Hypermedia and Telecommunications, at Denver, Colorado, USA, 1492–1497.

Cummings, C., J. Pacitto, D. Lauro & M. Forest (2015), 'Why people move: understanding the drivers and trends of migration to Europe,' working paper 430, London: Overseas Development Institute (ODI).

Dekker, R. & G. Engbersen (2014), 'How social media transform migrant networks and facilitate migration,' *Global Networks*, 14(4), 401–418.

Dekker, R., G. Engbersen & M. Faber (2015), 'The use of online media in migration networks,' *Population, Space and Place*, 22(6), 539–551.

Dekker, R., G. Engbersen, J. Klaver & H. Vonk (2018), 'Smart refugees: How Syrian asylum migrants use social media information in migration decision-making,' *Social Media+ Society*, 4(1), 1–11.

Diminescu, D. (2008), 'The connected migrant: An epistemological manifesto,' *Social Science Information*, 47(4), 565–579.

Faulkner, W. (2001), 'The technology question in feminism: A view from feminist technology studies,' *Women's Studies International Forum*, 24(1), 79–95.

Garrido, M., J. Sullivan & A. Gordon (2010), 'Understanding the links between ICT skills training and employability: An analytical framework', paper presented at the Proceedings of the 4th ACM/ IEEE International Conference on Information and Communication Technologies and Development, December.

Gelb, S. & A. Krishnan (2018), 'Technology, migration and the 2030 Agenda for Sustainable Development', Briefing paper, ODI, September.

Gillespie, Marie, Souad Osseiran & Margie Cheesman (2018), 'Syrian refugees and the digital passage to Europe: Smartphone infrastructures and affordances,' *Social Media+ Society*, 4(1), 1–12.

Gillespie, Marie, Lawrence Ampofo, Margaret Cheesman, Becky Faith, Evgenia Iliadou, Ali Issa, Souad Osseiran & Dimitris Skleparis (2016), *Mapping Refugee Media Journeys: Smartphones and Social Media Networks*, The Open University, accessed 12 March 2020. At https://www.open.ac.uk/ ccig/sites/www.open.ac.uk.ccig/files/Mapping%20Refugee%20Media%20Journeys%2016%20May %20FIN%20MG_0.pdf.

Groes-Green, C. (2014), 'Journeys of patronage: Moral economies of transactional sex, kinship, and female migration from Mozambique to Europe,' *Journal of the Royal Anthropological Institute*, 20(2), 237–255.

GSMA (2018), *The Mobile Gender Gap Report*, London: GSMA.

GSMA (2017), *The Importance of Mobile for Refugees: A Landscape of New Services and Approaches*, London: GSMA.

Gurung, L. (2018), 'The digital divide: An inquiry from feminist perspectives,' *Dhaulagiri Journal of Sociology and Anthropology*, 12, 50–57.

Hamel, J. Y. (2009), 'Information and communication technologies and migration,' Human Development Research Paper 2009/39, UNDP.

Hilbert, M. (2011), 'Digital gender divide or technologically empowered women in developing countries? A typical case of lies, damned lies, and statistics,' *Women's Studies International Forum*, 34(6), 479–489.

Horst, H. A. (2006), 'The blessings and burdens of communication: Cell phones in Jamaican transnational social fields,' *Global Networks*, 6(2), 143–159.

Inda, Jonathan Xavier & Renato Rosaldo (ed.) (2008), *The Anthropology of Globalisation*, Oxford: Blackwell Publishing.

International Organisation for Migration (IOM) (2019), *Supporting Brighter Futures: Young Women and Girls and Labour Migration in South East Asia and the Pacific*, Geneva: IOM.

Kaufmann, K. (2018), 'Navigating a new life: Syrian refugees and their smartphones in Vienna,' *Information, Communication & Society*, 21(6), 882–898.

Khoury, R. B. (2015), 'Sweet tea and cigarettes: A taste of refugee life in Jordan,' *Forced Migration Review*, 49, 93–94.

Kleif, T. & W. Faulkner (2003), 'I'm no athlete [but] I can make this thing dance!'—Men's pleasures in technology,' *Science, Technology and Human Values*, 28(2), 296–325

Kok, S. & R. Rogers (2017), 'Rethinking migration in the digital age: Transglocalization and the Somali diaspora,' *Global Networks*, 17(1), 23–46.

Lee, EunKyung (2013), 'Formation of a talking space and gender discourses in digital diaspora space: Case of a female Korean im/migrants online community in the USA,' *Asian Journal of Communication*, 23(5), 472–488.

Leurs, Koen (2012), *Digital Passages: Moroccan–Dutch Youths Performing Diaspora, Gender and Youth Cultural Identities across Digital Space*, Utrecht University.

Leurs, Koen (2015), *Digital Passages: Migrant Youth 2.0: Diaspora, Gender & Youth Cultural Intersections*, Amsterdam University Press.

Leurs, Koen & Sandra Ponzanesi (2013), 'Intersectionality, Digital Identities and Migrant Youth,' in Cynthia Carter, Linda Steiner & Lisa McLughlin (eds), *Routledge Companion to Media and Gender*, London: Routledge, pp. 632–642.

Liff, S., A. Shepherd, J. Wajcman, R. Rice & E. Hargittai (2004), 'An evolving gender digital divide?', Internet Issue Brief, 2, Oxford Internet Institute, July.

Lohan, M. (2000), 'Constructive tensions in feminist technology studies,' *Social Studies of Science*, 36(6), 895–917.

Lohan, M. & W. Faulkner (2004), 'Masculinities and technologies: Some introductory remarks,' *Men and Masculinities*, 6(4), 319–329.

Lutz, H. & E. Palenga-Möllenbeck (2012), 'Care workers, care drain, and care chains: Reflections on care, migration, and citizenship,' *Social Politics*, 19(1), 15–37.

MacKenzie, D. & Wajcman, J. (1999), *The Social Shaping of Technology*, 2nd edition. Milton Keynes: Open University Press.

Madianou, M. (2014), 'Smartphones as polymedia,' *Journal of Computer-Mediated Communication*, 19(3), 667–680.

McAuliffe, Marie & Martin Ruhs (2017), *World Migration Report 2018*, Geneva: IOM.

McAuliffe, Marie & Binod Khadria (eds) (2019), *World Migration Report 2020*, Geneva: IOM.

McAuliffe, Marie, Adrian Kitimbo, Alexandra M. Grossens & Ahsan Ullah (2017), 'Understanding migration journeys from migrants' perspectives,' in Marie McAuliffe and Ruhs Martin (eds), *World Migration Report 2018*, Geneva: IOM, pp. 171–190.

Mechefske, C. K., Wyss, U. P., Surgenor, B. W., and Kubrick, N. (2005), 'Alumni/ae surveys as tools for directing change in engineering curriculum', Proceedings of the Canadian Engineering Education Association.

Mixed Migration Centre (MMC) (2018), 'Monthly Migration Movement: Social Media,' MMC, August.

Monsutti, A. (2007), 'Migration as a rite of passage: Young Afghans building masculinity and adulthood in Iran,' *Iranian Studies*, 40(2), 167–185.

Nedelcu, M. (2012), 'Migrants' new transnational habitus: Rethinking migration through a cosmopolitan lens in the digital age,' *Journal of Ethnic and Migration Studies*, 38(9), 1339–1356.

O'Donnell, A. and C. Sweetman (2018), 'Introduction: gender, development and ICTs,' *Gender & Development*, 26(2), 217–229.

Parreñas, R. (2005), 'Long distance intimacy: Class, gender and intergenerational relations between mothers and children in Filipino transnational families', *Global Networks*, 5(4), 317–336.

Ponzanesi, S. (2016), 'Digital crossings in Europe: Gender, diaspora and belonging,' accessed 12 March 2020. At http://connectingeuropeproject.eu/wp-content/uploads/2016/12/ERC_CONNECTINGEUROPE_summary21.pdf.

Pujol, J. & M. Montenegro (2015), 'Technology and feminism: A strange couple', *Revista de Estudios Sociales*, 51, 173–185.

Sánchez-Querubín, N. & Richard Rogers (2018), 'Connected routes: Migration studies with digital devices and platforms', *Social Media+ Society*, 4(1), pp. 1–13.

Sassen, S. (2002), 'Towards a sociology of information technology', *Current Sociology*, 50(3), 365–388.

Schoellkopf, J. C. (2012), 'Gender: An infinite and evolving theory', *Lesbian Gay Bisexual Transgender Queer Centre*, Paper 29.

Sey, Araba & Nancy Hafkin (ed.) (2009), *Report of EQUALS Research Group, Led by the United Nations University*, United Nations University.

Sey, Araba and Nancy Hafkin (ed.) (2019), *Taking Stock: Data and Evidence on Gender Equality in Digital Access, Skills and Leadership*, United Nations University, Tokyo.

Sida (2015), 'Gender and ICT', Sida, Brief, March.

Smets, K. (2018), 'The way Syrian refugees in Turkey use media: Understanding "connected refugees" through a non-media-centric and local approach,' *Communications*, 43(1), 113–123.

Sun, J. (2016), 'Her voice in the making: ICTs and the empowerment of migrant women in Pearl River Delta, China,' *Asian Journal of Women's Studies*, 22(4), 507–516.

Thulin, E. & B. Vilhelmson (2014), 'Virtual practices and migration plans: A qualitative study of urban young adults', *Population, Space and Place*, 20, 389–401.

Tona, C. D. & A. Whelan (2009), '"Re-mediating" the ruptures of migration: The use of internet and mobile phones in migrant women's organizations in Ireland,' *Translocations: The Irish Migration, Race and Social Transformation Review*, 5(1), 1–20.

Uy-Tioco, C. (2007), 'Overseas Filipino workers and text messaging: Reinventing transnational mothering,' *Continuum*, 21(2), 253–265.

Venables, E. (2008), 'Senegalese women and the cyber café: Online dating and aspirations of transnational migration in Ziguinchor,' *African and Asian Studies*, 7(4), 471–490.

Vertovec, S. (2004), 'Cheap calls: The social glue of migrant transnationalism,' *Global Networks*, 4(2), 219–224.

Vu, K., P. Hanafizadeh & E. Bohlin (2020), 'ICT as a driver of economic growth: A survey of the literature and directions for future research,' *Telecommunications Policy*, 44(2). https://doi.org/10.1016/j.telpol.2020.101922.

Wajcman, Judy (1991), *Feminism Confronts Technology*, Cambridge: Polity.

Wajcman, J. (2006), 'The gender politics of technology', in Robert E. Goodin and Charles Tilly (eds), *The Oxford Handbook of Contextual Political Analysis*, Vol. 5. Oxford: Oxford University Press, pp. 707–721.

Wajcman, J. (2009), 'Feminist theories of technology', *Cambridge Journal of Economics*, 34(1), 143–152.

Wall, M., M. Otis Campbell & D. Janbek (2017), 'Syrian refugees and information precarity', *New Media & Society*, 19(2), 240–254.

Wellman, B. (2001), 'Physical place and cyberplace: The rise of personalized networking,' *International Journal of Urban and Regional Research*, 25, 227–52.

Witteborn, S. (2011), 'Constructing the forced migrant and the politics of space and place-making,' *Journal of Communication*, 61(6), 1142–1160.

Witteborn, S. (2015), 'Becoming (im)perceptible: Forced migrants and virtual practice,' *Journal of Refugee Studies*, 28, 350–367.

Witteborn, S. (2018), 'The digital force in forced migration: Imagined affordances and gendered practices,' *Popular Communication*, 16(1), 21–31.

Wood, Nancy & Russell King (2013), 'Media and migration: An overview', in Russell King & Nancy Wood (eds), *Media and Migration: Constructions of Mobility and Difference*, London: Routledge, pp. 11–32.

Zapata-Barrero, Ricard & Evren Yalaz (ed.) (2018), *Qualitative Research in European Migration Studies*, New York, NY: Springer.

Zijlstra, J. & I. van Liempt (2017), 'Smart (phone) travelling: Understanding the use and impact of mobile technology on irregular migration journeys,' *International Journal of Migration and Border Studies*, 3(2–3), 174–191.

20. Mobility of tech professionals in the world economy: the case of Indian entrepreneurialism in the United States

Binod Khadria and Ratnam Mishra

ENTREPRENEURIALISM: BOOST THROUGH BLENDING OF TECH AND BUSINESS EDUCATION

Engineering is a profession that prepares individuals for development and application of new technologies. Logically, engineers are innovators who apply their scientific knowledge in designing the process to develop a new product. However, the modern world is cross-functional and therefore modern engineers are also required to be well versed in communicating with business representatives for the commercial success of their product designs. Therefore, current engineering education has started combining marketing, distribution, pricing and finance with its existing curriculum leading a new way to entrepreneurialism. Schumpeter (1942:83) had called innovation a process of industrial mutation that revolutionized the internal structure of the economy, incessantly destroying the old one and creating a new one. He also believed that innovation was the centre of economic change, causing gales of what he termed 'creative destruction'. This is what was defined as entrepreneurship, or organization, or management – the fourth factor of production in textbook economics that determines 'risk-taking' in the choice and application of technology to combine the other three, viz. land, labour and capital for maximization of what is called profit.

In India, the main focus of traditional engineering education was to provide a technology-driven workforce to manage and operate the growing industry. But the dynamic industrial structure and its ever-changing nature gave rise to a subsequent mismatch between the oversupply of engineering graduates and their declining demand from hi-tech industry, which resulted in large scale unemployment among the engineering graduates. This prompted them to jump into new venture development and entrepreneurialism, either directly or after topping up their engineering qualifications with business schooling. Thus, to be qualified for the future needs of the industry, general or soft skills were also required along with the specific technical skills of engineering education (Mechefske et al., 2005; Schwieler, 2007). The objective of the contemporary engineering education, therefore, is to provide a blend of technical knowledge along with skills required for innovation and entrepreneurship. With the emergence of high dependence on information technology, and globalization of industries and services along with research and development (R&D), the traditional engineering education therefore started incorporating new interdisciplinary fields leading to amalgamation of management-led curricula in the discipline of engineering (Luryi et al., 2007).

In previous years, the courses on entrepreneurship and management were delivered only in management schools. Now, however, they are also available in the engineering and technical education curriculum. Standish-Kuon and Rice (2002) highlighted the cases of Massachusetts

Institute of Technology (MIT) and Stanford University, where engineering schools were offering business and entrepreneurial courses to engineering students, and management schools were established in the School of Engineering. The MIT programme now offers engineering and science courses, but these are located in its Sloan School of Management. Along with management courses, it provides technical courses oriented towards growing technology-based businesses available to engineering and science students. In contrast, Stanford University has a business school that belongs to the School of Engineering. In India, institutions like Indian Institutes of Technology (IITs), National Institutes of Technology (NITs) and various Regional Engineering Colleges (RECs), which were earlier engaged purely with providing tech-based education, also started incorporating management and entrepreneurship courses in their curricula. They started opening separate Schools of Management education: for instance, IIT Madras has an entrepreneurship cell that organizes programs to evangelize entrepreneurship among the engineering students and encourages potential entrepreneurs to start businesses of their own. Another example is Delhi Technological University (DTU), earlier Delhi College of Engineering, where, along with engineering, many management programmes are also delivered, such as Management in Innovation, Entrepreneurship and Venture Development (MBA-IEV), and Management in Family Business and Entrepreneurship (MBA-FBE).

Within this broad context of entrepreneurialism, the chapter begins with a background to skill formation that led to the rise of Indian tech-professionals and entrepreneurs in the foreground of their global mobility in the 21st century. In this context, it refers to the demand and supply mismatch of the first generation of highly skilled knowledge workers born and educated in India and being absorbed gainfully in the global labour market of the developed west, primarily in the United States. This is followed by a reference to the second generation of tech-professionals and entrepreneurs born in India but educated at the tertiary level in the US. Focusing on the US, the chapter provides a comparison of some of these aspects between India and China, both major countries of origin of hi-tech professionals and entrepreneurs migrating to the global markets of labour, goods and services. One of the objectives of the chapter is to highlight, in a historical perspective, both the paradigm shift and gradual change in the mobility pattern of tech-professionals, turning them into entrepreneurs and their specific skills into generic. The chapter concludes with a brief remark on their future trajectories, both negative and positive, arising respectively from the unanticipated shock of the COVID-19 pandemic and long-awaited avenues of hope opened up by the world coming together on the Global Compact for Migration (GCM).

ENTREPRENEURSHIP IN THE WORLD MOBILITY OF TECH PROFESSIONALS

Starting with the final decade of the last century, there has been a vast literature on competitiveness, growth and productivity that theoretically underpins the linkage of hi-tech professionals and entrepreneurs with the key role of technological innovations (Grossman et al., 1990), creation of scientific knowledge and opportunities (Shane and Venkataraman, 2000) and risk (Sarasvathy et al., 1998). Peri and Sparber (2009) took note of a large number of research studies based on the migration of skilled labour force to the countries with a shortage of high-skilled workers in science and technology. Most of these studies looked at migrant scientist and tech-professionals only as employees, whereas there is a shortage of literature that directly links skilled migrants to self-employment through entrepreneurialism. Nevertheless, there are a few studies emphasizing greater advantages to migrant entrepreneurs as compared with their native-born counterparts in terms of access to better opportunities (Florida, 2006), higher possibilities to get engaged in international operations (Wang and Liu, 2015) and advantages of representing self-selected groups due to personality traits (Akee et al., 2013; Hunt and Gauthier-Loiselle, 2010; Kerr and Lincoln, 2010).

In addition, there are arguments stating that there exist disadvantages to native-born entrepreneurs as compared with migrant entrepreneurs in terms of cultural differences, language barriers (Borjas et al., 2008), and lack of access to institutions and networks that facilitate knowledge transfer, recruitments and flow of financial capital (Fairlie et al., 2012). Based on the study of migrant-owned firms in the US high-tech sector, Brown et al. (2018) found that there was a higher rate of innovations in migrant-owned firms as compared to native-owned firms in terms of higher propensity to engage in R&D and filing of patents. These migrant entrepreneurs had the background of science and engineering and had distinct motives to start businesses. Based on the same data, Kerr and Kerr (2018) had a very different finding that migrant entrepreneurs formed a proportionately larger share of new firms than native firms, but created less employment, supporting the fact that migrant-owned start-ups and new ventures in food services, accommodation and retail trade etc. formed a larger industrial concentration. Vandor and Franke (2016) also support the argument that migrants are more entrepreneurial than the natives. They suggested that, to get a better understanding of economic significance of migrant entrepreneurship, it is necessary to distinguish between 'necessity-based entrepreneurship' and 'opportunity-based entrepreneurship'. In terms of this classification, the chapter supports a proposition that the entrepreneurialism of the Indian migrant tech-professionals started as the former when recession hit the world and the US before the turn of the century, but soon transformed into the latter in the new century. This is what also had a role in boosting the amalgamation of tech and business education in India discussed above.

GLOBALIZATION OF LABOUR: RISE OF INDIAN GENERIC HUMAN CAPITAL OVER SPECIFIC

In the context of liberalization and reforms that swept the world economies in the 1990s, 'globalization of capital' in terms of foreign investments by multinational corporations had become a familiar and accepted phenomenon. In contrast, 'globalization of labour' was slow to attract recognition under mode 4 of General Agreement on Trade in Services (GATS) in the World Trade Organization (WTO) as 'movement of natural persons' or 'presence of service providers' for temporary stay gradually substituting permanent residency for purpose of work in a foreign country (WTO, 1998). This was despite the projection that Peter Drucker had made in the 1960s, coining the terms 'knowledge workers' and 'service workers' to introduce a classification of work that would be discernible by the end of the 20th century (Drucker, 1993). According to him, globalized industries would be typically free to locate their processes anywhere in the world and the knowledge workers would therefore be swiftly globalized, welcomed in most countries, and their income would be determined by the world economy. Contrary to Drucker's prediction, however, even the service workers comprising the medium and unskilled workers also globalized by engaging in the 3D (dirty, dangerous, difficult) tasks in industries like fisheries, agriculture, construction, manufacturing, garbage disposal, etc. However, a particular category of occupation or profession that overtook all others in coming and staying closest to Drucker's description of 'knowledge workers' has been the tech-professionals. Their mobility worldwide surpassed that of any other in specific knowledge occupations, such as doctors, engineers, scientists, architects, nurses, teachers and so on, who were dominating the global mobility of professionals until then. This shift supported the conjecture that the world economy moved away from the specific occupations to generic professions, making the technology embodied in an IT professional applicable in a large number of fields in a knowledge society (Khadria, 2001).

India, the origin country of the world's largest diaspora of 17.5 million in 2019 (IOM, 2019) surpassing the over 15 million two years earlier (IOM, 2017), happens to be the main supplier of two kinds of skilled people in this generic field – IT professionals and students in computer sciences – to the world market of labour and education respectively, particularly in developed countries. By the close of the 20th century, IT professionals emerged as the most sought-after category of employment-related migrants from India (NASSCOM-McKinsey Report, 2005). In 1999/2000, in a 47-country ranking of brain drain, calculated on the basis of a survey on whether 'well-educated people emigrate or do not emigrate abroad', India was placed 42nd (or sixth from bottom) with a significance score of 3.291 on a 10-point scale (IMD, 2000). The low ranking and score implied that India had a high degree of brain drain in terms of well-educated and skilled persons who emigrated abroad. In the European Union, the 'Blue Card' visa was introduced in 2008/09, to attract IT professionals, although it was not very successful (Khadria, 2018).

There has been another major paradigm shift, the scale tilting from entry of specific knowledge workers as permanent immigrants to entry of tech-professionals with the status of temporary visitors or guest workers (like students), for example on 'non-immigrant' visa category H-1B in the case of the US. The distinction between permanent and temporary residency, which once separated the two specific categories of migrants (professionals and students) was losing its prominence in defining global talent mobility. In the year 2006, 54 percent, i.e. more than half of new recipients of the H-1B temporary visa-holders for work in the US were from India while only 9 percent were from China. The share of all Asian countries including India and China was more than three-quarters of all visas issued (Khadria, 2012). As per the USCIS report (2018), by October 2018, there were as many as 419,637 foreign nationals working in the US on H-1B visas and of these, 309,986 were Indians which made up to 73.8 percent, i.e. approximately three-quarters of the world's total.[1]

The trend of rising mobility of IT or software tech professionals has not been an overnight phenomenon. The practice of on-site software development in the US, popularly called 'body shopping', was predominant in the 1980s and 1990s because telecom infrastructure was not fully developed to undertake offshore online jobs like Business Process Outsourcing (BPO) in India at that time. The trend of offshore software development in India began when Indian companies (tech-giants like WIPRO, Infosys etc.) made their mark in executing large and complex projects and when the telecom and satellite links also improved. This resulted in an advantage to India, as foreign countries too started inviting Indian professionals to work for them on site, which was in a way a return of 'body shopping' in the 21st century in a new form. Given that most of these professionals were young unmarried males with parental families back in India to whom they would send the remittances, the surge in remittances to India from the US and other developed countries led to the volumes surpassing those coming from the relatively low- and medium-skilled Indian migrant workers in the Gulf until then

(Khadria, 2004).[2] Hence there was open encouragement by the Indian government to this outflow of high-tech professionals who, it was thought, would also bring back to the country better technologies and experience of working in a global environment. It was ironical that not many Indians came back; many adjusted their status to that of permanent residents in the host countries and those who returned did not choose to live and work permanently in India mainly due to the differences in standard of living between home and abroad. Ultimately, when the US-educated engineers started returning to India noticeably, they did it not for transfer of information or upgradation of India's technological infrastructure but primarily as roving ambassadors of their overseas multinational employer companies in India (Khadria, 2001:56–59).

HISTORICAL ROOTS OF TECH EDUCATION IN INDIA AND EMIGRATION OF SCIENTISTS & ENGINEERS

India had a moderate number of universities at the time of independence. However, it lacked in highly trained human resources with scientific and technical expertise. India was also far behind as an institutional base for Science & Technology, which was required for industrialization and modernization of the Indian economy during the Nehru era. Hence, Prime Minister Jawaharlal Nehru gave greater autonomy to the group of scientists including Homi J. Bhabha, S. Bhatnagar, D. S. Kothari and P. C. Mahalanobis to develop the scientific and technical base of higher education in India.

The systematic efforts in this direction started with the establishment of the Scientific Manpower Committee (SMC) in the year 1947 and recommendations of the Sarkar Committee for setting up of at least four higher technical institutions. The Parliament passed the Indian Institute of Technology Act in 1956 and the first Indian Institute of Technology (IIT) was established in the same year at Kharagpur. This was followed by IIT Bombay in 1958, IIT Madras in 1959, IIT Kanpur in 1960 and IIT Delhi in 1963 – with aid from the US, the Soviet Union, Germany and the UK respectively (Khadria, 2007). Closer to present times, when the tertiary level enrolments in India responded to the global mobility at 'breakneck speed' between 2006 and 2011, it was mainly driven by enrolment in Engineering, which grew by over 200 percent from 1.8 million in 2006/07 to 5.46 million in 2011/12 (see Table 20.1).

Traditionally, the UK was the main recipient of the Indian migrants including the highly skilled until the end of the 1960s due to the colonial ties and the English language as the medium of the higher professional and technical education in India. After 1970, large numbers of Indians in 'specific' knowledge and skill occupations (doctors, engineers, scientists, academicians etc.) were absorbed into the US labour market. By the close of the 20th century, however, approximately 80 to 90 percent of the US-bound Indians were 'generic' professionals skilled in IT professions, largely moving to the IT capital of the US, Silicon Valley (Saxenian, 2002: 22).

The 'New Indian diaspora' that was being formed through naturalization of these Indian legal permanent residents ('green card' holders in the US) to citizenship played an important role in the field of Science & Engineering (S&E) in the US. The standing of these Indians in the scientific community of the US and the world is best illustrated by the celebrated examples of Dr. Hargobind Khorana and Dr. S. Chandrashekhar among the first generation of Indian emigrants from Asia winning a Nobel Prize in Science (in medicine in 1968 and physics in

Table 20.1   *Growth of tertiary level enrolment in India by field of study: 2006–2011 (in millions and percent)*

Faculty	2006/07		2011/12	
	Total	Percent	Total	Percent
Engineering & Technology	1.80	13.00	5.46	25.00
Education	0.62	4.50	1.30	6.00
Medicine, Nursing and Pharmacy	0.59	4.30	1.20	5.50
Commerce and Management	2.28	16.50	3.43	15.80
Agriculture & Veterinary Science	0.09	0.70	0.12	0.60
Science	2.54	18.40	3.05	14.00
Arts	5.48	39.60	6.57	30.20
Law	0.30	2.20	0.34	1.60
Others	0.11	0.80	0.27	1.30
Total	13.85	100.00	21.78	100.00

Source: Planning Commission of India (2013), Twelfth Five Year Plan, 2012–17, Social Sector, Volume 3, Government of India, New Delhi: Sage.

Table 20.2   *An overview of the DST-sponsored brain drain studies on IITs*

	IIT Bombay	IIT Madras	IIT Delhi
Year of Study	**1987**	**1989**	**1997**
Period Covered	1973–77	1964–87	1980–90
Population size	1262	5942	2479
Sample Size	501	429	460
In India	179	184	316
Out of India	322	245	144
Magnitude of Brain Drain	30.8%	25-28%	23.1%
	(±2%)		(±1.5%)

Source: Deshmukh et al. (1997) as cited in Khadria (1999).

1983 respectively). In India, the emigration of such knowledgeable people was lamented by the home scientific community and policy makers as the problem of brain drain of highly qualified persons (HQP) (Krishna & Khadria, 1997). Although the outflow from the field of science and technology continued through the 1950s and 1960s, it was only from the late 1960s and early 1970s that it began to ring 'alarm bells'. There were a number of diverse reasons that drove this mobility, such as relative underdevelopment of specialized research communities, low intellectual climate in India coupled with limited opportunities, increasing bureaucratic apathy, absence of government policies to restrict the outflow of talent in a situation of soaring graduate unemployment, and low incomes. These push factors were complemented by the pull factors, not least a new demand pattern for highly skilled people in developed countries, especially manifested through the US Immigration and Nationality Act of October 3, 1965 (Khadria, 1999). This was a sequel to the American emphasis on space research at NASA in the 1950s and 1960s. Subsequently, the post-oil crisis recession of the early 1970s gave a new impetus to R&D in the US, particularly in the energy-saving technologies that were to usher the microelectronics revolution (Khadria, 2012).

A series of India's DST-sponsored (Department of Science and Technology) studies on brain drain from India provides the estimates of brain drain from a number of IITs in a comparative perspective (Table 20.2) (Deshmukh et al., 1997). The studies reveal a very high rate of brain drain from the top institutions of science and technology in India, viz. three IITs. The above exodus of one-fourth of tech-professionals accelerated America's IT-led economic boom in the 1990s, to which Indian IT professionals made a visible and widely recognised contribution. The resulting phenomenon of change from specifics to generics led to a shift of choices of subjects by the Indian students. The tilt was in favour of IT disciplines such as computers over pure sciences in anticipation of the possibility to migrate rather than having to face unemployment in India. Figure 20.1 shows the pathways of Indian talent to the US triggering this phenomenon.

According to industry projections made in 2005 by the NASSCOM-McKinsey report (2005), the demand for Indian software professionals was to rise tenfold over the next seven years in the new century. This led to an increase in the number of seats offered to students in the Indian Institutes of Technology (IITs); and many new institutions were opened up, such as National Institutes of Technologies (NITs), Indian Institutes of Information Technology (IIITs), and new IITs. The city of Bangalore, the so-called 'Silicon Valley of India', faced the crunch of such professionals. Indian IT giants like INFOSYS and WIPRO faced competition from abroad in recruitment of top-class tech-professionals (Khadria, 2012).

Choudaha and Roy (2015), citing recent research from the US NSF, have stated that India was the top country of birth of immigrant scientists and engineers in the US, which rose by almost 85 percent in the decade between 2003 and 2013, reaching 950,000 immigrants. This corroborated the fact that India sent a large number of skilled migrants to the US through the H-1B and L1 (for intra-company transferees) work visas and international students through F1 visa. Currently the flows of Science & Engineering (S&E) personnel migrating from India comprise students who pursue S&E education in the US and the Indian workers who seek S&E jobs in the labour markets there. Although both the categories have increased over time, the former has increased very rapidly in the past decade. According to them, 'While more than 100,000 Indian students are enrolled in U.S. higher education, ... 65% of them are enrolled in

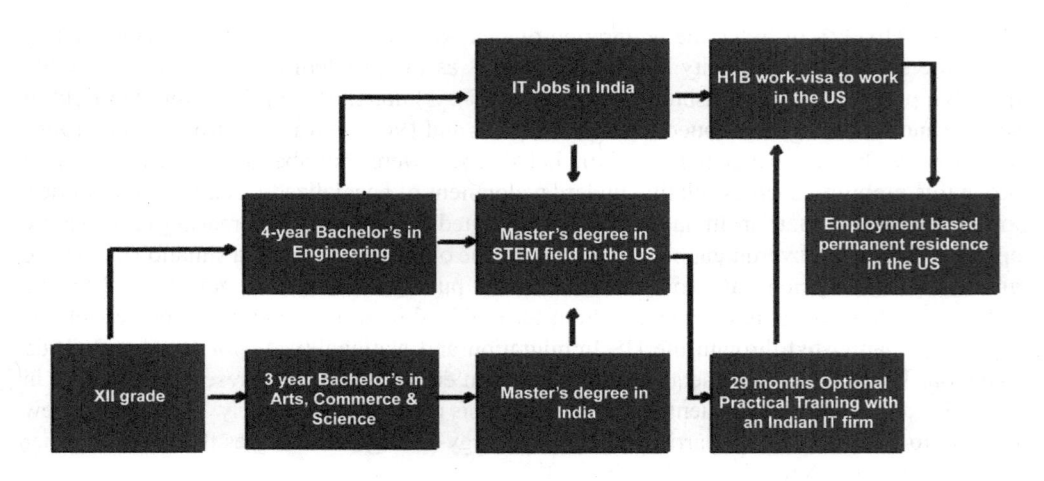

Source: Choudaha and Roy (2015).

Figure 20.1   Pathways of Indian Talent to the US

engineering and computer science programs as compared to 22% of international students in the U.S. more broadly' (Choudaha & Roy, 2015).

MOBILITY AND THE CHANGING FRONTIERS OF GLOBAL TECHNOLOGIES

In the global context, highly skilled managers, academicians, professionals and entrepreneurs form a privileged transnational migrant group. They migrate to advanced nations in pursuit of higher studies or to access the premier educational institutions with better laboratories and other research infrastructures not available in the home country (Hazen & Alberts, 2006). Portes et al. (2002) found that highly skilled professionals were motivated by entrepreneurial desires to migrate as evident from the transnational entrepreneurs forming a major chunk of self-employed persons among the immigrants. The study also found that transnational entrepreneurs were educationally and legally sound and their entrepreneurial activities gave them higher than average incomes than the salaried migrants.

In the case of tech-professionals, return migration of skilled labour force that is taking place from developed to developing countries mostly includes engineers and scientists. Sabharwal and Varma (2017) explored the experiences of faculty in Science & Engineering who decided to return to India from the US. They argued that, unlike push–pull factors, economic factors such as salary differential, which are more appropriately applicable in the case of unskilled, manual and semi-skilled labour force, are not the major reasons for scientists and engineers to migrate. The conventional migration theories perceive return migration as the outcome of failed migration experiences in destination countries. But there are several academic reasons, including ample research funding, less competition for grants, ability to work on theoretical topics, and freedom in research objectives, that prompted tech-professionals to return to India from the US. Additionally, there are various non-academic and sociological reasons of return home such as the availability of domestic help services and family network including mutual support from parents and grandparents etc. (Khadria, 2004).

In the case of Indian tech-professionals, business media started focusing on young NRI professionals returning to India as 'angels of venture capital' and 'financial sector MNC executives'. Large scale layoffs of Dot-Com professionals in the US triggered a wave of return migration to India, although these involved mainly short-term diploma holders in IT, not long-term degree holders (Khadria, 2001). Based on a qualitative survey of IT professionals in Bangalore, Khadria (2004) found that outmigration of IT professionals from India was mainly to grab the opportunity to work on various projects and to get work experience abroad. Return migration was mostly due to family issues like old age of parents and attachments to family members. 'India emerging as IT power in the world' has also been among such motivational factors for the to-and-fro migration of tech-professionals, turning them from one-time migrants to circulating transnationals.

Table 20.3   *Growth of international students in the US from all countries, by field of study: from 2017–18 to 2018–19 (numbers and percent)*

Field of Study	Total 2017–18	2018–19	% change
Agriculture	12,473	13,754	10.3
Business and Management	196,054	182,170	-7.1
Communications and Journalism	22,824	24,017	5.2
Education	17,615	16,786	-4.7
Engineering	232,710	230,780	-0.8
Fine and Applied Arts	63,795	63,097	-1.1
Health Professions	35,169	35,446	0.8
Humanities	17,040	17,013	-0.2
Intensive English	25,845	22,026	-14.8
Legal Studies and Law Enforcement	16,894	16,483	-2.4
Maths and Computer Science	186,003	203,461	9.4
Physical and Life Sciences	78,700	81,580	3.7
Social Sciences	83,708	84,320	0.7
Other Fields of Study	88,720	86,057	-3.0
Undeclared	17,242	18,309	6.2
Total	1,094,792	1,095,299	0.05

Source: Compiled from Open Doors, www.iie.org.

PRIMACY OF TECH-RELATED STREAMS IN US INTERNATIONAL EDUCATION

Professionals holding a degree or diploma in IT from India were not the only ones fuelling Silicon Valley. Many were educated in the US at the postgraduate level having emigrated as a generic category, as students with a first engineering degree (B.Tech/B.E.) from IITs, NITs, IIITs, Regional Engineering colleges or institutions of excellence. Similarly, graduates with Post-Graduate Diploma in Business Management came from the Indian Institutes of Management (IIMs) to pursue higher education abroad and enter the global labour market. Universities in countries of the global north started coming to India to hold 'education fairs' in Indian metros and cities to 'recruit' the hottest-selling generic human capital, the Indian students. The USEFI attributed this to Indian students being rated the highest in the international students' community.

Table 20.3 gives an overview of international students in the US in various subjects over a period from 2017/18 to 2018/19. The data show that, although there is a slight fall in the number of Engineering students migrating to the US from 232,710 in the year 2017/18 to 230,780 in 2018/19, the total number of students was still the highest in Engineering Education among all subjects in both years. The students in Maths and Computer Science formed the second-highest group in 2018/19 with an increase of 9.4 percent between the two years.

A notable fact is that Business and Management students migrating to the US for higher studies formed the third-largest group in 2018/19, much above the proportion of students in Physical and Life Sciences. This implies that not only the demand for high-tech professionals was at its peak but that for Business and Management professionals also continued to be substantially high despite the small decline in the number as compared with the previous year.

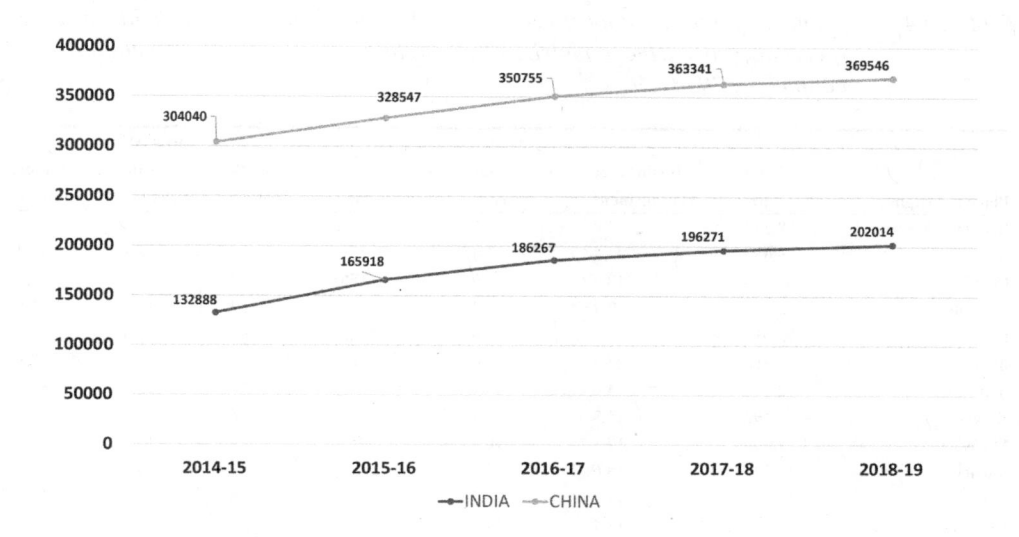

Source: Compiled from Open Doors, www.iie.org.

Figure 20.2   Stocks of students in the US (in numbers), from India and China: 2014–2019

Figure 20.2 shows a continuously increasing trend, both for India and China, in the number of students studying in the US. China has remained consistently ahead of India in terms of total number of students in all these years from 2014 to 2019. Tables 20.4 and 20.5 respectively present a comparative picture of developing and developed countries. China and India were the top two source countries in terms of total international students studying in the US in 2018/19. In the same year, as Table 20.4 shows, India was ahead of China in Maths & Computer Science (i.e., Software Technology), and Engineering, in three out of four STEM fields. In the fourth field, viz., Science too, India was ahead in Health Sciences, although China was ahead in Physical & Life Sciences.

As per Table 20.4, out of 369,546 students from China, 18.9 percent were in Business & Management whereas out of 202,014 students from India, 10.3 percent belonged to this field. This implies that, in entrepreneurialism, China had higher numbers of students in the US than India. But other neighboring countries of India had lower numbers of students studying in the US: 7,957 students from Pakistan, 8,249 from Bangladesh and 13,229 from Nepal, making India the largest source of international students emigrating from South Asia to the US.[3]

Figure 20.3 shows that, from 2014 to 2019, among all STEM fields, Engineering and Maths & Computer Science had the largest share of Indian students. It may be noticed that Indian students from Business & Management showed a higher proportion than the students in Health Professions and Physical & Life Sciences. However, it has to be kept in mind that a substantial proportion of students pursuing Business & Management were already degree holders in Engineering.

Table 20.4   *Number of international students in the US and their shares by fields of study (percent), from select countries of origin in the Global South (developing countries), 2018–19*

Place of Origin	Total Students	Business & Management	Maths/ Computer Science	Engineering	Health Sciences	Physical/ Life Sciences	Other Subjects
Bangladesh	8,249	7.8	17.7	40.4	3.4	15.3	15.5
Brazil	16,059	23.3	5.4	12.6	2.8	8.8	47.3
China	369,546	18.9	19.9	18.0	1.4	8.4	33.3
Colombia	8,060	19.3	5.6	14.2	2.3	10.0	48.5
India	202,014	10.3	37.0	34.2	3.2	5.6	9.6
Indonesia	8,356	27.4	10.0	16.7	1.5	6.5	37.8
Iran	12,142	4.6	13.5	52.1	2.4	12.4	15.1
Kuwait	9,195	12.3	1.4	63.8	1.6	3.7	17.3
Malaysia	7,709	17.3	10.7	25.8	2.4	12.5	31.5
Mexico	15,229	18.6	5.6	16.2	3.6	9.6	46.4
Nepal	13,229	11.3	28.5	19.0	5.4	20.7	15.0
Nigeria	13,382	13.4	11.3	21.5	13.7	13.4	26.7
Pakistan	7,957	17.6	16.7	24.6	2.9	8.0	30.3
Saudi Arabia	37,080	18.8	9.6	30.2	8.1	6.5	26.9
Taiwan	23,369	19.1	13.4	17.4	3.8	9.5	36.9
Turkey	10,159	14.1	12.0	26.1	1.2	7.9	38.8
Venezuela	7,760	25.3	5.3	18.2	3.4	4.8	42.9
Vietnam	24,392	28.5	14.2	10.9	4.6	9.6	32.2

Source: Compiled from Open Doors, www.iie.org.

Table 20.5   *International students in the US (numbers), by fields of study (percent), from select countries of origin in the Global North (developed countries), 2018–19*

Place of Origin	Total Students	Business & Management	Maths / Computer Science	Engineering	Health Professionals	Physical / Life Sciences	Other Subjects
Canada	26,122	16.3	5.3	8.9	14.0	9.8	45.8
France	8,716	22.6	6.4	13.9	1.3	5.7	50.1
Germany	9,191	26.1	5.1	9.4	1.7	7.4	50.3
Japan	18,105	17.9	4.7	4.8	2.7	4.6	65.3
South Korea	52,250	14.0	10.8	13.5	5.0	7.8	48.9
Spain	7,262	24.4	4.7	14.6	2.1	6.8	47.3
United Kingdom	11,146	18.1	4.7	4.7	3.3	8.1	61.1

Source: Compiled from Open Doors, www.iie.org.

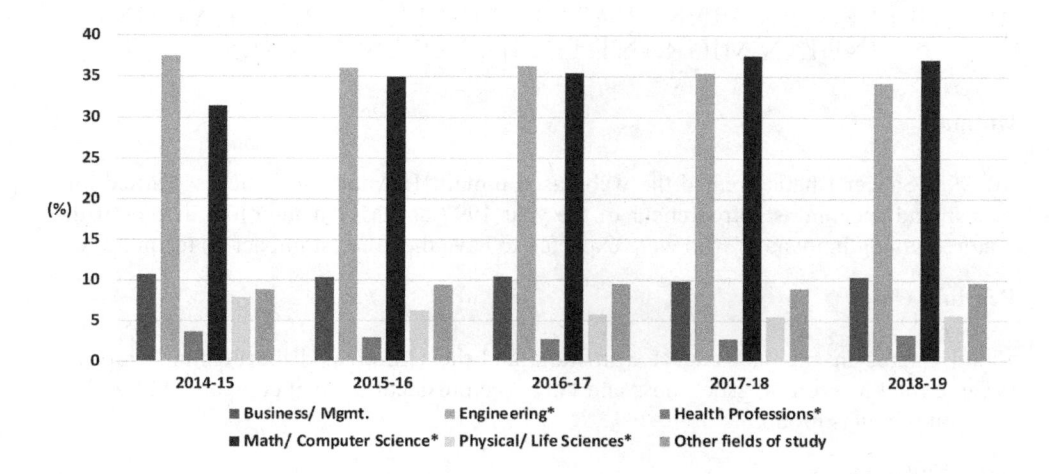

Note: Other fields of study include Humanities, Intensive English, Education, Fine Arts & Applied Arts, Social Sciences and some undeclared fields.
Source: Compiled from Open Doors, www.iie.org.

Figure 20.3   Indian students in the US (in %), by fields of study: 2014/15–2018/19

FROM TECH-PROFESSIONALS TO ENTREPRENEURS: ROLE OF THE INDIAN BUSINESS DIASPORA

Indian professionals-turned-entrepreneurs are those groups of Indian professionals who initially moved to the US as engineers but ended up being venture capitalists and executives in medium- and large-sized companies by the end of the 1990s. Many of these companies, where Indian IT professionals made up about a quarter of the total workforce during that decade, were based in Silicon Valley (Lal, 2006). A few tech-professionals of the first-generation Indian diaspora made significant contributions in path-breaking innovations in technology (see Box 20.1).

Subsequently, since the emergence of globalized economic activities in the 21st century, and the Indian entrepreneurial diaspora becoming a key player on the world stage, development of transportation and telecommunication made it easier for businessmen to operate virtually from any part of the world (McAuliffe et al., 2017: 150–55). Now, with higher returns being expected from foreign investments due to the liberalization of state regulations and diminishing constraints on the flow of transnational capital, these entrepreneurs have been a source of capital and foreign direct investment in the 'homeland', also to act as intermediaries enabling and facilitating the development of networks and connections abroad.

This was followed by the current generation of tech-professionals rising to leadership positions of leading multinational giants such as the CEOs or vice presidents of Microsoft, Google, Nokia, Adobe, IBM, etc. An outcome of such people coming together was the forming of many a diaspora entrepreneurial organization like the TiE, initially established in Silicon Valley as the Indus Entrepreneurs in 1992 (Lal, 2006).[4]

BOX 20.1 LEADING INNOVATIONS BY FIRST GENERATION INDIAN MIGRANT TECH-PROFESSIONALS

Hotmail

In 1995, Sabeer Bhatia created the web-based e-mail 'Hotmail'. His success earned him widespread acclaim as entrepreneur of the year 1997 and MIT named him as one of the hundred young innovators who were expected to have the greatest impact on technology.

Pentium Chip

Vinod Dahm, an engineer educated in India and the US, created the Pentium chip. He became Intel Microprocessors' boss and later vice president of rival company AMD computation products group.

Sun Microsystems

Vinod Khosla studied at IIT Delhi and Carnegie Mellon in the US and founded Sun Microsystems. He was among the first venture capitalists to realise that a combination of internet technology and fibre optics would lower the cost of communication at high speed while being easy to use.

Bose Audio Speakers

Amar Gopal Bose, an electrical engineer by education, designed a speaker that would emulate the symphony hall experience at home. Bose has become a symbol of quality and technical innovation in the world of audio systems.

Source: The Encyclopedia of the Indian Diaspora (Lal, 2006).

CONCLUDING REMARKS

Given the historical trajectory of tech professionals and entrepreneurialism among the Indian migrants abroad, particularly the US, the latest pandemic of COVID-19 will have significant implications for bringing about a paradigm shift in the future. There will be both negative and positive fallouts. Incomes are going to be dipped and there will be large-scale layoffs also. These will adversely affect the remittances. At the same time, new avenues for application of information technology, IA, Robotics and others to online services will emerge that will be taken up by technologists and entrepreneurs. R&D in the field of medical and health care industries would get priority. This will necessitate new technological and entrepreneurial support for inventions, innovations and experiments in the direction of life-saving applications for effective and long-term safeguards to both life and livelihood across the world. All this will influence the choices that the migrant STEM (Science, Technology, Engineering and Mathematics) professionals and students are likely to make in terms of re-ordering their preferences about their fields of education, career, occupation, employment, business and vocations, and most significantly their locations through migration leading even to possible re-emergence of the brain drain (Khadria, 2020a).

This is where the role of the Indian migrants and diaspora tech professionals and entrepreneurs in the US would have seamless potential and possibilities. It would also open up new avenues of global partnerships and collaborations among other countries of origin and destination, necessitating global cooperation in migration governance. In this endeavour, the principles and objectives of the Global Compact for Migration (GCM) could provide a new optimistic roadmap for countries to reflect upon their past conduct and make constructive and enduring amendments. A detailed analysis of such a roadmap, however, is outside the scope of this chapter.

ACKNOWLEDGMENT

The authors are grateful to an anonymous reviewer for the constructive comments and suggestions received on an earlier draft.

NOTES

1. More recently, however, in the wake of the COVID-19-driven unemployment and to save jobs for American citizens, the axe of Presidential Executive Orders had fallen on Green Cards, followed by H-1B and L-1 visas. This reflects the turn towards a stringent visa regime of the United States, which has since September 2017 suspended employment authorisation document (EAD) for spouses of H-1B visa holders, having adverse gender implications for the majority of qualified Indian women, who were receiving over 90 percent of the 120,000 H-4 visas issued since 2015 (Economic Times, 2019).
2. This surge in remittances also did not last long as the professionals aged, married, had families and then turned into entrepreneurs with more permanent roots in the destination countries. India being at the top of the league of remittance receiving countries with US$ 83 billion in 2019, there is now some new thinking on economic contributions of migrants and their relation with technology (McAuliffe et al., 2019, pp. 172–175). What is imminently required, however, is to revisit remittance analyses in view of substantial reduction anticipated in their volumes to India and worldwide caused by the COVID-19 pandemic (Khadria, 2020b). Further elaboration on this, however, is outside the scope of this chapter.
3. See Khadria and Mishra (2021) and Khadria (2020c), respectively, for trends in intra-subregional migration within South Asia and inter-subregional migration in Asia.
4. More recently, even smaller entrepreneurial fora have cropped up. One example is the setting up of a network called Global Assamese Entrepreneurship Forum (GAEF) in July 2021 with its head office located in New Jersey, USA. Its aim is "to unite the Assamese-origin business owners to help mentoring, networking, education, and incubation of Assamese origins business owners, startups and future entrepreneurs." See https://www.facebook.com/groups/440248043822192/ (accessed on 6 October 2021).

REFERENCES

Akee, R. K., Jaeger, D. A. & Tatsiramos, K. (2013), 'The persistence of self-employment across borders: New evidence on legal immigrants to the United States', *Economics Bulletin*, 33(1), 126–137.

Borjas, G. J., Grogger, J. & Hanson, G. H. (2008), Imperfect substitution between immigrants and natives: A reappraisal, Technical report, *National Bureau of Economic Research*.

Brown, J. D., Earle, J. S., Kim, M. J. & Lee, K. M. (2018), 'Immigrant entrepreneurship, job creation, and innovation', in *The Role of Immigrants and Foreign Students in Science, Innovation, and Entrepreneurship*, University of Chicago Press.

Choudaha, R. & Roy, M. (2015), 'Mobility Patterns and Pathways of Indian Engineers to the U.S.', WES Research & Advisory Services, WENR.

Deshmukh, S. K., Raju, R. & Rao, P.N. (1997), 'A Study on Brain Drain and Career Profile of IIT Delhi Graduate', project sponsored by Department of Science & Technology, Government of India.

Drucker, P. (1993), *Post-Capitalist Society*, New York, NY: Harper Collins.

Economic Times (2019), 'Reprieve for spouses of H-1B visa holders', 18 September, accessed on 13 August 2021 at https://economictimes.indiatimes.com/nri/visa-and-immigration/reprieve-for-spouses -of-h-1b-visa-holders/articleshow/71176908.cms?utm_source=contentofinterest&utm_medium=text &utm_campaign=cppst.

Fairlie, R. W. et al. (2012), 'Immigrant entrepreneurs and small business owners, and their access to financial capital', *U.S. Small Business Administration: Office of Advocacy*, accessed on 13 August 2021 at https://www.sba.gov/sites/default/files/rs396tot.pdf.

Florida, R. (2006), 'The flight of the creative class: The new global competition for talent', *Liberal Education*, 92(3), 22–29.

GOI (2020). *Ministry of Human Resource Development (MHRD)*, accessed on 20 February 2020 at https://mhrd.gov.in/iits.

Grossman, G. M. & Helpman, E. (1990), 'Trade, innovation, and growth', *The American Economic Review*, 80(2), 86–91.

Hazen, H. D. & Alberts, H. C. (2006), 'Visitors or immigrants? International students in the United States', *Population, Space and Place*, 12, 201–216.

Hunt, J. & Gauthier-Loiselle, M. (2010), 'How much does immigration boost innovation?', *American Economic Journal: Macroeconomics*, 2(2), 31–56.

IMD (2000), *The World Competitiveness Year Book 2000*, Institute of Management Development, Lausanne.

IOM (2017), *World Migration Report 2018* (M. McAuliffe and M. Ruhs, eds). Geneva: IOM, accessed on 13 August 2021 at www.iom.int/wmr/.

IOM (2019), *World Migration Report 2020* (M. McAuliffe and B. Khadria, eds). *International Organisation for Migration, Geneva*, accessed on 13 August 2021 at www.iom.int/wmr/

Kerr, S. P. & Kerr, W. R. (2018), 'Immigrant entrepreneurship in America: Evidence from the survey of business owners 2007–2012', technical report, National Bureau of Economic Research.

Kerr, W. R. & Lincoln, W. F. (2010), 'The supply side of innovation: H-1b visa reforms and US ethnic invention', *Journal of Labor Economics*, 28(3), 473–508.

Khadria, B. (1999), *The Migration of Knowledge Workers: Second-generation Effects of India's Brain Drain*, New Delhi: Sage Publications.

Khadria, B. (2001), 'Shifting paradigms of globalization: The twenty-first century transition towards generics in skilled migration from India', *International Migration*, 39, 5, Special issue 1/2001 on International Migration of the Highly Skilled, pp. 45–72.

Khadria, B. (2004), Migration of Highly Skilled Indians: Case Studies of IT and Health Professionals, STI Working Paper 2004/6, April, OECD, Paris. https://doi.org/10.1787/381236020703

Khadria, B. (2007), 'Tracing the Genesis of Brain Drain in India through its State Policy and Civil Society', in Nancy Green and François Weil, eds., *Citizenship and Those Who Leave: The Politics of Emigration and Expatriation*, University of Illinois Press, pp. 265-282.

Khadria, B. (2012), *India Migration Report 2010–11: The Americas*, Delhi: Cambridge University Press.

Khadria, B. (2018), 'Blue Card: A comment on the EU's preferential immigration visa for attracting human capital vis-à-vis its own multilateral conundrum', in I. Rajan (ed.), *India Migration Report 2018*, London and New York: Routledge, pp. 56–61.

Khadria, B. (2020a), 'STEMming brain drain in COVID-19 era', *Down to Earth*, 5 May, accessed on 13 August at https://www.downtoearth.org.in/blog/economy/stemming-brain-drain-in-covid-19-era -70873.

Khadria, B. (2020b), 'Migrants and borders: My wishlist in a post-COVID-19 world', *Geography and You*, 20(8), 146, 4–7.

Khadria, B. (2020c), 'Between the "hubs" and "hinterlands" of migration in South Asia: The Bangladesh-India Corridor', *International Journal of South Asian Studies*, Tokyo, 10, 1–10, accessed on 3 October 2021 at https://jasas.info/en/publication/ijsas_vol10/.

Khadria, B. & Mishra, R. (2021), 'Migration in Asia and its subregions: Data challenges and coping strategies for 2021', *Migration Policy Practice*, 11(1), Jan–Feb, IOM–UN Migration & EurAsylum, Geneva, pp. 14–20, accessed on 3 October 2021 at https://publications.iom.int/books/migration-policy-practice-vol-xi-number-1-january-february-2021.

Krishna, V. V. & Khadria, B. (1997), 'Phasing scientific migration in the context of brain drain and brain gain in India', *Science, Technology and Society*, 2(2), 347–385.

Lal, B. (2006), *The Encyclopedia of the Indian Diaspora*, Singapore: Editions Didier Millet.

Luryi, S. et al. (2007), 'Entrepreneurship in engineering education', 37th ASEE/IEEE Frontiers in Education Conference.

Mechefske, C. K., Wyss, U. P., Surgenor, B. W. & Kubrick, N. (2005), 'Alumni/ae surveys as tools for directing change in engineering curriculum', Proceedings of the Canadian Engineering Education Association.

Mejtoft, T. & Vesterberg, J. (2017), 'Integration of generic skills in engineering education: Increased student engagement using a CDIO approach', Umea University, Proceedings of the 13th International CDIO Conference, University of Calgary, Calgary, Canada, 18–22 June.

McAuliffe, M., Goossens, A.M. & Sengupta, A. (2017), 'Mobility, migration and transnational connectivity', in *World Migration Report 2018* (M. McAuliffe & M. Ruhs, eds), Geneva: IOM, pp. 149–169, accessed on 20 February 2020 at www.iom.int/wmr/chapter-6

McAuliffe, M., Kitimbo, A. & Khadria, B. (2019), 'Reflections on migrants' contributions in an era of increasing disruption and disinformation', in IOM, *World Migration Report 2020* (M. McAuliffe & B. Khadria), Geneva: IOM, pp. 161–184.

NASSCOM–McKinsey Report (2005), 'Extending India's Leadership of the Global IT and BPO Industries'.

NSF (2010), 'Science and Engineering Indicators 2010', Chapter 2, Higher Education in Science & Engineering, National Science Foundation, Discussion of Science Resources Statistics (NSF/SRS).

Peri, G. & Sparber, C. (2009), 'Task specialization, immigration, and wages', *American Economic Journal: Applied Economics*, 1(3), 135–169.

Planning Commission of India (2013), Twelfth Five Year Plan, 2012–17, Social Sector, Volume 3, Government of India, New Delhi: Sage.

Portes, A., Haller, W. J. & Guarnizo, L. E. (2002), 'Transnational entrepreneurs: An alternative form of immigrant economic adaptation', *American Sociological Review*, 67, 278–298.

Sabharwal, M. & Varma, R. (2017), 'Grass is greener on the other side: Return migration of Indian engineers and scientists in Academia', *Bulletin of Science, Technology & Society*, 37(1), 34–44.

Sarasvathy, D. et al. (1998), 'Perceiving and managing business risks: Differences between entrepreneurs and bankers', *Journal of Economic Behavior and Organization*, 33(2), 207–225.

Saxenian, A. (2002), 'Silicon Valley's new immigrant high-growth entrepreneurs', *Economic Development Quarterly*, 16(20), 20–31.

Schumpeter, Joseph A. (1942), *Capitalism, Socialism and Democracy*, New York, NY: Harper Torchbooks.

Schwieler, E. (2007), *Anställningsbarhet Begrepp, principer och premisser* [Employability Concepts, principles and premises], UPC-rapport 2007:2 (Stockholm: UPC Stockholmsuniversitet).

Shane, S. & Venkataraman, S. (2000), 'The promise of entrepreneurship as a field of research', *Academy of Management Review*, 25(1), 217–226.

Standish-Kuon, T. & Rice, M. P. (2002), 'Introducing engineering and science students to entrepreneurship: Models and influential factors at six American universities', *Journal of Engineering Education*, 91(1), 33–39.

USCIS Report (2018), 'H-1B petitions by gender and country of birth', accessed on 13 August 2021 at https://economictimes.indiatimes.com/nri/visa-and-immigration/three-fourths-of-h1b-visa-holders-in-2018-are-indians-us-report/articleshow/66289772.cms.

Vandor, P. & Franke, N. (2016), 'Why are immigrants more entrepreneurial?', *The Guardian*.

Wang, Q. & Liu, C. Y. (2015), 'Transnational activities of immigrant-owned firms and their performances in the USA', *Small Business Economics*, 44(2), 345–359.

WTO (1998), 'Presence of natural persons (mode 4)', Background Note by the Secretariat, Mimeo, Council for Trade in Service, World Trade Organization, Dec., pp.1–33.

21. Transnational families and technology: trends, impacts and futures
Jacqueline Bhabha, Abhishek Bhatia and Sam Peisch

INTRODUCTION

Although long-distance family relationships have existed for centuries, the fuel driving cross-border family communication has radically changed in the digital age. More than at any other time in human history, physical separation today does not present a barrier to the experience of family unity or the growth of transnational family identity. As social, economic, political and cultural players, these transnational social units occupy a central role in many aspects of contemporary life, including as drivers of distinct migration patterns.

Understanding the changing modalities through which transnational family life is conducted can complement our comprehension of family-based migration, long a key element of human mobility. This inquiry entails an exploration of the extent to which transnational family life is increasingly mediated by digital technologies developed in the information age. We propose a multi-faceted examination of the phenomenon, at once a transformative asset and a powerful threat, from the vantage point of both transnational families and institutional actors, state and non-state.

Transnational families have been defined as "familial groups with members living some or most of the time separated from each other, while nonetheless feeling a sense of collective welfare, unity, and familyhood across national borders" (Bryceson and Vuorela, 2002, p. 000). This broad definition is not migration specific, because it encompasses any combination of couples, parents, children and other relatives, living separately at "home" or abroad, whether erstwhile migrants or not (Fesenmeyer, 2014). But it helpfully highlights a key element of all migrant transnational family life – the sense of unity across national borders – a unity that may be increasingly mediated by digital technologies and the connections they enable.

The benefits of this digital mediation seem undeniable. Handwritten letters delivered by physical mail carriers have been replaced by the benefits of immediacy provided by emails or tweets; trunk phone calls have given way to FaceTime video exchanges; Instagram uploads have taken the place of postcards and photographs; telegrams have been rendered obsolete by emoji-laden text messages. With the change in media have come changes in the quantum of communication. Trunk calls and telegrams were expensive – necessarily brief and infrequent. Writing and posting letters was time consuming, subject to the vagaries of unpredictable delivery systems. By contrast, most modern information technology, at the fingertips of all transnational families, is virtually cost free and instantaneous. Taxi drivers in Berlin may carry out their work while chatting with relatives back home in Anatolia or Punjab; schedule permitting, grandparents in India may regularly do homework with children in America; migrant workers in Dubai may spend their down-time in dormitories connecting with their spouses or siblings back home in Kerala, Dhaka or the rural Philippines; unaccompanied Afghan youth en route to European destinations may seek advice on their migration route or legal strategy in real-time. Organizations like Médecins Sans Frontières now consider charging outlets in each domicile for mobile phones as critical as showers, health care, or food for forced migrants on the move, as their shift away from communal charging stations in Syrian refugee camps in Ritsona, Greece demonstrates (Latonero, Poole and Berens, 2018).

We argue, however, that, despite these dramatic benefits, advancing technology is not inherently beneficial for human lives any more than transnational migration is. Rather, like most complex social systems, both contemporary phenomena are, in and of themselves, value neutral. Their impact is context specific, determined by a host of contingent factors. As a result, generalizations are hazardous. Any examination of current trends and potential future impacts must take this Janus-faced migration/technology relationship into account.

This chapter examines transnational families and how they interact with technology at a time of growing anti-migrant xenophobia, which has spawned troubling and uncertain questions about the future of citizenship, asylum and migrant rights. Do more reliable and faster forms of international transport, internet communication and global banking – key migration-related aspects of technological advance relevant to the conduct of family life across borders – mitigate these concerning trends? Do they contribute to protecting and maintaining familial and communal relationships, expose transnational families to new, surveillance and exploitation related, dangers; or both?

THE PROMISE OF TECHNOLOGY IN THE DIGITAL AGE

According to the World Economic Forum,

> The use of apps to share information in real-time, including to support clandestine border crossings, together with the consolidation of social media platforms to connect geographically dispersed groups with common interests, has raised valid questions concerning the extent to which technology has been used to support irregular migration, as well as to enable migrants to avoid abusive and exploitative migrant smugglers. The "appification of migration" has become a topic of intense interest. (McAuliffe, 2016)

The rapid global proliferation of technology, combined with a sharp increase in smartphone penetration, has had a profound effect on both the processes that affect migration and the families affected by migration. The "appification of migration" has brought into the marketplace a multitude of products that have enabled the efficient exchange of information between groups, including transnational families across geographies, at a pace that seems primed to grow exponentially (McAuliffe, 2016). This enhanced family connectivity has undoubtedly strengthened transnational family relationships, enabling an intensity of regular communication and coordination that, as just noted, would have been unthinkable even a decade ago. Here, we explore the processes by which technology-enabled solutions have filled key institutional voids that have historically affected transnational families.

Information about Rights and Services

Historically, domestic migration policy has focused on the impact of migrant arrival on the institutional apparatus in place in the receiving country. Less attention has been paid to the corresponding impact of relocation on migrants and in particular on their needs for information. Integration assets available from already-migrated members of transnational families and their cultural and social support networks have also, until recently, received scant attention (Sørensen, 2012). Software solutions centered on providing up-to-date information about protection assistance are gradually addressing this dearth of support in integration practices appropriately addressed to newly arrived relatives.

Another example of state-driven attention to the impact of migrant arrival is the increasing collection and utilization of surveillance data related to migrants and transnational families for security purposes. The deployment of tools such as "geo-spatial mapping of overland flows, air and sea routes, crowd-sourced data from places with high concentrations of previous fatalities, or anthropogenic sources of light detectable from space (including electric lighting or cooking fires) from even the most remote region" provides states with substantial information about migrants, information that could be deployed to "prevent serious, even life-threatening harm", including to members of transnational families (Fatal Journeys, 2019, pp. 77–78). Again, however, these potentially valuable data are rarely shared with migrants and their families – both those who should have a proprietary right to such information and those who could leverage it for family reunification purposes, such as to trace "disappeared" family members, including children, they cannot locate (Fatal Journeys, 2019, p. 73). A recent UN General Assembly declaration reinforces the state obligation to make this information available to transnational families, because this is one way states can contribute to "fully protecting the human rights of all refugees and migrants, regardless of status" (United Nations General Assembly, 2016, para. 5). This state duty is particularly salient where children are concerned, because states have an obligation under the UN Convention on the Rights of the Child to facilitate access to information that could promote the best interests of children (CRC Art. 3(1)).

Data-driven tools can of course do more than locate individual migrants and groups of migrants en route. They can provide localized information of potential benefit to newly arrived family members. This includes targeted information on health centers, consulates and embassies, migration offices, and human rights protection centers as well as on immigrant community programs that offer advice, support and shelter. Examples of such digital resources include apps such as "InfoMigrants" and IOM's recently launched "MigApp", intended as a source of "reliable migration information on risks, visa regulations, health guidelines, migrants rights, government migration policies, access to services and programmes, that provide assistance on migrating safely, facilitate low-cost money transfers, and a secure space to communicate and share stories" (IOM, 2017). Global Positioning System (GPS) apps are also often integrated into the app arena. For example, My Route allows users to privately share their migration route in real-time with family members or anyone else they approve.

Connectivity and Sense of Community

Relationships and information exchanges between transnational family members are increasingly reliant on digital technology, telecommunications and air travel (Faist, 2000; Mazzucato et al., 2015). These resources have had dramatic impacts on transnational families' perceptions of both physical and social distance, perceptions that have only relatively recently been altered through ICT's transformative effects (Wilder and Wilding, 2006). These subjective transformations are complemented by significant objective changes that complement the experience of enhanced proximity. They include reductions in transportation and communication costs, both supported by better infrastructure and a globalized commons that facilitates uptake of services by transnational family members (Kilkey and Merla, 2014).

As a result, given contemporary conditions of polymedia and social network availability, it is fair to say that, unlike in the past, moral or emotional choice rather than financial cost drives the decisions of transnational families about how and when to be in touch (Madianou and Miller, 2012). With the co-presence of family members and social networks no longer limited to "in place" settings, and the exponential growth of ICTs generating a continuum of virtual practices of communication, support, gifts and remittance, we are witnessing the development of a new communication and exchange universe (Lee and Francis, 2009). Transnational family visits have, increasingly, become annual or biennial rather than once-in-a-lifetime events, regularly involving second or third generations, with day-to-day interactions via SMS, WhatsApp messages, Skype/Facetime calls, other voice-over-internet protocol (VoIP) services, with Facebook and Instagram filling the voids previously generated by transnational migration (Baldassar, 2001). These new, digitally mediated, conditions make it possible for separated generations to achieve a sense of satisfactory care through the co-presence of older or younger family members, with virtual proximity substituting for physical contact (Wilding and Baldassar, 2014; Baldassar and Merla, 2014).

Digitally mediated communication does not only benefit settled transnational families. Migrants still on the move increasingly rely on digital resources to maintain family contact and seek out information and advice (Wilding and Baldassar, 2014; Baldassar and Merla, 2014). Thousands of unaccompanied minors, stuck in harsh refugee camps or shelter facilities as they pursue their migration ambitions, rely on their smartphones to connect with relatives back home, and thus create a space of cultural familiarity and emotional support: the smartphone becomes their lived daily reality, partially shutting out the sense of deep alienation from the transit or host society reality outside that often colours the migrant experience (Digidiki and Bhabha, 2018).

Family Care Regimes

A central and constant element of family life, across time and space, is the activity of day-to-day nurture, the care and support of family members across different stages of the life cycle. Interruption of this "family care regime" has long been one of the harshest consequences of migration-induced family separation. Increasingly, however, digital technology, including some of the mechanisms already described, is enabling transnational families to participate in regular caregiving practices, whether as parents of young children, or children of aging relatives (Mahler and Pessar, 2001; Kilkey, Plomien, and Perrons, 2013; Foner and Dreby, 2011; Wang and Lim, this volume).

The profile of a "typical" transnational family has shifted dramatically in recent years. Nearly half the population of transnational migrants are women and girls – a trend that is particularly strong in South-East (SE) Asia (IOM, 2018b). In the past, women in SE Asia migrated for marriage, for family reunification, or to accompany a spouse traveling for work; they were considered dependent family members rather than heads of migrant households (Bhabha and Shutter, 1994). However, in recent years, growing numbers of women have been migrating independently, a phenomenon referred to as the feminization of migration. A range of motivations prompts these migrations. Some are driven by personal reasons, such as the desire to enter into or avoid marriage, or to access educational opportunities unavailable in the home context. Other female migrations are propelled by economic reasons, often closely tied to the needs of family members who remain in the country of origin (IOM, 2018b). According to a recent report on Asian and Pacific female labour migration, "the consequential effects of women's increased mobility have been significant. Sending countries have benefitted from higher inflows of remittances and changes in societal and family relations, particularly as they relate to gender roles and relations. For receiving countries, the welfare gains have been considerable, in the form of increased labour supply, opportunities for native women to enter the workforce and child and elder-care possibilities, especially in contexts where such services are limited" (Sijapati, 2015, p. 1).

Transnational migration inevitably generates logistical challenges for families dependent on intimate care regimes. While all families face complex negotiations over the division of labor needed to ensure that such regimes are effective, transnational families have to contend with the additional logistical challenges of prolonged physical separation. These challenges include the need to find replacement childcare or elderly care, and to address the possibility of high care dependency ratios. Replacement carers may be family or non-family members serving on a short- or long-term basis, including peripatetic nuclear or extended family members, friends and family members who move in and out on a need basis.

Many migrant workers are prevented, by the terms of their contract or by irregular immigration status, from making regular visits to family back home. Instead, they are forced to rely on frequent mobile phone and internet contact to achieve a meaningful off-site care role. More privileged workers, by contrast, may be able to supplement regular but remote contact with periodic visits. Often the spacing of children in transnational families records the periodicity of these precious periods of in person contact. Baldassar (2007) delineates different categories of transnational family care flows (emotional, practical, financial and accommodation-related) and separate forms of delivery (face-to-face or technology-mediated). Examples from Trinidad and Jamaica highlight how internet services, webcams and cell phones serve as social networks that migrant communities rely on to sustain intense family dynamics, including in moments of crisis (Horst and Miller, 2006; Miller, 2011). The increasing feminization of transnational labour migration has raised the issue of "care deficits" for so-called left-behind children, often compelled to grapple with momentous and potentially adverse decisions they had little or no control over (Lam and Yeoh, 2019). According to Lam and Yeoh, "children in migrant families positioned within a web of home care linked to their migrant parents' absence are simultaneously powerful and powerless", a conclusion that is consistent with earlier work (Lam and Yeoh, 2019, p. 4; also Dreby, 2007 and Hoang et al., 2015).

Case studies of transnational families in the Philippines highlight the complex impact that increasing globalization has on women's place in the household. In some cases, it "bestow[s] them with greater autonomy … slowly eroding men's authority at home" (Lam and Yeoh, 2019, p. 4; Osteria, 2011). In others, however, where South-East Asian women were left at home to care for their children while husbands travelled abroad, relatively high incidences of mental disorder were detected, suggesting hidden and gendered costs of international labour migration, so far neglected by public policy makers (Graham, Jordan and Yeoh, 2015). Rather than opting for a simple positive/negative dichotomy in evaluating the impact of transnational labour migration, the evidence calls for a more nuanced approach that acknowledges the complexities embedded in the lives of transnational families (Levitt, 2001).

The response to transnational migration of family members takes a wide range of forms. An interesting variant on digitally mediated family care regimes comprises the "transnational care collectives" established to generate support for elderly family members left behind. ICTs have been effectively leveraged by economic female migrants whose caregiving roles have been filled by other family members, or auxiliary social support to establish support networks that include members beyond the child–parent relationship. In both India and the Philippines these care collectives have revolutionized elderly persons' care through the use of VoIP and video conferencing (with a plethora of options to choose from based on the family member's digital literacy and comfort). The digital infrastructure in some of these contexts includes in-home camera-based monitoring of family members needing a higher degree of hands-on care, sharing of images and videos via group conversations, and general social media to maintain intimacy (Ahlin, 2018; Madianou and Miller, 2012).

A central goal of women's labor migration historically has been improvement of the long-term life chances of their children, especially with respect to educational and employment opportunity. A robust literature documents the complex impact of maternal migration on children who do not migrate but remain in the country of origin. Whereas some studies report improved child health, nutrition, educational performance and skill development compared to non-separated peers, other studies report negative impacts, including deteriorating mental health, and other negative consequences for the development of adolescents (Gamburd, 2008; Yiu and Yun, 2017). As with elder care in separated families, so with childcare, digital technology has started generating diverse family care regimes. Case studies from China, where close to 70 million children are left behind as their parents migrate from rural to urban centers, describe technology-mediated support systems proliferating between separated families (UNICEF, 2014). E-learning platforms developed by UNICEF and the Center for Child Rights and Corporate Social Responsibility have helped distant parents connect with their families and teach their children, the latter activity aided by a series of cartoon lessons and quizzes.

Another variant of these digitally mediated care regimes, this time focused on the migrant worker rather than the family member in the country of origin, is that recently developed through a collaboration between the International Labor Organization and the International Finance Corporation. A set of interventions to support the mental health of garment workers in Jordan highlights the importance of mental health awareness training for all stakeholders at the factory level. They include factory-level mental health services and protocols, a mental health information system for workers, mental health support across the migration timeline, and public support for improved national mental health policies in Jordan (Better Work, 2019).

Other tools in the transnational family space aim to tackle the issue of lack of educational support and bolster parent–child relationships and relationships between absent parents and education workers who address common parenting problems during the parents' absence (Shi, Nazer, Ma and Zhou, 2016). Non-profit groups like Stepping Stones have launched programs aimed at bridging gaps in digital literacy for students unable to effectively leverage technological resources at hand. The Hu+ Project developed by Hujiang EdTech has created online classroom content for distribution to rural schools for the benefit of "left-behind children" without access to tutoring or familial support (Collective Responsibility, 2017). The advent of Artificial Intelligence-based recognition systems across sectors in China has also contributed to the development and deployment of educational algorithms through various hubs in the country that read student body language and behavior in classrooms to identify students that may need more attention from teachers, with a focus on left-behind children (Pelley, 2019). While this initiative appears to have a constructive and child protective goal, the possibility of such in-depth and individually unauthorized scrutiny of personal behavior by remote sensors raises important and so-far under-studied questions about the protection of privacy and broader personal well-being.

Clearly ICT cannot substitute for in-person care for transnational families, especially where there is a need for hands-on caregiving. However, the plethora of options currently leveraged by transnational families, while placing onerous caring responsibility on new constituencies such as grandparents and state child welfare entities, and raising the possibility of uncharted privacy concerns, has undoubtedly mitigated some aspects of transnational family member abandonment and loneliness (Madianou and Miller, 2012).

Family Remittances

The remittance of funds by migrating family members to relatives in the home country has long been a central aspect of transnational family exchange. As another chapter in this volume explores in detail, financial contributions may be used for a wide range of purposes, including to offset migration debts, to provide for family needs, or to cover unexpected expenses generated by health or other emergencies (Flore, 2018). The influx of technology-enabled solutions has filled institutional voids that did not allow all migrant family members to send remittances to their families. The shift from paper vouchers to mobile-based wire transfers has reduced family reliance on agents, radically speeding up and rationalizing the costs incurred for the transfer of funds, with clear benefits for transnational family economies (Ratha, 2016). Digitally mediated financial transfers are particularly useful in supporting family members back home who are affected by large scale disasters, as was demonstrated by the transnational family response to the 2015 Nepal earthquake (Bettin, Presbitero and Spatafora, 2014; World Bank, 2018). Phone-based remittance services like M-Pesa, launched in Kenya in 2007 and designed around the African customer's experience transferring prepaid airtime between earning and non-earning family members, has leveraged the existing digital literacy and mobile phone penetration to radically improve cell phone access and usage by members of transnational families who would have lacked the resources to benefit from these innovations without external assistance (Hughes and Lonie, 2007). Other forms of remittance services that function through informal networks of brokers and rely on user trust and self-regulation, also aim to remove the cost barrier and introduce more families to formal channels for transferring money, trading off lower security protocols for a more affordable price (El Qorchi, Maimbo and Wilson, 2003). These consequences of digitally mediated financial assistance have dramatic impacts on the density and reliability of contemporary intra-family exchange, a sharp contrast to the opportunities available to previous generations of transnational migrants and their families.

Family Reunification

Migration specific factors are key to understanding some crucial impacts of technology for transnational families. For families with recent or ongoing migration histories, technology can be an increasingly critical evidentiary tool. This is the case for a key subset of transnational family migration – cases of family reunification. Evidence proving that the intending migrant is "related as claimed" to his or her sponsor is foundational – without such proof, family-based migration visas are likely to be denied. Documentary evidence is often unavailable and even where it exists, its authenticity may be questioned – for example marriage and birth certificates are easily forged (El Qorchi, Maimbo and Wilson, 2003). Alternative or supplementary information sources, once adduced as optional but increasingly expected or demanded by migration destination states are matching DNA samples for biological relatives (for example in Denmark, Sweden, Norway for biological relatedness of children, siblings, parents) (European Council on Refugees and Exiles, 1999). While this growing reliance on DNA evidence can facilitate proof of relationship, it can also pose challenges for transnational families unable to afford the cost of genetic testing, and unduly raise the evidentiary bar required to demonstrate family relatedness (Kritzman-Amir, in press).

Digital technology greatly eases the burden on sponsors expected to demonstrate dense communication to prove the continuing existence of "family life" to justify the grant of a family reunification visa despite geographic separation – texts, emails, video-chats are easy to retrieve (Home Office, 2020). Electronic records also make proof of ongoing financial support easier. In short, both for migrant parents seeking to facilitate the immigration of children left in the care of relatives back home and for grown children seeking to bring elderly parents who never migrated to join them during the latter's final years, technological advances have greatly reduced the evidentiary burden. Gathering scrappy Western Union receipts, saving for expensive and hard-to-access genetic testing, finding old phone bills or airmail letters proving a continuing relationship has in many cases given way to the production of much easier to access electronic evidence.

RISKS AND UNINTENDED CONSEQUENCES

The proliferation of new technologies in the daily life of transnational families is not without risks and unintended consequences. As we have already noted, technology in and of itself is inherently neither rights enhancing nor oppressive. Its potential for utility or harm is largely contingent upon the strength and legitimacy of governmental institutions and non-governmental actors to effectively monitor and regulate its use in society. Stimulating and supporting public education about the possibilities and pitfalls of technological innovation is critical for preventing harm and promoting social benefit. The excitement of technological advance has to be tempered, at our stage of development, with realistic acknowledgement that the present pace of technological innovation often far outpaces governmental (and societal) attempts to develop effective regulatory protocols. This poses a danger to transnational families, particularly those who are vulnerable to digitally mediated exploitation or discrimination. This section explores these threats to the rights and well-being of transnational migrants and their families.

Exploitation of Family Members

Transnational migration can afford a life-changing source of safety or income for migrating family members, generating previously unimaginable improvements for physically distant relatives. But it can also generate traumatic isolation or exposure to other forms of irreversible harm, with potentially far-flung family repercussions. The same is true of modern technology. The ubiquity of connectivity enables exploitative or threatening exchanges, just as it enables reassuring or enjoyable ones. Among the most extreme examples of this deleterious spin-off are the growing number of ransom cases where traffickers and other criminal actors seize on migrants as revenue sources, blackmailing distant family members reachable through the migrants' smartphones to hand over cash to prevent torture, rape or other forms of extreme violence against the captive migrant. Youth across Sudan, Egypt, Eritrea and Niger, and Rohingya refugees from Myanmar to Thailand and Bangladesh have all described such encounters (HRW, 2014; Fortify Rights, 2019; Info Migrants, 2019).

Other types of transnational family migration may also leverage technology in potentially exploitative ways. Undocumented, already-migrated family members seeking to facilitate the subsequent migration of relatives abroad are known to communicate online with migration professionals – smugglers or traffickers – to make requisite payments and supply contact information as a prelude to their relative's journey. Whereas some such technology-mediated relationships are pure business transactions – the exchange of monetary resources for a required mobility service – others deteriorate into highly exploitative, ransom-like interactions, where familial ties are manipulated to extract punitively high sums under threat of brutal retaliation. As official migration exclusion policies targeting vulnerable communities and distressed migrants accelerate so does the pressure on migrants to resort to these abusive migration facilitation strategies to secure the safety and mobility of loved ones (Perez and Stevenson, 2019; Al Jazeera, 2019).

State Surveillance Systems

Other harms impacting transnational migrants, and their families, include the growing technology-mediated use of surveillance in the context of migration. In the modern information age, a defining feature of quantitative data is how frequently it is generated by users themselves through the interaction with a system that is either online or mediated by a mobile application. Biometric identification technology and State access to personally identifiable information generated by migrants enables states to rapidly identify asylum applicants transiting from one country to another to reunify with family. European Union authorities routinely use this identification information generated by the Eurodac system, to block asylum seekers' onward journey, whatever its human benefit, and return them to the first "safe country" they have entered (Dernbach and Der Tagesspiegel, 2015). This is technology at the service of family separation not family reunification.

Technology does not just hinder migrant family reunification at transit points. Where migrants are already settled within a destination country, governments have been able to exploit family connectivity to identify persons they seek to target or exclude. Zapya, a file-sharing application developed and used by transnational Muslim families to download and share teachings from the Quran with their loved ones over distances, especially in areas with poor connectivity with each other, has been extremely popular with the Uighur population in China as a means of cultural connectivity. Data from this app have been closely monitored and used by officials in the Chinese government to identify, and systematically monitor Uighur individuals and their family members (BBC News, 2019). Similar surveillance-related uses of migrant-generated digital data have proliferated. Regrettably, this vast new databank has largely been deployed for the benefit of destination states' "security", to advance border control strategies. The security of vulnerable migrants, by contrast, including those forced to embark on perilous journeys to secure safety and family reunification, has yet to benefit. A case in point is the dearth of information about the whereabouts or remains of migrants who have disappeared. Relatives intent on tracing missing family members have not so far secured much sought after information from the plethora of potential data sources captured by state surveillance (Fatal Journeys, 2019, p. 73). There is, however, a welcome counter-example. Recent migrant caravans to the United States that originated in El Salvador have communicated with each other through social media, in particular Facebook and WhatsApp group chats (Sieff and Partlow, 2018; Verza and Aleman, 2019).

LOOKING FORWARD: BLOCKCHAIN, BIG DATA AND AI

The core functionality of blockchain technology allows for the creation of a distributed system of transaction databases (blocks) that serve as archives for all transactions, rendering them absolute and unchangeable through a guaranteed historical record. This technology has great potential for transnational family exchanges because it can enable speedy, low cost and private transactions controlled exclusively by family members and thus much less penetrable by outside surveillance systems including those utilized by states.

The advent of blockchain and cryptocurrency has spurred an entirely different transfer service, to tackle both high costs and blurry distribution channels (Iyke, 2017). In combination with the emerging forms of cryptocurrency in the market, this system has the potential to enable instant, high security, traceable transactions across distant spaces, bypassing the need for banks, agents, cash reserves and other intermediaries who have long held transnational families to ransom to different degrees. Though still nascent in the public sector, the private digital remittance market built on this framework is growing rapidly, with companies like BitSpark (Hong Kong), Abra (North America-Philippines), BitPesa (Africa) already catering to consumers in their respective markets. Specialized services like Mojaloop launched by the Bill and Melinda Gates Foundation, Rebit (Philippines), CoinPip (South and South East Asia) and Volabit (Latin America) have been designed to cater to overseas workers and individuals located in under-banked areas, a set of developments that clearly sets the scene for future adoption by transnational families globally (Flore, 2018).

As the global momentum for remittance shifts towards the use of blockchain, there are important gaps in policy and regulation that are yet to be addressed. These need urgent attention if the integrity of the process is to be protected and the benefit of this new technology for transnational family transfers is to be assured. Specifically, migrant families and their advocates need to take note of the lack of a robust formal financial infrastructure mediating these processes at present. They also need to take into account the risks of non-compliance with Counter-Terrorism Financing and Anti-Money Laundering regulations that can present dangers to exchanges between transnational migrant family members (Flore, 2018). The practicality of a blockchain system for digital remittance is heavily dependent on cost reduction, but for this system to be effective and secure, governments and regulators need to set up a multi-regional system to carefully manage and supervise these processes. In sum, though blockchain technology is poised to be of considerable practical utility for transnational families in the near future, optimal functionality will depend on a robust regulatory apparatus not yet fully in place.

Artificial Intelligence (AI) also has tremendous potential to influence the management of international migration for the benefit of transnational families. It is poised to replace tasks associated with human intelligence like border security and control, identity checks, and analysis of migrant data for decision-making (Chui et al., 2018). Data generated by migrants are vast and extremely dynamic. As data collection systems to collect and aggregate structured data improve, the influence of these big data in combination with AI algorithms could generate valuable resources for forecasting and protecting migratory flows (Rango, 2015; Pew Research Center, 2017; Beduschi, 2018; IOM, 2018a; Spyratos et al., 2018). But this benign application depends on robust rights-respecting political will and an attendant regulatory regime. The reason for this is evident: AI can also fuel anti-migrant decision making, streamlining exclusion procedures and accelerating the efficacy of xenophobic migration procedures. AI algorithms have already begun influencing decision-making: in Canada they are used for immigration and asylum determination, in Switzerland for refugee integration, and they are fueling the decision to revise the Schengen Information System (SIS) in the European Union (EU) to use DNA, biometrics and facial recognition data to facilitate migrants' return (Molnar and Gill, 2018; Bansak et al., 2018, Regulation 2018/1860/EU). The application of this body of digital technology to the protection and facilitation of transnational family migration remains to be studied. At its best, it could streamline cumbersome, time-consuming and expensive procedures to make transnational family migration for many as smooth and seamless as it is for the much smaller number of highly privileged families relocated through corporate transfers or other high-level mechanisms.

Here again, though, it is wise to sound a warning note. Within the scope of the potential of AI algorithms fueled by big data, lies uncertainty, and the potential for discrimination and harm for transnational families seeking to take advantage of this new technological frontier. Beduschi (2020) highlights the potential of AI to increase disparities in access to information and the power associated with that information.

A reliance on greatly increased migrant-generated data to drive border-control decision-making gives rise to additional concerns about new forms of "surveillance humanitarianism" leading to exclusion from protection (Latonero, 2019). These developments also raise the specter of biased data and black-box algorithms, together reinforcing stereotypes and resulting in unlawful discrimination, including against members of transnational families (Pasquale, 2015; Citron and Pasquale, 2014).

Transnational families, of course, have very varied characteristics. Not all such families will be at risk from the cybersecurity pitfalls just delineated. But some will be, particularly the families originating from less AI-capable geographies, minorities that have historically been subject to unlawful discrimination, and migrants who do not fall in the jurisdiction of a state party to treaties recognizing the right to data privacy (UN, 1985; UN, 2014). In this era of evolving global and regional jurisprudence, the implications of AI algorithms and big data use for not only these families but all subsets of migrant populations need to be comprehensively discussed and carefully considered.

CONCLUSION

The experience of belonging to a transnational family is changing. A major contributor to this transformation is the evolving digital technology mediating communication, both within families and between family members and states or other entities they engage with. Much of the transformation is positive – a reflection of dramatically lowered thresholds for regular, affordable and high-quality exchange. The enhanced transaction opportunities strengthen transnational family ties, reduce the emotional cost of separation, and disseminate more accurate and comprehensive information about migration-related variables, in origin, transit and destination contexts. These opportunities do not just deliver incremental change, as in a greater quantum of affordable communication time or higher resolution images of significant family events. They also deliver radically new engagement opportunities – shared daily collaboration of routine tasks (e.g., homework), real-time advice on legal or logistical migration challenges, unprecedented tracing opportunities to re-establish critical human contacts during emergencies.

But, as with so much technological innovation, there is a darker side to the story of digital transformation and its impact on transnational families. Contemporaneous communication offers exploiters and other nefarious actors powerful opportunities to capitalize on migrants' extreme vulnerability and deep affective ties to family by leveraging assets of distant family members – through ransom and other forms of extortion; online advertising links vulnerable migrant communities to service providers offering deeply sought-after reunification opportunities embedded in financial terms that often lack transparency and accountability; crowd-sourced data harvesting techniques provide unprecedented surveillance possibilities that can be deployed to deflect or otherwise impede humanitarian movement; and expanded, digitally mediated genetic tools can generate vast reserves of evidentiary data capable of being deployed to undermine migrant family reunification plans, as much as to enhance them. The newest forms of technological innovation – blockchain and a range of artificial intelligence applications – suggest expansive future deliverables for transnational families, but they have yet to be carefully scrutinized for their migrant-related utility.

With the prospect of seamless exchange comes the threat of predatory interception and misappropriation. Technology is a double-edged sword that can be wielded for human rights-based advocacy or political sabotage. As the world becomes ever more connected through a wider and more complex array of digital technologies, the need for forward-thinking regulation, intergovernmental collaboration and public–private partnership that facilitates both information exchange and accountable and transparent oversight of false, misleading or inaccurate dissemination will remain paramount. So will the political determination to secure digital technology as a public good, for the global commons, including the migrant families within it.

REFERENCES

Ahlin, T. (2018), "Only near is dear? Doing elderly care with everyday ICTs in Indian transnational families", *Medical Anthropology Quarterly*, 32(1): 85–102. doi:10.1111/maq.12404.

Al Jazeera (2019), "Rohingya refugees continue to risk lives to seek safety: UNHCR", 1 October, at https://www.aljazeera.com/news/2019/10/rohingya-refugees-continue-risk-lives-seek-safety-unhcr-191001075311416.html.

Baldassar, L. (2001), *Visits Home: Migration Experiences between Italy and Australia*, Melbourne: Melbourne University Press.

Baldassar, Loretta (2007), "Transnational families and the provision of moral and emotional support: The relationship between truth and distance", *Identities*, 14(4): 385–409.

Baldassar, Loretta, and Laura Merla, eds. (2013), *Transnational Families, Migration and the Circulation of Care: Understanding Mobility and Absence in Family Life*, 1st edn, London: Routledge.

Baldassar, L., and R. Wilding (2014), "Middle-class transnational caregiving: The circulation of care between family and extended kin networks in the Global North", in L. Baldassar and L. Merla (eds), *Transnational Families, Migration and the Circulation of Care: Understanding Mobility and Absence in Family Life*, New York and London: Routledge, pp. 235–51.

Bansak, K., et al. (2018), "Improving refugee integration through data-driven algorithmic assignment", *Science*, 359(6373): 325–9.

BBC News (2019), "Data leak reveals how China 'brainwashes' Uighurs in prison camps", 24 November, accessed on 17 September 2021 at https://www.bbc.com/news/world-asia-china-50511063.

Beduschi, A. (2018), "The big data of international migration: Opportunities and challenges for states under international human rights law", *Georgetown Journal of International Law*, 49(2): 982–1017.

Beduschi, Ana (2020), "International migration management in the age of artificial intelligence", *Migration Studies*. https://doi.org/10.1093/migration/mnaa003.

Better Work (2019), "Better Work Jordan Annual Report 2019: A Year in Review", Better Work, accessed on 17 September 2021 at https://betterwork.org/portfolio/19446-2/.

Bettin, Giulia, Andrea Filippo Presbitero and Nikola Spatafora (2014), *Remittances and Vulnerability in Developing Countries*, International Monetary Fund Working Paper.

Bhabha, J., and S. Shutter (1994), *Women's Movement: Women under Immigration, Nationality and Refugee Law*, Stoke-on-Trent: Trentham Books.

Bryceson, Deborah, and Ulla Vuorela, eds (2003), *The Transnational Family*, 1st edn, London: Routledge.

Chui, M. et al. (2018), *Notes from the AI Frontier. Applying Artificial Intelligence for Social Good*, Washington D.C.: McKinsey Global Institute.

Citron, D.K., and Pasquale, F. (2014), 'The Scored Society: Due Process for Automated Predictions', *Washington Law Review*, 89(1): 1–33.

Collective Responsibility (2017), "Educating China's Left-Behind Children", 9 June, accessed on 17 September at https://www.coresponsibility.com/educating-chinas-left-behind-children/.

Dernbach, Andrea, and Der Tagesspiegel (2015), "Eurodac fingerprint database under fire by human rights activists", *Euractiv*, 15 July, accessed on 17 September 2021 at https://www.euractiv.com/section/justice-home-affairs/news/eurodac-fingerprint-database-under-fire-by-human-rights-activists/.

Digidiki, Vasileia, and Jacqueline Bhabha (2018), "Sexual abuse and exploitation of unaccompanied migrant children in Greece: Identifying risk factors and gaps in services during the European migration crisis", *Children and Youth Services Review*, 92, 114–21. https://doi.org/10/gfdfjz.

Dilip, Ratha (2016), *The Impact of Remittances on Economic Growth and Poverty Reduction*, Migration Policy Institute, Washington, D.C.

Dreby, J. (2007), "Children and power in Mexican transnational families", *Journal of Marriage and Family*, 69(4): 1050–1064.

El Qorchi, Mohammed, Samuel Munzele Maimbo, and John F. Wilson (2003), *Informal Funds Transfer Systems: An Analysis of the Informal Hawala System*, International Monetary Fund, August, accessed on 6 November 2019 at http://blogs.reuters.com/great-debate/2015/09/16/amid-documentation-crisis-syrians-turn-to-forgeries/.

European Council on Refugees and Exiles (1999), *Survey of Provisions for Refugee Family Reunion in the European Union*, Note 1, London: Author.

Faist, Thomas (2000), "Transnationalization in international migration: Implications for the study of citizenship and culture", *Ethnic and Racial Studies*, 23(2), 189–222, DOI: 10.1080/014198700329024

Fatal Journeys (2019), "Fatal Journeys Volume 4: Missing Migrant Children", accessed on 6 November 2019 at https://publications.iom.int/books/fatal-journeys-volume-4-missing-migrant-children.

Fesenmeyer, L. (2014), 'Transnational families', in M. Keith and B. Anderson (eds), *Migration: The COMPAS Anthology*, Oxford: COMPAS, pp. 135–6.

Flore, Massimo (2018), "How Blockchain-Based Technology Is Disrupting Migrants' Remittances: A Preliminary Assessment", *European Commission*, accessed on 6 November 2019 at https://ec.europa.eu/jrc/en/publication/how-blockchain-based-technology-disrupting-migrants-remittances-preliminary-assessment.

Foner, N., and J. Dreby (2011), "Relations between the generations in immigrant families", *Annual Review of Sociology*, 37, 545–64.

Fortify Rights (2019), "Crimes Against Humanity, Mass Graves, and Human Trafficking from Myanmar and Bangladesh to Malaysia from 2012 to 2015", 27 March, accessed on 17 September 2021 at https://www.fortifyrights.org/reg-inv-rep-2019-03-27/.

Gamburd, M.R. (2008), *Breaking the Ashes: The Culture of Illicit Liquor in Sri Lanka*, Ithaca, NY: Cornell University Press.

Graham, E., L.P. Jordan and B.S.A. Yeoh (2015), "Parental migration and the mental health of those who stay behind to care for children in South-East Asia", *Social Science & Medicine*, 132, 225–235. https://doi.org/10.1016/j.socscimed.2014.10.060.

Harwell, Drew (2019), "FBI, ICE find state driver's license photos are a gold mine for facial-recognition searches", *Washington Post*, 7 July, accessed on 6 November 2019 at https://www.washingtonpost.com/technology/2019/07/07/fbi-ice-find-state-drivers-license-photos-are-gold-mine-facial-recognition-searches/.

Hoang, L.A., T. Lam, B.S. Yeoh and E. Graham (2015), "Transnational migration, changing care arrangements and left-behind children's responses in South-east Asia", *Child Geography*, 13(3): 263–77.

Home Office (2020), "Settlement: Refugee or humanitarian protection", accessed on 17 September 2021 at https://www.gov.uk/settlement-refugee-or-humanitarian-protection/family-applying-as-dependants.

Horst, H., and D. Miller (2006), *The Cell Phone: An Anthropology of Communication*, Oxford: Berg.

Hughes, Nick, and Susie Lonie (2007), *M-PESA: Mobile Money for the "Unbanked" Turning Cellphones into 24-Hour Tellers in Kenya, Innovations*, Winter/Spring.

InfoMigrants (2019), "No choice: Migrants trapped for ransom", *InfoMigrants*, 15 February, accessed on 17 September 2021 at https://www.infomigrants.net/en/post/15149/no-choice-migrants-kidnapped-for-ransom.

IOM (2017), MigApp, at https://www.iom.int/migapp.

IOM (2018a), *Big Data and Migration*, Geneva: IOM.

IOM (2018b), "Op-Ed: Understanding Modern Migrant Women", accessed on 6 November 2019 at https://thailand.iom.int/news/op-ed-understanding-modern-migrant-women.

Iyke, Aru (2017), "From flying money to blockchain: How remittance industry has reached turning point", *Cointelegraph*, 1 July.

Kilkey, M., A. Plomien and D. Perrons (2013), "Migrant men's fathering practices and projects in national and transnational spaces: Recent Polish male migrants to London", *International Migration*, 52(1), 178–191.

Kilkey, M., and L. Merla (2014), "Situating transnational families' care-giving arrangements: The role of institutional contexts", *Global Networks*, 14, 210–229.

Kritzman-Amir, T. (in press), "Swab before you enter: DNA collection and immigration control", *Harvard Civil Rights*, 56.

Lam, Theodora, and Brenda S.A. Yeoh (2019), "Parental migration and disruptions in everyday life: Reactions of left-behind children in Southeast Asia." *Journal of Ethnic and Migration Studies*, 45(16), 3085–114.

Latonero, M. (2019), 'Stop surveillance humanitarianism', *The New York Times*, 11 July.

Latonero, Mark, Danielle Poole and Jos Berens (2018), "Refugee connectivity: A survey of mobile phones, mental health, and privacy at a Syrian refugee camp in Greece", April, accessed on 17 September 2021 at https://hhi.harvard.edu/publications/refugee-connectivity-survey-mobile-phones-mental-health-and-privacy-syrian-refugee-camp.

Lee, H., and S. Francis (eds) (2009), *Migration and Transnationalism: Pacific Perspectives*, Canberra: ANU Press.

Levitt, Peggy (2001), "Transnational migration: Taking stock and future directions", *Global Networks*, 1(3), 195–216.

Madianou, M., and D. Miller (2012), *Migration and New Media: Transnational Families and Polymedia*, London and New York: Routledge.

Mahler, S.J., and P.R. Pessar (2001), "Gendered geographies of power: analyzing gender across transnational spaces", *Identities*, 7(4), 441–459.

Mazzucato, V., D. Schans, K. Caarls and C. Beauchemin (2015), "Transnational families between Africa and Europe", *International Migration Review*, 49(1), 142–172. https://doi.org/10.1111/imre.12153.

Miller, D. (2011), *Tales from Facebook*, Malden, MA: Polity Press.

Miller, D., and J. Sinanan (2014), *Webcam*, Cambridge, MA: Polity Press.

Molnar, P., and L. Gill (2018), *Bots at the Gate: A Human Rights Analysis of Automated Decision-Making in Canada's Immigration and Degree System*. Toronto, ON: University of Toronto.

Osteria, T. (2011), "Changing family structure and shifting gender roles: Impact on Child and Development", in F. Rosario-Braid, R.R. Tuazon and A.L.C. Lopez (eds), *The Future of Filipino Children: Development Issues and Trends*, Philippines: Asian Institute of Journalism and Communication and UNICEF, 2–23.

Pasquale, F. (2015), *The Black Box Society: The Secret Algorithms that Control Money and Information*, Cambridge, MA: Harvard University Press.

Pelley, Scott (2019), "Facial and Emotional Recognition; How One Man Is Advancing Artificial Intelligence", CBS News, CBS Interactive, 13 January, accessed on 6 November 2019 at www.cbsnews.com/news/60-minutes-ai-facial-and-emotional-recognition-how-one-man-is-advancing-artificial-intelligence/.

Perez, Sonia, and Mark Stevenson (2019), "Mexico's crackdown forces migrants to more dangerous routes", *Associated Press*, 25 April, accessed on 6 November 2019 at https://apnews.com/f4fd64d4ae1d47e19c86fedad479588f.

Pew Research Center (2017), *The Digital Footprint of Europe's Refugees*, accessed on 6 November 2019 at http://www.pewglobal.org/2017/06/08/digital-footprint-of-europes-refugees.

Rango, M. (2015), *How Big Data Can Help Migrants*, accessed on 6 November 2019 at https://www.weforum.org/agenda/2015/10/how-big-data-can-help-migrants/.

Regulation 2018/1860/EU on the use of the Schengen Information System for the return of illegally staying third-country nationals OJ L312.

Sieff, Kevin, and Joshua Partlow (2018), "How the migrant caravan became so big and why it's continuing to grow", *Washington Post*, 23 October, accessed on 6 November 2019 at https://www.washingtonpost.com/world/the_americas/how-the-migrant-caravan-became-so-big-and-why-its-continuing-to-grow/2018/10/23/88abf1a6-d631-11e8-8384-bcc5492fef49_story.html.

Sijapati, Bandita (2015), "Women's Labour Migration from Asia and the Pacific: Opportunities and Challenges", IOM/MPI *Issue in Brief*, issue no. 12, page 1.

Sorensen, Ninna Nyberg (2012), Revisiting the Migration–Development Nexus: From Social Networks and Remittances to Markets for Migration Control, 15 May. DOI: 10.1111/j.1468-2435.2012.00753.x.

Spyratos, S. et al. (2018), *Migration Data Using Social Media: A European Perspective*, Luxembourg: Publications Office of the European Union.

UN General Assembly (1985), *Resolution 40/144. Declaration on the Human Rights of Individuals who are not Nationals of the Country in which they live*, A/RES/40/144.

UN General Assembly (2014), *Resolution 68/167. The Right to Privacy in the Digital Age*, A/RES/68/167.

UNICEF (2014), *Children in China: An Atlas of Social Indicators*. Accessed on 17 September at https://www.unicef.cn/sites/unicef.org.china/files/2019-03/Atlas%202018%20ENG.pdf.

Verza, Maria, and Aleman, Marcos (2019), "How does a Central American migrant caravan form?" *AP*, 19 April, accessed on 6 November 2021 at https://apnews.com/77e346d48eee42dc9ffce43c92bb261f.

Wilding, R. (2006), "'Virtual' intimacies? Families communicating across transnational contexts", *Global Networks*, 6(2), 125–142.

Wilding, Raelene, and Loretta Baldassar (2018), "Ageing, migration and new media: The significance of transnational care", *Journal of Sociology*, 54(2), 226–35. https://doi.org/10/gdsrsw.

World Bank (2018), *Remittance Prices Worldwide*, issue 26, June.

Ying Shi, Simon Nazer, Jifang Ma and Elaine Zhou (2016), "It's good to talk: Technology for children connecting children and parents in China and Indonesia", UNICEF, 3 June, accessed on 6 November 2021 at https://www.unicef.org/stories/its-good-talk-technology-children.

Yiu, Lisa, and Luo Yun (2017), "China's rural education: Chinese migrant children and left-behind children", *Chinese Education & Society*, 50(4), 307–314. DOI: 10.1080/10611932.2017.1382128

PART V

MIGRATION, TECHNOLOGY
AND PUBLIC DEBATES

22. How online disinformation and far-right activism is shaping public debates on immigration

Eileen Culloty and Jane Suiter

INTRODUCTION

In February 2017, US president Donald Trump encouraged the crowd at the Conservative Political Action Conference to 'take a look at what's happening in Sweden', implying a cautionary link between Sweden's history of welcoming immigrants and an apparent upsurge in violence. His source was a Fox News interview with an obscure online activist who had created a YouTube film about the supposed increase in 'rape and violence … across Sweden due its immigration policies'. In the aftermath of Trump's speech, a panoply of media outlets replicated false claims about immigrant crime in Sweden; among these was a Russian TV crew who attempted to bribe immigrants into staging violence for the camera (Gramer 2017). This episode illustrates the contemporary dynamics of far-right activism and anti-immigrant disinformation. While Sweden's immigration policies have long been a focal point for disinformation by far-right activists, this disinformation is now re-packaged for different audiences. It flows through a media system that begins with unapologetically racist online activists, enters the mainstream through conservative media and populist politicians, and flows back to the online activists. As Fekete argues (2014), contemporary media facilitate a process of 'cumulative racism' as anti-immigrant disinformation travels from the fringe to the mainstream and back again.

Anti-immigrant disinformation is a mainstay of far-right communication because it reinforces the core ideology of exclusion and nativist supremacy. These attitudes have increasingly permeated public debates across the world including, for example, Brazil (see Saad-Filho and Boffo 2020), the Philippines (see Ramos 2020), and South Africa (see Kerr, Durrheim, and Dixon 2019). These attitudes have increasingly permeated public debates. This chapter focuses specifically on anti-immigrant disinformation by the far-right in Europe and North America, where anti-immigration sentiments intensified following the 2015 refugee crisis and attacks by Islamist terrorists. Since then, countries across these regions have experienced an upsurge in far-right and populist politics. In some countries, the electoral gains of far-right parties have brought harsher policies on immigration and normalised anti-immigrant rhetoric while transforming liberal democracies into autocracies. The key backdrop to these developments is the crisis of legitimacy within democracy (Bennett and Livingston 2018). Trust in democratic institutions and mainstream news media has declined in tandem with the rise of digital media, which has enabled alternative information sources and disinformation to flourish on an unprecedented scale. Of course, digital media are not the root cause of far-right activism, but they have enabled far-right activists and various political actors (encompassing parties, movements, and individuals) to capitalise on democratic weaknesses and to exploit and manipulate world events that put issues of immigration at the centre of public debate.

For comprehension, this chapter will primarily refer to the 'far-right' as a collective term for actors pushing anti-immigrant disinformation. However, in some instances, the actors we mention might be more accurately described as 'alt-right' or 'right-wing populists'. The following section briefly outlines the wider context for anti-immigration disinformation in Europe and North America and the panoply of actors that scapegoat and attack immigrants.

BACKGROUND

Over the past two decades, the public debate on immigration in Europe has been punctuated by a series of high-profile controversies. Prior to the 2015 refugee crisis, debates about immigration centred on Muslim populations including fears of Islamisation (e.g., debates about the visibility of mosques and burqas) and reactions to Islamist terror attacks (e.g., the attack on Charlie Hebdo). These and related events have been exploited by far-right activists in Europe and North America who spread disinformation about 'no go zones' and then appeal to public concerns about security. As such, anti-immigrant disinformation is entwined in narratives about the failure of multiculturalism, the security of national borders, the threat of terrorism, and the protection of (white) women and girls (Juhász and Szicherle 2017). These narratives are not new. In the US, for example, they have animated recent conservative politics from Pat Buchannan in the 1990s, the Tea Party in the early 2000s, and the Trump phenomenon in the 2010s (Benkler, Faris, and Roberts 2018).

Research on anti-immigrant disinformation intensified following the election of Donald Trump and the UK's Brexit referendum in 2016. Although disinformation about immigration was a feature of both campaigns, it is important to also acknowledge the real and perceived grievances that made anti-immigrant rhetoric appealing to large segments of both populations. In both countries, the decline of white working-class communities is linked to feelings of alienation and a visceral opposition to immigration (Gusterson 2017; Hobolt 2016; Goodwin and Heath 2016; Inglehart and Norris 2016). In the US, Trump exploited this sentiment with his simplistic promise to 'build a wall' to prevent immigrants from 'swarming' over the southern border. While much of Trump's fear mongering was directed at Mexicans, alternative media outlets led by Breitbart also introduced a fear of Muslims and Islamist terrorism (Benkler, Faris, and Roberts 2018; Kamenova and Pingaud 2017). In the UK, similar political and media rhetoric focused on immigration from Eastern Europe and refugees from the Middle East. Legal immigration within the EU was frequently confused with asylum seeking and the Vote Leave campaign stoked up fears of an imminent arrival of millions of Turks (Ker-Lindsay 2018). As outlined below, immigration from predominately Muslim countries is the common ground that supports transnational cooperation between far-right groups across Europe and North America, and this cooperation is made possible by digital media.

Social media platforms enable far-right activists to push their anti-immigrant narratives and these platforms play a key role in the normalisation of anti-immigrant attitudes. Nativist, racist, and xenophobic narratives that were previously marginalised on fringe far-right websites – where people had to actively seek them out – now reach a wider audience on popular social media platforms such as Facebook and Twitter (Ekman 2019; Farkas et al. 2018). Thus, it is important to recognise that anti-immigrant disinformation is part of a culture war in which an ecosystem of actors (far-right, alt-right, populist, and conservative) reinforce a common opposition to a pluralist worldview even though those actors frequently disagree on other ideological issues.

As noted, this chapter refers to the 'far-right' as a collective term for actors pushing anti-immigrant disinformation. Currently, there are no commonly accepted definitions to describe the various actors that animate right-wing extremism (Carter 2018; Davey and Ebner 2017). In part, the conceptual difficulty stems from the sheer diversity of actors within the extremist ecosystem. Moreover, the disinformation narratives and digital tactics of these actors are increasingly entangled in the media system making it difficult to draw clear distinctions between, for example, far-right activists, alternative and conservative media actors, and populist politicians and parties. The range of extremist actors identified by anti-racism and anti-fascist activists is notably diverse. For example, Hope Not Hate (2019) identified the key personalities influencing hate culture, which included far-right politicians, conspiracy theorists, and social media influencers.

The so-called 'alt-right' are perhaps more influential on social media. Hartzell (2018: 8) characterises the alt-right as the 'youthful, intellectual, pro-white' faction of the far-right, which acts as a bridge between 'mainstream public discourse and white nationalism'. However, the alt-right encompasses a wide range of views representing 'an amalgam of conspiracy theorists, techno-libertarians, white nationalists, Men's Rights advocates, trolls, anti-feminists, anti-immigration activists, and bored young people' (Marwick and Lewis 2017:3).

While acknowledging certain ideological diversity, most scholars agree that these actors may be grouped together for their shared espousal of nationalism, racism, and xenophobia. Importantly, it is the diversity of extremist actors that enables the fluidity of anti-immigrant disinformation within the ecosystem. In other words, the diversity of actors creates different points of exposure for audiences and different rhetorical strategies and tactics through which disinformation is packaged. Consequently, to conceptualise the relationship between far-right activism and anti-immigrant disinformation and the implications for public debate, it is necessary to consider three interrelated areas: online disinformation, far-right activism, and anti-immigrant narratives. In what follows, we examine the key issues and research findings relating to each of these areas.

ONLINE DISINFORMATION

When thinking about the relationship between disinformation and public debates about immigration, the first challenge is to develop a clear definition of disinformation that distinguishes it from other kinds of negative content. This is important because there is much confusion surrounding the term disinformation, which limits our ability to understand the problem (Wardle and Derakhshan 2017). Put simply, disinformation is false information that is created to cause harm. In other words, someone is deliberately trying to confuse the public by sharing false information, rumours, or manipulated content. In contrast, misinformation involves sharing false information without an intention to cause harm. For example, misinformation occurs when an official issues an incorrect statement by mistake and journalists then report that statement on the assumption it is true (Wardle and Derakhshan 2017). While these mistakes can distort public debate, they lack the intention to deceive.

In addition, we may distinguish disinformation from negative information such as public criticism of an organisation or critiques of public policy. This negative information does not cross over into disinformation unless it is accompanied by a deliberate effort to cause harm by spreading false claims. For example, criticism of the 2018 Global Compact for Safe, Orderly and Regular Migration (GCM) is not disinformation. Yet, distinct from criticism, there existed a network of far-right actors spreading disinformation about the GCM including false claims that it would outlaw criticism of migration (ISD 2019). This network of GCM opponents filled a vacuum in coverage and was responsible for almost half of the most popular YouTube videos about the GCM (ISD 2019).

Reduced to its basic components, disinformation is a process that engages different actors and different stages. First, we have the disinformation actors who strategically push false information into the network. These actors may be defined collectively for their common intention to deceive or manipulate the public, but of course they have a diverse range of goals and motivations. To date, much of the scholarly and journalistic attention has focused on state-based actors such as Russia's Internet Research Agency, but the network of far-right activists that push anti-immigrant disinformation encompasses a wide range of ideological movements and their media channels as well as partisan media outlets and individual social media influencers (Marwick and Lewis 2017).

The infrastructure of digital platforms facilitates disinformation by these actors in many ways; not least in the unprecedented speed and scale at which disinformation can travel. As Facebook and other platforms now play a central role in distributing political disinformation (Merrill and Åkerlund 2018), these platforms can be used by far-right activists to advance their xenophobic and anti-democratic ideology. Moreover, as Törnberg and Wahlström (2019) argue, social media provide multiple opportunity mechanisms for the far-right including: discursive opportunities that support anti-immigrant mobilisation by exploiting topical issues to spread disinformation; group dynamic opportunities that strengthen community ties by linking the local with the transnational; and coordination opportunities that allow diverse actors to cooperate in targeting key messages across different platforms and at different audiences. In this context, some argue that social media platforms have given rise to a digital hate culture (Ganesh 2018) augmented by the coordinated action of anonymous and automated accounts (Phillips 2015; Kessler 2018; Massanari 2017; Zannettou et al. 2017, 2018).

Receptive audiences are arguably the most important component of the disinformation process because disinformation only becomes a problem when it finds an audience willing to believe and share it. In general, disinformation is successful when it triggers negative emotions such as fear and anger (Bakir and McStay 2018; Wirz et al. 2018). Immigration has long been a potent source of fear mongering; after all, humans are a tribal species and fear of the other is easily ignited. As noted above, anti-immigrant disinformation often manipulates the anxieties of the white working class, whose communities have struggled to maintain their industrial identities in the age of global capitalism. Some researchers also identify a generational dimension arguing that the aging populations in the Global North may perceive immigration as a 'threat to their accumulated nest egg' (Crépeau 2018: 650). To those who feel marginalised, far-right movements offer strong communities that provide a sense of intimacy and belonging (Back 2002; Daniels 2009; Simi and Futrell 2015). Thus, the digital hate cultures mentioned above give rise to a 'networked white rage' (Daniels 2018) that migrates across websites and platforms and adopts an always-evolving coded language that makes it difficult to monitor and govern (Ganesh 2018).

It is difficult to draw a conclusive cause and effect relationship between disinformation and public attitudes; not least because there are so many factors that influence public attitudes. Nevertheless, anti-immigrant rhetoric has been a recurring trope among the far-right for many decades (Bajomi-Lázár 2019) and has intensified since the 2015 refugee crisis. Recent studies find evidence that these narratives have filtered down to ordinary social media conversations about immigration. At the same time, researchers investigating the process through which individuals come to accept extreme ideological beliefs identify online media as a pivotal point of exposure (Hamm and Spaaij 2017; McCauley and Moskalenko 2011).

Bakamo Social (2018) identified five major immigration narratives in their study of social media conversations across 28 European countries over a one-year period. A resounding anti-immigrant stance was evident across the online conversations with varying levels of intensity in individual countries. The humanitarianism narrative (49.9%) concerned moral obligations to support refugees, but included arguments for and against humanitarianism. The security narrative (25.9%) focused on the threat of immigration in terms of immigrant crime and terror attacks while the identity narrative (15.3%) concerned the threat of immigration to social cohesion and the traditional identity of European countries. Finally, the economic narrative (8%) and the demographics narrative (1%) focused on issues of sustainability.

At the country level, identity narratives were most prevalent in Germany, the Netherlands, and Slovakia while security narratives dominated in Hungary, Poland, Estonia, and Austria. Identity and security narratives also subverted discussions of humanitarianism. For example, in France the humanitarian narrative was undermined by those questioning whether refugees were genuinely in need of assistance while in Spain the humanitarian narrative was subverted by concerns that left-wing politicians would prioritise the needs of migrants over Spaniards. Consequently, those advocating humanitarianism were characterised as a threat to the welfare of European countries (Bakamo Social 2018).

FAR-RIGHT ACTIVISM

In terms of formal politics, the nationalist and xenophobic attitudes of the far-right are increasingly embraced by segments of the population in many regions of the world. Much of this is underpinned by a nativist ideology that upholds the value of a homogeneous nation-state and perceives non-native elements – whether people, customs, or ideas – as threatening (Mudde 2019). Populist and far-right political actors vary in the extent to which this ideology is made explicit, but a nativist superiority is almost always implied in their messaging (Mudde 2019). All populist rhetoric – whether on the left or the right – shares a fundamental distinction between 'we' the pure people and 'them' the corrupt elite. Far-right populism takes this further to identify an enemy other – Muslims, Jews, immigrants – on whom insecurities may be projected.

Gaining power by entering government is perhaps the ultimate success of far-right activism. Research indicates that the success of far-right populist parties is linked to their political communication which may combine anti-immigrant disinformation with dramatic, emotional appeals to the target audience (Wirz et al. 2018). In some cases, these communication strategies enable far-right actors to capture mainstream media coverage and thereby increase exposure for their anti-immigrant messages (Wirz et al. 2018). Anti-immigration disinformation has played a central role in the success of far-right parties in Europe. In Hungary, for example,

Prime Minister Viktor Orbán has framed immigration as an existential crisis for the country despite an absence of any evidence to support this view (Kiss and Szabó 2018). Nevertheless, immigration is conceived as a threat to Hungarian society and Orbán's government are cast in the role of national saviours while pro-immigration 'elites' such as the EU and George Soros are the anti-Hungarian villains (Bajomi-Lázár 2019). Orbán's politics has been bolstered by the government's gradual take-over of media outlets, which has enabled the normalisation of anti-immigrant attitudes and disinformation. In this context, Bajomi-Lázár (2019) argues that the impact of anti-immigrant disinformation is likely to be greater in countries that have limited levels of political and media pluralism.

Nevertheless, as argued in the following section, the digital tactics and strategies of far-right activists also seek influence in countries in which media freedom and pluralism is relatively high. For example, during the 2016 US presidential election, far-right perspectives influenced the agenda of mainstream news including *The New York Times* and *The Washington Post* (Benkler, Faris, and Roberts 2018). Moreover, far-right actors attack the legitimacy of the news media and promote the authenticity of their own media outlets. Asperholm Hedlund (2019) analysed how anti-immigration disinformation was utilised in Swedish online news media before the 2018 Swedish national elections. She analysed 123 articles from different online news media from ten days before the election up until election day. All articles from far-right news media contained anti-immigrant disinformation that was based on tactics of misappropriation, fabrication, and propaganda. However, the study found little evidence of anti-immigrant disinformation among mainstream outlets.

Apart from political actors, there is some evidence that the online activities of the far-right in Europe and North America have become more cooperative and transnational; at least in relation to common issues of opposition such as Muslim immigration. At the turn of the millennium, the online presence of far-right groups was heavily decentralised and centred on individual websites. Burris, Smith, and Strahm's (2000) study found a lack of connection between different groups in North America and Europe and the general absence of links to mainstream conservative movements. Almost two decades later – following the advent of social media and the protracted fallout of the War on Terror era – there is evidence of a move towards transnational far-right activism. A study by the Institute for Strategic Dialogue (Davey and Ebner 2017) analysed content from more than fifty online platforms. The authors identified a high level of transnational cooperation among far-right activists and the emergence of international campaigns such as Defend Europe, which was launched in May 2017. The campaign raised more than €75,000 to charter a ship with the intention of preventing search and rescue vessels from rescuing migrants in the Mediterranean Sea (Oppenheim 2017).

As noted, it can be difficult to draw clear lines of distinction between the online activities of different groups of far-right actors. Nevertheless, studies indicate that social media are utilised in different ways by far-right movements and far-right political parties. Klein and Muis (2018) examined far-right Facebook pages from France, Germany, the Netherlands, and the UK during 2014/2015. They found that far-right political parties primarily focused on opposing political 'elites' (i.e., established parties and the EU), while far-right groups focused on immigration and Islam. Perhaps unsurprisingly, non-institutional groups were also far more extreme in their discussion of immigration. The British National Party (BNP) in the UK was an exception in that the party focused heavily on immigration rather than formal politics; Klein and Muis attribute this to the fact that far-right parties in the UK have traditionally had fewer opportunities to exercise influence through formal politics.

In addition, there is evidence of far-right groups coordinating transnational activity ahead of national elections; for example, to bolster support for far-right parties such as AfD in Germany and the National Front in France (Avaaz 2019). Typically, activities are coordinated on lesser-known online platforms and their messages are then disseminated to a wider audience on popular platforms such as Twitter and Facebook (Davey and Ebner 2017; Marwick and Lewis 2017). In their analysis of far-right Twitter activity across France, Germany, Italy and the UK, Froio and Ganesh (2018) found that cross-border retweets between far-right groups were generally limited. However, as noted above, Twitter is a target platform rather than a coordinating one. Nevertheless, Froio and Ganesh did find evidence of cooperation in relation to the issue of Muslim immigration, which was commonly framed as a threat to Europe in terms of the continent's collective security, economy, and culture. Consequently, the authors concluded that the 'transnational glue of the far-right' is Islamophobia (Froio and Ganesh 2018: 19). In response, some Internet service providers have removed their technical support for platforms known to foster extremist ideologies and far-right activity has moved to new platforms with strict free speech policies (Zannettou et al. 2018).

MEDIA MANIPULATION TACTICS

Since the advent of the web, websites and online forums enabled far-right activists to foster a strong sense of community among individuals who are geographically distant (Meddaugh and Kay 2009). As noted above, social media platforms offer many benefits for far-right activists as these platforms facilitate opportunities for more direct and personal communication with target audiences and in ways that were not possible in the age of mass media (Ernst et al. 2017; Stier et al. 2017; Törnberg and Wahlström 2019). A growing number of researchers investigate the online manipulation and disinformation tactics of the far-right and how these tactics are employed to reach different audiences and achieve different communication goals. For example, some tactics aim to increase the exposure of the general public to anti-immigrant disinformation while other tactics aim to reinforce and bolster the cohesion of the far-right community (Törnberg and Wahlström 2019).

The advent of social media greatly expanded opportunities to reach the wider audience. In an analysis of far-right activity on Twitter, Graham (2016) identified three key disinformation tactics that enable the far-right to direct their messages to a wider, mainstream audience. The 'piggybacking' and 'backstaging' manipulation tactics infiltrate trending topics to push extremist content while the 'narrating' manipulation tactic inverts the meaning of trending topics to reframe the original meaning through irony. For example, hashtags such as #WhiteLivesMatter attempt to reframe the narrative of the Black Lives Matter movement.

This process of appropriating existing hashtags has also been characterised in terms of 'hijacking' (Jackson and Foucault Welles 2015) and the promotion of 'critical counter-narratives' (Poole et al. 2019). In essence, far-right activists exploit major news events to increase exposure for their anti-immigrant disinformation. For example, during the 2016 US presidential election far-right activists routinely used the hashtag #StopIslam in conjunction with pro-Trump and anti-Clinton hashtags as part of a broader effort to normalise an anti-immigration narrative and to introduce anti-immigrant disinformation into the election campaign (Poole et al. 2019). Other studies have identified the use of this manipulation tactic in relation to the refugee crisis (Siapera et al. 2018) and Brexit (Green et al. 2016). These hashtag campaigns support the formation of 'affective' (Papacharissi 2015) or 'ad hoc' (Dawes 2017) publics that facilitate the circulation (Groshek and Koc-Michalska 2017) and fermentation (Farkas et al. 2017) of far-right narratives and attitudes. While the ad hoc publics created by hashtag campaigns tend to be short lived (Poole et al. 2019; Dawes 2017), they have a 'liminal' power to disorientate and confuse public debate (Siapera et al. 2018).

Far-right activists have been adept at exploiting the irreverent currency of social media, fusing pop culture with extremist content. White power symbolism and Nazi iconography has been a traditional feature of far-right groups in Europe and North America. However, such overtly racist symbols are a target for censorship on popular social media platforms. To counteract this and to appeal to a broader, younger audience, the far-right has adopted new symbols and has made extensive use of memes, music videos, jokes, and irony (Beran 2017; Luke 2016; Marwick and Lewis 2017; Nagle 2017). The logic of this manipulation tactic is frequently discussed by far-right activists as an effort to attack young audiences and guide them towards more extremist content and ideas.

The playful and ironic use of pop culture creates coded in-jokes for far-right affiliates. These coded references help foster a collective identity for the far-right that is distinguished from ordinary people or 'normies' while also pushing the boundaries of what is considered acceptable (Nagle 2017). Ekman (2019) outlines how these manipulation tactics result in the gradual normalisation of previously unacceptable utterances; utterances that dehumanise immigrants and even designate them as legitimate targets of violence. The participatory culture of digital media is central to the success of this tactic. In their analysis of far-right activity, Marwick and Lewis (2017:4) found that far-right memes often function as image macros that are engineered to 'go viral' by conveying far-right ideology through humour. In other words, these memes encourage participation and appropriation by individuals who can participate in the meme without necessarily understanding what it promotes. Importantly, the use of ironic memes presents a challenge to our understanding of disinformation as the intentional promotion of false information. Memes are often presented as entertainment rather than information and the use of irony can make it difficult to ascertain intention.

As with disinformation generally, automated bots and fake accounts are frequently used to inflate the popularity of content. Avaaz (2019) investigated far-right disinformation on Facebook ahead of the European parliament elections. In response, Facebook removed 77 pages and 230 accounts from France, Germany, Italy, Poland, Spain, and the UK. Facebook estimated that this content reached 32 million people and generated 67 million 'interactions' through comments, likes, and shares (Avaaz 2019). Across the countries, fake and duplicate accounts artificially inflated the popularity of far-right content and unsuspecting audiences were targeted through the deceptive branding of Facebook pages. For example, Facebook pages were initially set up as lifestyle or music pages, but the followers then received a stream of far-right content.

Finally, we may consider the role of mainstream media within the manipulation tactics of the far-right. The news media are already predisposed to use fear as a framing device in news stories (Yadamsuren and Heinström 2011). Consequently, news stories about refugees and immigrants tend to focus on crime, public unrest, and violence resulting in a perpetual flow of 'bad news' about immigrants and refugees (Philo et al. 2013). As such, the news media provide the far-right with stories that can be repurposed and de-contextualised to emphasise their own agenda (Ekman 2019). De-contextualisation is a simple but effective disinformation tactic (Wardle and Derakhshan 2017). By omitting key explanatory factors, adding textual amendments, or adopting different naming standards, far-right activists can transform a relatively neutral news story into one that is imbued with racist or anti-immigrant disinformation (Ekman 2019). In contrast to 'fake news' that is entirely fabricated, these disinformation stories contain nuggets of truth that are corroborated by mainstream news sources.

In addition, Marwick and Lewis (2017) argue that far-right conspiracy theories appeal to the media logic of mainstream news. In the US, the network news media has often featured 'documentaries' investigating conspiracy theories without fully refuting them (Byford 2011). For example, the 'birther' conspiracy, prompted by Donald Trump, claimed Barack Obama was not a US citizen. At the height of this disinformation controversy, CNN and MSNBC devoted a higher percentage of their Obama coverage to birther conspiracy than Fox News (Rosenberg 2011). This is a significant problem because such media manipulation may contribute to decreased trust of mainstream media, increased misinformation, and further radicalisation. At the same time, the far-right exploits growing distrust towards the mainstream media, particularly regarding immigration reporting (Andersson 2017 cited in Ekman 2019). This declining trust reflects growing political polarisation (Suiter and Fletcher 2020) as well as growing use of social media for news (Kalogeropoulos et al. 2019) and declining news values including clickbait and unverified sources (Enli and Rosenberg 2018). Thus, it is important to recognise that the capacity for disinformation to distort public debates about immigration extends far beyond the activities of the far-right.

CURRENT AND FUTURE RESEARCH CHALLENGES

Although this chapter has focused on anti-immigration activism in Europe and North America, it is important to recognise that the resurgence of nationalism, racism, and xenophobia is a global phenomenon. In Europe and North America, anti-immigrant extremism is typically associated with white populations, but the same patterns of scapegoating minorities and fuelling or manipulating perceived grievances are found across power constructs. In South Africa, for example, xenophobic discourses echo the sentiments expressed by the far-right in Europe in so far as they accuse immigrants of undermining the socio-economic struggles of native South Africans (Kerr, Durrheim, and Dixon 2019).

Far-right and anti-immigrant activism is a challenging object of study and these challenges impede efforts to combat extremism. At a purely conceptual level, one difficulty lies in the ideological diversity of anti-immigrant activists and the lack of universal definitions to describe far-right extremism (Carter 2018; Davey and Ebner 2017; Holt et al. 2020). For example, the core ideological tenets of different far-right groups may encompass varying degrees of anti-globalist, anti-immigrant, and nativist attitudes. While far-right groups commonly scapegoat immigrants, greater conceptual clarity is needed to understand the specific role of anti-immigrant disinformation within the communication strategies of different groups.

At a practical, methodological level, efforts to monitor the online activities of the far-right are restricted by the speed of change in the online environment. As noted above, the online manipulation tactics of the far-right vary across different platforms and evolve in response to countermeasures. As accounts are closed, activists move to new platforms and reopen accounts under different names; when access to platforms is denied, activists regroup on new platforms with stricter free-speech policies; and when favoured phrases such as 'white genocide' are restricted or demoted by algorithms, new coded phrases take their place. The speed of these changes makes it difficult to develop a reliable method of data collection (Marwick and Lewis 2017). Moreover, the design of each platform gives rise to distinct forms of participation, which makes it difficult to operationalise a consistent set of indicators that can be used to identify far-right accounts (Crosset, Tanner, and Campana 2019). Finally, as with social media research generally, researchers are impeded by a lack of quality data from the social media platforms.

In terms of counteracting anti-immigrant disinformation, more research is needed to understand what makes different audiences receptive to anti-immigrant messages. Research shows that the negative framing of immigration can affect public attitudes and voting behaviour. However, quite apart from any bias on the part of the individual, repeated exposure can increase perceptions of credibility over time (Fazio et al. 2015). Thus, reducing exposure to disinformation and providing more supports to help audiences evaluate online content appear to be key for mitigating disinformation (Schleicher 2019). Various regulatory, legal, educational, and technological measures have been proposed to counteract disinformation, but to date we know little about the effectiveness of these measures in general and in the context of anti-immigrant disinformation specifically.

Proponents of critical thinking and information literacy argue for the importance of strengthening the capacity of individuals to evaluate content (Schleicher 2019). This is often accompanied by calls for technological approaches that filter out extremist content or flag disinformation. However, research remains limited and there are contradictory findings about the effectiveness of these approaches. It is likely that countering disinformation and helping audiences evaluate online content will require more systematic action. In this regard, Janda and Víchová (2019) call for a 'whole of society approach' that rests on cooperation between technology companies, governments, civil society organisations, and individuals.

The regulatory environment has failed to keep pace with the rapid evolution of digital platforms and their use for political campaigning and propaganda (Jones 2019). Some countries have introduced legislation or regulations against disinformation, but the implementation of new regulations is not without controversy. For authoritarian governments, concerns about online disinformation are an opportunity to exercise more control over the media landscape. For democratic governments, policies that aim to counter disinformation must also respect the values of freedom of thought and freedom of expression. However, this debate is often distorted by far-right activists who claim freedom of expression as a defence for the promotion of extremist agendas (O'Hagan 2018). As Jones (2019:50) argues, 'freedom of expression does not entail that there must be no restriction of online political content; rather, that any restriction must be properly tailored.' Thus, the major challenge for countering far-right and extremist disinformation rests on the wider issue of establishing normative frameworks for the online environment.

In addition, the ascendency of the far-right also requires attention from mainstream media and politicians in terms of how immigration is discussed and reported. While the online environment is flooded with disinformation, mainstream news media remain highly influential. It is vital that these outlets offer audiences a comprehensive and accurate understanding of issues relating to immigration and avoid platforming extremist views for sensational coverage. In other words, there is no magic bullet to counter anti-immigrant disinformation. It requires a 'whole of society' approach that engages top-down approaches to regulating and monitoring the information and security environments as well as bottom-up approaches to everyday media practices at organisational and individual levels.

REFERENCES

Asperholm Hedlund, Laura (2019), *Identifying and Understanding Anti-Immigration Disinformation: A Case Study of the 2018 Swedish National Elections* (dissertation), Swedish Defence University.

Avaaz (2019), *Far Right Networks of Deception*, London: Avaaz.

Back, L. (2002), 'Aryans reading Adorno: Cyber-culture and twenty-first century racism', *Ethnic and Racial Studies*, **25**(4), 628–651.

Bajomi-Lázár, P. (2019), 'An anti-migration campaign and its impact on public opinion: The Hungarian case', *European Journal of Communication*, **34**(6), 619–628.

Bakamo Social (2018), *Migration Narratives in Europe 2017–2018*, London: Bakamo Social.

Bakir, V. and A. McStay (2018), 'Fake news and the economy of emotions: Problems, causes, solutions', *Digital Journalism*, **6**(2), 154–175.

Benkler, Yochai, Robert Faris, and Hal Roberts (2018), *Network Propaganda: Manipulation, Disinformation, and Radicalization in American Politics*, London: Oxford University Press.

Bennett, W.L., and S. Livingston (2018), 'The disinformation order: Disruptive communication and the decline of democratic institutions', *European Journal of Communication*, **33**(2), 122-139.

Beran, D. (2017), '4chan: The skeleton key to the rise of Trump', *Medium.com*, 14 February 14, accessed 20 January 2020 at https://medium.com/@DaleBeran/4chan-the-skeleton-key-to-the-rise-of-trump-624e7cb798cb.

Burris, V., E. Smith, and A. Strahm (2000), 'White supremacist networks on the Internet', *Sociological Focus*, **33**(2), 215–235.

Byford, Jovan (2011), *Conspiracy Theories: A Critical Introduction*, New York: Palgrave Macmillan.

Carter, E. (2018), 'Right-wing extremism/radicalism: reconstructing the concept', *Journal of Political Ideologies*, **23**(2), 157–182.

Crépeau, F. (2018), 'Towards a mobile and diverse world: "Facilitating mobility" as a central objective of the global compact on migration', *International Journal of Refugee Law*, **30**(4), 650–656.

Crosset, V., S. Tanner, and A. Campana (2019), 'Researching far right groups on Twitter: Methodological challenges 2.0', *New Media & Society*, **21**(4), 939–961.

Daniels, J. (2009), 'Cloaked websites: Propaganda, cyber-racism and epistemology in the digital era', *New Media & Society*, **11**(5), 659–683.

Daniels, J. (2018), 'The algorithmic rise of the alt-right', *Contexts*, **17**(1), 60–65.

Davey, Jacob and Julia Ebner (2017), *The Fringe Insurgency: Connectivity, Convergence and Mainstreaming of the Extreme Right*, London: Institute for Strategic Dialogue.

Dawes, Simon (2017), '#JeSuisCharlie, #JeNeSuisPasCharlie and ad hoc publics', in Gavan Titley, Des Freedman, Gholam Khiabany and Aurélien Mondon (eds), *After Charlie Hebdo: Terror, Racism and Free Speech*, London: Zed Books, pp. 180–191.

Ekman, M. (2019), 'Anti-immigration and racist discourse in social media', *European Journal of Communication*, **34**(6), 606–618.

Enli, G. and L.T. Rosenberg (2018), 'Trust in the age of social media: Populist politicians seem more authentic', *Social Media + Society*, **4**(1), 1–11.

Ernst, N., S. Engesser, F. Büchel, S. Blassnig, and F. Esser (2017), 'Extreme parties and populism: An analysis of Facebook and Twitter across six countries', *Information, Communication & Society*, **20**(9), 1347–1364.

Farkas, J., J. Schou, and C. Neumayer (2017), 'Cloaked Facebook pages: Exploring fake Islamist propaganda in social media', *New Media & Society*, **20**(5), 1850–1867.

Farkas, J., J. Schou, and C. Neumayer (2018), 'Platformed antagonism: Racist discourses on fake Muslim Facebook pages', *Critical Discourse Studies*, **15**(5), 463–480.

Fazio, L.K., M.N. Brashier, B.K. Payne, and E.J. Marsh (2015), 'Knowledge does not protect against illusory truth', *Journal of Experimental Psychology: General*, **144**(5), 993–1002.

Fekete, L. (2014), 'Anti-fascism or anti-extremism?', *Race & Class*, **55**(4), 29–39.

Froio, C. and B. Ganesh (2018), 'The transnationalisation of far-right discourse on Twitter. Issues and actors that cross borders in Western European democracies', *European Societies*, **21**(4), 513–539.

Ganesh, B. (2018), 'The ungovernability of digital hate culture', *Journal of International Affairs*, **71**(2), 30–49.

Goodwin, M.J. and O. Heath (2016), 'The 2016 referendum, Brexit and the left behind: An aggregate-level analysis of the result', *The Political Quarterly*, **87**(3), 323–332.

Graham, R. (2016), 'Inter-ideological mingling: White extremist ideology entering the mainstream on Twitter', *Sociological Spectrum*, **36**(1), 24–36.

Gramer, R. (2017), 'Russian TV crew tries to bribe Swedish youngsters to riot on camera', *Foreign Policy*, 7 March, accessed 25 January 2020 at https://foreignpolicy.com/2017/03/07/russian-tv -crew-tries-to-bribe-swedish-youngsters-to-riot-on-camera-stockholm-rinkeby-russia-disinformation -media-immigration-migration-sweden/

Green, S.F., C. Gregory, M. Reeves, J.K. Cowan, O. Demetriou, I. Koch, M. Carrithers, R. Andersson, A. Gingrich, S. Macdonald, and S.C. Açiksöz (2016), 'Brexit referendum: First reactions from anthropology', *Social Anthropology*, **24**(4), 478–502.

Groshek, J. and K. Koc-Michalska (2017), 'Helping populism win? Social media use, filter bubbles, and support for populist presidential candidates in the 2016 US election campaign', *Information, Communication & Society*, **20**(9), 1389–1407.

Gusterson, H. (2017), 'From Brexit to Trump: Anthropology and the rise of nationalist populism', *American Ethnologist*, **44**(2), 209–214.

Hamm, Mark S. and Ramon Spaaij (2017), *The Age of Lone Wolf Terrorism*, New York: Columbia University Press.

Hartzell, S. (2018), '"Alt-White: Conceptualizing the "Alt-Right" as a rhetorical bridge between white nationalism and mainstream public discourse', *Journal of Contemporary Rhetoric*, **8**(2), 6–25.

Hobolt, S.B. (2016), 'The Brexit vote: A divided nation, a divided continent', *Journal of European Public Policy*, **23**(9), 1259–1277.

Holt, T.J., J.D. Freilich, and S. M. Chermak (2020), 'Examining the online expression of ideology among far-right extremist forum users', *Terrorism and Political Violence*, DOI: 10.1080/09546553.2019.1701446.

Hope Not Hate (2019), *State of Hate 2019*, London: Hope Not Hate.

Inglehart, R.F. and P. Norris (2016), 'Trump, Brexit, and the rise of populism: Economic have-nots and cultural backlash', *HKS Faculty Research Working Paper Series*, RWP16-026, accessed 20 January 2020 at https://www.hks.harvard.edu/publications/trump-brexit-and-rise-populism-economic-have -nots-and-cultural-backlash#citation

Institute for Strategic Dialogue (2019), 'ISD research featured in POLITICO about the trolling of the UN migration pact', *Institute for Strategic Dialogue*, accessed 20 January 2020 at https://www.isdglobal .org/isd-research-featured-in-politico-surroundingthe-trolling-of-the-un-migration-pact/.

Jackson, S.J. and B. Foucault Welles (2015), 'Hijacking #myNYPD: Social media dissent and networked counterpublics', *Journal of Communication*, **65**(6), 932–952.

Janda J. and V. Víchová (2019), 'What can we learn from Russian hostile information operations in Europe?', in Vasu Norman, Jayakumar Shashi, and Ang Benjamin (eds), *DRUMS: Distortions, Rumours, Untruths, Misinformation, and Smears*, Singapore: World Scientific, pp. 133–146.

Juhász, Attila and Patrik Szicherle (2017), *The Political Effects of Migration-Related Fake News, Disinformation and Conspiracy Theories in Europe*, Budapest: Friedrich Ebert Stiftung and Political Capital.

Jones, K. (2019), 'Online disinformation and political discourse: Applying a human rights framework', *Chatham House International Law Programme*, 6 November, accessed 20 January 2020 at https:// www.chathamhouse.org/publication/online-disinformation-and-political-discourse-applying-human -rights-framework

Kalogeropoulos, A., J. Suiter, and I. Udris (2019), 'News media and news consumption: Factors predicting trust in news in 35 countries', *International Journal of Communication*, **13**, 3672–3693.

Kamenova, D. and Pingaud, E. (2017), 'Anti-migration and Islamophobia: Web populism and targeting the "easiest other"', in Mojca Pajnik and Birgit Sauer (eds), *Populism and the Web: Communicative Practices of Parties and Movements in Europe*, London: Routledge, pp. 108–121.

Ker-Lindsay, J. (2018), 'Turkey's EU accession as a factor in the 2016 Brexit referendum', *Turkish Studies*, **19**(1), 1–22.

Kerr, P., K. Durrheim, and J. Dixon (2019), 'Xenophobic violence and struggle discourse in South Africa', *Journal of Asian and African Studies*, **54**(7), 995–1011.

Kessler, A. (2018), 'Who Is @TEN_GOP from the Russia indictment?', *CNN*, 17 February, accessed 20 January 2020 at https://www.cnn.com/2018/02/16/politics/who-is-ten- gop/index.html.

Kiss, B. and G. Szabó (2018), 'Constructing political leadership during the 2015 European migration crisis: The Hungarian case', *Central European Journal of Communication*, **11**(1), 9–24.

Klein, O. and J. Muis (2018), 'Online discontent: comparing Western European far-right groups on Facebook', *European Societies*, **21**(4), 540–562.

Luke, T.W. (2016), 'Overtures for the triumph of the tweet: White power music and the alt-right in 2016', *New Political Science*, **39**(2), 277–282.

Marwick, Alice and Rebecca Lewis (2017), *Media Manipulation and Disinformation Online*, New York, NY: Data & Society Research Institute.

Massanari, A. (2017), '#Gamergate and the fappening: How Reddit's algorithm, governance, and culture support toxic technocultures', *New Media & Society*, **19**(3), 329–346.

McCauley, Clark and Sophia Moskalenko (2011), *Friction: How Radicalization Happens to Them and Us*, Oxford, UK: Oxford University Press.

Meddaugh, P.M. and J. Kay (2009), '"Hate speech" or "reasonable racism"? The other in Stormfront', *Journal of Mass Media Ethics*, **24**(4), 251–268.

Merrill, S. and M. Åkerlund (2018), 'Standing up for Sweden? The racist discourses, architectures and affordances of an anti-immigration Facebook group', *Journal of Computer-Mediated Communication*, **23**(6), 332–353.

Mudde, Cas (2019), *The Far Right Today*, Oxford: Polity.

Nagle, Angela (2017), *Kill All Normies: Online Culture Wars from 4Chan and Tumblr to Trump and the Alt-Right*, Alresford: John Hunt Publishing.

O'Hagan, E. (2018), 'The far right is rising, and Britain is dangerously complacent about it', *The Guardian*, 7 May, accessed 20 January 2020 at https://www.theguardian.com/commentisfree/2018/may/07/far-right-britain-freedom-of-speech

Oppenheim, M. (2017), 'Defend Europe: Far-right ship stopping refugees ends its mission after a series of setbacks', *The Independent*, 21 August, accessed 20 January 2020 at https://www.independent.co.uk/news/world/europe/defend-europe-far-right-ship-stop-refugees-mediterranean-end-mission-c-star-setbacks-migrant-boats-a7904466.html

Papacharissi, Zizi (2015), *Affective Publics: Sentiment, Technology, and Politics*, New York: Oxford University Press.

Phillips, Whitney (2015), *This Is Why We Can't Have Nice Things: Mapping the Relationship Between Online Trolling and Mainstream Culture*, Cambridge, MA: MIT Press.

Philo, Greg, Emma Briant, and Pauline Donald (2013), *Bad News for Refugees*, London: Pluto Press.

Poole, E., E. Giraud, and E. de Quincey (2019), 'Contesting #StopIslam: The dynamics of a counter-narrative against right-wing populism', *Open Library of Humanities*, **5**(1), DOI: http://doi.org/10.16995/olh.406.

Ramos, C.G. (2020), 'Change without transformation: Social policy reforms in the Philippines under Duterte', *Development and Change*, **51**(2), 485–505.

Rosenberg, E. (2011), 'CNN, MSNBC ran more birther coverage than Fox News', *The Atlantic*, 28 April, accessed 20 January 2020 at https://www.theatlantic.com/politics/archive/2011/04/chart-cnn-msnbc-ran-more-birther-coverage-fox-news/350112/.

Saad-Filho, A. and M. Boffo (2020), 'The corruption of democracy: Corruption scandals, class alliances, and political authoritarianism in Brazil', *Geoforum*. https://doi.org/10.1016/j.geoforum.2020.02.003.

Schleicher, A. (2019), 'Distinguishing fact from fiction in the modern age', in Vasu Norman, Jayakumar Shashi, and Ang Benjamin (eds), *DRUMS: Distortions, Rumours, Untruths, Misinformation, and Smears*, Singapore: World Scientific, pp. 159–170.

Siapera, E., M. Boudourides, S. Lenis, and J. Suiter (2018), 'Refugees and network publics on Twitter: Networked framing, affect, and capture', *Social Media + Society*, **4**(1), 1–21.

Simi, Pete and Robert Futrell (2015), *American Swastika: Inside the White Power Movement's Hidden Spaces of Hate*, Lanham, MA: Rowman & Littlefield.

Stier, S., L. Posch, A. Bleier, and M. Strohmaier (2017), 'When populists become popular: comparing Facebook use by the right-wing movement Pegida and German political parties', *Information, Communication & Society*, **20**(9), 1365–1388.

Suiter, J. and R. Fletcher (2020), 'Polarisation and Inequality: Key drivers of declining trust in news', *European Journal of Communication*. In Press.

Törnberg, A. and M. Wahlström (2018), 'Unveiling the radical right online: Exploring framing and identity in an online anti-immigrant discussion group', *Sociologisk Forskning*, **55**(2–3), 267–292.

Wardle, Claire and Hossein Derakhshan (2017), *Information Disorder: Toward an Interdisciplinary Framework for Research and Policy Making*, Strasbourg: Council of Europe Report DGI(2017)09.

Wirz, D.S., M. Wettstein, A. Schulz, P. Müller, C. Schemer, N. Ernst, F. Esser, and W. Wirth (2018), 'The effects of right-wing populist communication on emotions and cognitions toward immigrants', *The International Journal of Press/Politics*, **23**(4), 496–516.

Yadamsuren, B. and J. Heinström (2011), 'Emotional reactions to incidental exposure to online news', *Information Research*, **16**(3), accessed 17 September 2021 at http://informationr.net/ir/16-3/paper486 .html.

Zannettou, S. et al. (2017), 'The Web centipede: Understanding how web communities influence each other through the lens of mainstream and alternative news sources', *Proceedings of the 2017 Internet Measurement Conference*, New York, NY: ACM, pp. 405–417.

Zannettou, S., B. Bradlyn, E. De Cristofaro, H. Kwak, M. Sirivianos, G. Stringini, and J. Blackburn (2018), 'What is gab: A bastion of free speech or an alt-right echo chamber?', in *Companion Proceedings of the Web Conference 2018*, New York, NY: ACM, pp. 1007–1014.

23. The role of networked publics in immigration debates

Markus Ojala

INTRODUCTION

The academic interest towards the impact of network technologies on the public debate has grown substantially over the past decade as digital networks and social media platforms have become increasingly central for how news and information is produced, consumed and shared. As with other phenomena involving technology, there are both optimistic expectations and sceptical warnings concerning the effects of online media. On the one hand, the digitalisation of communication enhances the opportunities for public participation in democratic processes. Through online platforms, people have new means to come together and influence the generation of public agendas both locally and nationally, and even across national boundaries. On the other, recent controversies across the world have demonstrated that digital networks can be used as efficient tools for spreading disinformation and mobilising hate, for example, against minorities.

This chapter focuses on the formation of 'networked publics' around the issue of immigration and the ways in which digital technologies are being used to shape public debates over immigration. Networked publics consist of groups discussing common concerns using digital media and online platforms. Scholarship has not only explored the nature of online communication but also sought to understand the ways in which networked publics can intervene in political processes and the generation of the public agenda. It appears that, in recent years, networked publics have become increasingly visible and influential players also in immigration policy. The tension between the democratic promises and worrying realities of online communication highlights the importance of examining the complex nature of networked publics and their role in shaping societal responses to the major trends of the twenty-first century, including immigration.

The first two sections of this chapter introduce the public and 'networked' public as theoretical concepts and research topics. Having established these conceptual grounds, the chapter turns to immigration as a public issue. The third section discusses why immigration has turned into a highly charged topic in many liberal democracies over the past two decades. The fourth section examines the roles of networked publics in contemporary immigration debates. It should be noted at the outset that the focus will be on the national public debates of countries of net *immigration* that reside mostly in the global north. The broader migration debates, whether in the countries that experience extensive emigration or at international forums, remain beyond the scope of this chapter.

PUBLICS, ISSUES AND CONFLICTS

As a term, the public has always been difficult to pin down and it continues to carry multiple meanings (Price 1992; Warner 2002). In contemporary research on political communication, we can roughly distinguish between two approaches. One of them, typically related to media and cultural studies traditions, derives its perception of the public from the notion of (media) audience (Varnelis 2008; boyd 2011): the public consists of those who read, watch and listen to media contents and are thus addressees of communication. The term public is used here to emphasise that, rather than being passive receivers, audiences actively interpret and remake the messages and further distribute them in their own communities and networks. In today's digital media environment, people shape the circulation of political communication, alternating between the roles of the receiver and the disseminator of opinions and information (Castells 2009).

The second approach builds on the liberal tradition of democratic theory and associates the public with civic practices (e.g., Habermas 1996; Dahlgren 2005). This tradition views the public as a loosely defined and unorganised aggregation of citizens who engage in argumentation over common affairs, critically scrutinising the conduct of state authorities, denouncing abuses of power and establishing collective views on the pressing matters of the day. It holds that citizens form a public when they participate in various kinds of civic activities and discuss current affairs in public spaces. The term public is thus used in a manner that somewhat diverges from the former perspective, which equates the public with audiences or the people as such. Against this all-encompassing notion, the liberal-democratic tradition stresses that people as users and consumers of media are not members of any public by default. They only become such members through active participation in the public debate.

The liberal approach to the public builds on an idea of democracy as a set of cultural practices through which people participate in the public debate and the political system becomes responsive to people's concerns. In other words, this perspective emphasises that the operation of democracy entails not only that the public expresses its consent to government legislation, ultimately granting legitimacy to the political system as a whole, but that the public also participates in political processes, actively informing decision-making institutions about its preferences. In its civic activities, the public embodies, reproduces and strengthens a democratic culture (O'Mahony 2013). Accordingly, when analysing various forms of political communication, this perspective focuses on the democratic qualities of public participation and its impacts on political decision-making. This is also why the notion of the citizen is central when talking about publics. It is mainly on the merit of being citizens with the corresponding rights and duties that people can participate in democratic processes. This obviously does not deny the possibility that non-citizens and even non-human agents such as online bots can, at times, have a notable impact on public debates.

Contemporary public spheres are commonly understood to be fragmented and differentiated, comprising a multitude of 'publics' that are overlapping and without fixed social and temporal boundaries (Habermas 1996; Friedland et al. 2006). Thematic publics form around specific issues that their participants perceive important enough to merit discussion. They are matters that cannot be dealt with privately and call for collective responses. Indeed, issues and publics cannot be separated; we would hardly recognise any shared problems in society unless there was a public, an aggregation of people engaged in discussing and promoting a given issue as a matter of collective discretion. Publics aim to bring the issue to the attention of decision-makers, forcing them to respond. Insofar as they are successful, the decision-makers themselves are drawn into the public.

Many critical scholars of the public sphere emphasise that, in addition to being divided, contemporary public spheres are highly conflictual. Different publics and 'counterpublics'— referring to publics formed by subordinated groups, representing marginalised views and positioning themselves in confrontation with dominant groups—regularly clash with each other in ideological and political contestation over the definition of public problems (Fraser 1992; Warner 2002; Fenton 2016). Publics also define themselves in relation to their political and ideological opponents. Instead of speaking about 'issues', the critical perspective thus emphasises the importance of conflicts in mobilising people and provoking them to public action.

As a result, the communication of publics has both internal and external dimensions. Internally, it constructs the sense of 'us versus them'—that is, collective identity and a sense of shared interests. Externally, the public aims to communicate to those who do not (yet) belong to it, including decision-making authorities, adversarial groups and the general audience. In contemporary heavily mediated societies, the public agenda of which is shaped in the interplay between major news media organisations and social media platforms, a public can bring an issue into the awareness of large numbers of people only when it is influential enough to penetrate through the gatekeeping of professional media organisations or can successfully harness the social media's operational logic.

NETWORKED PUBLICS AND POLITICAL CONTESTATION

The great curiosity—and concern—of late over the impact of digital information and communication technologies on various aspects of social life have also been reflected in studies of the public sphere, paving the way for analyses of how digital networks shape the dynamics of public debate (e.g., Castells 2009; Chadwick 2013; Fenton 2016; Iosifidis and Wheeler 2016). While the notion of the 'networked public' is not always explicitly present in this literature, one of the main issues addressed is how digital networks alter the public operation of actors and the formation of publicly acting collectives. This is exactly what the term 'networked public' aims to capture when exploring the impact of information and communication technologies on democratic politics and the public sphere. However, when talking about the networked *public* in this context, we should keep in mind that the notion carries certain liberal-democratic undertones as discussed in the previous section. That means it should be understood as having a different meaning from related terms such as online users, 'networked individuals' (Castells 2009) or online communication in general.

The notion of networked publics implies not only that publics increasingly take shape and act in and through digital networks but also that these communication technologies affect the nature and operation of publics in various ways (Varnelis 2008; boyd 2011). The generation of publics and their activities—exchanging views, organising and participating in public events, bringing in new members, reaching out to other people, and so on—is greatly enhanced by the availability of digital tools, which make it easy to record, archive, replicate and spread messages. Network technologies facilitate communication that is less bounded by geography and more inclusive of diverse views, more dialogical and less hierarchical in nature (Papacharissi 2004). However, while the digital platforms and associated technologies provide certain structural affordances and limitations, they (alone) do not determine the democratic quality of societal discussion, which ultimately depends on intentional action, values, interests and political purposes.

The participation in networked publics tends to be highly individualistic; they are formed by 'connected individuals' (Maireder and Schwarzenegger 2012) for whom it is easy to engage and disengage from the conversation. Being unorganised and largely spontaneous formations, networked publics lack clear leaders and authorities. The research of online participation often highlights the nature of networked publics as spaces for affective expression, the construction of identities and the formulation of common lines of action (e.g., boyd 2011; Papacharissi 2015).

While most online activities are oriented to diversion, entertainment and social networking, they always contain the possibility of becoming political (Dahlgren 2005; Fenton 2016; Roslyng and Blaagaard 2018). Any interaction among groups can develop into a formulation of public problems that require societal and political responses. When becoming political, networked publics often draw in various types of actors, including civil society actors, politicians, journalists, celebrities and experts (Wang and Chu 2019). The news media have an important role in the creation of networked publics who use professional journalism about current affairs as a fuel for discussion (Maireder and Schlögl 2014).

Despite the new capacities for transnational communication brought about by digital technologies, networked publics continue to be bounded by political systems as well as by language. Publics strive for political and cultural influence—that is, to shape the ways people live together—and that tends to lead them to orient towards the nation-state and its institutions. Accordingly, national public spheres form the principal field of operation for networked publics, and participation in the public debate typically requires the use of the official or dominant language. This is not to deny that many of today's central political issues, including climate change, immigration and digital surveillance, generate international discourses and forms of publicity. Similarly, some of the major recent protest movements and popular uprisings, including the Arab Spring, the Occupy movement and anti-austerity protests in Europe, were notable for their use of digital networks to mobilise publics across borders (Castells 2015; Papacharissi 2015; Iosifidis and Wheeler 2016). But even in those cases, most of the public mobilisation focused on effecting political change at the national level—with the partial exception of the anti-austerity movement that also targeted European Union institutions.

In online domains, people typically seek spaces where they can exchange ideas with people who share their attitudes, views and interests. As a result, networked publics often comprise primarily like-minded people. They tend to engage in advocacy rather than critical and self-reflective discussion, and their modes of operation resemble those of social movements or activists (Dahlgren 2005; Bennett and Segerberg 2013). Minorities, too, establish their own counterpublics with the help of digital networks (Renninger 2015; Kuo 2018). In this way, the internet further drives the fragmentation of the public sphere, leading to the diversification of political identities. Different views do not necessarily meet, at least not in a mutually respectful manner; divides between groups deepen; and the public debate becomes increasingly conflictual (Sunstein 2017; Dutton 2018).

THE POLITICAL INFLUENCE OF NETWORKED PUBLICS

Networked publics are often understood to mark citizens' increased salience and influence in the public sphere. Yet it is less than clear what this salience means in terms of the political debate more generally and how it influences political processes. One of the perceived impacts is the capacity of networked publics to shape the public agenda by bringing in underrepresented viewpoints. The news media, which have been traditionally oriented to institutional and elite sources, nowadays regularly tap into online conversations for new stories and reactions, enabling publics to bring their concerns into the broader public view (Chadwick 2013; Maireder and Schlögl 2014). In this reading, networked publics constitute an influential new actor in a media environment characterised by the interplay between online publics and professional news organisations.

A somewhat bolder argument is to see networked publics having a direct impact on political decision-making and government. According to this view, the new communication technologies enable people to increasingly monitor the conduct of powerful institutions and individual decision-makers, expose their wrongdoings, mobilise criticism against them and demand direct responses from them (Dutton 2009; Keane 2009). When managing to disrupt the public agenda, typically through scandals and other controversies, these activities can have significant outcomes, prompting authorities not only to publicly justify their decisions but to reassess their conduct whether in fear of losing public legitimacy or because feeling morally compelled to do so (Bovens 2007; Moore 2014).

Generally speaking, networked publics seem to have strengthened the importance of *publicity* in the everyday politics and government. When making decisions and communicating them, policymakers and officials must increasingly take into account the potential reactions of networked publics and plan ahead the public representation of policies (Louw 2010). However, the rise of networked publics does not signal a decisive shift of political power from established elites to citizens. In online spaces, attention is a scarce resource and political influence continues to be exerted by the few, not the many (Hindman 2009). Due to their large follower bases, elite opinion leaders are in a privileged position to get people involved in an issue, mobilise publics and direct their attention to new developments (boyd 2011).

Politicians, corporations and organised civil society actors have quickly learned to see new media technologies as tools that they can use to exert their power and influence in the public arena (Castells 2009; Chadwick 2013). They aim to harness the power of networked publics to achieve their strategic purposes. Moreover, most of today's popular online spaces for political communication are controlled by transnational corporations and engineered for profit creation (Allmer 2015). Established interests and profit-seeking motives thus continue to have a major impact on the public agenda even in online environments, shaping the formation and evolution of networked publics as well as their broader political impact.

We must also bear in mind that the public agenda is highly volatile and tends to be focused only on a few issues at a time. Therefore, the rising importance of publicity in policymaking should not be interpreted as a general tendency. Even as the role of the public debate in daily politics grows in some policy areas, most of the policymaking and administration continues to be carried out outside the glare of publicity. In recent decades, issues related to immigration have frequently become subject to public controversies, potentially contributing to the impact of networked publics on immigration policy.

IMMIGRATION AS A PUBLIC ISSUE

As discussed previously, the contemporary public spheres of liberal democracies tend to be divided into contesting and mutually hostile publics. Immigration is one of the most central issues that both mobilises publics and brings them into a conflictual relationship with one another. While social media are often accused of creating such a hostile public environment around immigration, it results from much more fundamental transformations in western societies and the global political economy (Žižek 2017). The number of people forced to migrate due either to violent conflicts or to the destruction of livelihoods has risen dramatically at a time when societies in the global north have suffered from economic crises, unemployment, precarisation, rising inequality and disinvestment in public services.

These developments can only be understood against the backdrop of newly emergent political and ideological cleavages in liberal democracies (Crouch 2011; Mouffe 2013). In the 1990s, the traditional confrontation between the social-liberal left and the market-liberal right was undermined by economic globalisation and the demise of socialism. The neoliberal consensus over market-led globalisation narrowed the political divergence of established parties, emptying the public debate over economic policy of any real alternatives and diminishing the capacity of the dominant political forces to mobilise publics. Concurrently, political conflicts began to move to the terrain of culture, lifestyle and social group relations. That gave a new lease of life to socially conservative groups whose approach to economic and social policies was often ambiguous but who articulated their political project in terms of defending the nation and national sovereignty, protecting traditional professions and ways of life—and restricting immigration (Eatwell and Goodwin 2018). In recent years, the national-conservative populist parties have been able to attract a sizable support base and make themselves visible in public debates in many if not most of the western countries. In doing so, they have presented immigration and the related issues of national values and lifestyles as the definitive questions for citizens to mobilise around publicly and politically.

This brief sketch may help explain why immigration has become a national question that recurrently shapes the agenda of parliaments and mainstream news media in many western countries. As such, immigration is a complex and multifaceted topic that includes a wide variety of issues (Gans et al. 2012; Swain 2018). Politicians, advocacy groups and citizens argue not only over such political and administrative considerations as naturalisation processes and immigrants' rights, but also over the costs of immigration to taxpayers, the impact of unskilled immigrants on the labour markets and other economic issues. Moreover, various social and cultural issues, including the social and cultural integration of immigrants to host societies, as well as multiculturalism and racism are often hotly debated themes related to immigration.

As the range of these issues of contestation suggests, publics tend to be highly divided on immigration-related matters. Indeed, even though immigration is often negatively portrayed and covered in mainstream news media (Allen and Blinder 2013; Georgiou and Zaborowski 2016), studies on public opinion indicate that people's perceptions are rather complex and divided on immigration (Freeman et al. 2013; Gilligan 2015). Perceptions alter depending on which group of immigrants the person is thinking about, and on the particular detail of immigration policy. Moreover, rather than questioning immigration itself, people may often have negative views of immigration policy or the way immigration is handled by the government. The political contestation over the framing of individual issues between those favouring liberal immigration policies and those striving for new restrictions on immigration is often key in swaying the public opinion (Sen and Mamdouh 2008).

Immigration is a transnational phenomenon by definition and as such an issue one might expect to be primarily discussed in various kinds of 'transnational public spheres' (Fraser 2007). Yet even within supranational polities such as the European Union, it has proven to be practically impossible to mobilise cross-border debates over immigration outside the very limited public spaces involving governments, international bureaucrats, experts and civil society organisations. Most people and the large majority of media outlets continue to orientate themselves around national political institutions. Often the language of communication also confines the immigration debate to the national level.

Language barriers also partly explain why immigrants themselves tend to be relegated to a marginal position within national public spheres. Immigrants are typically the objects of public claims made by others but seldom claims-makers themselves. Moreover, being a diverse and fragmented group, immigrants are not well placed to establish a public of their own around the issue of immigration. This is not to say that immigrants fail to participate in practices and forums of public communication. Ethnic minorities and diasporas, for instance, certainly look for ways to engage, and they may occasionally be highly involved in the immigration debates of their receiving countries. On the whole, however, immigration is mostly debated and contested in national public spheres by other people than immigrants.

IMMIGRATION AND NETWORKED PUBLICS

One of the main concerns in recent scholarship on immigration has been the rise of anti-immigrant parties and movements and the simultaneous prevalence of anti-immigrant voices and sentiments in public discourse. Many studies have focused on the online mobilisation of anti-immigrant users and its impact on the public debate and political processes. For instance, an analysis of US immigration debates on Twitter found that anti-immigrant groups took advantage of the social networking platform for mobilisation and were able to steer the overall debate in their favour (Flores 2017). More generally, digital networks have been found to make it easier for anti-immigrant activists and individuals to spread hostile messages against particular immigrant or minority groups (Ekman 2015; Kaján 2017; Merrill and Åkerlund 2018).

Such dissemination of anti-immigrant perspectives on social media may not alter users' views on immigration but can change their behaviour, inciting new people to join these public activities. Concurrently, incivility and hate speech often dominate online debates on immigration (Musolff 2015; Pöyhtäri et al. 2019; Santana 2015). Conversations tend to be highly confrontational, and users often denounce politicians and other public figures with whom they disagree. Online commentary is typically provoked by news media reports, but users often direct their criticism at the newsmakers themselves in addition to those expressing pro-immigrant views and positions (Bartlett and Norrie 2015; Kaján 2017).

Roger Eatwell and Matthew Goodwin (2018) have argued that anti-immigrant parties have tapped into a widespread discontent among citizens with the elitist nature of liberal-democratic politics and the increasing economic inequality, as well as into their concerns about the scale and pace of ethnic change. Demands for more restrictive immigration policies are often tied to a broader ideology of 'national populism' that does not simply address immigration as an economic burden and a threat to cultural values and ways of life but also denounces liberal institutions for their failure to represent the voices and interests of the less-educated and more conservative segments of the population. This overlapping of negative attitudes on immigration with anti-elitism can also be witnessed in online discussions about immigration, where anti-immigrant views are often linked to grievances about unemployment and the erosion of the welfare state (Gibson et al. 2018). Political and media attention to the plight of refugees is juxtaposed with the absence of similar alarm about crumbling social services and domestic inequalities (Ojala, Kaasik-Krogerus and Pantti 2019). Such claims provoke a sense of being a second-class citizen and paves the way for denouncing elites as those who have abandoned their own people.

The focus of much of the recent research on the spreading of anti-immigrant views online ought not to lead us to disregard the existence of strong liberal and pro-immigrant publics that also make efficient use of networked spaces (e.g., Siapera 2004; Burke and Goodman 2012; Grover et al. 2019; Macgilchrist and Böhmig 2012). Their discourses of solidarity, human rights and resistance to xenophobia and racism have at times had as significant an impact on the public debate as the public's demanding more restrictive immigration policies. This may be the case particularly during periods such as the so-called 'refugee crisis' in Europe in 2015, when the rapidly increasing number of migrant deaths in the Mediterranean and the surge of asylum seekers created a sense of great urgency among European policymakers and civil society. Digital networks were used by refugee organisations, activists and volunteers not only to organise aid to asylum seekers. With the help of opinion leaders, including left-wing politicians, human rights organisations and journalists, they also mobilised the public in demonstrations of solidarity and in defence of asylum seekers' rights, as well as to gain visibility for pro-immigrant views in the public debate to counter anti-immigrant discourses and disinformation (Holmes and Castañeda 2016; Siapera et al. 2018; Ojala, Pantti and Laaksonen 2019).

To be sure, we can speak of 'the' public debate on immigration only on a highly abstract level as online debates on immigration are not conducted similarly across digital networks. Studies of online debates during the so-called refugee crisis in Europe indicate that popular discussion forums and users' comment sections of online newspapers are often dominated by anti-immigrant views, whereas the blogosphere and platforms like Twitter and Facebook—highly fragmented spaces in themselves—tend to include a more even ratio of liberal-humanitarian and anti-immigrant discourses (Siapera et al. 2018; Ojala, Kaasik-Krogerus and Pantti 2019; Pöyhtäri et al. 2019). This suggests that the contribution of the internet to the multiplication of spaces of debate and to the fragmentation of the public sphere (Dahlgren 2005; Fenton 2016) can also be evidenced in the immigration debate.

In recent years, a lot of concern has been raised over the impact of disinformation, fake news and foreign information operations on national political debates (Bennett and Livingston 2018; Spohr 2017). While the spreading of lies, distortions and propaganda is by no means a new phenomenon in political communication, the networked media environment and targeted political messaging, enabled by the pervasive collection of user data, provide organised actors with new means to manipulate the public opinion. In addition, citizen actors themselves actively post online hearsay, rumours and speculation, which can rapidly spread with the help of various alternative and counter media sites, blogs and video blogs. The immigration debate is obviously vulnerable to these phenomena, and the dissemination of rumours and disinformation tends to become particularly extensive during emergencies or exceptional circumstances. For instance, Pöyhtäri and colleagues (2019) found that Finnish online users often spread negative rumours and false information about local asylum seekers during the 2015 'migration crisis'. A few anti-immigrant online discussion forums, blogs, video blogs and counter-media sites were effective in drawing nationwide attention to these often fabricated local 'news'. These events illustrate how individual users and minor media initiatives, through both intentional and unintentional spread of false information, may shape the public debate on immigration, often entirely in the absence of any organised or external information operation. Whether originating from foreign and organised sources or domestic and individual activities, fake news and disinformation obviously distort people's knowledge environment (Dahlgren 2018). As such, they may play a part in debilitating the public sphere as an arena of societal debate, for instance, by accelerating political polarisation.

There are indeed some indications that online platforms exacerbate the polarisation of the immigration debate into two opposing camps. As the consumption of news and discussion on current affairs increasingly takes place on social networking sites, the very networks that users form by being followed or following others may consist of relatively like-minded people who mostly confirm the existing opinions and beliefs they hold about immigration. Aside from users' voluntary networking choices, the recommendation algorithms of the platforms may enhance this effect, trapping people within 'echo chambers' or 'filter bubbles' that expose them to only one-sided views (Pariser 2012; Spohr 2017); although some studies suggest that the problem has been exaggerated and that most users in fact avoid echo chambers in online networking and news consumption (e.g., Dubois and Blank 2018; Bruns 2019). Nevertheless, even if most people continue to be exposed to contradictory information and views, the public debate on immigration appears to be driven by highly partisan actors who rarely seek the middle ground or reach out to those they oppose. As a result, even though online debates in the multitude of forums and platforms may feature a greater number of themes than what circulates in the mainstream news media (Nerghes and Lee 2019; Pöyhtäri et al. 2019), the diversity of viewpoints does not necessarily increase (Siapera et al. 2018).

Neither do the online platforms seem to significantly reallocate power and influence in the immigration debate. Even though the number of participants engaged in public argumentation over immigration has been greatly expanded by the digital means of communication, public attention continues to be highly unequally distributed. Instead of being levelling devices, social media platforms tend to concentrate agenda setting power and opinion leadership to very few actors who more often than not represent established elites from the domains of politics, media and the organised civil society (Ojala, Pantti and Laaksonen 2019; Siapera et al. 2018). On occasions, a non-elite individual may certainly make an impact on the immigration debate, for instance, by leaking information from within the immigration system or exposing malpractices of asylum officials; yet they usually need mainstream media professionals or public figures as message sponsors to gain broader attention for their cause.

What, then, can we say about the political outcomes that may result from the engagement of networked publics in immigration debates? Thus far, there have been very few studies that have addressed the relation between online debates and immigration policies. Nevertheless, one indication of the significance of publics in the politics of immigration can be found in the European Union, whose member states differ markedly in terms of the relative openness or restrictedness of their immigration policies. Marc Howard and Sara Goodman (2018) have argued that these differences cannot be explained based on the countries' diverging demography, economic indicators or other structural factors. The relevant factor explaining the differences appears instead to be the success of anti-immigrant groups in making immigration a salient issue in national public debates. In practically all EU countries, more restrictive immigration policies have been historically associated with the mobilisation of an anti-immigrant sentiment in the public arena by far-right groups or parties. Conversely, those member states that carried out significant liberalisation of immigration policies in the 1980s and 1990s were connected by the absence of a visible anti-immigrant party or movement. At least in liberal democracies, immigration policy seems to be partly responsive to shifts in the public debate.

Another indication of the political impact of networked publics can be discerned at the level of politicians' and officials' communicative practices. Ojala, Pantti and Laaksonen (2019) studied the public interactions between Finnish immigration officials and citizens during the so-called refugee crisis in 2015 and 2016. They discovered that the volume of online challenges and critiques directed at the government's immigration agency exploded following the surge of asylum seekers and were expressed in various online forums. In the early phase of the 'crisis', most of the criticism came from conservative anti-immigrant publics that protested against the opening of new refugee reception centres and suspected the immigration officials of being too lenient in their asylum decisions. However, after the acceptance rate of asylum applications dropped significantly in the spring of 2016, the agency faced mounting critiques from liberal pro-immigrant publics, as well as highly critical press coverage, which continued till the end of that year.

Moreover, Ojala and colleagues (ibid.) found that, throughout the period, the immigration agency repeatedly addressed its critics both on its Twitter account and through the national news media, attempting to explain and justify its actions and policies. While these efforts were hardly successful in quelling public discontent, this interaction created an unprecedented level of responsiveness between immigration officials and its publics while making the operation of the immigration system as a whole more transparent to large segments of the population. The heightened attention on and criticism of the immigration agency even compelled the government to conduct an external review of Finnish asylum processes. The authors concluded that in this case networked publics operated as 'agents of accountability', capable of forcing authorities to provide explanations and justifications of their use of public power. The publics thus contributed to what John Keane (2009) has dubbed 'monitory democracy', enforcing the social accountability of public authorities.

CONCLUSION

Much of the existing research on online debates over immigration focuses on the practices and discursive strategies of participants in advancing either fear and anger towards immigrants or feelings of empathy and solidarity. One of the main challenges is to look beyond such practices of shaping the debate and addressing the political impact of networked publics. As Chris Gilligan (2015) points out, it is not at all evident that governments always respond to the pressure of public opinion in immigration matters; instead, it often appears that political leaders themselves are influential shapers of public opinion. Moreover, the publics hold diverse attitudes and opinions regarding immigration, and there are various actors, institutions and contextual factors that mediate the impact of public opinion on the administration, including political parties, advocacy groups, media organisations, legal frameworks and immigration flows. There is thus much to be discovered in terms of 'the conditions under which the public influence policy-makers, and vice versa, as well as identifying the mechanisms through which this influence is exerted' (ibid., p. 1385).

Being in full agreement with that research agenda, I conclude by reiterating some useful starting points for future studies on the basis of the research reviewed in this chapter. First, owing to the long-term economic and political transformations, national populism is turning into a persistent phenomenon in the public life of liberal democracies. With their growing impact on public life, national populists are increasingly capable of defining the issues around which societal, political and cultural contestations are fought. Therefore, even as future migration trends are difficult to predict, immigration is unlikely to wither away as a source of political conflict any time soon in the global north. Both national-conservative and liberal publics have become increasingly confrontational in recent years in their attitudes and activities regarding the political system and immigration administration. The publics aim to shape the public debate and manipulate public opinion, and they engage in the scrutiny and critique of officials and policymakers.

Second, digital networks and social media platforms certainly play a role in the rise of contesting publics around immigration in the sense that both anti-immigrant parties and movements and liberal groups have taken advantage of the new media technologies and tools to spread their message and rally people around their cause. The kind of public activities that were traditionally the domain of activist groups and social movements are now available to technology-savvy individuals. Without any organisational resources, they can mine online databases for information and make it known to others, publish sensitive information, launch online addresses and target political leaders with critiques. When they successfully interact with the professional news media, networked publics can set into motion public controversies, derail the political system and force authorities to respond. Moreover, far-right activists, in particular, have adopted practices and techniques such as disinformation, fake news, trolling, online shaming and doxing to disrupt the public debate and harass those they consider as political opponents. They can make the life of politicians, officials and researchers increasingly difficult by instigating other users to attack them online and harass them offline.

Finally, while the capacity of networked publics to shape immigration debates, monitor the conduct of immigration authorities and even influence certain policies seems to be growing due to the increased salience of immigration in the public agenda, we should be cautious about jumping to any assumptions about their contribution to advancing the democratisation of policymaking or generating a more inclusive public debate on immigration. Even with the emergence of new forms of online communication and social media networking, established actors constantly find ways to exert their influence by manipulating the publics that form on these platforms. The elites largely maintain their grip on the public agenda to preserve existing power hierarchies.

Moreover, it is evident that some of the activities of the networked publics have deteriorated the public sphere and undermined the conditions for a democratic debate. The problem lies not only in anti-immigrant publics who spread hateful and racist discourses and aim to silence humanitarian voices; liberal pro-immigrant publics also often engage in derogatory and exclusionary rhetoric about their opponents (Burke and Goodman 2012; Gibson et al. 2018). The resulting polarisation of the immigration debate into highly antagonistic camps may be partly driven by the logics of digital networks. However, while the private online media platforms have certainly benefited from and contributed to the increasingly polarised political environment, the roots of these problems lie in the practices of the networked publics themselves and are not determined by the technologies used for their intra-public and inter-public communication.

REFERENCES

Allen, W. and S. Blinder (2013), 'Migration in the news: portrayals of immigrants, migrants, asylum seekers and refugees in national British newspapers, 2010–2012', Migration Observatory report, Oxford: University of Oxford.

Allmer, T. (2015), *Critical Theory and Social Media*, London; New York: Routledge.

Bartlett, J. and R. Norrie (2015), *Immigration on Twitter: Understanding Public Attitudes Online*, London: Demos.

Bennett, W. L. and A. Segerberg (2013), *The Logic of Connective Action: Digital Media and the Personalization of Contentious Politics*, Cambridge: Cambridge University Press.

Bennett, W. L. and S. Livingston (2018), 'The disinformation order: disruptive communication and the decline of democratic institutions', *European Journal of Communication*, **33** (2), 122–39.

Bovens, M. (2007), 'Public accountability', in E. Ferlie, L. E. Lynn, and C. Pollitt (eds), *The Oxford Handbook of Public Management*, New York: Oxford University Press, pp. 182–208.

boyd, d. (2011), 'Social network sites as networked publics', in Z. Papacharissi (ed.), *A Networked Self: Identity, Community, and Culture on Social Network Sites*, New York: Routledge, pp. 39–58.

Bruns, A. (2019), *Are Filter Bubbles Real?*, Cambridge: Polity Press.

Burke, S. and S. Goodman (2012), '"Bring back Hitler's gas chambers": asylum seeking, Nazis and Facebook – a discursive analysis', *Discourse & Society*, **23** (1), 19–33.

Castells, M. (2009), *Communication Power*, Oxford: Oxford University Press.

Castells, M. (2015), *Networks of Outrage and Hope: Social Movements in the Internet Age*, 2nd edn, Cambridge: Polity.

Chadwick, A. (2013), *The Hybrid Media System*, Oxford: Oxford University Press.

Crouch, C. (2011), *The Strange Non-Death of Neoliberalism*, Cambridge: Polity Press.

Dahlgren, P. (2005), 'The internet, public spheres, and political communication: dispersion and deliberation', *Political Communication*, **22** (2), 147–62.

Dahlgren, P. (2018), 'Media, knowledge and trust: The deepening epistemic crisis of democracy', *Javnost*, **25** (1–2), 20–27.

Dubois, E. and G. Blank (2018), 'The echo chamber is overstated: the moderating effect of political interest and diverse media', *Information, Communication & Society*, **21** (5), 729–45.

Dutton, W. H. (2009), 'The Fifth Estate emerging through the network of networks', *Prometheus*, **27** (1), 1–15.

Dutton, W. H. (2018), 'Networked publics: multi-disciplinary perspectives on big policy issues', *Internet Policy Review*, **7** (2). https://doi.org/10.14763/2018.2.795.

Eatwell, R. and M. Goodwin (2018), *National Populism: The Revolt against Liberal Democracy*, London: Penguin.

Ekman, M. (2015), 'Online Islamophobia and the politics of fear: manufacturing the green scare', *Ethnic and Racial Studies*, **38** (11), 1986–2002.

Fenton, N. (2016), *Digital, Political, Radical*, Cambridge: Polity Press.

Flores, R. D. (2017), 'Do anti-immigrant laws shape public sentiment? A study of Arizona's SB 1070 using Twitter data', *American Journal of Sociology*, **123** (2), 333–84.

Fraser, N. (1992), 'Rethinking the public sphere: a contribution to the critique of actually existing democracy', in C. Calhoun (ed.), *Habermas and the Public Sphere*, Cambridge: MIT Press, pp. 109–142.

Fraser, Nancy (2007), 'Transnationalizing the public sphere: on the legitimacy and efficacy of public opinion in a post-Westphalian world', *Theory, Culture & Society*, **24** (4), 7–30.

Freeman, G. P., R. Hansen and D. L. Leal (eds) (2013), *Immigration and Public Opinion in Liberal Democracies*. London: Routledge.

Friedland, L. A., T. B. Hove and H. Rojas (2006), 'The networked public sphere', *Javnost*, **13** (4), 5–26.

Gans, J., E. M. Replogle and D. J. Tichenor (eds) (2012), *Debates on U.S. Immigration*, Thousand Oaks, CA: Sage.

Georgiou, M. and R. Zaborowski (2016), 'Media coverage of the "refugee crisis": a cross-European perspective', Council of Europe Report No. DG1(2017)03, Brussels: Council of Europe.

Gibson, S., M. Crossland and J. Hamilton (2018), 'Social citizenship and immigration: employment, welfare, and effortfulness in online discourse concerning migration to the United Kingdom', *Qualitative Psychology*, **5** (1), 99–116.

Gilligan, C. (2015), 'The public and the politics of immigration controls', *Journal of Ethnic and Migration Studies*, **41** (9), 1373–90.

Grover, T., E. Bayraktaroglu, G. Mark and E. H. R. Rho (2019), 'Moral and affective differences in U.S. immigration policy debate on Twitter', *Computer Supported Cooperative Work*, **28** (3), 317–55.

Habermas, J. (1996), *Between Facts and Norms*, Cambridge: MIT Press.

Hindman, M. (2009), *The Myth of Digital Democracy*, Princeton, NJ: Princeton University Press.

Holmes, S. and H. Castañeda (2016), 'Representing the "European refugee crisis" in Germany and beyond: deservingness and difference, life and death', *American Ethnologist*, **43** (1), 12–24.

Howard, M. M. and S. W. Goodman (2018), 'The politics of citizenship and belonging in Europe', in C. M. Swain (ed.), *Debating Immigration*, Cambridge: Cambridge University Press, pp. 329–46.

Iosifidis, P. and M. Wheeler (2016), *Public Spheres and Mediated Social Networks in the Western Context and Beyond*, London: Palgrave Macmillan.

Kaján, E. (2017), 'Hate online: anti-immigration rhetoric in Darknet', *Nordia Geographical Publications*, **46** (3), 3–22.

Keane, J. (2009), 'Monitory democracy and media-saturated societies', *Griffith Review*, **24**, 47–69.

Kuo, R. (2018), 'Visible solidarities: #Asians4BlackLives and affective racial counterpublics', *Studies of Transition States and Societies*, **10** (2), 40–54.

Louw, E. (2010), *The Media and Political Process*, Los Angeles, CA: Sage.

Macgilchrist, F. and I. Böhmig (2012), 'Blogs, genes and immigration: online media and minimal politics', *Media, Culture & Society*, **34** (1), 83–100.

Maireder, A. and S. Schlögl (2014), '24 hours of an #outcry: the networked publics of a socio-political debate', *European Journal of Communication*, **29** (6), 687–702.

Maireder, A. and C. Schwarzenegger (2012), 'A movement of connected individuals: social media in the Austrian student protests 2009', *Information Communication and Society*, **15** (2), 171–95.

Merrill, S. and M. Åkerlund (2018), 'Standing up for Sweden? The racist discourses, architectures and affordances of an anti-immigration Facebook group', *Journal of Computer-Mediated Communication*, **23** (6), 332–53.

Moore, M. H. (2014), 'Accountability, legitimacy, and the court of public opinion', in M. Bovens, R. E. Goodin and T. Schillemans (eds), *The Oxford Handbook of Public Accountability*, Oxford: Oxford University Press, pp. 632–646.

Mouffe, C. (2013), *Agonistics: Thinking the World Politically*, London: Verso.

Musolff, A. (2015), 'Dehumanizing metaphors in UK immigrant debates in press and online media', *Journal of Language Aggression and Conflict*, **3** (1), 41–56.

Nerghes, A. and J.-S. Lee (2019), 'Narratives of the refugee crisis: a comparative study of mainstream-media and Twitter', *Media and Communication*, **7** (2), 275–89.

O'Mahony, P. (2013), *The Contemporary Theory of the Public Sphere*, Oxford: Peter Lang.

Ojala, M., S. Kaasik-Krogerus and M. Pantti (2019), 'Presidential speeches and the online politics of belonging: affective–discursive positions toward refugees in Finland and Estonia', *European Journal of Cultural Studies*, **22** (2), 164–79.

Ojala, M., M. Pantti and S.-M. Laaksonen (2019), 'Networked publics as agents of accountability: online interactions between citizens, the media and immigration officials during the European refugee crisis', *New Media & Society*, **21** (2), 279–97.

Papacharissi, Z. (2004), 'Democracy online: civility, politeness, and the democratic potential of online political discussion groups', *New Media & Society*, **6** (2), 259–83.

Papacharissi, Z. (2015), *Affective Publics*, Oxford: Oxford University Press.

Pariser, E. (2012), *The Filter Bubble*, New York: Penguin Press.

Pöyhtäri, R., M. Nelimarkka, K. Nikunen, M. Ojala, M. Pantti and J. Pääkkönen (2019), 'Refugee debate and networked framing in the hybrid media environment', *International Communication Gazette*. https://doi.org/10.1177/1748048519883520.

Price, V. (1992), *Public Opinion*. Newbury Park: Sage.

Renninger, B. J. (2015), '"Where I can be myself ... where I can speak my mind": networked counter-publics in a polymedia environment', *New Media & Society*, **17** (9), 1513–29.

Roslyng, M. M. and B. B. Blaagaard (2018), 'Networking the political: on the dynamic interrelations that create publics in the digital age', *International Journal of Cultural Studies*, **21** (2), 124–38.

Santana, A. D. (2015), 'Incivility dominates online comments on immigration', *Newspaper Research Journal*, **36** (1), 92–107.

Sen, R. and F. Mamdouh (2008), *The Accidental American: Immigration and Citizenship in the Age of Globalization*, San Francisco, CA: Berret-Koehler.

Siapera, E. (2004), 'Asylum politics, the Internet, and the public sphere: the case of UK refugee support groups online', *Javnost – The Public*, **11** (1), 79–100.

Siapera, E., M. Boudourides, S. Lenis and J. Suiter (2018), 'Refugees and network publics on Twitter: networked framing, affect, and capture', *Social Media + Society*, **4** (1), 1–18.

Spohr, D. (2017), 'Fake news and ideological polarization', *Business Information Review*, **34** (3), 150–60.

Sunstein, C. R. (2017), *#Republic: Divided Democracy in the Age of Social Media*, Princeton, NJ: Princeton University Press.

Swain, C. M. (ed.) (2018), *Debating Immigration*, Cambridge: Cambridge University Press.

Varnelis, K. (ed.) (2008), *Networked Publics*, Cambridge: MIT Press.

Wang, R. and K.-H. Chu (2019), 'Networked publics and the organizing of collective action on Twitter', *Convergence: The International Journal of Research into New Media Technologies*, **25** (4), 393–408.

Warner, M. (2002), 'Publics and counterpublics', *Public Culture*, **14** (1), 49–90.

Žižek, S. (2017), *Against the Double Blackmail: Refugees, Terror and Other Troubles with the Neighbours*, London: Penguin.

24. Using new media platforms for human rights advocacy in real-time: people seeking asylum in Nauru and Papua New Guinea

Cecilia Cannon and Shaminda Kanapathi

INTRODUCTION TO AUSTRALIA'S 'OFFSHORE PROCESSING'

On 14 November 2019 Behrouz Boochani, an Iranian journalist, walked through the arrival gates at Auckland airport, New Zealand, armed with a one-month visa to speak at a literary festival in Christchurch. His Twitter status, liked more than 22,000 times, read, 'So exciting to get freedom after more than six years' (Boochani 2019). Seven years earlier, Boochani had fled Iran following an Iranian government raid on the magazine he had co-founded. In 2013, he attempted to travel from Indonesia to Australia by boat, was intercepted by the Australian government and was transported to the Manus Island Regional Processing Centre. The Australian government had previously announced, on 19 July 2013, that every person arriving by boat in Australia would be subject to indefinite detention either on Manus Island, Papua New Guinea (PNG), or in the Republic of Nauru, under 'processing' arrangements with those Pacific states (Australian Department of Foreign Affairs and Trade [DFAT] 2013). Successive Australian governments have supported, managed and funded this offshore processing regime, where one of the co-authors of this chapter remains in 2020.

Boochani was one of 3,127 asylum seekers detained on Manus Island and Nauru as part of Australia's offshore processing policy between 2013 and 2020 (RCA 2019). While in detention, the asylum seekers used contraband phones and new media platforms to document and report on unsanitary conditions: 12 deaths caused by medical neglect, suicides and one killing by a security guard; violence, abuse and hunger strikes. They also documented several traumatic incidents, such as in October 2017, when the Australian government cut off electricity, food, water and other services for 25 days for more than 600 men on Manus Island who refused to go to a new, unfinished processing compound. Boochani wrote a book in Persian—*No Friend But the Mountains*—and transmitted it via WhatsApp to Omid Tofighian, an academic, who translated it into English and helped Boochani publish it in 2018. Boochani won several literary awards, including Australia's Victorian Prize for Literature. He gained a Twitter following of 50,000+ people and he participated in seminars, academic conferences and literary events and festivals remotely, via Skype.

Examining the use of communications technology and new media platforms by people seeking asylum on Manus Island and Nauru reveals that their experiences largely align with scholarly research on the topic. It also reveals additional insights not yet explored. In the first section of this chapter, we review the key themes that have emerged in existing research on the topic. In Section II, we use these themes as the framework for analysis for examining the experiences of people seeking asylum on Manus Island and Nauru utilizing new media platforms for human rights advocacy. In doing so, we hope to draw lessons for others who find themselves in similar situations, while contributing to the academic discourse. Finally, in Section III we reflect on the case of PNG and Nauru within the broader context of the global refugee system, highlighting what is unique to PNG and Nauru, and what can potentially be applied in other asylum processing settings.

We have written the first and third sections of this chapter in third person, while Section II, which is largely based on one of the author's own experiences of being detained on Manus Island and in Port Moresby between 2013 and 2020 under Australia's offshore asylum processing policy, is written in first person. Both authors have contributed to, and are responsible for, the entire chapter.

SECTION I: NEW MEDIA, HUMAN RIGHTS ADVOCACY AND EXISTING RESEARCH

A growing number of scholars have examined the topic of real-time human rights advocacy using new media platforms, with the following key themes emerging. New media platforms: (1) serve as a tool to facilitate freedom of expression, especially in environments of secrecy; (2) facilitate the documentation of, and awareness-raising about human rights abuses, including by engaging the public, key stakeholders, non-traditional partners and by facilitating organization within social movements; and (3) enable agency, self-representation and personal narratives for persons suffering human rights abuses. Select studies also examine or refer to (4) challenges and risks for those advocating in real time through new media platforms. We elaborate on each below, highlighting topics not yet examined.

1.   New Media Platforms as a Tool to Facilitate Freedom of Expression

Article 19 of the Universal Declaration of Human Rights stipulates that 'everyone has the right to freedom of opinion and expression; this right includes freedom to hold opinions without interference and to seek, receive and impart information and ideas through any media and regardless of frontiers' (United Nations General Assembly 1948). Scholars have shown that activists often face subtle censorship by the media because they are perceived as a threat to hegemonic power structures. Advocates of marginalized groups therefore communicate through alternative media platforms (Downing 2001; Rodriguez 2001 in Harlow 2017, pp. 318-319). Unofficial censorship is compounded when government legislation directly censors advocates, or subtly silences them through ambiguous legislation, high legal costs and surveillance laws, which provoke 'uncertainty, fear and self-censorship among writers and journalists' (Townend 2017, p. 73).

Scholars largely see the internet as a disrupter to traditional media power structures, creating an accessible 'alternative space where information and counter-information can easily and cheaply circulate, uninhibited by the gatekeepers of the traditional press' (Bennett 2004 in Harlow 2017). Digitalization of public communication offers new opportunities for individual and collective expression, while raising concerns about privacy, data usage, online harassment and blackmailing, national sovereignty and control in digital communication, as well as government and commercial surveillance (Tumber and Waisbord 2017, p. 6). Social media activism is also viewed as digital citizenship, particularly for those denied political participation, such as people seeking asylum (Nikunen 2019; Stavinoha 2019).

As we will demonstrate in this chapter, several people seeking asylum in PNG and Nauru utilized new media platforms not only for advocacy through social media status updates, but also to express themselves through their arts and skills, increasing their social media followings and establishing themselves as high-profile advocates and media spokespersons, which paved the way to resettlement for a small number of them.

2.   Documenting and Publicizing Human Rights Abuses Using New Media Platforms

The widespread use of smartphones, with built-in cameras and recording devices, has greatly facilitated the documentation, monitoring, reporting and publicity of human rights abuses in real time by citizens, witnesses, victims and even offenders. Marginalized groups and persons suffering human rights abuses are found to use social media to generate empathy and garner support/protection among human rights bodies, NGOs, activists, journalists and the general public (Coddington and Mountz 2014; Rae et al. 2019). This type of advocacy has been observed in various settings, including police violence towards marginalized groups; indigenous rights; sexual assault and abuse of women and other women's rights issues; documented and undocumented immigrant rights; and forceful/violent migrant deportation proceedings (Harlow 2017; Rollins 2019).

Scholars have debated the utility and effectiveness of this type of advocacy. Among the advantages cited by scholars are that it facilitates: awareness-raising, particularly amidst intentional state secrecy when NGO and media access is restricted (Martin 2019); the mobilization of publics and civil society networks; community building, civic and political engagement; and access to justice (Gregory 2010; Harlow 2017; Rollins 2019). Rollins explains, 'deportations become harder the moment more people know about them, and communications platforms make this instantaneously and enduringly possible' (Rollins 2019, 54).

Persons subjected to human rights abuses are found to use social media to organize within social movements by facilitating networking, connecting people with resources, sharing information and promoting solidarity (Peuchaud 2014; Bishop 2019; Costanza-Chock 2014). Examples cited by scholars include undocumented migrants in the US using online platforms to connect with one another and to learn more about each other's activities; LGBTQIA+ refugees accessing information on Facebook and other websites about their rights; and Egyptian women using an app to report sexual harassment, which triggered the deployment of citizen groups to provide physical protection against further abuse (Peuchaud 2014).

Awareness-raising and other forms of social media activism are also found to facilitate asylum seekers' contact with the media, particularly when media access is restricted. Once linkages with journalists are established, asylum seekers' messages are found to spread beyond their own social media followers (Rae et al. 2018). Social media can also help marginalized groups counter the mainstream media's biased representation of a situation or population, as analyzed in the case of black feminist activism on Twitter (Williams 2015).

Social media helps those affected by human rights abuses to raise awareness about key issues impacting their community and to mobilize community members unaware of the issues, as occurred among youth members of the Congolese diaspora and Brazilian favela residents using social media to publicize and warn their communities about police brutality (Costanza-Chock 2014; Godin and Donà 2016; Talbot and Carter 2018). Social media can also strengthen solidarity within a social movement by helping those affected to share their stories with other survivors, such as in the case of women and teenage girls using social media to condemn and share personal accounts of sexual harassment (Keller et al. 2018; Russell and Rae 2019).

Scholars have examined mobile applications designed to aid the documentation and sharing of human rights abuses. For example, AppCID is designed to help immigration detainees understand their rights and to report abuse (Harlow 2017; Rollins 2019, p. 55); and EyeWitness to Atrocities is designed to enable people to report war crimes (McAuliffe 2016, p. 320). However, Rollins shows that such apps can create a false sense of security for people suffering human rights abuses, and that their effectiveness is hindered because many migrants lack access to phones, power to charge phones and the internet (Rollins 2019, p. 56).

Some scholars warn against overstating digital technology's capacity to reduce human rights abuses or to bring perpetrators to justice, with great variation in usage by different groups and in different locations (Harlow 2017; Rollins 2019). Smugglers generally prohibit migrants from using phones during journeys for fear of detection by authorities, and places of immigration detention frequently ban phones because migrants can use them to bring attention to inhuman or illegal detention conditions, lack of due process, violence and human rights abuses (Rollins 2019, pp. 25–37, 54). International organizations, governments and NGOs can also pressure social media platforms to shut down migrant web pages, arguing that they facilitate irregular migration and smuggling, despite serving as spaces where migrants can share their migratory experiences, including human rights abuses (Rollins 2019, pp. 23–24). The gendered digital divide also leaves women less likely to have access to smartphones to engage in social media activism (Mancini et al. 2019).

Scholars have not yet examined the way this sort of advocacy can transform people seeking asylum and others suffering human rights abuses into high-profile, semi-celebrity advocates, attracting large followings on social media, becoming sought-out experts and media spokes-persons, and sometimes publicity fronts for NGOs and human rights and refugee organiza-tions, who at times use refugees to champion their causes, to raise funds and to gain publicity. Similarly, scholars have not yet examined in detail the retribution that can be brought upon migrants utilizing new media platforms to raise awareness of the human rights abuses they suffer, although this has been examined in social science research outside migration.[1] In this chapter, we will show how this sort of advocacy can have dire physical, procedural and psychological consequences for migrants and people seeking asylum while they are being processed.

Social media and mobile applications, such as WhatsApp, are also found to facilitate refugees' connections with family and friends while in transit, detention or upon arrival in destination countries (Coddington and Mountz 2014; Mancini et al. 2019). However, scholars have not yet examined the way this can both strengthen mental health by offering support to advocates enduring human rights abuses, while simultaneously causing anxiety and guilt among advocates, who fear they are causing worry to their families.

3. Self-representation, Agency, and Re-humanization

Several scholars, outlined in this section, have examined digital technology and new media as enablers of self-representation and agency for people suffering or at risk of human rights abuses. Citizens, and not just human rights organizations or mainstream media, can use per-sonal stories and user-generated videos as testimonial evidence for reporting human rights abuses and 'help to change the conversation, raise awareness and mobilize supporters in the field of human rights,' which also facilitates organizations in their monitoring of human rights violations (Harlow 2017, p. 320). This provides immigration detainees with 'oft-denied agency' (Rollins 2019, p. 55). Sharing personal narratives can also be used as a political strategy, particularly by those participating in social movements that affect them personally, such as undocumented migrants advocating to improve their daily lives in the US, or asylum seekers to Australia exposing detention conditions (Bishop 2019). These personal narratives can also serve as a strategy of humanization and to connect with general publics (Rae et al. 2018).

Several scholars question the power dynamics and symbolism of personal narratives when shared by the subjects themselves vis-à-vis other actors. Personal narratives are shown to be important in countering dominant, inaccurate representations about a population such as refugees (Leurs et al. 2020; Williams 2015). Sharing personal narratives via social media is seen as a form of reclamation of the ownership of their image, which can be homogenized, essentialized, stereotyped or otherwise misrepresented by other actors (Godin and Donà 2016; Surma 2018). Individuals can use their own images and stories to challenge dominant narratives often perpetuated by the media and state. Scholars contrast refugees' use of 'the selfie'—understood as an act of self-representation and reclamation of individual identity when they are misrepresented, such as was observed during the influx of refugees to Europe in 2015/2016—with the media's use of images depicting refugees taking selfies, such as occurred in European media in 2015/2016 (Chouliaraki 2017; Nikunen 2019; Risam 2018; Georgiou 2018; Leurs et al. 2020). Scholars have debated whether other actors can ethically use refugees' selfies and personal narratives, and discuss the media's co-optation and integration of refugees' images, recordings or stories to anti-migrant rhetoric in reporting (Chouliaraki 2017; Georgiou 2018; Risam 2018).

Many scholars outline the limitations and questionable ethics of using stories of people who have suffered rights abuses, questioning whether this plays into a 'discourse of deserving', whereby only refugees who behave admirably or contribute to society are depicted as worthy of meriting protection, as opposed to receiving protection as a human right (Leurs et al. 2020; Martin 2019; Nikunen 2019; Stavinhoa 2019). Several scholars question whether victimhood can ever be ethically portrayed, referring to 'victim voyeurism', such as sharing the horrific treatment of asylum seekers in Australian offshore detention.

Migrants and refugees themselves question the use, selectivity and framing of their personal stories by the media, international organizations and NGOs, which usually depict them as pitiful victims, reinforcing stereotypes. An Afghan refugee explains, 'While the journalist apologized for the "heavy editing", this humiliating experience taught me that despite my background as a citizen journalist and an academic, for some I will forever be a traumatized Syrian refugee whose primary role is to evoke sympathy and tears' (Tammas 2019). Others highlighted the limitations of human rights advocacy by refugees themselves, arguing that the audience accessing or interacting with refugee-created content may be self-selecting (Rae et al. 2019).

The broader advocacy literature has shown that shock messages and emotional appeals can be effective when employed by advocates using social media (Auger 2013), and that personal testimonies are vital for advocacy networks and campaign success to help audiences to connect and empathize with affected people (Keck and Sikkink 1997). A recent study about the language used by asylum seeker advocates in Australia found that persuadables, the bulk of the population whose minds can be changed, are more frequently convinced by messages that lead with values like family, freedom, fairness; offer two parts solution, one part problem; name the cause of harms to attribute blame to specific actors; and seize the moral high ground, rather than refer to laws and costs, among others (ASRC 2015).

There does not appear to exist any scholarly analysis on the effectiveness of different message framing techniques and advocacy strategies adopted by individuals suffering human rights abuses.

4.   Challenges/Risks

Scholars raise concerns about the safety, consent and ethics around filming and documenting others experiencing human rights abuses, particularly when this can lead to issues of representation, re-victimization, retaliation through online social networks, and doctored and manipulated recordings and online documentation (Gregory 2010). Several articles also refer to risks of backlash against refugees using social media for human rights advocacy (Nikunen 2019; Wroe 2018), and restrictions placed on internet usage when affected people speak out (Rae et al. 2018). However, scholars have not yet examined the anxiety, paranoia and insomnia that people speaking out about the human rights abuses they are suffering endure, particularly concerning their families becoming aware of their situation through their social media advocacy; and/or fear of retribution from authorities and local populations they speak out against.

SECTION II: LESSONS FROM AUSTRALIA'S OFFSHORE PROCESSING IN PNG AND NAURU

Using the themes identified in the previous section, we now examine human rights advocacy utilizing new media by people seeking asylum and refugees held in Australia's offshore processing camps in PNG and Nauru. By doing so, we uncover new insights about this topic.

1.   New Media Platforms as a Tool to Facilitate Freedom of Expression: Only after 2016

My experience and the experience of my fellow detainees in PNG of utilizing new media platforms to document, circulate and publicize the inhumane conditions, denial of basic rights, such as medical care, and other human rights violations aligns, to a degree, with the analysis of scholars mapped earlier, but only after 2016. Prior to 2016, the majority of us had no access to smartphones. The few phones that did wind up in the hands of my fellow detainees prior to 2016 had been smuggled in, or had been received as bribes from security guards. They were used secretly and discreetly.

We were transferred from Christmas Island in 2013 to detention on Manus Island without any consultation. We had no idea where we were being taken. We were each given 5 minutes to phone our families to let them know we were alive. On arrival we were given two options regarding dormitories: air-conditioned dormitories with phone access twice weekly, but without internet access, leaving no way to contact the outside world; or non-air-conditioned dormitories, with access to a poor internet connection twice weekly, 45 minutes each time. The choice caused us to feel tension and stress. Many of the men were from Iran and found living in a hot, humid environment difficult. They opted for the air-conditioned rooms without internet access. Those of us who chose internet access could explore our options for leaving Manus Island, which gave us a small, temporary sense of empowerment over our situation each week.

To be detained in suffocating heat with little-to-no access to the outside world sent a clear message that we who had sought safety were to be punished. The time limits on our access to the outside world seemed cruel and punitive. A sense of fear and isolation spread among us as our fundamental rights to freedom of expression and access to information were systematically denied. We saw this as a process of de-humanization and a sort of torture as we were arbitrarily detained with no access to courts to challenge our detention. At times, cutting our access to communication was used as punishment, such as following the 2014 Manus Island detention centre riots. The men involved were placed in solitary confinement and denied access to information and communication technology. They were entirely uncommunicative for one month. The phone and internet time allocated to all of us was further reduced.

Mobile phones were forbidden between 2013 and May 2016. Our main communication platforms during our limited internet access times were Skype and Facebook. I tried, unsuccessfully, to reach out to NGOs and advocates due to the slow and limited internet access. Most of us prioritized contacting our families during those precious minutes.

In May 2015, the Australian government introduced legislation – the Australian Border Force Act 2015 – that silenced us and furthered censored information about our situation. Section 42 of the legislation, entitled 'Secrecy', provides that anyone contracted to work on Manus Island and Nauru commits an offence if he or she makes a record of, or discloses 'protected information' (Parliament of the Commonwealth of Australia 2015). Our only option for making complaints was to the Australian Border Force, and these complaints largely went un-registered. This created a sense that we were criminals, when our only supposed crime was exercising our right to seek asylum in Australia. For many of us, it led to attempted suicides.

Some social workers did document and speak out about the inhumane conditions on Manus Island and Nauru. They were charged, and faced court and a potential of up to two years in prison. Speaking about the censorship that we faced, Boochani (2018) explained:

> In a country like Iran you do not have the freedom to write what you wish, but in a liberal democracy you are relatively free .... However, the system can ... silence you in ways that even if you were to shout no one would hear. And even if they hear you, they do not have the capacity to understand. This has been my experience on Manus Island .... For nearly five years they [the Australian political parties and media] have conducted a program of systematic censorship against me and the other imprisoned refugees on Manus Island and Nauru .... Sometimes, I am ignored; other times I am insulted; and other times the censorship is accompanied by political propaganda ... the distortion becomes so intense that the Australian people cannot receive any correct information.

On 26 April 2016, the PNG Supreme Court ordered the Australian and PNG governments to end the detention of asylum seekers at the Manus Island Regional Processing Centre on the grounds that it was a breach of our right to personal liberty under the PNG constitution. The Australian government subsequently built new facilities on Manus Island—the East Lorengau Refugee Transit Centre. We were not permitted to leave the facility between 6pm and 6am. Outside those hours we could move freely on Manus Island and we were finally permitted to have mobile phones. Mobile phones cost between 800 kina and 1000 kina (approx. AUD $300–400). We received a weekly allowance in points (50 points per week) that we could use to buy items in the canteen. We would therefore sell our cigarettes to the guards in order to buy a phone or Australian friends would sometimes send us money or mobile phones.

Social media and our time accessing communications were vital for connecting with our families. Doing so reminded us of their love and support, and strengthened us in our advocacy work. Once we were permitted to have smartphones in 2016, we were able to call our families anytime, for birthdays and other special events. We could also download movies, songs and games that helped us pass our time in detention. We also had access to information on the internet and online courses, enabling us to stay abreast of current affairs and politics in Australia and around the world. This helped inform us as we began to speak out about our situation, and we began to use social media, including Facebook, Twitter, Telegram, Skype, Messenger and WhatsApp, in addition to email and direct calls.

2.   Documenting, Publicizing and Mobilizing around Human Rights Abuses: Again, Only after 2016

Having smartphones restored to us our right to freedom of expression, access to information and enabled us to reach out to stakeholders that could amplify our complaints and serve as our legal and mental support. It was important for us to work together as detainees and to support one another as we began to speak out. We were stronger and more courageous when we supported and encouraged one another. We quickly re-established our social media profiles and we slowly reached out to and got to know many people in Australia and around the world and alerted them to our situation. Prior to that, not many people knew or understood our situation. We began to daily expose the denial of freedom and the human rights abuses we faced. We filmed traumatic incidents when they occurred, such as abuse by the Australian security guards, and the 2017 siege when more than a hundred shots were fired at us, leaving nine wounded. Later, we used that footage to produce a 10-minute documentary about the siege. Mobile phones became our main source of strength in our daily lives, helping us show others what was happening.

When we first spoke out, few people paid attention. Facebook was our social media platform of choice because most of the refugee advocates used Facebook. I only began to use Twitter in early 2017 and it took one year to begin to gain a following. Communicating and connecting with people in the outside world was vital for the amplification of our complaints, for the spread of our stories and for the creation of a social movement that fought hard to win us back many of the rights we had been denied, such as a little more access to medical care (though this has never been adequate). We engaged key stakeholders, such as NGOs, human rights organizations, sympathetic government representatives, international organizations, supportive individuals, influencers on social media, scholars, other non-traditional partners, and the media. Together, all of these stakeholders played an important role in our advocacy efforts utilizing new media platforms.

Following introduction of the Australian Medevac Bill in 2018 that enabled the transfer of refugees and people seeking asylum to Australia if they required urgent medical treatment, I helped many of the men on Manus Island to complete their Medevac applications. This gave me a first-hand view of the dire physical and mental medical crisis unravelling among us, and I wrote about this frequently on social media. These social media writings were regularly used by the Asylum Seeker Resource Centre (ASRC) in their 'Save Medevac' campaign and were frequently quoted in the media. We additionally set up a secure online mobile application

Telegram channel late in 2016, through which we alerted 400+ NGOs, journalists and advocates about incidences and abuses as they occurred.

From 2016 onwards, I was among a small number of activist refugees on Manus who shared our stories on social media and raised awareness about the conditions, abuses and injustices we faced. Several detainees also developed and shared their skills via social media, establishing/re-establishing themselves as artists, academics, journalists, writers and film producers with an ever-increasing social media audience. They were subsequently to speak as experts remotely via online conferencing technology at literary and artistic events, workshops, academic seminars and in media interviews.

With perseverance in their crafts while in detention and over several years, a small number of refugees were granted temporary and/or permanent resettlement visas through their respective sectors, such as the arts and academia. Like Boochani, Ali Dorani/Eaten Fish, an Iranian cartoonist and political activist, gained global recognition for his cartoons depicting his experiences in Australia's offshore processing centre on Manus Island, including the abuse, rape and mental illness he endured. Cartoonists Rights Network International raised awareness of Dorani's situation, and he was granted an artist's residency in Norway through International Cities of Refuge (Wilcox 2017). Similarly, Abdul Aziz Muhammat sent over 4,000 voice messages to Melbourne based journalist Michael Green, who used the voice messages to produce a podcast *The Messenger*. In 2018, Muhamat was granted a temporary visa to travel to Switzerland to receive the Martin Ennals Award for Human Rights Defenders, where he applied for and was granted asylum.

When these high-profile advocates and community leaders leave Australia's offshore processing, we are happy for them. Their stories are inspirational and illustrate that human rights advocacy using new media platforms can lead to freedom and permanent resettlement. However, their stories are not common and their absence magnifies the desolation of the majority who remain in limbo in refugee camps or in temporary host communities around the world, escalating depression and suicide attempts. It becomes increasingly difficult for those of us who remain to speak out, causing us to often retreat into our rooms and into ourselves. With an uncertain future, we are afraid we will be forgotten in limbo in PNG, Nauru, and in camps and shelters around the world.

Their individual successes suggest that there are differences regarding social media campaigns for individual cases versus advocacy for a group, such as groups of people seeking asylum or refugees. The latter seems far more challenging. These few, individual stories of success resemble those of other high profile, intense, social media campaigns that saw international and human rights organizations and the general public pressure governments in 2019 to improve the human rights conditions of both Rahaf-al-Qunun, who was granted permanent resettlement to Canada after tweeting about the physical and mental abuse she faced from her father in Saudi Arabia; and Hakeem al-Araibi, who was granted leave to return to Australia—the country where he had previously been granted asylum. Intense, online social media campaigns pressured international organizations and governments to grant these individuals freedom.

3. Representation and Agency

Our experience working with NGOs and international organizations to advocate for our human rights has been mixed, reflecting the opportunities and concerns of representation and agency discussed by scholars. Since 2016, access to mobile phones and through them social media gave us a platform to tell our own stories and to report and share information on abuses as they happened. These new media channels have helped us connect with NGOs and international organizations who have sometimes amplified our messages to wider audiences throughout Australia and the world, countering misinformation often reported in the Australian and international media about refugees.

From Manus Island and Nauru we worked on advocacy campaigns with several nonprofits, including GetUp, Refugee Action Coalition, ASRC and Amnesty International. The more successful campaigns communicated our personal narratives and showed the human side of our stories, while resonating with the Australian community by focusing on things we have in common with the Australian public, such as the #BringThemHere, #KidsOffNauru and #SaveMedevac campaigns we worked on with GetUp (ASRC 2015; Shenker-Osorio 2019). We have worked with organizations to produce several video campaigns, including Amnesty International's 2019/2020 #GameOver testimonial campaign.

While many of these campaigns are hailed as wins in our struggle, their successes should not be over-stated. While they have gained for us small wins with respect to guaranteeing some of our human rights, none of them has significantly challenged the underlying policies that systematize the deprivation of liberty and human rights violations of people seeking asylum in Australia. Even following the successful introduction of the Medevac Bill, for example, refugees transferred to Australia for urgent medical care have been detained indefinitely in Australian hotels, aggravating mental illness and suicides, such as the tragic suicide of a 32-year-old Afghan doctor who spent four years on Manus Island, followed by two years in a Brisbane hotel in Australia (Davidson 2019). Some organizations have a political agenda and only contact us when they need us to serve as the face of their own issues and for their own publicity and fundraising purposes, rather than actively helping us find solutions or working to relieve our situation. Most nonprofits and IOs only tell a portion of our story that suits their own narratives and objectives, which is difficult for us to accept.

We also work closely with journalists, who frequently interview us and then amplify our messages in their reports. Some newspapers, such as *The Guardian* and *The Saturday Paper*, also gave some of us the opportunity to write our own articles about our experience on Manus Island. Journalists also came to rely on us as experts and spokespersons on the deteriorating health conditions and our situation, such as Radio New Zealand and Australia's SBS news and ABC.

4.   Challenges and Risks of Speaking Out

Not everyone was willing to speak out for fear of retribution by the Australian government. During my first year in detention on Manus Island, I was afraid and kept quiet, but I realized that our only hope of stopping the abuses was to let the world know what was happening. When we spoke out, we faced attacks and abuse in person and online. The local guards and the Australian contracted employees taunted us for our criticism of the PNG and Australian authorities. I sustained injuries that required hospitalization when we were forcibly moved out of the Lombrum detention facility and for punishment I was placed in the incomplete accommodation in West Lorengou Transit Center. My belongings were stolen and destroyed. My complaints to the Manus Island Police went unheard. In late 2019 and early 2020, I spoke out through social media about the 50+ asylum seekers arbitrarily incarcerated in inhumane conditions in Bomana Prison, with no access to lawyers, courts, or the outside world. Local service providers abused me because they feared that our exposure of the inhumane conditions would lead to the closure of the offshore processing centres, and the subsequent loss of their jobs. This intensifies their anger towards us because our plight is profitable for them.

Online trolls, including bots and anonymous accounts, incessantly flood our social media postings with hate speech, and with messages and images that undermine the credibility of what we write about. Over the years, we have come to see online hate and trolls as a sign that what we are saying matters and is having impact. Yet there is a stark difference between the online attacks and hate speech people seeking asylum face as compared with celebrities and average people on social media. While the majority of online attacks to celebrities are mere talk, and can be 'shaken off', the threats levelled against people seeking asylum and refugees by local people can and frequently do translate into actual retribution and physical attacks. For example, on 7 March 2020, refugees residing in Australia's offshore processing residence in Port Moresby faced an attack from a group of intoxicated, angry locals. Refugees from inside the accommodation under attack sent a video and text message via the 'Manus Alert' Telegram channel showing the attackers throwing stones, wielding knives and baseball bats, and screaming abuse at security guards: 'Open the gate. We need to kill these people... this is not your country.' Several people tweeted about this event, and it was later picked up by some news agencies, but such attacks are everyday life for refugees in PNG.

What the literature does not talk about is the incessant fear refugees are tormented with. Now in Port Moresby, I am recognized as one of the few remaining outspoken refugees that exposed the inhumane offshore processing policy. In September 2019, I published an article in the *Saturday Paper*, describing Port Moresby as 'one of the most dangerous cities in the world'. This was shared widely on social media and I was interviewed on radio. Subsequently I received death threats and threats of attack from locals via social media, such as 'We will get you. Look out', and phone calls from anonymous callers threatening to hunt me down and kill me. These threats fill me with fear and I, like many other refugees, remain confined to my room, afraid to leave my accommodation. Despite informing UNHCR and Amnesty International, I am offered no protection. This has negative consequences on mental and physical health that scholars do not discuss. I fear I will be abandoned in limbo in PNG forever, and will be attacked and eventually killed here.

Similarly, the literature fails to discuss the profound depression and post-traumatic stress that most people in our situation experience. The effects of these threats aggravate the trauma we have already experienced and the uncertainty of our situation, leading to anxiety, withdrawal, depression and insomnia, with no access to mental health services. This also leads to a withdrawal of social media use, because social media messages, even by our supporters, incessantly taunt us about our hopeless situation that has gone on for seven years and that has no end in sight. Even if the threats are not carried out, they silence us to a degree, even if not permanently. An added stress has been the worry and shame that we were causing to our families, especially our parents, leading many of us to lie to them and instead tell them we made our way to Australia. But our online activism and advocacy make this difficult, and our families come to know our situation, which makes them worry and unwell and creates in us a sense of guilt, pain and failure.

The co-optation of refugees' images referred to by scholars is not only carried out by the media. Online trolls frequently use our images in their online attacks against us. They select images we have posted on social media that depict us smiling or sharing positive messages and post them in reply to our comments about deplorable conditions and human rights abuses, arguing that we look happy and are living in paradise. Our enormous efforts to remain positive are then turned against us as our online tormentors claim we are lying. For example, Ezatullah Kakar—a refugee from Pakistan—has fought tough mental and physical conditions to become a renowned boxer while detained on Manus Island. His selfies and positive messages are frequently used against other refugees' posts about the deplorable conditions.

SECTION III: LESSONS FOR HUMAN RIGHTS ADVOCACY USING NEW MEDIA

The first-hand experience of one of the authors of this chapter and that of his fellow detainees on Manus Island and Nauru illustrates the complexity of experiences of access to and use of information and communication technologies and new media platforms. When reflecting on lessons that can potentially be drawn from the human rights advocacy utilizing new media platforms adopted by people seeking asylum in PNG and Nauru, many of them could potentially be adopted by other refugees and people seeking asylum around the world facing human rights abuses in camps, shelters, open centres and even by those living in the community of temporary host societies.

These ten lessons presented below offer insights into how human rights advocacy utilizing information and communications technology and amplified through new media platforms can offer improvements to individuals and groups' human rights situations and the formation of social movements and transnational advocacy networks:

1. Access to smartphones, power and the internet is a necessary foundation for refugees and people seeking asylum wishing to engage in human rights advocacy utilizing new media platforms. This access to information and communication technology can be blocked by legislation or standards of practice by government and security officials in host communities; or it may stem from other barriers, such as financial difficulty, lack of connectivity and infrastructure, or the demands of smugglers and traffickers.
2. Working together to support and encourage fellow refugees and asylum seekers will give you the strength and courage you need to speak out, and will create a unified voice among people with a shared objective.
3. When you begin to reach out to people and begin to document and report on the conditions you are being subjected to and the human rights abuses you might be facing, it may seem at first that no one is listening. Keep speaking anyway. It was only through perseverance and persistence, and after some years, that the people seeking asylum and refugees on Manus Island and Nauru had their voices heard, which has led to some improvements for them. Keep the pressure up, even and especially when it seems that you are getting nowhere, or are going backwards.
4. Reach out to people strategically. Think about what you are trying to achieve and think about which people, NGOs, human rights activists, influencers, international organizations, government representatives and media might be willing to listen to your reports and stories and amplify your messages. Build a network of supportive individuals and organizations that, over time, can amplify your messages.

5. Speak for yourself, tell your own story and if other individuals and organizations speak on your behalf, ensure that they tell your story accurately and show your human side.
6. Think about your target audiences and what will connect with them. Take time to understand the political and social context of your host communities or the communities that can act on your behalf and/or offer you protection, guarantee your rights and/or ideally offer you resettlement. You can only connect with them if and when you understand them and their situations and settings.
7. If you have skills and you are able to continue developing them and working on them, do so and, if appropriate, try to use them to amplify your story and your voice, and to connect with wider groups, organizations and communities.
8. When you advocate and speak out on your own behalf or on behalf of others utilizing new media platforms, you will likely face many challenges. These obstacles may come in the form of retribution from representatives of government, security guards, international organizations or local host communities. You may be attacked verbally, online, over the phone, or in person, and in some cases even physically attacked. You may suffer incessant fear and mental health issues. Your family and friends who will follow your online advocacy closely may also suffer mental health challenges and may worry a great deal. Remember point 2 above—talk to your fellow asylum seekers and refugees who are going through the same experiences as you and work together and support one another. It is often there that you will find the strength and encouragement you each need, which may also help you decide when it might be safest and more strategic to remain silent for a while, or when it is important to keep speaking out. Ultimately, it is everyone's unique decision as to whether they wish to advocate about the human rights abuses they suffer, and thereby subject themselves to these risks, or not.
9. Small wins for individual human rights, such as access to adequate food, medical care and due process, are good, but don't forget the long game of advocating to stop the deprivation of liberty of asylum seekers and refugees, for equality, and for more permanent resettlement visas so that you can not only survive and live with human dignity, but thrive and reach your potential and contribute to society.
10. When you are free or permanently resettled, don't forget your fellow brothers and sisters who remain in shelters, camps, detention or in limbo in the community—keep speaking out on their behalf.

While the above ten points were possible in the case of people seeking asylum who had reached the waters of Australia—a liberal democracy that receives a relatively small number of asylum seekers each year—it is not so easy for people who find themselves for years and sometimes decades in refugee camps, in open refugee shelters/centres, and even those living in temporary host communities in developing countries, such as in Chad, Kenya, Ghana, Tunisia, Ethiopia, Uganda, Jordan, Lebanon, Bangladesh, Syria, Turkey, among other countries. The points will need to be adapted to local contexts and host communities, with particular attention paid to points 4, 5 and 6—think carefully about what your objectives are and who your target community is. Understand your rights, as well as the political, social and economic contexts in which you are advocating. Understand what, if any, potential opportunities you have to permanent resettlement, and which host communities may be able to offer you assistance, protection, support and potential resettlement.

Finally, it is worth reiterating that the successful advocacy of a small number of people seeking asylum in PNG and Nauru, while immensely important for raising awareness of the deplorable conditions and abuse, has not shaken the underlying policies in Australia or globally that propagate systematic human rights abuses against people seeking asylum and refugees. And this responsibility should not rest on their shoulders or on the shoulders of any refugees. This responsibility is one that only governments around the world can resolve by ceasing to support conflicts and by offering a sufficient number of permanent resettlement visas at a time when only 1% of refugees have access to them (UNHCR 2020).

At the time of concluding this chapter in May 2020, Australia's Acting Immigration Minister Alan Tudge proposed legislation that would grant the Australian Border Force officials search and seizure powers, and would permit the ban of certain items from immigration detention facilities, including mobile phones (Farhart 2020). The proposed laws have been widely condemned by human rights groups, who argue that the Australian public would not have been made aware of instances of abuse and mistreatment inside Australia's immigration detention facilities, including several high-profile cases that enabled legal action. They also argue that the proposed laws would drastically restrict the access immigration detainees have to their legal representatives, their families and loved ones, as well as reducing their ability to document and report in real-time the abuse, conditions and mistreatment they face while in immigration detention (ASRC 2020).

NOTE

1.   See, for examples, K. Kenny, F. Marianna and S. Scriver, "Mental health as a weapon: whistle-blower retaliation and normative violence", *Journal of Business Ethics*, 160 (2019), 801–815; Human Rights Watch, *We could disappear at any time: Retaliation and abuses against Chinese petitioners*. Volume 17 (2005), 11 (C).

REFERENCES

ASRC (2020), 'Expanded border force powers desperate distraction from Government's inhumane immigration detention', accessed 12 May 2020 at www.asrc.org.au/2020/05/14/expanded-border -force-powers-desperate-distraction-from-governments-inhumane-immigration-detention/.

ASRC (2015), 'Words that Work: making the best case for people seeking asylum', accessed 5 October 2019 at www.asrc.org.au/wp-content/uploads/2016/05/ASRC-Words-that-Work-4pp.pdf.

Auger, G. (2013), 'Fostering democracy through social media', *Public Relations Review*, **39** (4), 369–376.

Australian DFAT (2013), 'Regional resettlement arrangement between Australia and PNG', accessed 6 March 2020 at www.dfat.gov.au/geo/papua-new-guinea/Pages/regional-resettlement-arrangement -between-australia-and-papua-new-guinea.aspx.

Bishop, S. C. (2019), *Undocumented Storytellers: Narrating the Immigrant Rights Movement*, Oxford, UK: Oxford University Press.

Boochani, B. (2019), 'So exciting to get freedom after more than six years', 14 November, accessed 6 March 2020 at https://twitter.com/BehrouzBoochani/status/1194918470703443968.

Boochani, B. (2018), 'Silenced in ways that no-one hears your shouts and screams', *Sydney Pen Magazine*, May 2018.

Chouliaraki, L. (2017), 'Symbolic bordering: the self-representation of migrants and refugees in digital news', *Popular Communication*, **15** (2), 78–94.

Coddington, K. and A. Mountz (2014), 'Countering isolation with the use of technology: how asylum-seeking detainees on islands in the Indian Ocean use social media to transcend their confinement', *Journal of the Indian Ocean Region*, **10** (1), 97–112.

Costanza-Chock, S. (2014), *Out of the Shadows, Into the Streets: Transmedia Organizing and the Immigrant Rights Movement*, Cambridge, MA: MIT Press.

Davidson, H. (2019), 'Afghan man dies in Brisbane two years after medical transfer from Manus Island', *The Guardian*, 17 October, accessed 17 September 2021 at www.theguardian.com/australia-news/2019/oct/17/afghan-man-dies-in-brisbane-two-years-after-medical-transfer-from-manus-island.

Farhart, C. (2020), 'Government pushes for increased Border Force powers inside immigration detention facilities', *SBS News*, 14 May.

Georgiou, M. (2018), 'Does the subaltern speak? Migrant voices in digital Europe', *Popular Communication*, **16** (1), 45–57.

Godin, M. and G. Doná (2016), '"Refugee voices," new social media and politics of representation: young Congolese in the diaspora and beyond', *Refuge*, **32** (1), 60–71.

Gregory, S. (2010), 'Cameras everywhere: ubiquitous video documentation of human rights, new forms of video advocacy, and considerations of safety, security, dignity and consent', *Journal of Human Rights Practice*, **2** (2), 191–207.

Harlow, S. (2017), 'Live-witnessing, slacktivism and surveillance', in H. Tumber and S. Waisbord (eds) 2017, *The Routledge Companion to Media and Human Rights*, New York and Oxford: Routledge, pp. 318–326.

Keck, M. and K. Sikkink (1997), *Activists Beyond Borders*, Ithaca, NY: Cornell University Press.

Keller, J., K. Mendes and J. Ringrose (2018), 'Speaking "unspeakable things": documenting digital feminist responses to rape culture', *Journal of Gender Studies*, **27** (1), 22-36.

Leurs, K., I. Agirreazkuenaga, K. Smets and M. Mevsimler (2020), 'The politics and poetics of migrant narratives', *European Journal of Cultural Studies*, **23** (5), 1-19.

Mancini, T., F. Sibilla, D. Argiropoulos, M. Rossi and M. Everri (2019), 'The opportunities and risks of mobile phones for refugees', *PLOS One*, **14** (12), 1–24.

Martin, G. (2019), 'Turn the Detention Centre Inside Out: Challenging State Secrecy in Australia's Offshore Processing of Asylum Seekers', in P. Billings (ed.), *Crimmigration in Australia: Law, Politics, and Society*, Singapore: Springer, pp. 327–352.

McAuliffe, M. (2016), 'The appification of migration', accessed 25 May 2020 at www.policyforum.net/the-appification-of-migration/.

Nikunen, K. (2019), 'Once a refugee: selfie activism, visualized citizenship and the space of appearance', *Popular Communication*, **17** (2), 154–170.

Parliament of the Commonwealth of Australia (2015), *Australian Border Force Bill 2015*.

Peuchaud, S. (2014), 'Social media activism and Egyptians' use of social media to combat sexual violence: an HiAP case study', *Health Promotion International*, **29** (1), 113–120.

Rae, M., E. K. Russell and A. Nethery (2019), 'Earwitnessing detention: carceral secrecy, affecting voices, and political listening in *The Messenger* podcast', *International Journal of Communication*, **13**, 1036–1055.

Rae, M., R. Holman and A. Nethery (2018), 'Self-represented witnessing: the use of social media by asylum seekers in Australia's offshore immigration detention centres', *Media, Culture, and Society*, **40** (4), 479–495.

RCA (2019), 'Offshore processing statistics', accessed 6 March 2020 at www.refugeecouncil.org.au/operation-sovereign-borders-offshore-detention-statistics/.

Risam, R. (2018), 'Now you see them: self-representation and the refugee selfie', *Popular Communication*, **16** (1), 58–71.

Rollins, T. (2019), 'Physical fences and digital divides', Global Detention Project, December 2019.

Russell, E. K. and M. Rae (2019), 'Indefinite stuckness: listening in a time of hyper-incarceration and border entrapment', *Punishment and Society*, **22** (3), 281–301.

Shenker-Osorio, A. (2019), 'People Seeking Asylum – Australia', *Brave New Words*.

Stavinoha, L. (2019), 'Communicative acts of citizenship: contesting Europe's border in and through the media', *International Journal of Communication*, **13**, 1212–1230.

Surma, A. (2018), 'In a different voice: "a letter from Manus Island" as poetic manifesto', *Continuum*, **32** (4), 518–526.

Talbot, A. and T. F. Carter (2018), 'Human rights abuses at the Rio 2016 Olympics: activism and the media', *Leisure Studies*, **37** (1), 77–88.

Tammas, R. (2019), 'Refugee stories could do more harm than good', *Open Democracy*, 1 November 2019.

Townend, J. (2017), 'Freedom of expression and the chilling effect', in H. Tumber and S. Waisbord (eds) (2017), *The Routledge Companion to Media and Human Rights*, New York and Abingdon: Routledge, pp. 73–82.

Tumber, H. and S. Waisbord (eds) (2017), *The Routledge Companion to Media and Human Rights*, New York and Abingdon: Routledge, pp. 318–326.

UNHCR (2020), 'Resettlement', accessed 6 March 2020 at www.unhcr.org/resettlement.html.

United Nations General Assembly (1948), 'Universal Declaration of Human Rights', General Assembly Resolution 217A.

Wilcox, C. (2017), 'When cartoonists are in the line of fire', *Sydney Pen*, May 2018.

Williams, S. (2015), 'Digital defense: black feminists resist violence with hashtag activism', *Feminist Media Studies*, **15** (2), 341–344.

Wroe, L. E. (2018), '"It really is about telling people who asylum seekers really are, because we are human like anybody else": negotiating victimhood in refugee advocacy work', *Discourse and Society*, **29** (3), 324–343.

PART VI

DIGITAL MIGRATION FUTURES

25. Technological transformations in migration processes: spatiality, temporality and agency

Huub Dijstelbloem

INTRODUCTION

Just like other areas of life, such as work, health care, financial transactions and entertainment, migration is nowadays steeped in technology. Whether applying for a visa, undergoing border checks, arranging transportation, traveling by land, sea and air or maintaining relationships with relatives, migrants use technology at every stage of their journey. The same applies to states that monitor the mobility of people and perform border controls, and organizations that conduct humanitarian tasks. From the issuance of passports by governments to the establishment of refugee camps by non-state actors, and from the registration of personal data to verify people to the provision of medical care to migrants, technologies come into view, ranging from databases and the collection of fingerprints to tents, food supplies and medical devices.

Although human migration and technology have long been interconnected, there are two aspects in modern history that have brought about a major change. The first is the simultaneous rise of nation-states, the exercise of sovereignty over territory and the delimitation of borders, in relation to the growth of international governance (Mazower 2012). Modern states have the authority to determine who can circulate within and cross their borders. They control 'the legitimate means of movement' (Torpey 1998: 4). The invention and expansion of trains, railways and telegraphs provided modern states with more and more means to keep an eye on the population, to monitor the borders and to organize the movement of people (Maier 2016). With the advent of nation-states, the international governance of migration also developed, particularly in the second half of the twentieth century after World War II, ranging from the 1951 Refugee Convention and the 1967 Protocol to more recent international migration policies, border control management and negotiated agreements such as the Global Compact for Safe, Orderly and Regular Migration prepared under the auspices of the United Nations in 2018.

The second development concerns the meaning of technology. Ever since the industrial revolution, technology has been seen as a productive force that enables the organization of labor and the conversion of natural resources into energy, materials, capital and mobility, with all the beneficial and disruptive consequences that this entails, and impact on societies and the environment. Today, however, technology can increasingly be seen as a disseminating force. Instead of providing a background for human action or a machine that makes production and transportation possible, technologies of all sorts deeply penetrate all facets of society. Technology acts as a distribution mechanism that organizes the relationships between people, things and their environment – or disrupts them. Technology is not a mechanism that only works within discrete institutions, organizations or situations, such as the hospital, laboratory or city. Technologies connect different spheres of society and bring about a variety of relations. Infrastructures for transportation, for instance, have already refined forms of coordination, such as clock alignment, the issuance of tickets, the streamlining of guidelines and taxes, the organization of labor and the development of commercial activities (Allen and Hecht 2001). As a result, technologies affect the very notion of territory and the delineation with borders. Territory, as Elden (2013: 323) argues, 'is not simply land … nor is it a narrowly political strategic question that is closer to a notion of terrain. Territory comprises techniques for measuring land and controlling terrain.' The rise of information technology, the internet, the organization of large-scale data systems, the growth of social media platforms such as Facebook, algorithmic operations and the widespread use of communication technology all make technologies even more invasive.

What are the consequences of the changing nature of technologies and their relationship to migrants, states and humanitarian organizations? How do technologies transform migration and the various stages it involves? What new issues arise through the use and diffusion of technology in migration, in particular now migration, security and border control policies increasingly intermingle? And which new political questions have to be addressed to evaluate the relation between technology and migration? This chapter argues that a conceptual change in the understanding of technology is required to grasp the dissemination capacity of technologies and the way in which technologies affect all types of relationships in migration processes. The chapter will discuss various examples from the study of migration and technology to demonstrate what this diffusion capacity of technologies entails. The first part of the chapter describes three ways to conceive technologies that appear in migration processes: (1) technology as an object: a tool, a device, a machine, an instrument; (2) technology as an infrastructure, system or (part of a) network; and (3) technology as a world view and a way of 'world making'. These different notions of technology are not entirely distinct. They overlap, coincide and vary in time.[1] The second part of the chapter analyzes the intermingling and mutual interaction between these conceptions of technology and explains how technologies *transform* the categories of place, time and action from the inside out. The final part of the chapter discusses these transformations and concludes they, first, produce hybrid relations of technologies with people and with natural surroundings; second, implicate different temporal regimes for various groups of migrants, but also other groups of actors; and third, give rise to a multiplicity of actors and institutions related to migration issues. The chapter suggests transformations in spatiality, temporality and agency and the resulting hybridity and multiplicity that arise through technology in migration processes have a pervasive influence on the kind of questions to be asked in the study of migration.

THE DISSEMINATION CAPACITY OF TECHNOLOGY

The study of migration and technologies has seen various approaches that study technologies meticulously without reducing them to mere instruments. In order to understand the way technology functions in migration processes, whether as an instrument, an infrastructure or a world view, two approaches deserve specific attention as they are widespread and have guided the literature into many interesting directions: approaches that built on the oeuvre of Foucault and on science and technology studies, in particular the work of Latour.

Foucault's analyses, for example, denote a particular relationship between technology, power and knowledge and the ways 'subjects' and 'objects' are created by particular ways of surveillance. Although Foucault refrained from explicitly discussing migration and border control, his view of 'governmentality' and the ways policy apparatuses govern the life forms of populations and secure territories have spoken to a wide range of scholars and stimulated analyses of citizenship, the humanitarian government of refugees and international mobility regimes (Walters 2015). Moreover, these notions allow for regarding policies themselves as kinds of technologies, that organize the governance of migration.

Science and technology studies, stimulated by the work of scholars such as Callon, Haraway, Latour, Mol, Stengers and Law, have demonstrated that living with technologies does not only mean that some things, such as long-distance communication or curing diseases, have become easier or more efficient, but also that the content of these actions – how we communicate, what we treat as disease and what not – is altered by the presence of technology. In the study of international mobility, this view has generated attention for specific technologies, their genealogies, and the way they connect and disconnect in networks and chains of associations (e.g., Salter 2015; 2016). The 'material turn' in border studies, international relations and political theory fuels this interest in the various modes of technology even further (Barry 2013; Best and Walters 2013).

In short, a rich and variegated conceptual and methodological ecosystem has arisen to understand technologies in different shapes, situations and meanings. The study of borders, migration and security has seen a rise of approaches that regard technology as a form of situational acting, thinking, seeing and arranging. An early overview given by Haggerty and Ericson in 'the surveillant assemblage' (2003) discusses many of these approaches and identifies a growing tendency to relate the oeuvres of Deleuze and Foucault to a Latourian lens. Notions such as 'dispositive,' 'apparatus' and 'assemblage' take the intermingling of knowledge and power and the immanent constellation of people, politics and technologies as a starting point. This angle has a specific sensitivity for the machinery of governance and the way this machinery performs technologically even without being related to specific technologies. Technological increasingly is regarded as a repertoire of actions, interventions and representations by which issues related to border control, mobility management, surveillance and security may be articulated.

This conception allows us to rethink the relation between humans and technology in a socio-technical matter. Technologies are not reducible to mere instruments. Instead, technologies are indispensable linkages in the socio-material world. Thinkers such as Haraway and Latour argue that one should not see humans as opposed to technology. Technology is intrinsic to how humans build societies, collapsing the means-end analysis completely. For instance, communication through telephone does not replace an authentic form of direct communication, but replaces already-mediated practices of communication through writing. How essential technology in society may be, it has at the same time the tendency to disappear in the background as our interaction with interfaces makes us forget what working lies behind the vast range of technological objects we have at our disposal, e.g. consuming news media through Facebook usually does not make us question the underlying algorithm. How then can we illuminate the effects and meaning of technology if it is so ubiquitous and yet fleeting?

Despite the differences between approaches that regard technologies as mere instruments versus approaches that suggest a more intimate and co-constitutive entanglement between humans and technology, the distinctions are gradual and vary in time. Body scanners at airports, for instance, can be considered as specific tools that are relatively unconnected to the surveillance environment as the data are not saved or shared with other security systems. The scanners present the person as colored spots on a grey outline of a human figure on a computer screen. There are no images that can be saved and no raw data are saved either. Despite this relatively isolated function, body scanners evoke particular questions about privacy but also about the gendering of border control as border guards or officials distinguish between 'male' and 'female' bodies and represent with distinct colors (blue and pink). At the same time, body scanners can be said to be part of the security and migration management environment of the airport. As a tool, body scanners are part of the 'smart border' at the airport. Airport scanners were introduced as *body* scanners. Following inquiries and allegations of privacy violation, they became *security* scanners (Bellanova and Fuster 2013; see also Schouten 2014). As an object, a body scanner is not static, but a shifting ensemble in which modifications are made so as to insert it more fully into law, space, use. Body scanners become part of the smart border environment of international airports that combine mobility, security and surveillance systems – visa systems, ticket registration systems, all kinds of border control technologies from iris scanners to cameras. Finally, body scanners are part of a technological world view and a way of world making. Airports are crucial nodes in the global mobility regime – or more precise, the visa regime that allows people from specific countries to travel with a visa and places restrictions on others, requiring them to take different routes.

A key feature of approaches that understand technologies not as mere objects or tools, is that they conceive technologies as things that affect the doing, thinking, seeing and feeling of the people involved and are an inherent aspect of realities that undergo transformations of all sorts. A study of Jeandesboz (2016) on border security and the rise of 'smart borders' in Europe is an example that illustrates the transition to such an approach. Jeandesboz criticizes the 'decisionist' position in security political theory favoring a less agency-centered perspective that acknowledges the transformations and mediations technologies bring in the context of smart borders. According to Jeandesboz (2016: 306) security politics is 'grounded in a hard labor of association, translation and enrolment.' Jeandesboz points to 'the non-human component of border control policymaking' (2016: 304) as he states 'the "smartening" of EU-borders is held together by multiple translations and enrolments through which the technical side … has become associated with the processes of policymaking on border security.'

The mixing of technologies with ways of knowing, seeing and governing has important consequences for the conceptualization of technologies. One of the implications is that it is hard to individualize technologies as they are often part of networks of actors and are deeply related to ways of thinking. As Latour (1999: 182) once stated: 'B-52s do not fly, the US Air Force flies.' And with the air force comes the machinery of organizing, governing, strategic thinking and acting (see also Andersson 2016). The following will argue a perspective that posits an intimate entanglement between humans and things and between migration and technology affects the spatiality, temporality and agency of migration.

TECHNOLOGY AND THE DISTRIBUTION OF ACTIONS OVER SPACE AND TIME

The previous examples draw attention to how technologies work as a dissimination device that relates, connects and disconnects various actors, institutions and technologies at different places in different moments in time. As a result, technological issues do not leave the categories of time, space and action unaffected. In particular in migration processes and the specific stages and situations they consist of, technologies redesign the inner layout of these categories and their mutual relationships. The following subsection will discuss how technologies in migration processes affect the categories of space, time and action from the inside out.

Spatialities of Migration

Migration is unmistakably a spatial phenomenon, as people move from one place to another. Migration and 'place' influence each other mutually if places are given a different population or affect cultures and societies, or when migration areas arise where reception takes place or an extraordinary amount of human mobility, for instance in refugee camps in conflict areas or regions affected by disasters. Transportation technologies of all sorts function as instruments to facilitate migration.

Migration is a spatial phenomenon, but so is migration control. As Lahav and Guiraudon (2000) argue, migration control has moved 'upwards, downwards and outwards.' The national and international connection of registering systems and databases as well as the policies of different countries to restrict irregular migrants' access to health care, housing and education has moved border control inwards. The externalization of border control to other countries such as conducted by the European Union in Africa and by the United States in Mexico and other countries in the South has moved migration control outwards. 'The border is everywhere,' Lyon (2005) famously concluded.

For these reasons, the research into migration has drawn attention to all kinds of places and spaces that migrants encounter. Geographical perspectives have analyzed the ways migration and technologies interrelate with landscapes and seascapes (Boyce 2016; Squire 2015), and how materialities of all sorts intermingle with practices of care and control, of security policies and humanitarian aid (Pallister-Wilkins 2017). Meanwhile, the material and infrastructural turn in political theory and international relations has analyzed the technological machinery that is required for states and migrants in process of migration, but also in containment and deportation policies (Walters 2018).

Besides identifying all kinds of novel technological and infrastructural places and spaces related to migration, studies show space and place as a category are affected by technologies. First of all, instead of making borders 'virtual', digital technologies create specific relations with concrete material places. To execute the Dublin regulations, that require migrants to apply for asylum in the country of their first arrival, the Eurodac database that stores fingerprints of migrants is closely connected with the fingerprinting machines in the reception centers of the hotspots the EU has initiated during the so-called 'refugee crisis' of 2014–2016, such as in Greece and Italy. The gathering of biometric information thus connects the bodies of migrants at specific locations with databases and regulations.

The intermingling of data processing, controlling migrants and providing care allows for detecting the logistics of specific locations, such as hotspots. Hotspots, in such a view, can be considered as infrastructures that align 'staff from different organizations with databases, devices, and migrants all in one place and organizes mundane practices such as filling out forms, taking fingerprints, signing, and entering datasets along a chain' (Pollozek and Passoth 2019: 606). Hotspots combine 'data infrastructures, knowledge practices, and bureaucratic procedures through which populations unknown to European actors are translated into "European-legible" identities' (Pelizza 2019: 262).

Secondly, digital technologies affect the visualization and representation of migration. Borders are not just instruments that reach across different geographies. They intertwine with them in spatial and material ways. This intertwinement is conceptualized by Bigo (2015). He argues that borders on land, sea and 'in the air' can be solid, liquid or gaseous. He further distinguishes between three 'fields of action' in border control practices. The first, *solid* form of control is the border as a physical barrier, most often on land. The second, *liquid* form concerns border checks and practices of 'policing and surveillance' involving processes of identifying, authenticating and filtering. The third, *gaseous* form is 'the universe of the transnational database', connected to 'the digital and the virtual, to data doubles and their cohorts, to categorizations resulting from algorithms, to anticipations of unknown behaviors, to the prevention of future actions' (Bigo 2015: 59). As such, technologies transform the nature of the border and turn borders at land, sea or in the air into entities that allow for particular ways of representing and intervening in migration processes.

Temporalities of Migration

In his essay 'Waiting: Keeping Time' Khosravi (2014) discusses the relationship between migration and time by focusing on the aspect of 'waiting'. 'Waiting', according to Khosravi, is a way to regulate social interactions and to manipulate time. 'Large numbers of displaced people – undocumented migrants, refugees and asylum seekers – spend extended periods waiting in camps, in transit lands, or in search of papers. Lack of information on how long they have to wait, or what exactly they have to do to get their permits, makes their lives unpredictable and uncertain. This is most palpable in the case of asylum seekers, and in detention centers where migrants can often be kept indefinitely before deportation' (Khosravi 2014: 1).

Technologies in migration processes affect time in different ways. They may intensify and instrumentalize the process of waiting by denying people access to countries or receiving them in reception centers or add procedures so as to expand the possibilities to manipulate, control and discourage migrants, for instance by influencing the 'processing time', the time between signing an asylum application and receiving the decision from the immigration services. As such, technologies create new doubts, hopes and uncertainties (see also Janeja and Bandak 2018).

On the other hand, technologies accelerate the traveling of people, not only by creating faster ways of transportation but also by speeding up the processing of passengers and visa applications. The circulation of people at airports, digital platforms where people can register as trusted travelers, the use of biometrics for purposes of identification and verification all add to the intensification of time. For that reason, Sontowski (2018) speaks of 'tempo politics,' a concept that refers to the ways the EU Smart Borders Package and the EU's bio databases problematize border control primarily on the level of its temporalities.

But instead of making traveling and migration processes faster or slower, it is probably more accurate to say technologies create different regimes of temporality and distinguish not only between different forms of traveling, but between different kinds of travelers. As Van der Ploeg and Sprenkels (2011) have argued, biometrics tend to create 'machine readable bodies.' The body thus becomes a kind of passport (Dijstelbloem and Meijer 2011; see also Longo 2020). Meanwhile, migration procedures distinguish between different kinds of passports and different kinds of bodies, and add a temporal dimension to processes of 'social sorting' (Lyon 2003). Risk categories, Jones (2009: 177) argues, create power by 'obscuring difference by forcing the multiple into manageable units (categories) with solid separations (boundaries) between them.'

Technologies not only affect the ways migration is experienced or how states control and select people, but also how human mobility is being monitored. In particular the field of irregular migration has seen an increase of the means that are used to track and trace migrants, to monitor migration routes, and to identify risky situations, for instance when migrants cross the Mediterranean to reach Europe.

The monitoring of migration connects and disconnects visualization and action in different ways, and concerns the past, the present and the future. The concept of 'situational awareness' has entered migration policies, in particular the field of surveillance and search and rescue operations at sea (Suchman 2015, Dijstelbloem et al. 2017, Jeandesboz 2017; Walters 2017). The real-time monitoring of critical events by satellites radar equipment on vessels of border guards aims to make the present 'present' and to transform representations of events into actionable events. Meanwhile, monitoring re-directs time's arrow backwards and forwards. As Tazzioli (2018: 272) argues, 'the detection of migrants "on the spot" coexists with both a future-oriented temporality and an archival one.' Migration mapping monitoring devices create 'temporalities of visibility.' Tazzioli calls the resulting data sets and monitoring devices 'track-and-archive gazes' that detect migrants in (nearly) real time and, elaborating on the data collected, craft spaces of intervention that are future oriented (Tazzioli 2018: 272). The monitoring system Eurosur, in that sense, is as much a tool to anticipate on future migration routes ·as well as an archive of the logistics of migration crossings (Tazzioli 2018: 278). Amoore (2013) emphasizes that anticipating on future situations contains a 'politics of possibilities.' Profiling people and making risk assessments of various categories of travelers and migrants requires the gathering together of 'multiple elements of what is thought to be known about a person—each element, in the singular, a mere possibility … in order to give appearance to an emergent subject. … Contemporary border controls mobilize correlations between fragments of a human subject, incomplete but, once assembled, used to define the borderline itself and who may cross it' (Amoore 2013: 81). Both the border and the subject that aims to cross it take on the shape of a mosaic, the outcome of a process of 'piecing together.'

Technologies in migration processes not only accelerate or slow down processes of traveling or of passenger management; they also re-organize the relationship between the past, the present and the future by reconstucting events, anticipating future ones and creating real-time awareness.

Actions, Interactions, Transactions

As a dissemination device, technologies not only affect the categories of time and place. They also connect and disconnect different groups of actors and affect the notion of 'action' itself. The introduction of technological systems in the monitoring of migration and the management of border passages has been accompanied by the deployment of all kinds of technical specialists, consultants, trainers and officials. As Broeders and Dijstelbloem (2015: 243) argue, 'the state's perception of reality becomes more technologically and statistically mediated and "datafied."' Migration policies are conducted through 'stretched screens' (Lyon 2009) that relate public border, immigration and law enforcement officials and private 'unofficials'. For instance, the introduction of an entry–exit system and biometric documents in Senegal is as much driven by the circulation of knowledge about biometrics as through all kinds of professional practices (Frowd 2017).

Technologies not only implicate actions and interactions, but also generate 'transactions.' The concepts of 'mediation' and 'translation' as developed in science and technology studies are increasingly applied to the study of technologies in migration processes to emphasize that transformation takes place. Latour (1994), for instance, distinguishes between acts of delegation, composition and 'black-boxing'. Delegation means a redistribution of tasks attributed to human actors, institutions and technologies takes place. Composition implies a novel socio-technical configuration comes into being. The act of 'black-boxing' entails that all kinds of ideas, motives and intentions that underlie the application of a certain technology disappear in the technical construction of a technology – and become hard to trace.

The acts of delegation, composition and 'black-boxing' circulate in approaches that conceive the role of technology in migration processes in terms of the emergence of 'infrastructures' (Dijstelbloem and Van der Veer 2019; Pelizza 2019; Pollozek and Passoth 2019). Infrastructures can consist of data and knowledge infrastructures, of border infrastructures, of professional infrastructures but also of migration infrastructures (Xiang and Lindquist 2014). Large-scale infrastructures as developed by states and industries require substantial investments. But infrastructures do not have to be conceived solely as 'grand projects.' Infrastructures also arise out of the numerous socio-technical interactions migrants conduct. The notion of 'migration infrastructures' introduces a less state-oriented perspective on migration and takes the labor, housing, financial and transportation infrastructures migrants create as a starting point. Infrastructures, in that sense, are created through actions, interactions and transactions and emergence via the movements and modifications of migrants.

Approaches to arrive at a less state-centric perspective are motivated by the idea to 'bring the migrant back in.' Attending to technologies, in that respect, is not just a way to do justice to the 'missing masses' (Latour 1992) of migration by pointing at the material and technical lay-out of migration processes. Following the technologies that figure in migration processes and opening up the networks to identify the associations technologies have with different groups of actors, is also a way to bring the variety of actors into light.

When Dijstelbloem, van Reekum and Schinkel (2017), for instance, study 'surveillance at sea,' they are interested in the use of Eurosur and the cameras of border control vessels that visualize migration at the Aegean Sea between Greece and Turkey. But they are also interested in how migrants claim an active role and transform themselves from the 'objects' of migration and surveillance policies into 'subjects' by acts of 'recalcitrance.'

The use of technologies by migrants is not restricted to overcoming an oppositional subject–object relationship. In particular social media and cell phones have empowered migrants and provided them with important instruments. The IOM's (2017: 157) *World Migration Report 2018* mentions the 'appification' of migration, a development that makes 'migration processes fundamentally different in specific but important ways' (IOM 2017: 158). The report refers to apps such as InfoAid, which 'has been set up by a Hungarian couple to provide real-time advice on how to cross borders.' Other examples concern *Refugermany* and *Arriving in Berlin*. Unmistakably, this appification gives migration processes and migration policies a twist. For instance, according to the IOM (2017: 158), 'large numbers of refugees and other migrants in transit and host countries such as Turkey are not sitting and waiting for resettlement or return. They are taking matters into their own hands, principally because they can. Information, advice and money can be shared quickly, and the constraints of geography more easily overcome.' As a result, social media networks are indispensable for migrants. Leurs (2015) refers to them as the 'digital passage' of migrants whereas Smets (2018: 115) calls smartphones of migrants their 'mobile homes.' For that reason, according to Gillespie et al. (2018: 1), 'smartphones are lifelines, as important as water and food' (see also Dijstelbloem 2020).

Meanwhile, technologies empower not only migrants but also states and corporations. An example is the use of data and the multiple purposes data serve. Data that are required to connect migrants to labor markets or to analyze the demographic and social composition of societies may support the integration of migrants in societies but can also become part of a political economy of migration data and be applied as a surveillance tool (e.g., Zuboff 2019). Tracking mobility using mobile phone data, for instance, affects the privacy and social organization of migrant groups and further empowers states (Taylor 2016). Another example concerns financial transactions. On the one hand, providing migrants and refugees with digital identities and mobile money supports their capacity to act and protects them. On the other hand, the introduction of migrants and refugees in the credit card economy shapes new forms of technological participation with new obligations and vulnerabilities (e.g., Tazzioli 2017). The use of humanitarian data and 'digital humanitarianism' in general can help migrants and humanitarian organization, but also strengthens the position of states and companies that are able to track and trace migrant mobility and the transactions they conduct by analyzing data (see also Macias 2020: 343).

As migration and in particular irregular migration becomes increasingly contested, technologies become part of power struggles and conflicts. On the one hand, there has been greater disclosure by states on specific aspects, such as arrival numbers and demographics. The provision of information on migration by state and non-state actors to the public at large has seen an enormous increase. Websites with infographics, visual maps, representations of migrant routes, and updates on conflicts and risky situations flourish. State and non-state actors have the task to provide the public with information and to ground policies on analyses, evidence and knowledge. On the other hand, there are examples of states refusing to share data with NGOs about dangerous situations for migrants, for instance at the Mediterranean. This neglect threatens their well-being and survival of migrants and the ability of other actors to provide assistance. Technologies, for that reason, are also applied when actions are inadequate, attention is missing, support is lacking, and/or policies lack sufficient cooperation (see Dijstelbloem 2017). Various groups of academics, artists, NGOs and non-state actors have constructed registers, lists, databases, maps and visual representations of 'border deaths,' i.e., mortality-related migration (see Cuttitta and Last 2019). The reasons for such databases vary, from starting investigations after violations of people's 'right to life', and/or holdings states and other actors responsible, creating public awareness of the 'human costs of border control' (Spijkerboer 2007) to sharing information with family members, the recognition of the casualties and returning their bodies. In general, it is important that the provision of information does not 'technologize' the monitoring of migration by mainly focusing on the representation of human mobility while paying less attention to the reasons, intentions and aspirations that underlie migration.

Technologies in migration processes, in short, not only affect the nature of actions, interactions and transactions by delegating and transforming the roles of humans and technologies; studying technologies is also a means to bring in certain actors again, such as migrants, or to advocate for attention for actions not taken.

CONCLUSION

As a phenomenon, migration is by definition based on movement. Human mobility in a bordered world forms the ground of modern migration. Technologies mediate human migration by the deployment of all kinds of instruments and infrastructures to support movement, and by the way migration is monitored, surveilled and controlled. The entanglement between migration and technology and between movement and mobility control turns it into a phenomenon that causes spatial and temporal movements that affect the agency of migrants, states and non-state organizations.

This chapter distinguished between three conceptions of technology: as object or instrument, as network or infrastructure, and as world view. The analysis showed these notions are hard to separate. Technologies may be conceived differently at different moments in time as they evolve and the networks they function in increase or decrease. Moreover, technologies can operate as instrument, network or world view at the same time, depending on the specific context and configuration of actors and institutions. Despite the blurring of the boundaries between these conceptions, it is worthwhile to apply the different notions of technology in migration processes as a lens to detect the particular issues that are at stake. Analyzing technologies in migration processes with a variegated and relational conceptual repertoire prevents reducing them to mere instruments.

Reducing technologies to their instrumental use is a common pitfall. Even when technologies in migration processes are thoroughly discussed, something might escape from attention. The intensified international management of migration goes accompanied with the deployment of technologies of all sorts. Whether it concerns migration data and the monitoring of human mobility, border control management, humanitarian aid and asylum and integration procedures, states, non-state actors such as the various UN-bodies, NGOs and industries collaborate intensively. While this cooperation can improve international policies and support for migrants, the use of technologies runs the risk of creating a 'migration machinery' in which policies take the form of socio-technical programs and pilot projects that disguise state policies and humanitarian aid as technological projects. If debates are only focused on specific technologies and applications, the more gradual, long-term process of 'technologization' of migration and migration policies can escape attention. This complicates the ways technologies can be discussed, evaluated, regulated and addressed as political, policy, humanitarian and existential issues in a context of mobility aspirations, humanitarian needs, state responsibilities, social interactions, power and conflict.

For that reason, studies of migration and technology are often hesitant about the capabilities of states and NGOs of fully taking the consequences and the meaning of technologies into account. Many studies show existing tensions or emerging issues fail from being attributed to actors, agencies or forums that ought to consider their impact. Whereas the instrumental applications of technologies can be captured in evaluator frameworks, the more intimate entanglement between technology and ways of governing and ways of knowing is more complicated to address. As a result, the risk is that technologies in migration processes are reduced to instrumental relations so as to make them fit for debate.

An entangled relationship with technology suggests that we as humans cannot simply use technology for our own purposes or even delegate our wishes to a thing that functions on our behalf. Our intimate relationship with technology affects our ways of being, thinking and living. Technology transforms the reality of migration and its representation. It affects the spatiality of migration, the temporality and the relationship between actions and actors of all sorts. Meanwhile, technologies are unequally distributed and often work in asymmetric ways. But technologies, the above explained, not just generate beneficial or adverse effects, but affect the categories of time, space and action in migration processes from the inside out.

Three technological transformations in migration processes deserve specific attention: the hybridization of technology; the variation of temporal regimes; and the multiplication of the actors and institutions involved in migration processes. First, technological transformations in migration processes not only affect actors and institutions; they also affect the spaces where migration takes place. Landscapes and seascapes undergo particular *interventions* as well as particular *representations* by the migrants who cross them and the states that aim to govern them. For instance, as argued above, technologies turn borders at land, sea or in the air into 'solid', 'liquid' or 'gaseous' entities that allow for particular ways of representing and intervening in migration processes. Meanwhile, technologies work differently in different spaces: search and rescue operations or apps run by migrant organization operate differently in landscapes such as the Sonoran Desert than in seascapes such as the Mediterranean. The interaction between physical border management tools such as vessels and vehicles, monitoring and visualization techniques, and communication devices connects migrants, states, technologies and territories in particular ways. In such cases it is not only the growth of socio-technical relationships that needs to be attended to, but also the *hybrid entanglements* between technology, territory and geography that require further reflection. Not only human relations are increasingly mediated by technologies, but also the terrains and territories they relate to.

Second, technological transformations in migration processes produce a variety of temporal regimes. Time management through technology may favor some groups of migrants and disadvantage others, such as migrants who are allowed to enter countries by plane and granted a visa versus migrants who have to take other routes to apply for asylum. More generally, technologies produce different temporal regimes simultaneously. This variety not only concerns migrants, but also policy-makers and professionals as they have to combine analyses of past, present and future migration movements. The coming into being of different temporal regimes of migration emphasizes migration is *mediated* by all kinds of information and communication technologies. As an international political phenomenon, migration already is highly driven by hope and expectations and fueled by anxieties and fears. The representation of migration by all kinds of media, by states and migrants themselves, is infused with looking back and looking ahead.

Third, technological transformations generate a multiplication of actors and institutions. Technology is difficult to limit to a clearly defined object, instrument or device. The chapter showed that technologies operate in networks of actors and institutions, spread over various landscapes and seascapes. Technology itself can become a 'world view', a way of looking at the world in order to understand, to represent, to intervene and to govern. This shrinkage and growth of networks not only concerns the different modalities of technologies but also the chains of relations between actors, institutions and technologies. Technologies require investments and maintenance and come with all kinds of additional services, varying from applications to consultancy. For that reason, technological projects such as smart border management, large-scale databases, surveillance projects or the installment of registration centers come about with all kinds of actors, varying training sessions to intervention teams and from software developers to legal services. The deployment of technologies in migration processes not only strengthens the intensity of human mobility; it also connects migration to an increasing number of actors and institutions.

Governing migration, for these reasons, requires taking the role of technology into account – without being focused on technologies solely. Humans are not the only ones that travel. Technologies travel through migration processes as well, affecting them at every stage in every aspect. The same applies to the governance of migration. Technologies operate in all facets of state policies concerning migration, sometimes turning policies into technologies themselves, when policies are regarded as an apparatus to manage and monitor human mobility – and 'technology' becomes a substitute for 'politics.' Where the politics of migration and migration policy can enjoy extensive attention, technology in migration processes also deserves a treatment that does justice to all its forms and aspects and to the transformations it undergoes. A focus on how technologies affect spatiality, temporality and agency whilst technologies themselves transform from objects into networks and worldviews and vice versa not only creates more awareness for the widespread and variegated role of technologies in migration processes, but also sharpens the view on the novel political questions with regard to human mobility, borders, security and fundamental rights that arise and how these political questions appear or disappear in other guises.

NOTE

1.   The distinction between technology as instrument, infrastructure and world view is discussed in detail in Huub Dijstelbloem (2021), *Borders as Infrastructure: The Technopolitics of Border Control*, Cambridge, MA: The MIT Press.

REFERENCES

Allen, Michael Thad and Gabrielle Hecht (2001), *Technologies of Power. Essays in Honor of Thomas Parke Hughes and Agatha Chipley Hughes*, Cambridge, MA: The MIT Press.
Amoore, Louise (2013), *The Politics of Possibility: Risk and Uncertainty Beyond Probability*, Durham, NC: Duke University Press.
Andersson, Ruben (2016), 'Hardwiring the frontier? The politics of security technology in Europe's "fight against illegal migration"', *Security Dialogue* **47** (1): 22–39.
Barry, Andrew (2013), 'The translation zone: between actor–network theory and international relations', *Millennium: Journal of International Studies*, March 20, 1–17. https://doi.org/10.1177/0305829813481007.
Bellanova, Rocco and Gloria Gonzalez Fuster (2013), 'Politics of disappearance: scanners and (unobserved) bodies as mediators of security practices', *International Political Sociology* **7** (2): 188–209.
Best, Jacqueline and William Walters (2013), '"Actor–network theory" and international relationality: lost (and found) in translation', *International Political Sociology* **7** (3) (September 1): 332–34. https://doi.org/10.1111/ips.12026_1.
Bigo, Didier (2015), 'Death in the Mediterranean Sea: the results of the three fields of action of European Union Border Controls', in Jansen Yolande, Robin Celikates and Joost de Bloois (eds), *The Irregularization of Migration in Contemporary Europe*, London and New York: Rowman & Littlefield, pp. 55–70.
Bourne, Mike, Heather Johnson and Debbie Lisle (2015), 'Laboratizing the border: the production, translation and anticipation of security technologies', *Security Dialogue* **46** (4): 307–25.
Boyce, Geoffrey A (2016), 'The rugged border: surveillance, policing and the dynamic materiality of the US/Mexico frontier', *Environment and Planning D: Society and Space* **34** (2): 245–62.

Broeders, Dennis and Huub Dijstelbloem (2015), 'The datafication of mobility and migration management: the mediating state and its consequences', in Irma van der Ploeg and Jason Pridmore (eds), *Digitizing Identities: Doing Identity in a Networked World*, New York and London: Routledge, pp. 242–60.

Cuttitta, Paolo and Tamara Last (eds) (2019), *Border Deaths Causes: Dynamics and Consequences of Migration-related Mortality*, Amsterdam: Amsterdam University Press (AUP).

Dijstelbloem, Huub (2017), 'Migration tracking is a mess', *Nature*, 543, 2 March.

Dijstelbloem, Huub (2020), 'Borders and the contagious nature of mediation', in Kevin Smets, Koen Leurs, Myria Georghou, Saskia Witteborn and Radhika Gajjala (eds), *The Sage Handbook of Media and Migration*, London: Sage, pp. 311–20.

Dijstelbloem, Huub and Albert Meijer (eds) (2011), *Migration and the New Technological Borders of Europe*, Basingstoke: Palgrave Macmillan.

Dijstelbloem, Huub and Lieke van der Veer (2019), 'The multiple movements of the humanitarian border: the portable provision of care and control at the Aegean Islands', *Journal of Borderlands Studies* [10.1080/08865655.2019.1567371]

Dijstelbloem, Huub, Rogier van Reekum and Willem Schinkel (2017), 'Surveillance at sea: the transactional politics of border control in the Aegean', *Security Dialogue* **48** (3): 224–40.

Elden, Stuart (2013), *The Birth of Territory*. Chicago, IL: Chicago University Press.

Frowd, Philippe (2017), 'The promises and pitfalls of biometric security practices in Senegal', *International Political Sociology* **11** (4): 343–59.

Gillespie, M., S. Osseiran and M. Cheesman (2018), 'Syrian refugees and the digital passage to Europe: Smartphone infrastructures and affordances', *Social Media + Society* **4** (1): 1–12.

Haggerty, Kevin and Richard V. Ericson (2003), 'The surveillant assemblage', *The British Journal of Sociology* **51**: 605–22. doi:10.1080/00071310020015280

IOM (2017), *World Migration Report 2018*, Geneva: IOM.

Janeja, Manpreet and Andreas Bandak (2018), *Ethnographies of Waiting: Doubt, Hope and Uncertainty*, New York: Bloomsbury.

Jeandesboz, Julien (2016), 'Smartening border security in the European Union: an associational inquiry', *Security Dialogue* **47** (4): 292–309.

Jeandesboz, Julien (2017), 'European border policing: EUROSUR, knowledge, calculation', *Global Crime* **18** (3): 256–85.

Jones, Reece (2009), 'Categories, borders and boundaries', *Progress in Human Geography* **33** (2): 174–89.

Khosravi, Shahram (2014), 'Waiting: keeping time', in B. Anderson and M. Keith (eds), *Migration: A COMPAS Anthology*, Oxford: COMPAS. https://compasanthology.co.uk/explore/keeping-time/ (accessed August 18, 2021).

Lahav, Gallya and Virginie Guiraudon (2000), 'Comparative perspectives on border control: away from the border and outside the state', in Peter Andreas and Timothy Snyder (eds), *The Wall around the West: State Borders and Immigration Controls in North America and Europe*, Lanham, MD: Rowman & Littlefield Publishers, pp. 55–77.

Latour, Bruno (1992), 'Where are the missing masses? The sociology of a few mundane facts', in Wiebe Bijker and John Law (eds), *Shaping Technology / Building Society: Studies in Sociotechnical Change*, Cambridge, MA: The MIT Press, pp. 151–80.

Latour, Bruno (1994), 'On technical mediation', *Common Knowledge* **3** (2): 29–64.

Latour, Bruno (1999), *Pandora's Hope: Essays on the Reality of Science Studies*, Cambridge, MA: Harvard University Press.

Leurs, Koen (2015), *Digital Passages: Migrant Youth 2.0: Diaspora, Gender and Youth Cultural Intersections*, Amsterdam, The Netherlands: Amsterdam University Press.

Longo, Matthew (2020), 'Your body is a passport', at https://www.politico.eu/article/future-passports -biometric-risk-profiles-the-codes-we-carry/ (accessed August 18, 2021).

Lyon, David (2003), *Surveillance as Social Sorting: Privacy, Risk and Digital Discrimination*, New York: Routledge.

Lyon, David (2005), 'The border is everywhere: ID cards, surveillance and the other', in Elia Zureik and Mark Salter (eds), *Global Surveillance and Policing: Borders, Security, Identity*, Collumpton: Willan Publishing, pp. 66–82.

Macias, Lea (2020), 'Digital humanitarianism in a refugee camp', in Kevin Smets, Koen Leurs, Myria Georghou, Saskia Witteborn and Radhika Gajjala (eds), *The Sage Handbook of Media and Migration*, London: Sage, pp. 343–5.

Maier, Charles. S. (2016), *Once Within Borders: Territories of Power, Wealth and Belonging since 1500*, Cambridge, MA: Harvard University Press.

Mazower, Mark (2012), *Governing the World: The History of an Idea*, New York: Penguin Books.

Pallister-Wilkins, Polly (2017), 'Humanitarian borderwork: actors, spaces, categories', in Reece Jones, Corey Johnson, Wendy Brown, Gabriel Popescu, Polly Pallister-Wilkins, Alison Mountz and Emily Gilbert (2017), 'Interventions on the state of sovereignty at the border', *Political Geography* 59: 1–10. https://www.researchgate.net/publication/313829670_Interventions_on_the_state_of_sovereignty_at _the_border.

Pelizza, Annalisa (2019), 'Processing alterity, enacting europe: migrant registration and identification as co-construction of individuals and polities', *Science, Technology, & Human Values*, **45** (2): 262–288.

Pollozek, Silvan and Jan Hendrik Passoth (2019), 'Infrastructuring European migration and border control: the logistics of registration and identification at Moria hotspot', *EPD: Society and Space*, **37** (4): 606–624.

Salter, Mark B (ed.) (2015), *Making Things International 1: Circuits and Motion*, Minneapolis, MN and London: University of Minnesota Press.

Salter, Mark B (ed.) (2016), *Making Things International 2: Catalysts and Reactions*, Minneapolis, MN and London: University of Minnesota Press.

Schouten, Peer (2014), 'Security as controversy: reassembling security at Amsterdam Airport', *Security Dialogue* **45** (1): 23–42.

Smets, Kevin (2018), 'The way Syrian refugees in Turkey use media: understanding "connected refugees" through a non-mediacentric and local approach', *Communications*, **43** (1), 113–123.

Sontowski, Simon (2018), 'Speed, timing and duration: contested temporalities, techno-political controversies and the emergence of the EU's smart border', *Journal of Ethnic and Migration Studies* **44**: 16. DOI: 10.1080/1369183X.2017.1401512

Spijkerboer, Thomas (2007), 'The human costs of border control' (June 22, 2009), *European Journal of Migration and Law*, **9**: 127–39. Available at SSRN: https://ssrn.com/abstract=1423779

Squire, Vicki (2015), *Post/Humanitarian Border Politics between Mexico and the US: People, Places, Things*, New York: Palgrave Macmillan.

Suchman, Lucy (2015), 'Situational awareness: deadly bioconvergence at the boundaries of bodies and machines' *Media Tropes* **V** (1): 1–24.

Taylor, Linnet (2016), 'No place to hide? The ethics and analytics of tracking mobility using mobile phone data', *Environment and Planning D: Society and Space* **34** (2): 319–36.

Tazzioli, Martina (2018), 'Spy, track and archive: the temporality of visibility in Eurosur and Jora', *Security Dialogue* **49** (4): 272–88.

Tazzioli, Martina (2017), 'The circuits of financial-humanitarianism in the Greek migration laboratory', at https://www.law.ox.ac.uk/research-subject-groups/centre-criminology/centreborder-criminologies/ blog/2017/09/circuits (accessed August 18, 2021).

Torpey, John (1998), *The Invention of the Passport: Surveillance, Citizenship and the State*, Cambridge: Cambridge University Press.

Van der Ploeg, Irma and Isolde Sprenkels (2011), 'The migration machine and the machine-readable body', in Huub Dijstelbloem and Albert Meijer (eds), *Migration and the New Technological Borders of Europe*, Basingstoke: Palgrave Macmillan, pp. 68–104.

Walters, William (2014), 'Drone strikes, dingpolitik and beyond: furthering the debate on materiality and security', *Security Dialogue* **45** (2): 101–18.

Walters, William (2015), 'Reflections on migration and governmentality', in *Movements. Journal for Critical Migration and Border Regime Studies* **1** (1). http://movements-journal.org/issues/01 .grenzregime/04.walters--migration.governmentality.html.

Walters, William (2017), 'Live governance, borders, and the time–space of the situation: EUROSUR and the genealogy of bordering in Europe', *Comparative European Politics*, **15** (5): 794–817.

Walters, William (2018), 'Deportation as air power', at https://www.law.ox.ac.uk/research-subject -groups/centre-criminology/centreborder-criminologies/blog/2018/05/deportation-air   (accessed January 24, 2020).

Xiang, Biao, and Johan Lindquist (2014), 'Migration infrastructure', *International Migration Review*, **48** (1): 122–48.

Zuboff, Shoshana (2019), *The Age of Surveillance Capitalism: The Fight for a Human Future at the New Frontier of Power*, New York: Hachette Book Group.

26. Migration forecasting using new technology and methods

Arkadiusz Wiśniowski

INTRODUCTION

International migration has become a key driver of population change in developed countries, where immigration is often considered a remedy to shrinking labour force and ageing population (Buettner and Muenz 2016; IOM 2018). In developing countries, emigration and diaspora living abroad are becoming an important source of income in the form of remittances. Thus, there is a need for detailed and accurate migration forecasts to aid decisions in policymaking, planning, supplying particular social services and allocating resources in other areas, such as humanitarian aid to asylum seekers and displaced populations (e.g., United Nations Population Division 2011; Azose & Raftery 2015; Lanzieri 2019; Sohst et al. 2020).

However, international migration is perhaps the most difficult-to-forecast component of population change. It is difficult to conceptualise and define in a unique way. Migration is highly susceptible to factors and drivers that are themselves barely predictable, such as economic, environmental, social and political crises and upheavals (e.g., Massey et al. 1993; Jennissen 2003; Castles et al. 2013). A recent example is a conflict in Syrian Arab Republic, which led to around 6.6 million refugees (under the UNHCR mandate) displaced abroad (UNHCR 2020).

In this chapter, the focus is on international migration, that is, a relocation across country borders to change a person's usual residence (UN 1998; Lanzieri 2019), rather than on migration within a country (internal migration), or short-term mobility (travelling or moving without changing residence).[1] Migration flows are events of crossing international border and are counted over a period of time, usually a year; whereas stocks are counts of persons (migrants) at a given point in time, e.g. on a census date. Further, the terms 'forecast' and 'prediction' are used interchangeably to denote unconditional statements about most plausible future migration developments, whereas a 'projection' refers to scenarios conditional on narratives and sets of assumptions about the future (cf. Bijak 2010; Bijak and Czaika 2020).

BACKGROUND: SITUATING FORECASTING IN THE EXISTING LITERATURE

Theories of Migration

One can base forecasts of migration on theories of migration, past data, expert opinions, or combinations of these three elements. Regarding migration theories, on the one hand, literature suggests that the existing theoretical frameworks are not able to explain migration decisions nor provide a causal relationship between various factors influencing the numbers of various types of migrants moving between countries (see, e.g., Massey et al. 1993; Arango 2000; Lutz and Goldstein 2004). Rather, the existing theories provide selective *ex post* explanations to observed migration patterns and they are too specific to prove useful in migration forecasting (Bijak 2010).

On the other hand, recently Courgeau et al. (2017), Burch (2018) and Willekens (2018) propose a shift towards causal migration forecasting; Willekens (2018) additionally refers to the demographic theory of socioeconomic change proposed by Lutz (2012). In principle, it assumes that the processes of making a decision about relocation vary between individuals. For instance, those seeking education set different objectives and utilise entirely different resources than those escaping conflicts. Thus, Willekens (2018) advocates that causal forecasting should rely on individual (micro-level) biographies, as opposed to aggregate cohort biographies traditionally analysed by demographers (macro-level). Data on biographies could then be combined with the cognitive models of decision-making and a model of interactions between migration and other events in the life course of individuals (Klabunde et al. 2017).

From the macroeconomic perspective, Docquier (2018) develops long-term migration forecasts by using a general equilibrium model of the world economy. The model explains past migration and forecasts future trends conditional on the input variables. The inputs are the production technology, i.e. differences amongst countries in wages and Gross Domestic Product, as well as migration technology, i.e. decisions on migration based on education and proxies for networks in the destination countries.

Finally, it is not essential for forecasts to be valid and useful even if the causal mechanism remains undiscovered (Keyfitz 1982). For example, Bijak and Wiśniowski (2010 cf. Sohst et al. 2020) use 'theory-free' time series models augmented with expert opinion to forecast immigration to seven European countries. These approaches are explained in more detail in the Methods section.

Uncertainty

The uncertainty about future migration stems from the uncertainty about individual decisions (micro level) on migration and uncertainty about migration drivers (macro level). Castles et al. (2013) list six main drivers: demographics, geography, migrant networks, culture, economy and political environment. These aggregate factors affect individual decisions about whether to migrate, when and to which destination. The decision-making processes are complex, and typically made under uncertainty and limited information (Klabunde and Willekens 2016; Hinsch and Bijak 2019; Bijak and Czaika 2020). The main reasons for migration include seeking employment, reunion with family, education, and fleeing persecution (Castles et al. 2013).

The uncertainty about migration decisions and the complexity of individual factors and external drivers of migration underlie an inherent uncertainty of future migration (Keilman 1990; Bijak 2010; Bijak and Czaika 2020). Individuals with the same characteristics, such as age, sex and the level of education, can make very different decisions about whether to migrate, when and where to. This is so-called aleatory uncertainty, as compared to the epistemic uncertainty, which is due to a lack of knowledge about the phenomena under study (Gray et al. 2017; Bijak and Czaika 2020).

The uncertainty at the macro level is rooted in the fact that there is no unique definition of a migrant (IOM 2019; Lanzieri 2019). International migration involves a relocation across a border and a change of usual residence for a given period of time. However, the exact details are informed by legal, political, geographical and other factors (IOM 2020). Subsequently, these factors affect both the data used as input to forecasts (e.g., UN 2019) and a baseline definition specific to a forecasting exercise, which adds another layer of uncertainty, especially to cross-country predictions.

The (epistemic) uncertainty in the forecasts also stems from the imperfect measurement of migration as well as gaps in the data. Stock data are available for 232 countries and areas (UN 2019), but only 45 countries provide flow data (UN 2015). Underlying definitions may exclude, e.g., returning nationals such as Romania. Data are often biased due to the underreporting of those who left a country and failed to deregister or register in the destination country (Raymer et al. 2013). For example, according to the Eurostat database, in the years from 2002 to 2011 around 500,000 people notified the Polish population register about their departure, whereas Census 2011 counted 1.9 million people living abroad (Wiśniowski 2017). Errors and biases in the data used as input to a forecasting model are yet another source of uncertainty (Kupiszewska and Nowok 2008).

Irrespective of the definition, migration flows are typically characterised by far greater volatility than the migrant stocks, which subsequently leads to much larger uncertainty of flow forecasts. Also, the uncertainty about specific types of migration is driven by the uncertainty of the underlying drivers behind them. For instance, a general unpredictability of conflicts (cf. Chadefaux 2017) and natural disasters, especially in the long term, renders forecasts of asylum seekers more uncertain than, say, of returning nationals (e.g., Bijak et al. 2019). Finally, smaller geographies are typically characterised by larger volatility than national and supra-national measures (Wilson and Shalley 2019).

Quantification of Uncertainty

In demographic forecasting, two approaches to uncertainty can be distinguished: deterministic projections and probabilistic (stochastic) forecasts (Keilman 1990; Bijak 2010; Sohst et al. 2020). Projections are based on what-if scenarios, typically including a principal, that is, the most plausible, and low and high. They offer no information on the chances of the actual migration falling between the low and high variant (Lutz and Goldstein 2004), while the probability of any single scenario is zero under any continuous probability measure (Bijak 2010). However, the scenarios are useful for long-term planning purposes and offer a narrative about the future developments that would lead to a projected migration situation (Sohst et al. 2020).

Probabilistic forecasts provide a range of possible outcomes and a probability attached to that range (Keilman 2001). The users of population forecasts typically require quantified measures of uncertainty attached to the forecasts (Keyfitz 1981; Keilman 2001; Dunstan and Ball 2016; Dunstan 2019; Wilson and Shalley 2019). Frequentist and Bayesian inferential framework can be used to quantify the uncertainty and present it in terms of, e.g., confidence (in frequentist) or credibility (in Bayesian inference) intervals. Bayesian approach also permits the explicit inclusion of expert opinion in the form of prior probability distributions that are combined with the observed data via Bayes theorem (Bijak and Bryant 2016).

Now-casting of Migration, New Forms of Data and New Technologies

A separate challenge are short-term forecasts of sudden changes in migration patterns. Early detection and timely information thereof are required for short- and medium-term planning and policymaking, as well as deploying humanitarian aid in the case of forced displacements (Bijak et al. 2017). Such short-term forecasting is often referred to as now-casting or prediction of the present (Giannone et al. 2008).

A recent example of a sudden shift in migration patterns is an influx to Europe of asylum seekers during an ongoing (at the time of writing this chapter) Syrian Civil War that began in 2011. The IOM's Displacement Tracking Matrix[2] (DTM) is a system for systematic tracking and monitoring displacement and population mobility (IOM 2020) and provides daily reports on the numbers of arrivals to the European countries through monitored key entry points. Another case is an influx of migrants after the EU enlargements in 2004 (mostly from Poland to the UK and Ireland) and 2007 (from Romania to the UK). The exact numbers of migrants have been a subject of forecasts (e.g., Dustmann et al. 2003) and long debates on the exact size of the post-enlargement migration (cf. Wiśniowski 2017).

Technological innovation, especially in the domain of information and communication technologies, provides access to and allows the exchange of information on an unprecedented scale. International Telecommunications Union estimates that 4.1 billion individuals (54 per cent of the world population) were using the Internet in 2019; 93 per cent were covered by at least 3G mobile network.[3] Technologies enable not only migration per se (McAuliffe 2018), but also new possibilities of measuring it. The digital traces left as a by-product of using the Internet (Cesare et al. 2018) constitute new forms of data[4] (see, e.g., Hughes et al. 2016 and Righi 2019 for reviews of new forms of data in migration and mobility studies). These data, despite their limitations in terms of the bias resulting from the lack of representativeness of the general population, contain timely and high-fidelity information unavailable in traditional sources (Cesare et al. 2018; Gendronneau et al. 2019). Thus, they can provide useful input to migration forecasting models, especially those aiming at very short-term forecasts and detection of fluctuations in migration trends and patterns. However, due to the biases, they may not be useful in predicting the exact levels of migration.

As examples of using Big Data for migration now-casting, Zagheni et al. (2017) utilised data obtained from the Facebook Advertising Platform to predict current migrant stocks in the U.S.; Gendronneau et al. (2019) did so for the EU migrant stocks. In tracking the displacement, Alexander et al. (2019) used Facebook data to measure the displacement induced by Hurricane Maria in Puerto Rico in September 2017, whereas Palotti et al. (2020) monitor the number of Venezuelans living in other South American countries. Zagheni et al. (2014) supplemented Twitter data with facial recognition algorithms to infer demographic characteristics of migrant populations. Mobile phone data have also been used to track population displacement (e.g., Bengtsson et al. 2011; Blumenstock 2012); Lai et al. (2019) combined call detail records from the mobile phone operators to estimate internal migration patterns using gravity-type spatial interaction models. Finally, Melachrinos et al. (2020) used Big Data on disruptive events to create 'push factor' indices that are highly correlated with observed data on irregular migration.

The other sources of data underutilised in migration forecasting are administrative sources routinely collecting data on populations. These data include tax records, such as the Internal Revenue System in the U.S., which cover 95–98 per cent of the taxpayers or 87 per cent of households and their dependants (Hauer and Byars 2019) and allow studying migration at a very detailed geography level. Another source is border crossings and apprehensions, such as those collected by Frontex[5] on detected crossings of the external EU borders. These data, similar to the ones collected through the DTM, can be combined with control models that rely on time series methods, and subsequently used in early-warning systems that produce now-casts of sudden changes in mobility and migration patterns (Bijak et al. 2017).

METHODS

Analogously to population forecasting, there are two main approaches for producing migration forecasts. In the first one, so-called 'random scenario' method, forecasts are derived from the opinions elicited from experts (Lutz et al. 1997; Lutz and Scherbov 1998). The second one uses time series models to extrapolate time series data (Bijak 2010 and Billari et al. 2014 provide reviews of those methods and their generalisations). Additionally, in this section, methods for integrating data from various sources as well as simulation methods are presented. Other approaches used for quantifying uncertainty and relying on time series data include extrapolation of empirical errors of the past forecasts (Stoto 1983; Keilman 2008; Bijak et al. 2019) and applications of a fuzzy set theory (Demirel and Basak 2019).

Expert-based Forecasts

Expert-based migration forecasts have the potential to incorporate any envisaged future developments in policies regarding migration and integration, economic situation, environmental change and technological progress that cannot be captured in the past data (Cohen 2003). Experts can also base their opinions on theoretical arguments and mental models of migration (De Beer 2008; Willekens 2019). On the other hand, experts are known to be overly confident and underestimate migration uncertainty, especially when recent trends remain relatively stable (Keilman 1990; Bijak 2010).

Deterministic trajectories are commonly based on such elaborations, especially about net migration (Billari et al. 2012; Cappelen et al. 2015; see Case Study box). They are also used by a 'vast majority of developed countries' (Bijak, 2010:54; see also UNECE Database on Population Projections Metadata[6]), as well as in population forecasts prepared by international organisations, such as Eurostat for the European Union (Lutz et al. 2019) and the United Nations (2019) for all countries. Sohst et al. (2020) present a recent overview of methods used in preparing migration scenarios.

Lutz et al. (1997) and Lutz and Scherbov (1998) demonstrate how the deterministic low and high variants can be used to create probabilistic forecasts. First, the experts provide the value of future migration in low and high variants, denoted by L_F and H_F, for some forecast horizon F such that they are 90 per cent certain of future migration level falling within that interval. Thus, the range between L_F and H_F contains 90 per cent of future migration paths. Then, assuming that the distribution of values is symmetric, a Gaussian distribution can be fit so that the 90 per cent probability mass falls between L_F and H_F.

This method has been extended to allow for correlation structure between, for example, fertility and migration or between migration of males and females (Billari et al. 2012, 2014). In particular, Billari et al. (2014) provide a co-called supra-Bayesian approach that allows eliciting expert opinions about future values of population components, including emigration and immigration, from experts and treating them as data in a population forecasting model. Each expert provides a migration trajectory and values it will reach at horizon *F*. By conditioning the variance of the distribution at *t+1* on the variance at *t*, the uncertainty of the forecasts increases with the horizon of the forecast. To take into account the heterogeneity of expert opinion and future migration values, the method utilises a mixture distribution over several groups of experts.

BOX 26.1 CASE STUDY: NET MIGRATION

Net migration – a difference between immigration and emigration – is a measure commonly used in population projections (Cappelen et al. 2015; Raymer and Wiśniowski 2018). It is often assumed to be constant, set to zero, or extrapolated based on past trends and expert opinions. It is the main measure used by, e.g., United Nations in their population forecasts for all countries in the world (United Nations 2015).

However, this measure conflates two entirely different processes of emigration and immigration (Rogers 1990). For instance, zero net migration might be a result of minimal movements to and from a country, or very large in- and outflows that cancel each other. It also conceals different population characteristics of immigrants and emigrants, which is often illustrated with emigration of highly skilled workers ('brain drain') who are 'replaced' with low-skilled immigrants (Stark 2004; Shiff and Özden 2005).

Finally, net migration forecasts are often based on a demographic residual method that uses the population balancing equation: Net Migration = P(t + 1) − P(t) − B + D, where P(t) is population at t, B and D are births and deaths, respectively, occurring between t and t+1. Any measurement errors in these components are propagated in net migration forecasts (Hyndman and Booth 2008).

Time Series Forecasting

Time series models extrapolate historical migration trends (Bijak 2010; Azose and Raftery 2015; Bijak et al. 2019). The methods rely on autoregressive integrated moving average (ARIMA) models, originally proposed by Box and Jenkins (1976).

The basic idea of the autoregressive models is that the future values of migration can be predicted by using past data. The simplest prediction is a naïve prediction, in which the level of migration in year *t*, denoted y_t, predicts y_{t+1}:

$$y_{t+1} = y_t + \xi_{t+1} \tag{26.1}$$

ξ_t is an error term, typically assumed to be Gaussian with expectation zero and variance σ^2. This model is also called a random walk, and it exhibits irregular patterns and ever-increasing variance over time. Adding a drift parameter φ_0 to the model,

$$y_{t+1} = \varphi_0 + y_t + \xi_{t+1} \tag{26.2}$$

introduces an increasing (if $\varphi_0 > 0$) or decreasing (if $\varphi_0 < 0$) trend.

If the series exhibit no systematic change in its mean nor its variance over time, then they are called stationary. In the model, this can be achieved by introducing an autoregressive parameter $\varphi_1 \in (-1,1)$ to create an AR(1) model:

$$y_{t+1} = \varphi_0 + \varphi_1 y_t + \xi_{t+1} \tag{26.3}$$

In cases where parameter $\varphi_1 > 1$ or $\varphi_1 < -1$, series exhibit explosive behaviour.

Examples of forecasts of the annual total immigration to Sweden for the years 2014 to 2018, based on data for 1998 to 2013 and the three models, (1) random walk, (2) random walk with drift, and (3) autoregressive model, are shown in Figure 26.1. We observe that the median forecasts from the random walk (RW) and autoregressive (AR) models underpredict the reported trends, comparing with the random walk with drift (RWD). Random walk models are characterised by the ever-increasing uncertainty, while in the autoregressive model the uncertainty is relatively smaller. Nevertheless, the observed immigration falls within the 95 per cent Credible Intervals (CI) in all three models.

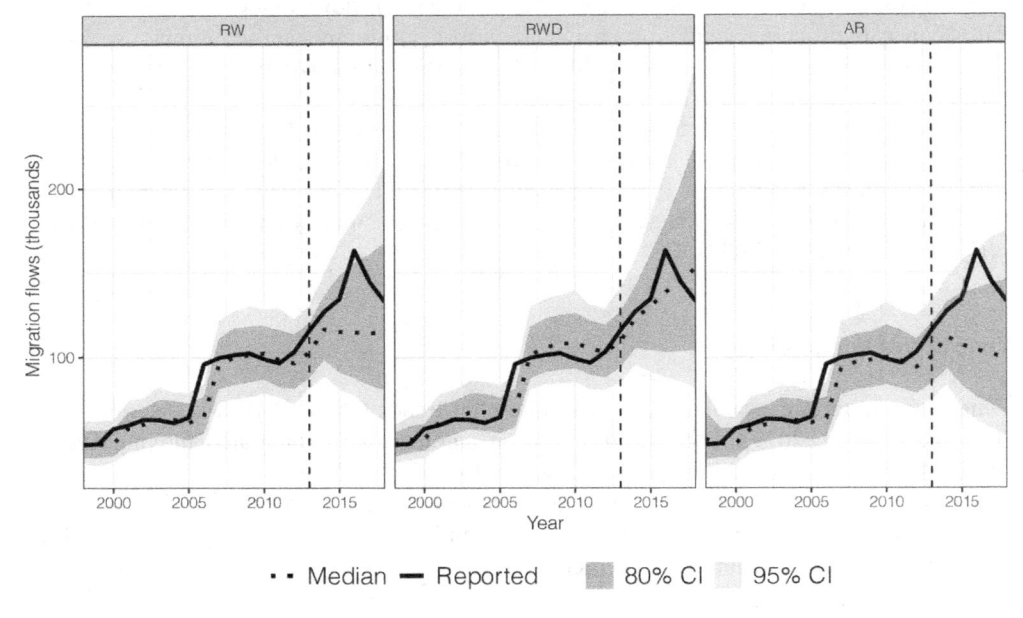

Notes: RW: random walk, RWD: random walk with drift, AR: autoregressive model specified as $y_t = \mu + \varphi_1 (y_{t-1} - \mu) + \xi_t$. All models fitted to the natural logarithm of reported immigration. X% CI denotes a Credible Interval with X% probability that the forecasted value will lie within that interval. Forecasts for 2014 to 2018 based on training sample 1998–2013. Prior distributions are weakly informative: $\mu \sim Uniform(-\infty,\infty)$, $\varphi_1 \sim Normal(0,1)$, $\varphi_1 \in (0,1)$, and $\sigma \sim Normal(0,1)$, $\sigma > 0$.
Source: Own calculations based on Eurostat data.

Figure 26.1   *Forecasts of immigration to Sweden, 2014–2018*

The models in equations (1)–(3) can be extended to include further lags of observed series, lags of the error terms, and series can be differenced.[7] More advanced extensions include local trend and state-space models (Bryant and Zhang 2017). It is also a common practice to work with logarithms of migration, as this ensures positivity of the forecast results (levels should be used when modelling net migration which can be negative).

Finally, economic variables and other covariates can be used in conjunction with the lagged values of migration to create econometric models (Dustmann et al. 2003; Bijak 2010; Cappelen et al. 2015; Bijak et al. 2019). In particular, information on Gross Domestic Product, unemployment or conflicts, as well as the access to technology, such as indicators measuring access to Internet or television, can be included depending on the case under study. The resulting forecasts are conditional on the trajectories of the covariates. Further, if the time series are sufficiently long, long-term fundamental relationship between the variables (so-called cointegration) can be tested and used to make joint forecasts for the covariates and migration (see, e.g., Bijak 2010: 121). Here, multivariate time series approaches, such as vector autoregressive models with or without imposed constraints, can be utilised (idem).

Delphi Surveys and Time Series with Expert Opinion

An interesting method of eliciting expert opinions is a Delphi method, originally developed for supporting decision-making in military applications (Dalkey 1969). It is an anonymous survey carried out amongst a heterogeneous group of experts. The method allows for the exchange of knowledge and opinions by deploying the survey in rounds and providing anonymised and aggregated results from the preceding round as input in the following round. Delphi survey has been applied in migration forecasting to predict emigration from the former USSR in the period 1992–1997 (Vishnevsky and Zayonchkovskaya 1994; after Bijak 2010: 55), as well as migration from East to West of Europe (Drbohlav 1996). At the time of writing this chapter, a large-scale Delphi study in migration projections is undertaken within the Cross Migration project (Acostamadiedo et al. 2020).[8]

Time series models can be combined with expert opinions. For example, Bijak and Wiśniowski (2010) forecast immigration to selected European countries based on past data and information elicited from the experts by using a Delphi survey. The expert opinions related to the type of the underlying model that 'generated' migration data: stationary or non-stationary, as well as the trajectory of reaching the levels at the forecast horizon. Then, the forecasts under different models were combined using Bayesian model averaging (cf. Abel et al. 2013). In another example, Wiśniowski et al. (2014) elicited expert opinions on the size of future migration to and from Scotland under the independence and remaining in the Union scenarios. The final forecasts were averaged over the two scenarios and the trajectory of reaching the target.

Models Explicitly Forecasting Age Schedules of Migration

Age schedules are known to exhibit regularities over age and across space (Rogers and Castro 1981). The largest peak in migration by age usually occurs for the graduates and young working-age population, then for new-borns and young children migrating with their parents, and at the retirement ages. Rogers and Castro (1981) provide a multiexponential model, useful for smoothing migration age schedules.

Another approach is based on multiplicative component (also known as log-linear) models. They decompose large arrays of migration data cross-classified by age and other dimensions, such as sex, origin, destination and time, into easy-to-interpret parameters that capture the main patterns in the data and can be extrapolated (Willekens and Baydar 1986; Raymer et al. 2006). A simple model for the logarithm of migration by age (denoted by x) and time (t) is

$$\log y_{x,t} = \alpha_x + \kappa_t + \xi_t, \quad \xi_t \sim Normal\left(0, \sigma_y^2\right) \tag{26.4}$$

where parameter α_x captures the average age profile and κ_t reflects the size of migration over time. Average age schedule is then forecast by using a time series model for κ_t, as described in the previous section.

The log-linear model can be extended to a model developed by Lee and Carter (in 1992) and widely used to forecast mortality (Booth 2006). The Lee–Carter approach decomposes migration by age and time into the age schedule averaged over time (parameter α_x), average changes of that schedule over time (β_x), with the size of changes over time captured by time parameter κ_t, (for which we again assume a time series model):

$$\log y_{x,t} = \alpha_x + \beta_x \kappa_t + \xi_t, \quad \xi_t \sim Normal\left(0, \sigma_y^2\right) \tag{26.5}$$

Model parameters can be estimated using frequentist (Hyndman and Booth 2008) or Bayesian inference (Raymer and Wiśniowski 2018). To capture correlations between emigration and emigration for males and females, Wiśniowski et al. (2015) and Raymer and Wiśniowski (2018) also use multivariate vector autoregressive (VAR) models for the time effects of each of four series.

A particular challenge of using those methods is the availability of detailed data. Sufficiently long series of migration data cross-classified by various characteristics are limited mostly to developed countries. Further, changes of definitions, such as administrative boundaries (in the case of internal migration) or criteria qualifying migrants, impact the applicability of these methods. Due to these limitations, migration forecasts are often produced for the main indicators, such as net migration or total immigration, which are then distributed to assumed age schedules (Azose et al. 2016; UN 2019).

Data Integration

Migration forecasts can be improved by integrating data from various sources. The integration approach is especially promising in utilising the new forms of data described in Background above. As accuracies of data collection vary between sources, integration allows a more trustworthy description of the forecast uncertainty, and overcoming the biases due to the lack of representativeness. The multiple data sources used simultaneously may provide a wider picture of future migration than a single data set alone.

Raymer et al. (2013) and Wiśniowski et al. (2016) develop a modelling framework for combining data reported by both sending and receiving countries, as well as information elicited from experts on migration, to estimate 'true' unobserved migration flows harmonised to a common definition and with measures of uncertainty. The model corrects the main inadequacies of the data: (i) different definitions of the duration of stay used by countries, (ii) undercount of migrants in population registers, (iii) accuracy of measurement, and (iv) coverage.

Gendronneau et al. (2019) adopt and extend the model of Raymer et al. (2013) to predict migrants stocks by using both data from traditional sources and new forms of data. The approach exploits the advantages of both types of data to reduce the biases present in them. Data used in the model include migration data officially reported to Eurostat by the EU member states, Census, Labour Force Survey (LFS), and data obtained from the Facebook Marketing API. In the model, each data source is assumed to produce biased measurements of migrant stocks; e.g., undercount in the population register or poor coverage of Facebook data. Also, each source has its own accuracy, i.e., censuses are assumed to be the most accurate, whereas LFS data suffer from sampling error. Informative prior distributions are then used to identify model parameters, and hierarchical time series models based on autoregressive models (Eq. 3) are applied to forecast migration.

Micro-level Approaches

The previously described methods produce forecasts of migration by using aggregate, macro-level data. These aggregates are the outcomes of the actions of individuals. Micro-level approaches analyse the life histories of individuals by using multi-state models (Willekens 2014). These models assume that, throughout lifetime, individuals move between various states (e.g., between countries or marital statuses). By analysing biographies, we can learn about the rates of transitions between the states. To forecast migration, one can then use these rates to simulate synthetic life histories and aggregate them to the quantity of interest, such as total number of migrants.

The simulation approaches, here broadly defined as microsimulation models, simulate a person's actions, which are then aggregated over many individuals (Zagheni 2015; Klabunde and Willekens 2016). The actions result from assuming pre-determined rates or probabilities, but without taking into account the decision-making process of individuals. The differences in actions between two individuals of the same characteristics are governed by 'chance and chance only' (Klabunde and Willekens 2016).

Microsimulations are applied across a wide range of disciplines (cf. Lomax and Smith 2017). In demography, they are used primarily for population forecasting. For example, they underlie the official population forecasts, including of international and internal migration, prepared by Statistics Canada (Morency et al. 2017). Lomax and Smith (2017) forecast population by age, sex and ethnicity for small areas in selected London boroughs. Bélanger et al. (2019) provide a framework for examining the future developments of population diversity measured in terms of language preference, religion and country of birth.

A sub-class of microsimulation approaches, called agent-based models, relies on causality by explicitly representing the behaviours and decisions of individuals ('agents'), who can interact with each other and with, for example, institutional agents (Klabunde and Willekens 2016). Such behavioural rules can be included in, for example, the multi-state models for biographies (Klabunde et al. 2017). The agent-based models allow an explicit inclusion of the diffusion of information and/or technology of accessing it amongst agents, as well as the rules on how it influences their decision-making about relocation (Gray et al. 2017). This is an important advantage over traditional approaches that utilise aggregate drivers of migration and conceal the complex decision-making process of individuals comprising the total number of migrants. The caveat of the agent-based models for their use in forecasting is that they typically explore possible theoretical explanations, rather than provide quantitative predictions about the future.

In an agent-based model, it can be assumed that agents do not have perfect foresight and acquire information about, say, wages in potential destinations, from their immediate networks, television or the Internet. Thus, agents with otherwise the same characteristics can make different decisions depending on their current information and not 'purely' at random. The ability of making use of technology by an individual can also be taken into account. For example, based on the method proposed by Kashyap and Villavicencio (2016) for simulating the dynamics of son preference in South Korea, a simple logistic function of technology diffusion amongst migrants can be formulated as:

$$Ability(t) = \frac{e^{\rho(t-\varphi)}}{1+e^{\rho(t-\varphi)}} \tag{26.6}$$

where $Ability(t)$ denotes an increasing with time t the probability of having access to technology, e.g. mobile Internet, ρ is a rate of increase and φ is a point in time when the increase becomes ever slower. In another example on forced migration, Hinsch and Bijak (2019) provide preliminary insights into how the access to limited information determines the choice of routes to destination when agents are forced to relocate because of conflict. The lack of full information makes the choices about the routes suboptimal and, thus, difficult to predict.

Microsimulations have the ability of overcoming the limitations of the models based only on macro-level data (such as log-linear models described in previous section, or cohort-component multi-state models, Rogers 1980), by explicitly modelling heterogeneous sub-populations and permitting interactions amongst individuals (Bélanger and Sabourin 2017; Bélanger et al. 2019). These approaches allow the inclusion of the decision rules, access to technology and information, which are typically too complex to be meaningfully incorporated in macro-level models. Finally, data from various sources, including expert opinion, can be used to inform model parameters, transition rates and decision rules (Burch 2018).

The main limitation of the micro-based approaches is, however, a frequent mismatch between the explanations of migration behaviours and the empirical patterns observed at a macro level, that is, in indicators such as total immigration or emigration (Billari 2015). Attempts are undertaken to bridge that gap in population forecasting. Willekens (2005) and Gampe et al. (2009) present the approach that combines the analysis of individual biographies with the outputs of a cohort-component model for projecting populations by age and sex with aggregated inputs. Unfortunately, migration in these projections has not been treated as comprehensively as fertility and mortality. Also, the agent-based models with explicit causal behavioural rules (cf. Hinsch and Bijak 2019), are yet to be developed and assessed in terms of their predictive power in the real world.

FURTHER READING

An excellent overview of probabilistic methods for forecasting migration within the Bayesian paradigm, as well as migration theories is offered by Bijak (2010). Bayesian inference in demographic estimation and forecasting is also at the core of the book by Bryant and Zhang (2017), and they provide examples of using log-linear models in demographic analysis. Log-linear models are also used in analysing contingency tables (or cross-classified data). Readers with interest in technical details of these models and other methods described in this chapter can reach to Agresti (2003, Ch. 8). A comprehensive and accessible introduction to time series analysis is given by Chatfield and Xing (2019). Further, Gelman et al. (2013) is perhaps the most often cited textbook on Bayesian inference, whereas Kaplan (2014) offers an accessible introduction for social scientists. For the simulation-based approaches, Klabunde and Willekens (2016) provide a review of causal theories of migration based on decisions taken by individual agents, while Zagheni (2015) offers an introduction to microsimulations for a non-technical audience; Bélanger and Sabourin (2017) provide a more technical text.

NOTES

1. The UN defines an international migrant as a person changing his or her country of usual residence for at least 12 months (United Nations 1998). The United Nations High Commissioner for Refugees (UNHCR) distinguishes between 'refugees' as those fleeing wars, persecution or natural disasters, from 'migrants' who 'choose to move not because of a direct threat of persecution or death, but mainly to improve their lives [...]' (Edwards 2016). From a demographic and statistical perspective, asylum seekers and refugees are migrants.
2. See https://dtm.iom.int/, accessed 27/07/2020.
3. See https://www.itu.int/en/ITU-D/Statistics/Pages/stat/default.aspx, accessed 28/07/2020.
4. Some of them are called Big Data (cf. Gandomi and Haider 2015).
5. The European Agency for the Management of Operational Cooperation at the External Borders of the Member States of the European Union.
6. See   https://statswiki.unece.org/display/PP/UNECE+Database+on+Population+Projections+ Metadata, accessed 08/08/2021.
7. Notice in Eq. (1) that $y_{t+1} - y_t = \xi_{t+1}$ has a mean zero and a constant variance σ^2 – that is, it is stationary.
8. See https://migrationresearch.com/migration-scenarios, accessed 08/08/2021.

REFERENCES

Abel, G. J., Bijak, J., Forster, J. J., Raymer, J., Smith, P. W. & Wong, J. S. (2013), 'Integrating uncertainty in time series population forecasts: an illustration using a simple projection model', *Demographic Research*, 29, 1187–1226.

Acostamadiedo, E., Sohst, R., Tjaden, J., Groenewold, G. & de Valk, H. (2020), 'Assessing immigration scenarios for the European Union in 2030 – relevant, realistic and reliable?', International Organization for Migration, Geneva, and the Netherlands Interdisciplinary Demographic Institute, The Hague.

Agresti, A. (2003), *Categorical Data Analysis*, Second Edition. Chichester: John Wiley & Sons.

Alexander, M., Polimis, K. & Zagheni, E. (2019), 'Leveraging Facebook's advertising platform to monitor stocks of migrants', *Population and Development Review*, 45(3), 617–630.

Arango, J. (2000), 'Explaining migration: A critical view', *International Social Science Journal*, 52(165), 283–296.

Azose, J. J. & Raftery, A. E. (2015), 'Bayesian probabilistic projection of international migration', *Demography*, 52(5), 1627–1650.

Azose, J. J., Ševčíková, H. & Raftery, A. E. (2016), 'Probabilistic population projections with migration uncertainty', *Proceedings of the National Academy of Sciences*, 113(23), 6460–6465.

Bélanger, A. & Sabourin, P. (2017), *Microsimulation and Population Dynamics*, Cham: Springer.

Bélanger, A., Sabourin, P., Marois, G., Van Hook, J. & Vézina, S. (2019), 'A framework for the prospective analysis of ethno-cultural super-diversity', *Demographic Research*, 41, 293–330.

Bengtsson, L., Lu, X., Thorson, A., Garfield, R. & Von Schreeb, J. (2011), 'Improved response to disasters and outbreaks by tracking population movements with mobile phone network data: a post-earthquake geospatial study in Haiti', *PLoS Med*, 8(8), e1001083.

Bijak, J. (2010), *Forecasting International Migration in Europe: A Bayesian View*. Dordrecht: Springer.

Bijak, J. & Wiśniowski, A. (2010), 'Bayesian forecasting of immigration to selected European countries by using expert knowledge', *Journal of the Royal Statistical Society: Series A (Statistics in Society)*, 173(4), 775–796.

Bijak, J. & Bryant, J. (2016), 'Bayesian demography 250 years after Bayes', *Population Studies*, 70(1), 1–19.

Bijak, J., Forster, J.J. & Hilton, J. (2017), 'Quantitative assessment of asylum-related migration: a survey of methodology', Report for the European Asylum Support Office. ESRC Centre for Population Change, Southampton.

Bijak, J., Disney, G., Findlay, A. M., Forster, J. J., Smith, P. W. & Wiśniowski, A. (2019), 'Assessing time series models for forecasting international migration: Lessons from the United Kingdom', *Journal of Forecasting*, 38(5), 470–487.

Bijak, J. & Czaika, M. (2020), 'Assessing uncertain migration futures: A typology of the unknown', QuantMig Project Deliverable D1.1. Southampton / Krems: University of Southampton and Danube University Krems.

Billari, F. C. (2015), 'Integrating macro- and micro-level approaches in the explanation of population change', *Population Studies*, 69(Suppl.), S11–S20.

Billari, F. C., Graziani, R. & Melilli, E. (2012), 'Stochastic population forecasts based on conditional expert opinions', *Journal of the Royal Statistical Society: Series A (Statistics in Society)*, 175(2), 491–511.

Billari, F. C., Graziani, R. & Melilli, E. (2014), 'Stochastic population forecasting based on combinations of expert evaluations within the Bayesian paradigm', *Demography*, 51(5), 1933–1954.

Blumenstock, J. E. (2012). 'Inferring patterns of internal migration from mobile phone call records: evidence from Rwanda', *Information Technology for Development*, 18(2), 107–125.

Booth, H. (2006), 'Demographic forecasting: 1980 to 2005 in review', *International Journal of Forecasting*, 22, 547–581.

Box, G. E. P. & Jenkins, G. M. (1976), *Time Series Analysis: Forecasting and Control*, San Francisco, CA: Holden-Day.

Bryant, J. & Zhang, J. L. (2018), *Bayesian Demographic Estimation and Forecasting*, London: Chapman and Hall/CRC.

Buettner, T. & Muenz, R. (2016), 'Comparative analysis of international migration in population projections', KNOMAD Working Paper 10. Washington DC.

Burch, T. K. (2018), *Model-based Demography: Essays on Integrating Data, Technique and Theory*, Cham: Springer.

Cappelen, Å., Skjerpen, T. & Tønnessen, M. (2015), 'Forecasting immigration in official population projections using an econometric model', *International Migration Review*, 49(4): 945–980.

Castles, S., De Haas, H. & Miller, M. J. (2013), *The Age of Migration: International Population Movements in the Modern World*, New York: Macmillan International Higher Education.

Cesare, N., Lee, H., McCormick, T., Spiro, E. & Zagheni, E. (2018), 'Promises and pitfalls of using digital traces for demographic research', *Demography*, 55(5), 1979–1999.

Chadefaux, T. (2017), 'Conflict forecasting and its limits', *Data Science*, 1(1–2), 7–17.

Chatfield, C. and Xing, H. (2019), *The Analysis of Time Series: An Introduction with R*, Seventh Edition. London: Chapman and Hall/CRC.

Cohen, J. (2003), 'Human population: the next half century', *Science*, 302, 1172–1175.

Courgeau, D., Bijak, J., Franck, R. & Silverman, E. (2017), 'Model-based demography: towards a research agenda', in Grow, A. and Van Bavel, J. (eds), *Agent-based Modelling in Population Studies*, Cham: Springer, pp. 29–51.

Dalkey, N. C. (1969), 'The Delphi method: an experimental study of group opinion', Research Memorandum RM-5888-PR. RAND, Santa Monica.

De Beer, J. (2008), 'Forecasting international migration: time series projections vs argument-based forecasts', in Raymer, J. & Willekens, F. (eds), *International Migration in Europe: Data, Models and Estimates*, Chichester: John Wiley & Sons, pp. 283–306.

Demirel, D. F. & Basak, M. (2019), 'A fuzzy bi-level method for modeling age-specific migration', *Socio-Economic Planning Sciences*, 68, 100664.

Docquier, F. (2018), 'Long-term trends in international migration: lessons from macroeconomic model', *Economics and Business Review*, 4(1), 3–15.

Drbohlav, D. (1996), 'The probable future of European East–West international migration – selected aspects', in Carter, F.W., Jordan, P. & Rey, V. (eds), *Central Europe after the Fall of the Iron Curtain; Geopolitical Perspectives, Spatial Patterns and Trends*, Wiener Osteuropa Studien No. 4, Frankfurt: Peter Lang & Co., pp. 269–296.

Dunstan, K. (2019), 'Stochastic projections: the New Zealand experience', Economic Commission for Europe / Eurostat Conference of European Statisticians, Work Session on Demographic Projections, Belgrade, 25–27 November 2019.

Dunstan, K. & Ball, C. (2016), 'Demographic projections: User and producer experiences of adopting a stochastic approach', *Journal of Official Statistics*, 32(4), 947–962.

Dustmann, C., Casanova, M., Fertig, M., Preston, I. & Schmidt, C. M. (2003), 'The impact of EU enlargement on migration flows', Research Development and Statistics Directorate, No. 25/03, Home Office, London.

Edwards, A. (2016), 'UNHCR viewpoint: "Refugee" or "migrant" – Which is right? United Nations Higher Commissioner for Refugees. At https://www.unhcr.org/en-us/news/latest/2016/7/55df0e556/unhcr-viewpoint-refugee-migrant-right.html (accessed 22/07/2020).

Gampe, J., Zinn, S., Willekens, F., Van Der Gaag, N., De Beer, J., Himmelspach, J., & Uhrmacher, A. (2009), 'The microsimulation tool of the MicMac project', in 2nd General Conference of the International Microsimulation Association, Ottawa, 8–10 June.

Gandomi, A. & Haider, M. (2015), Beyond the hype: big data concepts, methods, and analytics. *International Journal of Information Management*, 35(2), 137–144.

Gelman, A., Carlin, J. B., Stern, H. S., Dunson, D. B., Vehtari, A. & Rubin, D. B. (2013), *Bayesian Data Analysis*, Third Edition. London: Chapman and Hall/CRC.

Gendronneau, C., Wiśniowski, A., Yildiz, D., Zagheni, E., Fiorio, L., Hsiao, Y., Stepanek, M., Weber, I., Abel, G. & Hoorens, S. (2019), 'Measuring Labour Mobility and Migration Using Big Data: Exploring the potential of social-media data for measuring EU mobility flows and stocks of EU movers', Luxembourg: Publications Office of the European Union. DOI: 10.2767/474282.

Giannone, D., Reichlin, L. & Small, D. (2008), 'Nowcasting: The real-time informational content of macroeconomic data', *Journal of Monetary Economics*, 55(4): 665–676.

Gray, J., Hilton, J. & Bijak, J. (2017), 'Choosing the choice: Reflections on modelling decisions and behaviour in demographic agent-based models', *Population Studies*, 71(sup1), 85-97.

Hauer, M. & Byars, J. (2019), 'IRS county-to-county migration data, 1990–2010', *Demographic Research*, 40, 1153–1166.

Hinsch, M. & Bijak, J. (2019), 'Rumours lead to self-organized migration routes', Paper for the Agent-based Modelling Hub, Artificial Life conference, ALife 2019. At http://abmhub.cs.ucl.ac.uk/2019/papers/Hinsch_Bijak.pdf (accessed 8 August 2021).

Hughes, C., Zagheni, E., Abel, G. J., Wiśniowski, A., Sorichetta, A., Weber, I. & Tatem, A. J. (2016), 'Inferring migrations: Traditional methods and new approaches based on mobile phone, social media, and other big data', Luxembourg: Publications Office of the European Union. DOI: 10.2767/61617.

Hyndman, R. J. & Booth, H. (2008), 'Stochastic population forecasts using functional data models for mortality, fertility and migration', *International Journal of Forecasting*, 24: 323–342.

International Organisation for Migration (IOM) (2018), World Migration Report 2018. At https://www.iom.int/wmr/2018 [accessed 08/08/2021].

International Organisation for Migration (IOM) (2019), Glossary on Migration. International Migration Law No. 34, IOM, Geneva. At www.iom.int/glossary-migration-2019. https://www.iom.int/key-migration-terms#Migrant [accessed 20/07/2020].

International Organisation for Migration (IOM) (2020), World Migration Report 2020: Chapter 2 Migration and Migrants: A Global Overview. At https://www.iom.int/wmr/2020 [accessed 08/08/2021].

Jennissen, R. (2003), 'Economic determinants of net international migration in Western Europe', *European Journal of Population/Revue Européenne de Démographie*, 19(2), 171–198.

Kaplan, D. (2014), *Bayesian Statistics for the Social Sciences*, New York: Guilford Publications.

Kashyap, R. & Villavicencio, F. (2016), 'The dynamics of son preference, technology diffusion, and fertility decline underlying distorted sex ratios at birth: A simulation approach', *Demography*, 53(5), 1261–1281.

Keilman, N. (1990), *Uncertainty in National Population Forecasting: Issues, Backgrounds, Analyses, Recommendations*, Amsterdam: Swets & Zeitlinger.

Keilman, N. (2001), 'Uncertain population forecasts', *Nature* 412, 490–491.

Keyfitz, N. (1981), 'The limits of population forecasting', *Population and Development Review*, 7(4), 579–593.

Keyfitz, N. (1982), 'Can knowledge improve forecasts?', *Population and Development Review*, 8(4), 729–751.

Klabunde, A. & Willekens, F. (2016), 'Decision-making in agent-based models of migration: state of the art and challenges', *European Journal of Population*, 32(1), 73–97.

Klabunde, A., Zinn, S., Willekens, F. & Leuchter, M. (2017), 'Multistate modelling extended by behavioural rules: an application to migration', *Population Studies*, 71(sup1), 51–67.

Kupiszewska, D. & Nowok, B. (2008), 'Comparability of statistics on international migration flows in the European Union', in Raymer, J. & Willekens, F. (eds), *International Migration in Europe: Data, Models and Estimates*, Chichester: John Wiley & Sons, pp. 41–71.

Lai, S., Erbach-Schoenberg, E.z., Pezzulo, C. et al. (2019), 'Exploring the use of mobile phone data for national migration statistics', *Palgrave Communications*, 5, art. no. 34.

Lanzieri, G. (2019), An alternative view on the statistical definition of migration: Meeting of the Work Session on Migration Statistics, UNECE, Geneva, 29–31 October.

Lomax, N. & Smith, A. (2017), 'Microsimulation for demography', *Australian Population Studies*, 1(1), 73–85.

Lutz, W. (2012), 'Demographic metabolism: a predictive theory of socioeconomic change', *Population and Development Review*, 38 (supplement): 283–301.

Lutz, W., Sanderson, W. & Scherbov, S. (1997), 'Doubling of world population unlikely', *Nature*, 387(6635), 803–805.

Lutz, W. & Scherbov, S. (1998), 'An expert-based framework for probabilistic national population projections: the example of Austria', *European Journal of Population/Revue Européenne de Démographie*, 14(1), 1–17.

Lutz, W. & Goldstein, J. R. (2004), 'Introduction: how to deal with uncertainty in population forecasting?', *International Statistical Review*, 72(1): 1–4.

Lutz, W. (ed.), Amran, G., Bélanger, A., Conte, A., Gailey, N., Ghio, D., Grapsa, E., Jensen, K., Loichinger, E., Marois, G., Muttarak, R., Potančoková, M., Sabourin, P. & Stonawski M. (2019), 'Demographic scenarios for the EU – Migration, Population and Education – executive summary', EUR 29739 EN, Luxembourg: Publications Office of the European Union.

Massey, D. S., Arango, J., Hugo, G., Kouaouci, A., Pellegrino, A. & Taylor, J. E. (1993), 'Theories of international migration: a review and appraisal', *Population and Development Review*, 19(3), 431–466.

McAuliffe, M. (2018), 'The link between migration and technology is not what you think', Agenda, World Economic Forum, Geneva, 14 December. At www.weforum.org/agenda/2018/12/socialmedia-is-casting-a-dark-shadow-over-migration/ [accessed 28/07/2020].

Melachrinos, C., Carammia, M. & Wilkin, T. (2020), 'Using big data to estimate migration "push factors" from Africa', in International Organization for Migration (eds), *Migration in West and North Africa and across the Mediterranean*, Geneva: International Organization for Migration, pp. 98–116.

Morency, J. D., Malenfant, É. C. & MacIsaac S. (2017), 'Immigration and diversity: Population projections for Canada and its Regions, 2011 to 2036. Statistics Canada = Statistique Canada. Catalogue No. 91-551-X. At https://www150.statcan.gc.ca/n1/pub/91-551-x/91-551-x2017001-eng.htm [accessed 08/08/2020].

Palotti, J., Adler, N., Morales-Guzman, A., Villaveces, J., Sekara, V., Garcia Herranz, M. ... & Weber, I. (2020), 'Monitoring of the Venezuelan exodus through Facebook's advertising platform', *PloS one*, 15(2): e0229175.

Raymer, J., Bonaguidi, A. & Valentini, A. (2006), 'Describing and projecting the age and spatial structures of interregional migration in Italy', *Population, Space and Place*, 12(5): 371–388.

Raymer, J., & Wiśniowski, A. (2018), 'Applying and testing a forecasting model for age and sex patterns of immigration and emigration', *Population Studies*, 72(3), 339–355.

Raymer, J., Wiśniowski, A., Forster, J. J., Smith, P. W. & Bijak, J. (2013), 'Integrated modeling of European migration', *Journal of the American Statistical Association*, 108(503), 801–819.

Righi, A. (2019), 'Assessing migration through social media: a review', *Mathematical Population Studies*, 26(2), 80–91. https://doi.org/10.1080/08898480.2019.1565271.

Rogers, A. (1980), 'Introduction to multistate mathematical demography', *Environment and Planning A*, 12(5), 489–498.

Rogers, A. (1990), 'Requiem for the net migrant', *Geographical Analysis*, 22(4): 283–300.

Rogers, A. & Castro, L. J. (1981), 'Model migration schedules', Research Report 81-30, Laxenburg, Austria: International Institute for Applied Systems Analysis.

Sohst, R., Tjaden, J., de Valk, H. & Melde, S. (2020), *The Future of Migration to Europe: A Systematic Review of the Literature on Migration Scenarios and Forecasts*. International Organization for Migration, Geneva, and the Netherlands Interdisciplinary Demographic Institute, The Hague.

Stark, O. (2004), 'Rethinking the brain drain', *World Development*, 32(1): 15–22.

Stoto, M. A. (1983), 'The accuracy of population projections', *Journal of the American Statistical Association*, 78: 13–20.

United Nations (1998), 'Recommendations on statistics of international migration', Statistical Papers Series M, 58, Revision 1. Department of Economic and Social Affairs, Statistics Division, United Nations, New York.

United Nations (2015), 'World population prospects: The 2015 revision, methodology of the United Nations population estimates and projections', Working Paper No. ESA/P/WP.242, Department of Economic and Social Affairs, Population Division, United Nations, New York.

United Nations (2019), 'World Population Prospects 2019: Methodology of the United Nations population estimates and projections', ST/ESA/SER.A/425, Department of Economic and Social Affairs, Population Division, United Nations, New York.

United Nations Higher Commissioner for Refugees (UNHCR) (2020), *Refugee Statistics*. At https://www.unhcr.org/refugee-statistics/download/?url=4t7R [accessed 21/07/2020].

United Nations Population Division (2011), 'World population prospects: The 2010 revision', New York: United Nations.

Vishnevsky, A. & Zayonchkovskaya, Z. (1994), 'Emigration from the former Soviet Union: The fourth wave', in H. Fassmann & R. Muenz (eds), *European Migration in the Late Twentieth Century: Historical Pattens, Actual Trends, and Social Implications*, Aldershot, UK and Brookfield, VT, USA: Edward Elgar Publishing, pp. 239–259.

Willekens, F. (2005), 'Biographic forecasting: bridging the micro–macro gap in population forecasting', *New Zealand Population Review*, 31(1): 77–124.

Willekens, F. (2014), *Multistate Analysis of Life Histories with R*, New York: Springer.

Willekens, F. (2018), 'Towards causal forecasting of international migration', *Vienna Yearbook of Population Research*, 16: 1–20.

Willekens, F. (2019), 'Evidence-based monitoring of international migration flows in Europe', *Journal of Official Statistics*, 35(1): 231–277.

Willekens, F. & Baydar, N. (1986), 'Forecasting place-to-place migration with generalized linear models', in R. Woods and P. Rees (eds), *Population Structures and Models: Developments in Spatial Demography*, London: Allen & Unwin, pp. 203–244.

Wilson, T. & Shalley, F. (2019), 'Subnational population forecasts: Do users want to know about uncertainty?', *Demographic Research*, 41: 367–392.

Wiśniowski, A. (2017), 'Combining Labour Force Survey data to estimate migration flows: The case of migration from Poland to the UK', *Journal of the Royal Statistical Society Series A*, 180: 185–202.

Wiśniowski, A., Bijak, J. & Shang, H. L. (2014), 'Forecasting Scottish migration in the context of the 2014 constitutional change debate', *Population, Space and Place*, 20(5), 455–464.

Wiśniowski, A., Smith, P. W., Bijak, J., Raymer, J. & Forster, J. J. (2015), 'Bayesian population forecasting: extending the Lee–Carter method', *Demography*, 52(3), 1035–1059.

Wiśniowski, A., Forster, J. J., Smith, P. W., Bijak, J. & Raymer, J. (2016), 'Integrated modelling of age and sex patterns of European migration', *Journal of the Royal Statistical Society: Series A (Statistics in Society)*, 179(4), 1007–1024.

Zagheni, E. (2015), 'Microsimulation in demographic research', in Wright, J. D. (ed.), *International Encyclopedia of Social and Behavioral Sciences*, 2nd Edition, Vol. 15, Oxford, UK: Elsevier, pp. 343–346.

Zagheni, E., Garimella, V. R. K., Weber, I. & State, B. (2014), 'Inferring international and internal migration patterns from Twitter data', in *Proceedings of the 23rd International Conference on World Wide Web*, New York, NY: ACM Press, pp. 439–444.

Zagheni, E., Weber, I. & Gummadi, K. (2017), 'Leveraging Facebook's advertising platform to monitor stocks of migrants', *Population and Development Review*, 43(4), 721–734.

27. Ahead of the policy curve: migrants harnessing tech to survive

Emre Eren Korkmaz

INTRODUCTION

While recent developments in frontier technologies have brought many opportunities, they also raise significant concerns and challenges for societies. Giant steps taken in artificial intelligence (AI), the blockchain, robotics and big data have had tremendous impacts on the daily lives of people and destabilized traditional approaches to work and life, and indeed individuals' sense of self and the future (Floridi, 2007).

Governments are investing vast sums in developing new technologies, and tech companies are at the forefront of commercializing various tech-based solutions. However, it is not difficult to observe growing concerns about the challenges that the new frontier technology is heralding. The general optimism in the first decade of the twenty-first century has turned into a more pessimistic outlook as a result of the actual and potential outcomes that arise when corporations and governments deploy these new technologies. For instance, when social media platforms began to take off, they were met with optimism. It was said these platforms would allow new digital public spheres to emerge, global online communities to form and the democratization of online space as a result of increased connectivity of societies across borders (Floridi, 2015). However, today such enthusiasm has evaporated as social media platforms have consolidated their role as vectors of fake news and misinformation, augmenting social polarization and undermining public confidence in elections and other democratic institutions (Zuboff, 2019). The vast quantum of data that global tech corporations were able to leverage in the first decade of Web 2.0 saw these firms heralded as vanguards of progressive globalization and the development of new digital global societies (Floridi, 2014). Today, "big tech" increasingly serves as a synonym for "big brother" and these firms stand accused of being monopolies hungry for ever more consumer data. "Surveillance capitalism" thus emerges, where we offer our most intimate selves in exchange for convenience and the tantalizing prospect of previously undreamed-of goods and services (Zuboff, 2019). For instance, in many countries, the internet is synonymous with Facebook, which stands accused of failing to prevent all manner of adverse content spread on the platform. According to a UN investigation into the recent ethnic cleansing in Rakhine state in Myanmar, Facebook was found to be the main arena in which hate speech and mobilization against Rohingya Muslims was spread (Solon, 2018; Beniger, 2009; Kanter and Fine, 2018).

Such a critical change in the public mood towards new technologies has, therefore, shifted the frame from a utopian vision of human progress to a picture of an emerging dystopian tomorrow in which humankind is subsumed into the digital realm. Dystopic stories and cinema and television productions like Black Mirror have fuelled these concerns by helping people to visualize in concrete terms the potential disruptions these technologies may bring. Once a staple of the tech-dystopian image of some distant tomorrow, now thoughts of robots and AI enslaving humanity have been eclipsed by very current fears about the power of corporations and governments to manipulate elections in the US and the UK, spread fake news, and "deep fake" images and videos. China's social credit system is no Hollywood imaginary or Orwellian allegory, but rather a very real and present attempt at almost total surveillance of an entire population of human beings. Thus, technology has moved to the middle of the debate over the growing power of the dominant forces in society, supplementing and indeed catalysing their existing tools of hegemony, control and oppression. Technology offers ever more ways to manipulate and control the citizen-as-consumer with the help of non-human smart products (Meier, 2011).

Despite the obvious power imbalances, technology remains an arena of political struggle. The same technologies can be used to promote peace and spark conflict. For this reason, the political struggle to make corporations and governments accountable can still change the course of technological development. For instance, if workers are concerned about automation or the use of AI by corporate HR departments, trade unions and various political parties and organizations can raise these issues in the public square and political decision-makers can take action to upgrade skills, guide the automation process and grant unions new rights to monitor the decisions of algorithms. For people worried about automated weapons and military AI algorithms, various organizations devote their energy to controlling or even banning these weapons to maintain the peace. Anxiety over fake news and manipulative campaigns undermining democracy and the trustworthiness of elections is spurring many NGOs and political movements to advocate for democratic principles and make political parties accountable for their social media advertising. Finally, those worried about tech corporations' ethical missteps and commercial irresponsibility can lobby governments to take the steps necessary to write regulations to ensure accountability and transparency (Madianou, 2019; UNCTAD, 2018; Cottray and Larrauri, 2017; Cummings et al., 2018).

In sum, various means to challenge the dystopian vision of the future of humanity remain. However, migrants and refugees do not have the same access to the arena of political contestation to protect and advance their interests. For anyone wishing to observe the likely shape of a dystopian future of a tech-oriented society in which people lose their political autonomy, a close look at the daily experiences of migrants and refugees vis-à-vis the large tech providers will provide sufficient clues (Madianou, 2019).

This chapter contends that, although migrants and refugees rely heavily on social media, WhatsApp groups and mobile phone applications to cross borders and express solidarity with each other, the power of governments and corporations to deploy surveillance technologies to control, manage, manipulate—and, indeed, halt—migratory movements is far greater. The chapter will analyse how new technologies are used to surveil migratory movements and how they are used by migrants/refugees to resolve the challenges they encounter during and after the journey.

CONTEXT, CONCEPTS AND DEVELOPMENTS

The disparity reflects the limited political strength and bargaining/negotiation/representation power of people on the move. Although citizens can be victimized by the development of new technologies—for instance, through job losses to automation or the diminution of the public sphere—they can still find some ways to raise their concerns and apply pressure to governments and corporations. Refugees and undocumented migrants lack these opportunities. Who will defend migrants/refugees against the power of governments and corporations, which have expanded their capabilities of surveillance and oppression because of new technologies (Maxmen, 2019; Bigo, 2002)?

Against this backdrop, justifying or legitimizing anti-migration discourses presents few challenges. In almost all countries, it is not difficult to unite most citizens and political parties to support action against migrants and increased security measures. Governments are not generally under pressure to recognize the fundamental rights of migrants and refugees. Instead, they are criticized by opposition parties for not being sufficiently tough against migrants. Media narratives of migrants and refugees as invaders is a widespread phenomenon. Such a political climate creates a suitable environment for states and corporations to employ or experiment with new technologies vis-à-vis migrants and refugees. Many people who would baulk at the deployment of facial recognition systems or confiscation of mobile phones by security forces in the ordinary course of events remain silent when these are applied to vulnerable migrants and refugees (Burns, 2014).

Nevertheless, despite the vast power imbalance, migrants and refugees do benefit from information and communication technologies (ICTs) to cross borders and find solutions to their daily problems. Also, social media could help migrants and refugees to document and record unlawful practices they face; and this might attract political attention and help pro-migrant groups to organize solidarity campaigns. Migrants rely on information gained through social media groups and mobile phone applications. Such crucial information and necessary contacts and networks guide people through their decisions to migrate and prepare for the journey. The flow of information from networks is vital en route and in making choices after arriving in the country of destination. ICTs support migrants to connect with UN agencies and humanitarian organizations and are used to stay in touch with families and loved ones back home. ICTs also help migrants and refugees to contact smugglers and access legal and other assistance from private and public agencies (Hernandez and Roberts, 2018). Therefore, comparing and contrasting the ways that new technologies are used to surveil migratory movements with how they are used by migrants/refugees to resolve the challenges they encounter during and after the journey will provide insights about the role of technologies in migration studies. For instance, mobile phones have become an inseparable part of the lives of people on the move and ICTs are very valuable to them. However, it is important to eschew the ready romanticizing and dramatization of migrants and refugees' use of mobile phones, for two main reasons. The first is that mobile phone data, including the content of conversations, can be tracked and analysed by governments. Moreover, it is not difficult to infiltrate Facebook or WhatsApp groups to surveil and manipulate migrant communities. Therefore, that migrants can cross borders successfully does not necessarily imply that states cannot detect them— indeed, there is a high possibility that such movements are detected and tracked through information technologies. However, such surveillance is likely partial. A state, for instance, may turn a blind eye to some mass movements towards neighbours to signal potential leverage in international negotiations. Another may find it impossible to patrol the entire border all of the time, producing a "migrant lottery" in which some are able to cross while others are detained.

The second reason is that, while migrants and refugees heavily rely on social media and mobile phone applications, states, the UN agencies and NGOs charged with controlling, managing or even halting migratory movements and providing various services to migrants and refugees have access to a broader portfolio of technologies to deal with human movement, such as AI, satellites, drones, the blockchain and big data analysis. Such disparities in access to the benefits of new technologies intensify when the politics of migratory movements are considered. However, the use of ICTs by migrants and refugees could be considered as a form of resistance against a system dominated by governments, agencies, NGOs and corporations. ICTs allow migrants and refugees to understand challenges and opportunities they face during and after their journeys. Based on the information gained from ICTs, they try to find their way by relying on existing rights and allies and using contradictions among these actors, be it a contradiction among two neighbouring states.

These debates are not only related to those migrants who cross borders without prior authorization from destination countries since legal mobility through visa applications covers the majority of migration activity. However, for the people who have to apply for visa application processes due to their citizenships as well—including tourist and business visa applications—applicants must share almost all salient data about their financial, economic and political situation, and accept biometric registration and consulates evaluating whether to grant a visa. For legal migration, in addition to the collection of so much private data and biometrics, governments can—as can be seen in the Canadian case—employ AI algorithms in rendering the final decision. Looked at from another angle, potential migrants rely on the information they gather via ICTs in deciding when and where to migrate to and what to do once they arrive.

TECHNOLOGY AND MIGRATION THROUGH THE LENS OF INTERNATIONAL RELATIONS

Mainly the states, but also the UN agencies, and NGOs collect data from migrants and refugees continuously. The data is necessary to deliver services and make long-term plans. However, concerns arise about how migrant and refugee data are being collected, where they are being stored, whether and how people in vital need of assistance give their consent, and what measures are taken to prevent people's data from being used against them. These serious political, ethical and social questions still have not received a satisfying response. However, looking at the issue through the lens of International Relations (IR) offers a global perspective of great power competition that affords analysis of the issue of migration and technology/data.

In addition to natural disaster displacements, many cases of humanitarian emergencies are observed in regions of the globe beset by conflicts and civil wars triggering mass movements, such as in Syria, Iraq and Yemen today. Combatants in the field are often supported by—or proxies of—states, and under these conditions, accessing data and controlling mass movement through data analysis gains more importance for conflicting parties. Therefore, who can access the collected data in the field becomes a salient question.

The UN could be considered a neutral alliance of global powers and the UN agencies are investing new technologies. The UN agencies are working with technology corporations to register people on the move and deliver services to them, and the analysis of the resulting data is very valuable. What would be the potential risks of working with a US tech company providing digital infrastructure to a UN agency and storing the data in its cloud? The chapter will discuss the close relationship between military/intelligence bureaucracy and global tech corporations in the next section. Briefly, if these corporations earn huge profits from military and intelligence sectors, we cannot consider them as only tech corporations: they are military corporations as well. Given such an intense relationship with military and intelligence agencies, these corporations' role in the humanitarian sector and collaboration with the UN agencies should be questioned as well. Another thought experiment that comes to mind is to imagine how Western powers would react were Huawei to provide such services. The recent debates in the UK about Huawei's wish to market its infrastructure for 5G networks there gives us some idea (Benner, 2020). Therefore, a humanitarian exercise in registering people to distribute aid might pose other consequences when analysed through the lens of IR.

This trend of the convergence among global tech corporations and military/intelligence bureaucracy could not be considered as only a client–supplier relationship. This trend is uniting the political and economic interests of both parties, resulting in long-term contracts and hiring staff from each other. This is also related to debates on the future of warfare/evolution of warfare where cyberspace has become an arena of warfare that includes social media dis-information campaigns, deep fake, hacking and cyber espionage, and these tech corporations are at the heart of this conflict (Kello, 2017). Therefore, it is not only Huawei that has strong and undeniable ties with the Chinese government; the US tech corporations construct close relationships with the US state. Such a perspective is crucial to examine global technology companies, which are increasingly coming to play in dealing with humanitarian aid and migratory movements.

The Role of Tech Corporations and the Military-Intelligence Sector

In recent years, global technology companies have begun to pay more attention to migratory movements. Various tech corporations offer services and products to UN agencies, INGOs, states and migrants/refugees on a commercial basis (i.e., not as part of corporate social respon-sibility projects). Financial institutions, telecoms companies and tech firms seek to earn profits from their investments. Serving migrants and refugees has become a profitable business.

A question arises, then, about the role of global technology corporations such as Facebook, Amazon, Google, Microsoft and Palantir in dealing with migratory movements. As a result of their technical superiority and financial power, they are not merely positioning themselves as suppliers for UN agencies or NGOs but also introduce themselves as critical stakeholders. These corporations work with states and agencies to manage and control migratory move-ments. However, the impact of these corporations has so far been largely underreported and analysed.

Silicon Valley's Palantir's collaboration with the World Food Programme was criticized by many humanitarian actors (Parker, 2019) because this company produces anti-terrorism soft-ware that could be readily used with migrants and refugees, and the CIA is a client. However, is Palantir an isolated case? From Microsoft to IBM and Google, giant tech companies have become part of the military intelligence industry. For instance, Amazon is considered the largest military supplier in the US in terms of sales. There had been intense competition between Amazon, IBM and Microsoft to provide the US Department of Defence's cloud services and Microsoft was in the winning tender position in the last months of 2019. Google employees protested that company's participation in a military project in 2018. The MIT Tech Review even speaks of the military intelligence sector as having become dependent on Silicon Valley (Delcan, 2019; Weinberger, 2019).

Against this backdrop, it becomes next to impossible to separate the interests of these giant American technology corporations with US national interests. Palantir's CEO's recent state-ment at the World Economic Forum defending US interests is therefore hardly surprising, but this is an issue that goes well beyond one company (Feuer, 2020). Other tech companies are competing for the mighty "military dollar". Similar arguments are made against Huawei as well. For this reason, special attention is required when refugees are fleeing war/conflict zones or migrants crossing borders, as their use of social media platforms and mobile phone applications creates important data for tech companies with close relationships with the military and intelligence sectors in their home states.

In this vein, it behoves UN agencies and NGOs to think twice before inviting tech companies to bid to provide their services. After all, many global tech corporations are not merely technology providers—they are also part of the military and intelligence sector, and UN agencies and NGOs have a duty (under the "no harm" principle) to protect people in need. In the near future, similar issues are likely to arise, wherein UN agencies will face challenges to act neutrally as a result of their cooperation with tech companies and their policies on data collection. Here, in-house production could be one of the solutions to avoid these problems or, if this is not possible, the due diligence process shall require tech providers not to have any commercial relation with military and intelligence industries.

Even if an organization has strict internal rules for data privacy and data collection, problems loom. Organizations that lack the independent means to store data—especially where the data are uploaded to a proprietary cloud system or when they are shared with private financial organizations—face a massive question mark over the control that these private service providers obey its principles and have capacity against cyber-attacks.

Military Industry and "Smart Borders"

The second debate about the role of tech corporations concerning migratory movements is about so-called "smart border" products. New technologies are increasingly intersecting with border management, moving to the centre of the debate, particularly in forced migration studies. States increasingly rely on digital and frontier technologies to manage borders and control migratory movements, and the defence industry and military intelligence sector provide high-tech tools to manage borders. A recent report details that Europe's largest arms sellers are also now marketing "smart" border management tools (Akkerman, 2016). "The smart border" constructed by the US at its southern frontier with Mexico (Ghaffary, 2019) and the EU's digitalized border management systems, including Eurosur and Eurodac, are examples of such technology in action.

As satellites, drones and "big data" analyses push physical borders outwards, biometrics push them inwards. The use of algorithms and AI to decide on asylum applications and confiscation of asylum-seekers' mobile phones—ostensibly to check their social media accounts at the border to verify their identities—has become a daily routine of border management in the US and the EU (Franco, 2019; Ghaffary, 2019; Fussell, 2019).

Artificial intelligence is an important component of smart border solutions and these algorithms assist administrative tribunals, immigration officers, border agents, legal analysts and other key players in the process (Angwin et al., 2016). Using technology can be perceived to be neutral, logical and consistent, thus affording weaker human accountability. However, technology has the potential to inherit our discriminatory biases and be political in its decisions. Therefore, automated decision-making processes can exacerbate pre-existing vulnerabilities by adding on risks such as bias, error, system failure and theft of data, all of which can result in greater harm to migrants and their families. A rejected claim formed on an erroneous basis can lead to persecution on the basis of an individual's race, religion, nationality, membership in a particular group or political opinion (Keung, 2017).

There are also projects to develop AI algorithms to predict displacement of populations. For instance, UNHCR's AI programme aims to predict displacement from Somalia to Ethiopia, considering various datasets and economic variables, including fluctuations in the market price of agricultural products. The UNHCR asserts that the aim of this algorithm is to prepare for new waves of displacement and ensure arrivals are adequately taken care of. However, can we be sure that an AI algorithm that predicts migratory movements or displacements would not be used by neighbouring states to stop migration movements instead of investing in better services for newcomers?

Using the Blockchain for Identification and Financial Transactions

Identifying people is one of the main challenges both in developing countries and in the humanitarian context. Over 1 billion people are estimated to be without an official identity, and their lack of such identification hinders their access to basic services such as health and education as well as rendering them unable to open bank accounts and access the SIM cards necessary to benefit from new technologies. The "right to a legal identity" is implied by the Universal Declaration of Human Rights and achieving this for all included in the UN's sustainable development agenda in SDG 16.9. A legal identity promises to help individuals become "bankable" and to access social and medical care, police protection, and other services. In the context of forced migration, identification is a primary means of population management.

The utilization of blockchain and biometrics promotes globally verifiable information that can be used for the registration of births, voters and refugees. Nevertheless, a question remains about what the best approaches to such digital identities are, given the risk of appropriation of control by more powerful actors. Additionally, focusing on technology to provide identification is problematic. Do over 1 billion people not have identification because of the lack of sufficient technologies or is it the political decision of states not to grant citizenship to specific communities within their territories? It is clear that, under the current nation-state system, it is not possible to dictate any solution to sovereign states and all arguments concentrating on the benefits of a particular technology for identification carry the risk of missing the main point— namely, to find political solutions to political crises (Coppi and Fast, 2019).

Blockchain technology has been popular in dealing with refugees and migrants in recent years, and it is not limited to personal identification. Various pilot projects are starting to make use of the blockchain to support refugees. A particular focus of such initiatives is to provide services rapidly, cheaply facilitate payments and data protection, and support local businesses through cryptographic security without intermediaries—while also offering transparency and accountability. For instance, the WFP has introduced a pilot project called "Blockchain against Hunger", deploying blockchain technology in Jordan's Azraq camp to make cash-based transfers to 10,000 refugees. WFP relies on biometric registration data provided by the UNHCR and refugees can shop from local supermarkets using iris scans. The system confirms the identity of the refugee in this way, checks their account balance and confirms the purchase (Juskalian, 2018).

However, despite these claims, these projects are criticized as being relatively secretive— since they are not open source, they stand accused of a lack of transparency. Also, as Coppi and Fast (2019, p. 1) note,

[T]he removal of intermediaries and altered power dynamics may result in unintended consequences that could endanger end-users as well as future projects. Accountability may also be undermined through a lack of meaningful consent and engagement in project design from end-users, which could exacerbate power imbalances between aid organisations and their beneficiaries and facilitate a surveillance-type system that could be used to harm, rather than help, vulnerable populations.

Experimenting on Refugees and Migrants

In many cases, there is no clarity about the outcomes of these technologies. Pilot projects are essential for learning from these technologies. Most importantly, the dissemination of project findings, impact assessment, transparency and collaboration are necessary if progress is to be made. These projects should be transparent, and the results of the pilot projects must be clear, measurable and open to independent review. Short-term projects and high turnover of staff also affect the fate of the collected data. There are various projects where collected data stay in the cloud for years without any auditing of the companies hosting the data (Burns, 2015).

Many academics argue that technologies—such as the blockchain—used in several recent pilot projects were unnecessary (Coppi and Fast, 2019). In a recent interview of the author of the paper with the founder of a tech company (10/02/2019) that provides digital infrastructure for NGOs it was shown that a single company held everyone's keys—from technical service providers to the agencies implementing the project and beneficiaries (i.e., tens of organizations and thousands of people)—in blockchain-based projects in Kenya, Jordan and Thailand. The reason for the insistence on the blockchain when there are other technologies available has not been explained in many of these projects.

These projects also raise concerns about whether these new technologies are being tested on refugees who are unable to object and whether those living in camps are being used in such experiments without their knowledge. Sophisticated data mining techniques can be used to advance humanitarian aims. However, as Letouzé, Meier and Vinck (2013, p. 5) underline,

> [R]elying primarily on biased-and-messy-data analysis by number crunchers who may have never set foot in the field to inform sensitive policy and programmatic decisions in conflict-prone or -affected contexts would indeed be like pouring hot oil on burning ashes. In other words, the concern is that rushing to apply Big Data in such volatile and dangerous environments without fully understanding and addressing the associated risks and challenges—the non-representativeness of the data, the difficulty in separating "the signal from the noise", the larger challenge of modelling human behaviour, even the risk of misuse of such tools by oppressive regimes—may well end up spurring rather than preventing the spread of conflict.

ICTs in the Hands of Refugees and Migrants

Despite the enormous power imbalance, we cannot ignore the significance of ICTs for migrants and refugees. It is not just about the physical obstacles and borders they must overcome. They also need digital infrastructure to receive vital information for the journey. This includes social media platforms, translation services, digital maps, and various mobile applications that allow them to message and call others or access services. Migrants can also use phone cameras to record any incidents of harassment and violence and share these with solidarity networks (Gillespie et al., 2016; Latonero et al., 2018).

Migrants need information as much as they need food and shelter. However, as the mainstream media preserves their anti-migrant stance, migrants also come to rely on alternative sources of news. For instance, research projects conducted with Afghan migrants demonstrate that they trust the information provided digitally by smugglers, family members and close social networks. However, these digital traces can also threaten them as a result of surveillance techniques. As touched on above, confiscation of asylum-seekers' smartphones allows authorities to check their identities, review personal data archives and reach a decision on their applications based on evaluation of photos and documents (Gillespie et al., 2016; Latonero and Kift, 2018; Mercy Corps, 2016).

One argument in favour of tracing and analysing the data of migrants and refugees comprises the benefits that accrue to their welfare—the so-called "surveillance for good" approach. According to this approach, social organizations and humanitarian NGOs should be encouraged to analyse the data of people on the move in order to support them and defend their rights. Many academic research projects aim to serve this purpose (Gillespie et al., 2016). However, it is crucial to avoid creating new hierarchies over migrants and not to ignore or downplay their personal agency. It would clearly not be possible to negotiate with people crossing borders during their journeys. While states are unlikely to recognize those crossing borders as entitled to "free" movement, governments and civil society organizations could collaborate more with their own migrant communities and grassroots organizations to support those who have already arrived. For example, collaboration could be a way to understand the kinds of policy responses needed to help new arrivals settle and adjust. Also, digital ethnographies or qualitative research that complements "big data" analysis could help refugees to express themselves more effectively.

This is also valid for hundreds of mobile phone applications created to support migrants and refugees. The majority of these apps have failed to provide expected services and so are not widely adopted by refugees and migrants. The experiences and expectations of migrants and refugees should be considered in a deliberative manner to work on genuine tech-based solutions for people on the move.

It is also necessary to analyse the digital divide within migrant and refugee communities themselves. It is highly skilled, male and literate refugees and migrants that appear to have disproportionate access to smartphones and mobile applications. Thus, gender and socio-economic conditions have an impact on who can benefit from the advantages offered by smartphones (Gillespie et al., 2016; Mercy Corps, 2016).

Smartphones and ICTs help migrants obtain vital information, communicate with their families and learn from their networks. From another perspective, they create an ecosystem where migrants from different nationalities meet and act together to cross borders. The emergence of an internationalist spirit can be observed when people from various countries such as Iran, Iraq, Syria, Somalia and Afghanistan get on the same boat or follow the same path on their way to their destinations. Therefore, it is not only their existing networks that provide much-needed information. Indeed, migrants and refugees from different countries with different networks can use ICT to learn from each other. For instance, mobile phone apps for translation may help them to communicate with each other en route. Such a process of building trust among disparate groups on the move can be enabled through ICT.

The emergence of such an internationalist spirit bringing together migrants from Africa, Asia and Latin America to combine their advantages to overcome the challenges they face during and after their journeys also provides the seed for the creation of grassroots organizations and networks that migrants and refugees use to decide on tactics and actions they need to take before, during and after their journeys. The digital infrastructure provided by ICTs for the emergence of such grassroots networks is worth further analysis.

All these mobile applications used by refugees and other vulnerable populations are the products of global tech corporations seeking a profit and they are not designed for solidarity purposes. These are commercial products whose aim is to collect data from users for analysis and to be sold on to other businesses to advertise their products and services to target audiences. Therefore, the particular needs of migrants and the risks associated with migratory movements are not considered during the design processes (Latonero and Kift, 2018). Many corporations offer specialized solutions for refugees and migrants by collaborating with UN agencies and NGOs as a part of their corporate social responsibility projects. However, the majority of them are targeted projects applied in camp settings. The most well-known tech brands and apps—WhatsApp, Skype, Viber, Facebook and mobile banking—are not among these specific migrant applications. The popular commercial apps—designed to garner "likes" and convince users to purchase more "in-app" services—can actually do more harm than good for refugees and vulnerable migrants.

CONCLUSION: ICTs AS A RESISTANCE TOOL FOR PEOPLE ON THE MOVE

While frontier ICTs provide vital information for refugees and vulnerable migrants, the technological superiority of states and corporations vis-à-vis people on the move is the essential part of the story. Neglecting this power imbalance leads to an inevitable romanticization of the relationship between migrants and their smartphones. The tech-based power to track people, analyse the "big data" produced by migrants and refugees, as well as the use of algorithms and AI, and the employment of drones and robotics provide immense opportunities for states to manage migratory movements and for corporations to expand their customer base. This means both states and corporations can make choices between "desirable" and "undesirable" migrants, allowing some of them to be liberated and others to languish or die. Tech corporations are at liberty to select some of them to be employed and accredited as clients while others are totally excluded from access. This is clearly a demonstration of the power wielded by dominant actors against people on the move. In stark contrast to such displays against citizens, this power play against migrants and refugees can proceed openly and publicly, undermining the political agency of these vulnerable people.

At the same time, migrants and refugees are not "power-less", drawing mightily on ICTs to challenge the status quo. This is a matter of life and death for them, and any mistake or misjudgement about the next step might be perilous. Such a tactical approach to reach out the strategy to arrive in the country of destination is valid for migrants who seek prior permission to cross borders and who aim to pass without prior confirmation of authorities. Therefore, this chapter emphasizes the use of ICTs by migrants and refugees to cross borders as *a tool of resistance against the superiority and hegemony of states and corporations.*

Migrants leave digital traces behind them, so they should act cautiously. They should distinguish fake news and manipulative information from helpful data. For instance, in Afghanistan, some countries such as Australia advertise on YouTube showing armed men urging Afghans not to consider any migration to their lands. Most official and mainstream information aims to change minds and convince prospective migrants to stay where they are. However, many of them rely on the information they gain from their own networks, family members and smugglers, and set off on their journeys to a better life.

Migrants can readily access WhatsApp and Facebook groups, but they should be aware of the possible infiltration of these systems by authorities and other people with malevolent intent. They should find helpful smugglers, collect valid documents, and invest as much time and energy as they need for the journey. They should plan each step of the journey very carefully, ensuring they have back-up plans, and should check out the mobility patterns of security guard patrols, taking into account important factors like the weather and public holidays. They should use ICT to measure the shortest possible, secure distances, and to retain the addresses and phone numbers of people for emergencies and when they need help. They ought to keep some money spare and maintain a list of contacts for when they arrive in their destination country and continue living there.

All this requires a review of hundreds of webpages, talking with many people, evaluating the messages of those who have travelled previously, drawing insight from their experiences and bargaining with authorities, smugglers and various agencies. Any mistake or error might imperil the journey, the chance of being accepted as a legal refugee, or even his or her life. Teamwork is also necessary, and as mentioned above vis-à-vis the emerging internationalist spirit, this might require collaboration with people from other nationalities. Earning trust is thus critical for people on the move to achieve their objectives.

All told, then, access to a smartphone is just the beginning. Smartphones and mobile applications provide several possible pathways to success, but also many hundreds of potential dead ends and perilous routes. Relying on secure networks and acting tactically is crucial. Millions of people can succeed in arriving at their desired destinations through luck, of course, but also through use of information systems and network technology to carefully analyse the political situation (such as Merkel's embrace of migrants after 2015, the change in public mood after Alan Kurdi's tragic drowning and Turkey's decision to use migratory movements as a bargaining chip in pursuit of foreign policy goals), the weather (spring, summer) and all the other factors mentioned above. With some luck, those on the move have more opportunity than ever to use ICTs to challenge the status quo, navigate their way through their arduous journeys, and achieve their dreams.

REFERENCES

Akkerman, M. 2016. *Border Wars: The Arms Dealers Profiting from Europe's Refugee Tragedy*, Transnational Institute and Stop Wapenhandel.

Angwin, Julia et al. 2016. "Machine Bias", ProPublica. At https://www.propublica.org/article/machine-bias-risk-assessments-in-criminal-sentencing (accessed on 20 August 2020).

Beniger, J. 2009. *The Control Revolution: Technological and Economic Origins of the Information Society*. Cambridge, MA: Harvard University Press.

Benner, T. 2020. "Britain Knows It's Selling Out Its National Security to Huawei", *Foreign Policy*, 31 January.

Bigo, Didier. 2002. "Security and Immigration: Toward a Critique of the Governmentality of Unease", *Alternatives* 27, no. 1 (February): 63–92.

Burns, R. 2014. "Moments of closure in the knowledge politics of digital humanitarianism", *Geoforum*, 53, 51–62.

Burns, R. 2015. "Rethinking big data in digital humanitarianism: Practices, epistemologies, and social relations", *GeoJournal*, 80(4), 477–490.

Coppi, G. and Fast, L. 2019. "Blockchain and Distributed Ledger Technologies in the Humanitarian Sector," *HPG Commissioned Report*, Overseas Development Institute, February 2019.

Cottray, O. and Larrauri, H. P. 2017. *Technology at the Service of Peace*, 26 April. At https://www.sipri.org/commentary/blog/2017/technology-service-peace (accessed on 10 September 2019).

Cummings, M. L., Roff, H. M., Cukier, K., Parakilas, J. and Bryce, H. 2018. "Artificial Intelligence and International Affairs. Disruption Anticipated", *The Royal Institute of International Affairs*, London.

Delcan, P. 2019. "Meet America's newest military giant: Amazon", *MIT Technology Review*, November.

Feuer, W. 2020. "Palantir CEO Alex Karp defends his company's relationship with government agencies", *CNBC*, 23 January.

Floridi, L. 2007. "A look into the future impact of ICT on our lives", *The Information Society*, 23(1), 59–64.

Floridi, L. 2014. *The 4th Revolution: How the Infosphere is Shaping Human Reality*, Oxford University Press.

Floridi, L. 2015. *Onlife Manifesto: Being Human in a Hyperconnected Era*, Cham: Springer.

Franco, Marisa. 2019. "Democrats Want a 'Smart Wall': That's Trump's Wall by Another Name", *The Guardian*, 14 Feb. 2019, at www.theguardian.com/commentisfree/2019/feb/14/democrats-wall-border-trump-security (accessed on 23 September 2019).

Fussell, Sidney. 2019. "The Increase in Drones Used for Border Surveillance", *The Atlantic*, 11 October, at www.theatlantic.com/technology/archive/2019/10/increase-drones-used-border-surveillance/599077/ (accessed on 11 November 2019).

Ghaffary, Shirin. 2019. "The 'Smarter' Wall: How Drones, Sensors, and AI Are Patrolling the Border", Vox, 16 May, at www.vox.com/recode/2019/5/16/18511583/smart-border-wall-drones-sensors-ai (accessed on 8 March 2020).

Gillespie, M., Ampofo, L., Cheesman, M., Faith, B., Illiadou, E., Issa, A., Osseiran, S. and Skleparis, D. 2016. *Mapping Refugee Media Journeys Smartphones and Social Media Networks*, The Open University / France Médias Monde, 13 May.

Hernandez, K. and Roberts, T. 2018. "Leaving No One Behind in a Digital World", *The K4D Emerging Issues Report Series, Institute of Development Studies*, November.

Juskalian, R. 2018, "Inside the Jordan refugee camp that runs on blockchain", *MIT Technology Review*, 12 April, at https://www.technologyreview.com/s/610806/inside-the-jordan-refugee-camp-that-runs-on-blockchain/ (accessed on 21 January 2018).

Kanter, B. and Fine, A. 2018. "Civil Society in the Age of Automation: Understanding the Benefits and Risks of Artificial Intelligence, Machine Learning, and Bots", *Alliance for Peacebuilding and Toda Peace Institute Policy Brief*, No. 26, November.

Kello, Lucas. 2017. *Virtual Weapon and International Order*, New Haven, CT: Yale University Press.

Keung, Nicholas. 2017. "Canadian immigration applications could soon be assessed by computers", Toronto Star, at https://www.thestar.com/news/immigration/2017/01/05/immigration-applications-could-soon-be-assessed-by-computers.html (accessed on 15 October 2020).

Latonero, M. and Kift, P. 2018. "On Digital Passages and Borders: Refugees and the New Infrastructure for Movement and Control", *Social Media + Society*, January–March, 1–11.

Latonero, M., Poole, D. and Berens, J. 2018. *Refugee Connectivity: A Survey of Mobile Phones, Mental Health and Privacy at a Syrian Refugee Camp in Greece*, The Signal Program at the Harvard Humanitarian Initiative and the Data & Society Research Institute, March 2018.

Letouzé, E., Meier, P. and Vinck, P. 2013. "Big Data for Conflict Prevention: New Oil and Old Fires", in *New Technology and the Prevention of Violence and Conflict*, ed. Francesco Mancini, New York: International Peace Institute, April.

Madianou, M. 2019. "Technocolonialism: Digital Innovation and Data Practices in the Humanitarian Response to Refugee Crises", *Social Media + Society*, July–September, 1–13.

Maxmen, A. 2019. "Can tracking people through phone-call data improve lives?", *Nature*, 29 May, at https://www.nature.com/articles/d41586-019-01679-5 (accessed on 17 September 2021).

Meier, P. 2011, *Do "Liberation Technologies" Change the Balance of Power between Repressive States and Civil Society?*, unpublished PhD Thesis, Faculty of the Fletcher School of Law and Diplomacy.

Mercy Corps. 2016. *How technology is affecting the refugee crisis: Afghanistan, Greece, Iraq, Jordan, Syria*, 9 June.

Parker, B. 2019. "New UN deal with data mining firm Palantir raises protection concerns", *The New Humanitarian*, 5 February, at https://www.thenewhumanitarian.org/news/2019/02/05/un-palantir-deal-data-mining-protection-concerns-wfp (accessed on 28 September 2021).

Solon, O. 2018. "Facebook struggling to end hate speech in Myanmar, investigation finds", *The Guardian*, 16 August. At https://www.theguardian.com/technology/2018/aug/15/facebook-myanmar-rohingya-hate-speech-investigation (accessed on 10 October 2019).

UNCTAD. 2018. Technology and Innovation Report 2018. *Harnessing Frontier Technologies for Sustainable Development*, United Nations Conference on Trade and Development, Geneva.

Weinberger, Sharon. 2019. "Meet America's newest military giant: Amazon", *MIT Technology Review*, 8 November, at https://www.technologyreview.com/2019/10/08/75349/meet-americas-newest-military-giant-amazon/ (accessed on 10 January 2020).

Zuboff, Shoshana. 2019. *The Age of Surveillance Capitalism: The Fight for a Human Future at the New Frontier of Power*, London: Profile Nooks.

28. Migration, mobility and digital technology in a post-COVID-19 world: initial reflections on transformations underway

Marie McAuliffe and Jenna Blower[1]

INTRODUCTION

International migration is intrinsically linked to social, economic and political global transformations, and can be seen as an important aspect of globalization processes (Castles, 2010; de Haas et al., 2020; Held et al., 1999). Along with other international phenomena, migration has historically been affected by seismic geopolitical events, such as the two World Wars, the Cold War and large terrorist attacks such as 9/11, which can mark turning points in migration governance, as well as in broader discourse and sentiment (Faist, 2004; Newland et al., 2019). The COVID-19 pandemic is the latest seismic geopolitical event, stemming on this occasion from a global health emergency. COVID-19 has highlighted risks and impacts in very stark terms – through, for example, the numbers of confirmed cases and deaths (over 33.5 and 1 million respectively by the end of September 2020 (WHO, 2020)) – pointing to the vulnerability of populations to a highly infectious virus underpinned by high levels of international and internal mobility. With effective vaccines having been developed and due to be rolled out globally in coming weeks and months, attention is turning to global vaccination programming and responding to the global economic downturn.

Recognising that COVID-19 is by no means over, and that its impact on the world will be felt for years to come, has already profoundly affected aspects of migration governance. The widespread imposition of unprecedented movement restrictions by governments (national and subnational) are underpinned by emergency powers (Ponta, 2020, para. 22) as part of a suite of extraordinary measures designed to limit coronavirus transmission and infection during this global public health emergency (WHO, 2020). COVID-19 immobility is underpinned by technology, such as track and trace apps and other (im)mobility monitoring that has far-reaching implications for how migration is managed. Just as we saw the reconfiguring of mobility systems to accommodate increased security processing following 9/11, the implications of COVID-19 on systems in the medium term is yet to be fully apparent; however, systemic changes are underway.

In this final chapter, extraordinary COVID-19 measures related to migration and mobility impacts are examined, with particular reference to the use of digital technology related to health-proofing mobility systems and the measures being taken by States on restricting mobility as part of the ongoing securitization of migration. This has implications for the application of technology within migration and mobility systems, but, just as importantly, also for the COVID-19 intensification of tech solutions for key industries (such as management consulting, agriculture, social care, services). The risks of relying on highly mobile international workforces have been elevated by the pandemic, spurring on the pre-COVID forays into automation underpinned by machine learning and other AI streams. This chapter then takes an early look at how digital technology is being utilised to monitor and limit mobility. In doing so, the chapter also examines the implications for human rights and mobility beyond the acute global health crisis. Before turning to COVID-19 (im)mobility, we situate the analysis in the framework of the ongoing securitization of migration before outlining the current data on mobility restrictions. We then examine some of the new mobility data being generated before highlighting examples from industry on the use of digital technology in sectors previously reliant on very high international mobility regimes.

SECURITIZATION OF MIGRATION: A BRIEF RECAP

The security risks posed by migration have long been acknowledged by leaders throughout the ages, as societies have sought to protect themselves from threats, while at the same time seeking increased prosperity through trade, finance and cultural exchange underpinned by migration (Watson, 2009).

In the modern era, it is clear that migration directly affects some of the defining elements of a State – that is, a permanent population and a defined territory.[2] The regulation of migration (entry and stay) is therefore considered a prerogative of sovereign States, supplemented by international cooperation on migration governance (Ferris and Martin, 2019; McAuliffe and Goossens, 2018). Over many decades migration has increasingly become an issue not only of migration management and border control, but also of national security (Waever, 1995). Further, many argued that one of the effects of the events of September 11, 2001 was that it reinforced the trend towards securitising migration, which directly resulted in increased migration control, significant investment in border intelligence systems and substantial institutional responses, most notably in the United States, but more generally throughout the Western world (Faist, 2004). To that end, conceptualisation of migration (especially "irregular" migration) has increasingly become State-centric and also beyond the earlier bounds of economic national interest to encompass national security. Some argue that the intensification of border controls is an overt demonstration of the securitization of migration processes that is especially apparent with the deployment of military resources to manage sea borders (Miggiano, 2009). Others have argued that the securitisation of migration in the twenty-first century is on course to intensify (Abebe, 2019; Humphrey, 2013; McAuliffe and Mence, 2017).

Health (in)security has been linked to migration and migrants throughout history, especially as it relates to potential public health threats (Greenaway and Gushulak, 2017). Border closures and travel restrictions have been enacted over the centuries in response to pandemics and other health crises (Schloenhardt, 2006). That said, with the increase in prevalence of global health emergencies over recent years (e.g., the SARS, MERS, H1N1, Zika and Ebola crises), there remains a tendency to stigmatise migrants during times of pandemics (Greenaway and Gushulak, 2017). In relation to COVID-19, we have also seen specific countries and peoples being blamed, spurring xenophobic racist behaviours that, while illogical, have been instrumentalized for political gain in some domestic political as well as geopolitical contexts (Hvistendahl, 2020). COVID-19 has increasingly framed migration as a national security issue within a global health crisis (Ferhani and Rushton, 2020; McAuliffe, 2020b; Nunes, 2020). Yet, *international travel* is widely recognised as being a key driver in the emergence and rapid spread of infectious diseases globally, most notably since developments in transportation technologies in the 1970s (Goldin, 2014; Greenaway and Gushulak, 2017). The massive increase in international air travel and the heightened risks of zoonotic viruses due to an increase in human–animal interactions has led some to call airports the "super-spreaders" of infectious disease (Goldin, 2014).

COVID-19 MOBILITY RESTRICTIONS: THE STORY SO FAR[3]

COVID-19 has been the most severe pandemic in recent decades, resulting by 30 September 2020 in 33.3 million confirmed cases and over one million deaths in 188 countries in the nine months following it, having being declared a global public health crisis by WHO on 30 January (Johns Hopkins University, 2020). The combination of high transmission and severity means that this pandemic is forcing policymakers into uncharted territory. Despite numerous warnings over many years about the growing risks of a severe pandemic and the need to strengthen preparedness, there has been significant pressure to develop and implement effective responses to COVID-19 in real time (Goldin, 2014; McAuliffe et al., 2020). As a result, governments around the world have implemented various measures to limit the spread of the virus. These measures have included severe restrictions on movements (international and internal) as well as on assembly, such that businesses have been forced to close, public and private transport systems shut down, and social activities severely discouraged or made illegal.[4] Some countries, such as El Salvador, Israel and Qatar, quickly imposed significant international movement restrictions in early to mid-March, while others took action weeks or months later.

By end-September 2020, a total of 219 countries, territories and areas had imposed over 90,000 specific travel-related restrictions (IOM, 2020c), reflecting an unprecedented situation globally. Some countries have stopped all entry of foreign nationals (including those who are seeking international protection), some have banned citizens of specific countries, while even further, some countries have completely closed borders to stop departure and entry of all people, including their own citizens. Quarantine measures have also been imposed by some countries, requiring that passengers entering a country are in quarantined isolation for a minimum period (typically 14 days) immediately upon arrival. There have also been significant internal movement restrictions imposed within some countries (see figure below). Notwithstanding the unprecedented pandemic context, there remain serious questions as to whether these extreme measures limiting freedom of movement are proportionate and temporary. This will continue to be an intense area of research and enquiry for years to come.

Overall, the movement restrictions put in place by governments (national and subnational) around the world in response to the pandemic have been swift, severe and unprecedented. They have also been introduced despite the WHO's recommendations not to implement travel restrictions, as per the International Health Regulations (IHR) as agreed by WHO Member States (Ferhani and Rushton, 2020). The IHR state a preference for borders remaining open and for controls being put in place in only very limited circumstances (Ferhani and Rushton, 2020; Greenaway and Gushulak, 2017). The IHR are consistent with existing human rights law, which provides for the right to leave any country and return to one's own country, as per the 1948 Universal Declaration of Human Rights and the 1966 Covenant on Civil and Political Rights. However, the right to leave one's own country under international law does not come with a corresponding right to enter another country, and the decision to let people in (usually

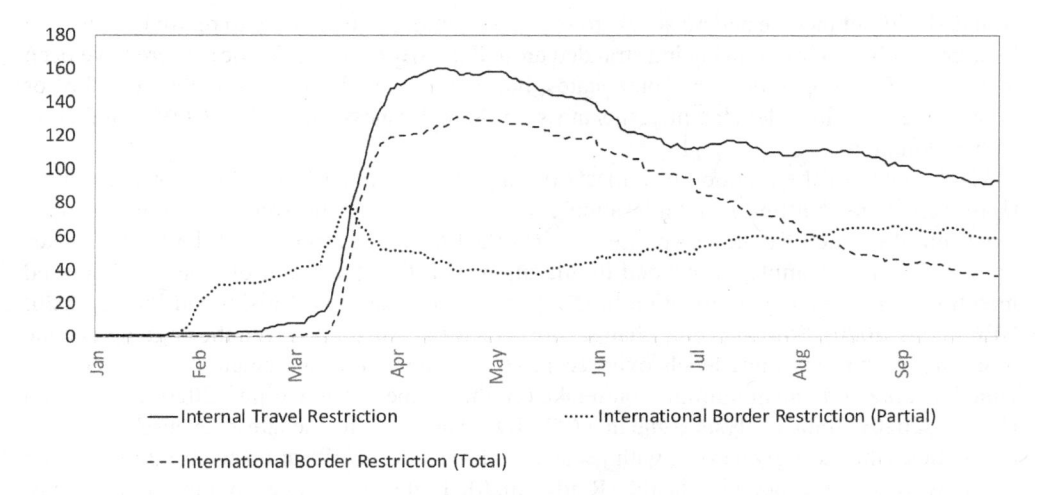

Note: Period is 1 January to 30 September 2020 (latest complete data at the time of writing).
Source: Hale et al., 2020.

Figure 28.1   Border and travel restrictions, number of countries

exercised in the form of entry visas) is made at the State level on the basis of bilateral relations in the context of political, economic and other considerations (Stringer, 2004); a major exception relating to the right to seek asylum is underpinned by the principle of non-refoulement (UNHCR, 2020). Restrictions on freedom of movement are limited to those necessary to protect national security, public order, public health or morals, or the rights and freedoms of others, such as in emergency situations (Ponta, 2020). COVID-19 has resulted in some governments using executive powers never before relied upon to declare a "state of emergency", for example, the use of the Australian state of Victoria's powers under the *Public Health and Wellbeing Act 2008* (Justice Connect, 2020).

Immobility: The Ultimate Migration Disrupter

As a result of COVID-19 movement and border restrictions, migration journeys, programming and services are being disrupted right along the migration cycle, with significant implications for human rights.[5] Over the course of the pandemic, many migrants have been unable to depart on planned migration journeys, such as for work, study or family reunion; further, people needing to seek asylum or otherwise depart unstable and highly insecure countries have been prevented from doing so, leaving them at risk of violence, abuse, persecution and/ or death. Impacts on migrants in destination countries have also been profound, especially for the most vulnerable in societies, who are without access to social protection and health care, and have also faced job loss, xenophobic racism and the risk of immigration detention, while being unable to return home. Further, refugees and internally displaced persons in camps and camp-like settings are subject to cramped, poor living conditions not conducive to COVID physical distancing and other infection control measures. Throughout the pandemic, border closure announcements in some countries have caused mass return to origin for fear of being stranded without income and no access to social protection. The inability to return has resulted in large numbers of migrants being stranded around the world (IOM, 2020c). There have been mass repatriation operations by some states, but many others have been unable to afford or organise repatriations, leaving migrants at risk of homelessness, starvation, COVID infection and exploitation.

While much of the immobility impacts on migrants stemmed from restrictions on travel, there were signs that limits on "migration", rather than movement, were also being incorporated into the suite of policy responses to COVID-19. The government of the United States of America, for example, suspended immigration with the stated aim of stopping imported infections *and* reducing competition to US-born workers for jobs (Chishti and Pierce, 2020; BBC News, 2020). Similar policy changes are forecast in Australia, given the ongoing decline in immigration programme levels over recent years as a result of incremental policy shifts to wind back the permanent immigration intake (Gothe-Snape, 2019; Collins, 2020; Pandey and Holmes, 2020). Some analysts argue that COVID-19 has become the latest "excuse" as migration policies increasingly harden, with yet another aspect of securitisation and migration called into play, this time concerning health (Reidy, 2020). That said, changes to migration policies, including how they relate to the entry and stay of migrants (including refugees, asylum seekers and visitors), need to uphold international law as well as national and subnational regulatory regimes. However, as highlighted by Ponta (2020, para. 22) there remain significant risks in some settings that

> a state of emergency can sometimes be used as a pretext for abuses, such as arbitrary detention, censorship, or other authoritarian measures .... There are increasing concerns that some governments might capitalize on emergency powers to undermine democratic principles, eliminate dissent, and violate the principles on necessity and proportionality. Most problematic are expansions of executive powers and repressive measures, which might continue after the national emergency in the respective countries.

A pronounced manifestation of the securitisation rhetoric has also emerged in relation to xenophobic racism being directed at migrants, with increases especially in relation to anti-Asian sentiment (Gamlen, 2020). Trend increases have been witnessed in Australia, the United States, Hungary and Italy among others (Coates, 2020; IOM, 2020b; Vertovec, 2020). Negative depictions of groups of migrants as posing health security threats during the pandemic have been articulated in multiple contexts, often tied in with other threats, and can be related to preparing the ground for tolerance of human rights infringements (Eves and Thedham, 2020). In other words, characterising groups of migrants in communities as public health security threats enables (successfully or otherwise) an indirect justification of human rights abuses contrary to prevailing norms – a form of "vice signalling" that is being deployed publicly on a number of issues related to COVID-19 (Berlatsky, 2020).

Alongside the immobility impacts on migrants and the evidence of deepening securitisation linkages, we have also seen significant advances in the use of AI in migration and mobility, which has intensified because of the pandemic. The figure below shows how the use of AI is being implemented through the migration cycle as well as how COVID-19 has resulted in new adaptations of AI-related digitalisation throughout the migration cycle.

While AI has increasingly been deployed in migration management settings since at least the 1990s, initially in visa entry and border processing systems but increasingly throughout the entire migration cycle (Beduschi and McAuliffe, 2021), the pandemic has intensified the race

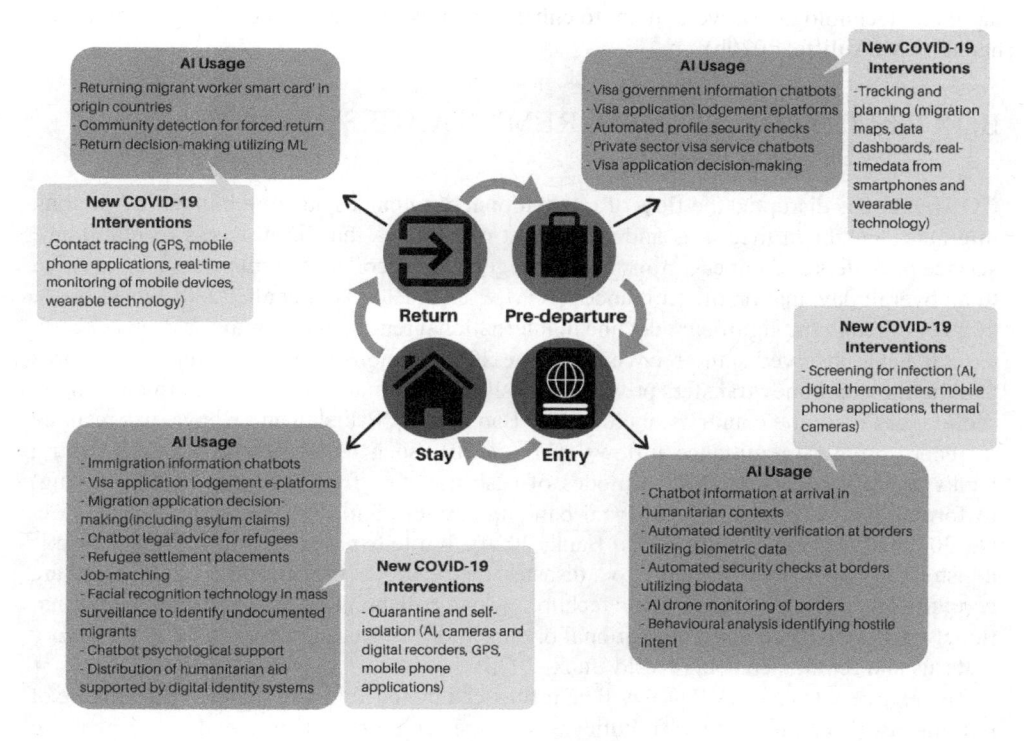

Source: Authors' elaboration, drawing upon Whitelaw et al., 2020.

Figure 28.2 COVID-19 AI developments through the migration cycle

for AI solutions to the COVID-19 rolling crisis. Contact tracing, population surveillance and quarantine tracking have quickly developed as fit-for-purpose digital tools, building on recent developments in machine learning such as those related to facial recognition and biometric analysis. The implications for privacy during and after the pandemic are a topic of intense scholarly and policy interest (Humer and Taylor, 2020; Privacy International, 2020; see also Molnar in this volume).

With the rise of xenophobic racism and human rights abuses, the threat to mobility has left millions of migrants in limbo, vulnerable to the social and economic shocks of the pandemic (ILO, 2020). The restrictions to global movement have made one thing clear: "mobility is fundamental to economic livelihood worldwide" (Barker et al., 2020, 1). The disruption to regular migration patterns creates challenges for economic development across the globe. On the one hand, the return and immobility of migrants have had direct implications in their pursuit for work and economic advancement, stemming into their capacity to provide support for family and community members in their countries of origin (McAuliffe, 2020b). International remittances are projected to decrease by 20 per cent in 2020 (World Bank, 2020), which will have dire consequences for communities in origin countries. While there is a robust digital infrastructure to support the transfer of remittances (Aron and Muelbauer, 2019), COVID-19 may incite technological development to enhance connectivity and accessibility to remittance channels (McAuliffe, 2020b).

BOX 28.1 INTERNATIONAL REMITTANCES AND DIGITAL TECHNOLOGY

COVID-19 has disrupted the flow of international remittances due to job insecurity among migrants, lockdown measures and movement restrictions that limit access to remittance service providers and impede informal exchange, and the collapse in oil prices that have led to an overall devaluation of remittances (IOM, 2020c; Rühmann et al., 2020; World Bank, 2020). Although the significant decline in international remittances remains a deep concern, experts have observed some recovery in June 2020, demonstrating how human networks and the flow of money transfers prove to be resilient (Rühmann et al., 2020). An upswing in remittances for some countries, including to Bangladesh, Pakistan and Kenya, has resulted in record monthly remittance inflows (IMF, 2020). Such recoveries reported by central banks are attributed to a shift in modes of cash transfers from informal (hand-carrying) to formal channels such as traditional banking services and digital platforms (Bisong et al., 2020; McAuliffe, 2020b; World Bank, 2020). While experts warn that this increase is a historically short-term response to a disaster, the pressures brought on by COVID-19 are certainly incentivising innovation in technology and collaboration among central banks, national statistical offices and international organizations for further exploration of migratory patterns and remittance data (World Bank, 2020).

The recovery from COVID-19 will be influenced by the ability to increase the volume of remittances (McAuliffe, 2020b). Policymakers are urging remittance sending countries to help facilitate the transfer of remittances by protecting migrants from job losses, relaxing stringent identification requirements that currently function as a barrier to accessing formal channels of exchange, reducing or eliminating transaction fees, and promoting the use of digital platforms (Bisong et al., 2020). Digital transfer platforms only account for approximately 30 per cent of global remittance transactions (Orozco et al., 2020). Access to digital platforms, online banking and the legal identification required to enable such accounts may pose a barrier to migrants (Bisong et al., 2020). Additionally, data privacy, regulatory concerns and the cost of remittances via new technology will pose as emergent challenges to technological developments. In any case, scholars urge global efforts to push beyond the status quo and enhance capabilities for international remittances in the long term (Jomo and Chowdhury, 2020).

With the restrictions for movement affecting a range of migrant groups including business travel, international students and tourism, communities in destination countries are facing unique economic challenges. Among these challenges are critical labour shortages in the immediate future, across key sectors including health care, agriculture, construction and service work (Gelatt, 2020). Without population growth to support the current labour force, there is growing concern among some countries that are not able to meet immigration targets. In Canada, for example, the Royal Bank of Canada anticipates a decline of about 100,000 people, reaching only 70 per cent of the originally targeted 341,000 new permanent residents (Canadian Press, 2020). Coldiretti, Italy's largest farmers' association, comments on labour shortages in the agricultural sector alone, reporting that they will be missing approximately

370,000 workers this year, primarily from Romania, Bulgaria and Poland (Liberti, 2020). Even as travel restrictions subside, migrants' decisions on where to migrate are unpredictable as they may seek to avoid common destination countries, such as the US, Italy and Spain due to the high number of COVID-19 cases (Smith and Wesselbaum, 2020). With the demographic crises already looming in advanced economies and the halt of incoming foreign workers, governments are being forced to swiftly respond to structural challenges. One response in the current scenario of immobility and border restrictions is the acceleration of the development and adoption of new technologies (Ting et al., 2020; Mitaritonna and Ragot, 2020). From big data to monitor the activity of COVID-19 to "virtual clinics", digital technology is being enhanced to respond to the direct pressures brought on by the pandemic (Ting et al., 2020).

MOBILITY, TECHNOLOGY AND BIG TECH: NEW TOOLS BUT NEW PROBLEMS TOO

As discussed in this volume (Yildiz and Abel; McAuliffe and Sawyer), there were some 271.6 million migrants in the world in 2019 (UN DESA, 2019). The number of internal migrants globally is much more difficult to estimate, and the most recent global UN estimate was more than a decade ago (740 million) and remains contested (UNDP, 2009). These estimates are of limited use when analysing mobility during acute health crises, such as recent Ebola outbreaks, SARS, MERS and COVID-19. Instead of migrant stock data, real-time information on mobility over both short and long distances, particularly on local interactions and on short-term movements, are needed. Accordingly, mobility restrictions began to be monitored and reported at the international level, particularly by the University of Oxford (see figure above), the International Air Transport Association (IATA) and (within Europe) the European Commission (Hale et al., 2020; IOM, 2020c; Iacus et al., 2020). However, while "mobility" includes short-term *international* mobility such as circular migration and international business and tourist movements, it also encompasses *internal* commuting and other forms of short-distance, more localised movements (IOM, 2019; Skeldon, 2018). COVID-19 highlighted this by drawing further attention to how mobility underpins and enables different forms of migration, and that immobility can act as a major disrupter to migrants right the way through the "migration cycle" (McAuliffe, 2020b).

In the context of COVID-19, some of the big tech companies, such as Google and Facebook, started to release mobility data early in the pandemic that was based on users' mobility records. The reporting is, therefore, limited to users, and for both Google's "community mobility reports" and Facebook's "movement range trends", both companies emphasise that the data are aggregated and anonymised and that it is based on users who have opted in to the respective "location history" functions, which are switched off by default (Herdagdelen et al., 2020; Google LLC, 2020). Concerns about this type of mobility data revolve around two main points: (i) its quality and questionable utility; and (ii) the significant privacy and human rights issues it raises. While noting that there has been a "scramble" by tech companies, governments, research institutes, employers and others to gather data on a wide range of aspects related to COVID-19 – such as symptoms, transmission, clinical outcomes, local movements, remote working – some of this desperation has resulted in compromised quality (Humer and Taylor, 2020; Privacy International, 2020).

In respect to the Google and Facebook "mobility" data, which may sound appealing on first glance, there are significant quality issues that limit its potential (and stated) utility, foremost the lack of clarity on the total number of people included in the dataset (which could be quite limited) as well as the high likelihood that the dataset is not representative of the underlying population but over-represents sub-populations according to location (e.g., urban, suburban, rural), age, sex/gender, and that it even represents "users". We know, for example, that there has been a proliferation of mobile phone and app user accounts, such that many people have several mobile phones and multiple user identities (GSMA, 2020). This makes it impossible to understand what and who are being counted exactly, making utility highly questionable, and potentially even misleading – there is no information available on the numbers of "users" involved in the Google or Facebook outputs, or of any weights applied to account for demographic or other biases (Toh, 2020).

Further, the basic aspects underlying key data variables are not understood by the data scientists producing the outputs, leading to misrepresentation of what the data cover and can therefore tell us. This is highlighted in a Facebook article on how the site protects the privacy of users, which refers to (inaccurately) "social distancing measures"[6] when the more likely meaning is movement restrictions during "lockdown"; it also refers to terms such as "populations", strongly implying that the data reflect populations when they cannot or do not (Facebook, 2020):

> These data sets are intended to inform researchers and public health experts about *how populations* are responding to *physical distancing measures*. In particular, there are two metrics, Change in Movement and Stay Put, that provide a slightly different perspective on movement trends. Change in Movement looks at how much people are moving around and compares it with a baseline period that predates most *social distancing measures*, while Stay Put looks at the *fraction of the population* that appear to stay within a small area during an entire day. [emphases added]

In addition to the big data sets that monitor mobility and population movement, digital technology is notably integrated into a range of national responses to the pandemic spanning the activities of planning, surveillance, testing, contact tracing, quarantine, and health care provision (Whitelaw et al., 2020). A swift range of technological interventions into policy and health care may effectively assist in combatting COVID-19, however, such advances are met with a serious concern for breaches in privacy among other issues. Innovative responses to the onset of COVID-19 include Microsoft's partnership with Swedish Health Services to create an app that monitors COVID-19 patients and hospital capacity, providing real-time data on the volume of patients, ventilator usage, and protective equipment for staff (Drees, 2020). To curb COVID-19 South Korea has adopted strict measures for contact tracing, which include the coalition of security camera footage, global positioning system (GPS) data from vehicles and mobile phones, facial recognition data and bank card records to detail people's travel (Fisher and Sang-Hun, 2020). Similarly, Taiwan has implemented a monitoring system to track the GPS location of home-quarantined individuals through government-issued mobile phones. The monitoring system will send messages to the individual's phone and impose a fine in the occurrence of a breach (Wage, Ng and Brook, 2020). Virtual care platforms have also proliferated in response to COVID-19, with the use of video conferencing and digital monitoring to facilitate clinician-to-patient encounters and deliver remote health care services (Whitelaw et al., 2020).

COVID-19, Digital Technology and the Future of Work

The proliferation of AI and digital technologies has been bolstered by COVID-19 directly impacting the future of work, the livelihoods of migrants and migration patterns. Human workers will have to harness social and managerial skills as they compete with machines and computers that will increasingly take over repetitive and routine tasks (Hertog, 2019). While automation will influence how work is done across all sectors including "white-collar" jobs (Hanke, 2017), the segment of the population that works in low-skilled, routine occupations will undergo a drastic change. In the agricultural sector, the search for automated crop harvesting solutions utilising robotics is merely one example where routine occupations and migrant workers in particular are impacted (Mitaritonna and Ragot, 2020). The text box below highlights how COVID-19 has accelerated technological responses in the sector.

Moreover, in a report focusing on Bahrain, Egypt, Kuwait, Oman, Saudi Arabia and the UAE, researchers highlight how the majority of migrants in these countries are low skilled and will be impacted by the automation of routine work. We may see the annual inflow of mass migration to the UAE disrupted, for example, where more than 93 per cent of automation potential concerns jobs held by migrant workers (aus dem Moore et al., 2018). The intersections of migration and technology become even more challenging due to political and legal implications that may follow. This is displayed by Saudi Arabia who bestowed citizenship to Sofia robot in 2017, emphasizing the country's inclination for AI initiatives by giving preferential status to a robot, a status unachievable to migrant workers (McAuliffe, 2018). Researchers, policymakers and activists alike will be tenaciously monitoring the development of technology and the impact automation will have on human mobility, migrants, and the future of work.

BOX 28.2 AUTOMATION IN THE AGRICULTURAL SECTOR: RESPONSES TO THE IMMOBILITY OF MIGRANTS

The closing of borders demonstrates the significance of in- and out-migration on the balancing of national economies. The ability to address labour shortages and unemployment rates, and stabilise wages and prices depends on the movement of people (Greenaway-McGrevy et al., 2020). With the inability to secure migrant workers, advanced economies are turning to artificial intelligence, automation and digital technologies to sustain the production of goods and services. The pandemic is accelerating the development and adoption of new technologies, across several sectors including retail services, communications, transportation, health care and finance. Most notable are the investments and technological advancements in the agricultural sector. For example, in Singapore, a new $30M grant was launched in the spring to accelerate local food production, with direct efforts to advance the country's 30 by 30 goal. Currently, less than 10 per cent of food consumed is grown locally; the 30 by 30 goal aims to produce enough food locally to meet 30 percent of the country's nutritional needs by 2030 (Liu, 2020). Similarly, the Gulf is turning to agritech, the use of technology for agriculture, horticulture and aquaculture for improved efficiency and production (Radanliev et al., 2020), to enhance domestic food production amid the pandemic and beyond (Buller, 2020). The large quantities of fresh produce accumulating on farms without sufficient manpower for harvesting is a direct consequence of the disruption to the flow of seasonal migrant workers in agriculture (FAO, 2020). These circumstances have forced many countries to rely on their resources and rethink their strategy for food security (Fedotova et al., 2020). While the advancement of agritech may be slow, transformations in the industry focus on advancing connectivity, infrastructure and digital literacy (FAO, 2020). Experts caution that rural areas and emerging economies with limited infrastructure and information and communication technology (ICT) education be considered, promoting an inclusive, hybrid approach (digital and human cooperation) to the development of digital technologies (FAO, 2020). Automation in the agricultural sector will surely impact migration patterns and the nature of work; writ large, however, civil society emphasises the consequences for migrants in the immediate (PwC, 2020; Liberti, 2020). Reflecting on the hardships brought on by COVID-19 in Italy, Romano Magrini, head of labour policies at Coldiretti suggests considering the applications from asylum seekers to upshore the workforce in the agricultural sector. Magrini describes how thousands of asylum seeker applications have been denied during this time, forcing many to live and work informally when they could be legalised workers (Liberti, 2020). Researchers also observe how the disruption to food production leads to food insecurity. In this case, there could be increased migration both internationally and internally from rural to urban settings despite barriers to mobility (Smith and Wesselbaum, 2020). The rippling effects of COVID-19 are dire, with many framing these current challenges as a "migration crisis" (Chetail, 2020) and a "global food crisis" (Fedotova et al., 2020) within the current pandemic. The impact will be monitored for decades to come.

CONCLUSION

In this chapter we have provided initial analysis of the impacts of COVID-19 on mobility and migration globally as well as reflected upon the key strategic implications for migration and mobility systems. In doing so, we recognize that in many ways we are only at the start of a long-term shift in how migration will occur (and be managed) into the future with the increasing use of digital technology; at the time of writing a second wave was emerging as a major challenge in many parts of the world, with increasing discussions focusing on vaccination programming, COVID-19 mutations/strains and future devastating zoonotic viruses looming due to ever-increasing human–animal interactions.

Initial data indicate that COVID-19 has resulted in mass immobility, with countries enacting unprecedented restrictions on freedom of movement as well as a surge in population surveillance technologies for contact tracing measures, including through the use of AI systems. While there is no doubt that this reaction and response has been focused overwhelmingly on tackling COVID-19 and reducing its harmful health impacts, there are also signs that some sectors and economies are seeking to utilise digital technologies to a much greater degree in a post-pandemic world and beyond immediate pandemic impacts. COVID-19 has, for example, further intensified the migration securitisation rhetoric while enabling the uptake of digital technology for both pandemic and non-pandemic purposes. We are also starting to see AI and other forms of automation advance in key sectors in the wake of the shock of the first wave, with major implications for long-term migration corridors globally.

While technological innovations can help mitigate the impact of COVID-19 considerable concerns are being raised by researchers and civil society organisations on the privacy and human rights implications of the data collection and usage. Not to mention the use of technological interventions and the risk for malfunction, misinformation, limitations in connectivity, the high cost of management and upkeep, and unequal access to digital technology (Whitelaw et al., 2020). The increased use of technology for monitoring and surveillance could have grave implications for migrants' health and safety. Migration and migrants have been stigmatised as a threat to public health during the pandemic, leading to increased occurrences of assault and harassment (Guadagno, 2020). These instances may influence a migrant's willingness to get tested and access health care services (Guadagno, 2020), prompting inquiry into the extent to which technological advances can support vulnerable populations. Further, the uptake of big data in the context of COVID-19 may be manipulated and used as a divisive tool to support anti-immigrant narratives under the auspices of the securitisation rhetoric. The breadth of digital technology being developed in response to COVID-19 is rapid and expansive. The full implications for migration and mobility, individual privacy and security are as yet unknown, however, alarming developments regarding the extent to which human rights principles have been so rapidly set aside during the pandemic are cause for major concern.

NOTES

1. The opinions, comments and analyses expressed in this chapter are those of the authors and do not necessarily represent the views of any of the organisations or institutions with which the authors are affiliated.
2. As per Article 1 of the 1933 Montevideo Convention on the Rights and Duties of States.
3. As at 30 September 2020.
4. In addition, public health measures such as mandatory quarantine, including in designated facilities, have been implemented.
5. Impacts of the COVID-19 pandemic on migration and migrants has been extensively covered in the IOM's COVID-19 Analytical Snapshot series, which commenced in March 2020, comprising over fifty snapshots on different topics.
6. Physical distance measures relate to the distance needed between people to reduce the risk of transmission (e.g. 1.5 metres), whereas 'stay put' and 'change in movement' relate to movement restrictions (e.g. stay at home requirements).

REFERENCES

Abebe, T.T. (2019), 'Securitisation of migration in Africa: The case of Agadez in Niger', Africa Report No. 20. Institute for Security Studies, Dakar. Available from https://issafrica.s3.amazonaws.com/site/uploads/ar20.pdf (accessed on 2 December 2020).

Aron, J. and J. Muellbauer (2019), 'The economics of mobile money: harnessing the transformative power of technology to benefit the global poor', OMS Policy Paper. Available from: https://www.oxfordmartin.ox.ac.uk/downloads/May-19-OMS-Policy-Paper-Mobile-Money-Aron-Muellbauer.pdf (accessed on 2 December 2020).

aus dem Moore, J. P., V. Chandran and J. Schubert (2018), 'The future of jobs in the Middle East', World Government Summit in collaboration with McKinsey and Company. Available from https://www.mckinsey.com/~/media/mckinsey/featured%20insights/middle%20east%20and%20africa/are%20middle%20east%20workers%20ready%20for%20the%20impact%20of%20automation/the-future-of-jobs-in-the-middle-east.ashx (accessed on 2 December 2020).

Barker, N., A. C. Davis, P. López-Peña, H. Mitchell, M. A. Mobarak, K. Naguib, M. Emy Reimão, A. Shenoy and C. Vernot (2020), 'Migration and the labour market impacts of Covid-19'. WIDER Working Paper 2020/139, Helsinki: UNU-WIDER.

BBC News (2020), 'Coronavirus: Immigration to US to be suspended amid pandemic, Trump says', 21 April, BBC News. Available from www.bbc.com/news/world-us-canada-52363852 (accessed on 2 December 2020).

Beduschi, A. and M. McAuliffe (2021), 'AI, migration and mobility: Implications for policy and practice', in M. McAuliffe and A. Triandafyllidou (eds), *World Migration Report 2022*, IOM: Geneva (forthcoming).

Berlatsky, N. (2020), 'As Bethany Mandel's "grandma killer" tweet proves, vice-signaling is the right's newest and most toxic trend', *The Independent*, 7 May. Available from www.independent.co.uk/voices/bethany-mandel-grandma-killer-tweet-coronavirus-lockdown-protest-a9504391.html (accessed on 2 December 2020).

Bisong, A., E. Ahairwe and E. Njoroge (2020), 'The impact of COVID-19 on remittances for development in Africa', ECDPM Discussion Paper no. 269. European Centre for Development Policy Management. Available from https://ecdpm.org/wp-content/uploads/Impact-COVID-19-remittances-development-Africa-ECDPM-discussion-paper-269-May-2020.pdf (accessed on 2 December 2020).

Buller, A. (2020), 'How the Gulf is turning to technology to boost food security ambitions', *Arabian Business*. Available from: https://www.arabianbusiness.com/technology/451717-how-the-gulf-is-turning-to-agritech-to-boost-food-security-ambitions (accessed on 2 December 2020).

Canadian Press (2020), 'RBC report says immigration slowdown threatens Canadian economy', *Investment Executive*. Available from: https://www.investmentexecutive.com/news/research-and-markets/rbc-report-says-immigration-slowdown-threatens-canadian-economy/ (accessed on 2 December 2020).

Castles, S. (2010), 'Understanding global migration: A social transformation perspective', *Journal of Ethnic and Migration Studies*, **36** (10):1565–1586.

Chetail, V. (2020), 'Covid-19 and the transformation of migration and mobility globally—Covid-19 and human rights of migrants: More protection for the benefit of all', IOM: Geneva. Available from https://publications.iom.int/system/files/pdf/covid19-human-rights.pdf (accessed on 2 December 2020).

Chishti, M. and S. Pierce (2020), 'The U.S. stands alone in explicitly basing coronavirus-linked immigration restrictions on economic grounds', Migration Policy Institute, Washington DC, 29 May. Available from www.migrationpolicy.org/article/us-alone-basing-immigration-restrictions-economic-concerns-not-public-health (accessed on 2 December 2020).

Coates, M. (2020), 'Covid-19 and the rise of racism', *British Medical Journal Letters*. Available from https://doi.org/10.1136/bmj.m1384 (accessed on 2 December 2020).

Collins, J. (2020), 'Immigrants don't take Australian jobs. They create jobs for others', *The Guardian*, 4 May. Available from www.theguardian.com/australia-news/2020/may/04/immigrants-dont-take-australian-jobs-they-create-jobs-for-others (accessed on 2 December 2020).

de Haas, H., S. Castles and M.J. Miller (2020), *The Age of Migration: International Population Movements in the Modern World*, sixth edition, London: Palgrave.

Drees, J. (2020), 'Swedish Health Services taps Microsoft to build app that tracks COVID-19 patients, hospital capacity', *Beckers Hospital Review*. Available from: https://www.beckershospitalreview.com/healthcare-information-technology/swedish-health-services-taps-microsoft-to-build-app-that-tracks-covid-19-patients-hospital-capacity.html (accessed on 2 December 2020).

Eves, L. and J. Thedham (2020), 'Applying securitization's second generation to COVID-19', *E-international Relations*, 14 May. Available from www.e-ir.info/2020/05/14/applying-securitizations-second-generation-to-covid-19/ (accessed on 2 December 2020).

Facebook (2020), 'COVID-19 Interactive Map & Dashboard'. Available from https://dataforgood.facebook.com/covid-survey/?date=2021-08-12&dates=2021-06-12_2021-08-07&region=WORLD (accessed on 2 December 2020).

Faist, T. (2004), 'The migration–security nexus: International migration and security before and after 9/11', Willy Brandt Series of Working Papers in International Migration and Ethnic Relations. Malmo University, Sweden.

Fedotova, G.V., N.I. Mosolova, Y.I. Sigidov and N.N. Kulikova (2020), 'COVID-19 as a factor in the global food crisis', *IOP Conference Series: Earth and Environmental Science*.

Ferhani, A. and S. Rushton (2020), 'The international health regulations, COVID-19, and bordering practices: Who gets in, what gets out, and who gets rescued?' *Contemporary Security Policy*, **41** (3):458–477. Available from https://doi.org/10.1080/13523260.2020.1771955

Ferris, E. and S. Martin, (2019), 'The global compacts on refugees and for safe, orderly and regular migration: Introduction to the special issue', *International Migration*, **57** (6):5–18.

Fisher, M. and C. Sang-Hun (2020), 'How South Korea flattened the curve', *The New York Times* [online], 23 March. Available from: https://www.nytimes.com/2020/03/23/world/asia/coronavirus -south-korea-flatten-curve.html (accessed on 2 December 2020).

Food and Agriculture Organization of the United Nations (FAO) (2020), *Food systems and COVID-19 in Latin America and the Caribbean*. Available from: http://www.fao.org/policy-support/tools-and -publications/resources-details/en/c/1294676/ (accessed on 2 December 2020).

Gamlen, A. (2020), *Migration and mobility after the 2020 pandemic: The end of an age?* Geneva: IOM. Available from https://publications.iom.int/books/covid-19-and-transformation-migration-and -mobility-globally-migration-and-mobility-after-2020 (accessed on 2 December 2020).

Gelatt, J. (2020), *Immigrant Workers: Vital to the U.S. COVID-19 Response, Disproportionately Vulnerable, Migration Policy Institute*. Available from: https://www.migrationpolicy.org/research/ immigrant-workers-us-covid-19-response#:~:text=March%202020- (accessed on 2 December 2020).

Goldin, I. (2014), Princeton University Press Author Q&A with Ian Goldin on *The Butterfly Defect: How Globalization Creates Systemic Risks, and What to Do about It*. Princeton University Press, New Jersey. Available from https://iangoldin.org/books/the-butterfly-defect/ (accessed on 2 December 2020).

Google LLC (2020), 'Mobility Report CSV Documentation', COVID-19 Community Mobility Reports. Available from https://www.google.com/covid19/mobility/data_documentation.html?hl=en (accessed on 2 December 2020).

Gothe-Snape, J. (2019), 'Permanent migrant intake set for another cut under Coalition, according to experts', *ABC News*, 22 August. Available at www.abc.net.au/news/2019-08-23/permanent-migration -cut-looms-coalition-home-affairs/11438240 (accessed on 2 December 2020).

Greenaway, C. and B. Gushulak (2017), 'Pandemics, migration and global health security'. In: *Handbook on Migration and Security* (P. Bourbeau, ed.). Cheltenham, UK and Northampton, MA, USA: Edward Elgar Publishing, pp. 316–336.

Greenaway-McGrevy, R., A. Grimes, T. Maloney, A. Bardsley and P. Gluckman (2020), 'New Zealand's economic future: COVID-19 as a catalyst for innovation'. Available from: https://apo.org .au/node/309364 (accessed on 2 December 2020).

GSMA (2020), *Mobile Identity Enabling the Digital World*, GSMA, London. Available at https://www .gsma.com/idx/wp-content/uploads/2020/05/Mobile-Identity-enabling-the-digital-world-report-Final -1.pdf (accessed on 2 December 2020).

Guadagno, L. (2020), 'Migrants and the COVID-19 pandemic: An initial analysis', *IOM Online Bookstore*. Available from: https://publications.iom.int/books/mrs-no-60-migrants-and-covid-19 -pandemic-initial-analysis (accessed on 2 December 2020).

Hale, T., S. Webster, A. Petherick, T. Phillips and B. Kira (2020), Oxford COVID-19 Government Response Tracker, Blavatnik School of Government. Available from www.bsg.ox.ac.uk/research/ research-projects/coronavirus-government-response-tracker (accessed on 2 December 2020).

Hanke, P. (2017), 'Artificial Intelligence and Big Data—An Uncharted Territory for Migration Studies?', National Center of Competence in Research—The Migration–Mobility Nexus. Available from: https://blog.nccr-onthemove.ch/artificial-intelligence-and-big-data-an-uncharted-territory-for -migration-studies/ (accessed on 2 December 2020).

Held, D., A. McGrew, D. Goldblatt and J. Perraton (1999), *Global Transformations Reader: Politics, Economics and Culture*, Cambridge: Polity Press.

Herdagdelen, A. D., B. State, P. Mohassel and A. Pompe (2020), 'Protecting privacy in Facebook mobility data during the COVID-19 response', Facebook Research, 3 June. Available from: https://research .fb.com/blog/2020/06/protecting-privacy-in-facebook-mobility-data-during-the-covid-19-response/ (accessed on 2 December 2020).

Hertog, S. (2019), 'The future of migrant work in the GCC: Literature review and a research and policy agenda', Fifth Abu Dhabi Dialogue Ministerial Consultation. Available from http://eprints.lse.ac.uk/ 102382/ (accessed on 2 December 2020).

Humer, C. and M. Taylor, (2020), 'Retracted COVID-19 studies expose holes in vetting of data firms', *The Jakarta Post*, Agence France-Presse. Available from: https://www.thejakartapost.com/news/

2020/06/10/retracted-covid-19-studies-expose-holes-in-vetting-of-data-firms.html (accessed on 2 December 2020)

Humphrey, M. (2013), 'Migration, security and insecurity', *Journal of Intercultural Studies*, **34** (2):178–195.

Hvistendahl, M. (2020), 'As Trump and Biden trade anti-China ads, hate crimes against Asian Americans spike', *The Intercept*, 11 May. Available at https://theintercept.com/2020/05/11/china-trump-biden -asian-american-hate-crimes/ (accessed on 2 December 2020).

Iacus, S. M., C. Santamaria, F. Sermi, S. Spyridon, D. Tarchi and M. Vespe (2020), 'How human mobility explains the initial spread of COVID-19: a European regional analysis', Luxembourg: Publications Office of the European Union. Available from: https://www.econbiz.de/Record/how -human-mobility-explains-the-initial-spread-of-covid-19-a-european-regional-analysis-iacus-stefano -maria/10012294467 (accessed on 2 December 2020).

International Labour Organization (ILO) (2020), 'ILO warns of COVID-19 migrant "crisis within a crisis"'. Available from: https://www.ilo.org/global/about-the-ilo/newsroom/news/WCMS_748992/ lang--en/index.htm (accessed on 2 December 2020).

International Monetary Fund (IMF) (2020), 'Supporting migrants and remittances as COVID-19 rages on', IMF Blog. Available from: https://blogs.imf.org/2020/09/11/supporting-migrants-and -remittances-as-covid-19-rages-on/ (accessed on 2 December 2020).

International Organization for Migration (IOM) (2019), *World Migration Report 2020*. Available from https://publications.iom.int/books/world-migration-report-2020 (accessed on 2 December 2020).

International Organization for Migration (IOM) (2020a), *COVID-19 Analytical Snapshot #16: International Remittances Understanding the migration & mobility implications of COVID-19*. At https://www.iom.int/sites/default/files/documents/covid-19_analytical_snapshot_16_-_international _remittances.pdf (accessed on 2 December 2020).

International Organization for Migration (IOM) (2020b), *COVID19 Analytical Snapshot #33: Combating Xenophobia and Racism*, Geneva. Available from www.iom.int/migration-research/covid -19-analytical-snapshot (accessed on 2 December 2020).

International Organization for Migration (IOM) (2020c), *Travel Restrictions Related to Travel Routes*. Available from https://migration.iom.int/ (accessed 15 October 2020).

Johns Hopkins University (2020), Coronavirus Resource Centre. Available from https://coronavirus.jhu .edu/ (accessed on 2 December 2020).

Jomo, K.S. and A. Chowdhury (2020), *COVID-19 Pandemic Recession and Recovery. Development*. https://doi.org/10.1057/s41301-020-00262-0.

Justice Connect (2020), *How the Victorian Government's emergency restrictions on COVID-19 (corona- virus) work*. Available from https://justiceconnect.org.au/resources/how-the-victorian-governments -emergency-restrictions-on-coronavirus-covid-19-work/ (accessed on 2 December 2020).

Liberti, S. (2020), 'COVID-19 and Closed Borders: Italy's Agriculture at Risk'. Available from: https:// worldcrunch.com/coronavirus-1/covid-19-and-closed-borders-italy39s-agriculture-at-risk (accessed on 2 December 2020).

McAuliffe, M. (2020a), *Digital currency and remittances in the time of Covid-19 Center for Strategic and International Studies Webinar*, Geneva: IOM. Available from: https://www.iom.int/sites/default/ files/our_work/ICP/MPR/csis_remittances_africa_webinar_final_mcauliffe.pdf (accessed on 2 December 2020).

McAuliffe, M. (2020b), *Immobility as the ultimate "migration disrupter": An initial analysis of COVID-19 impacts through the prism of securitization*, Migration Research Series, MRS No 64, Geneva: IOM. Available from https://publications.iom.int/books/mrs-no-64-immobility-ultimate -migration-disrupter (accessed on 2 December 2020).

McAuliffe, M. (2018), 'The link between migration and technology is not what you think', *Agenda*, World Economic Forum, Geneva. Available from https://www.weforum.org/agenda/2018/12/social -media-is-casting-a-dark-shadow-over-migration/ (accessed on 2 December 2020).

McAuliffe, M. and A. Goossens (2018), 'Regulating international migration in an era of increasing inter- connectedness'. In: *Handbook on Migration and Globalisation* (A. Triandafyllidou, ed.). Cheltenham, UK and Northampton, MA, USA: Edward Elgar Publishing, pp. 86–104.

McAuliffe, M. and V. Mence (2017), 'Irregular maritime migration as a global phenomenon'. In: *A Long Way to Go: Irregular Migration Patterns, Processes, Drivers And Decision Making* (M. McAuliffe and K. Koser, eds.). Canberra: ANU Press, pp. 11–47.

McAuliffe, M., C. Bauloz and A. Kitimbo (2020), 'The challenge of real-time analysis: Making sense of the migration and mobility implications of COVID-19', *Migration Policy Practice*, **X** (2). Available at https://www.disinfobservatory.org/wp-content/uploads/2020/05/Covid-19-and-minsinformation-in-the-field-of-migration.pdf (accessed on 2 December 2020).

Miggiano, L. (2009), *States of Exception: Securitisation and Irregular Migration in the Mediterranean.* Geneva: UNHCR.

Mitaritonna, C. and L. Ragot (2020), 'After Covid-19, will seasonal migrant agricultural workers in Europe be replaced by robots?' CEPII Policy Brief. Available from: http://www.cepii.fr/CEPII/fr/publications/pb/abstract.asp?NoDoc=12680 (accessed on 2 December 2020).

Newland, K., M. McAuliffe and C. Bauloz (2019), 'Recent developments in the global governance of migration: An update to the *World Migration Report 2018*'. In: *World Migration Report 2020.* Geneva: IOM.

Nunes, J. (2020), 'The COVID-19 pandemic: Securitization, neoliberal crisis, and global vulnerabilization', *Cadernos de Saúde Pública*, **36** (5), 8 May. Available from http://dx.doi.org/10.1590/0102-311x00063120.

Orozco, M., K. Klaas and N. Ledesma (2020), 'The Remittance marketplace in 2019: The growing role of digital payments', *The Dialogue*. Available from https://www.thedialogue.org/analysis/the-remittance-marketplace-in-2019-the-growing-role-of-digital-payments/ (accessed on 2 December 2020).

Pandey, S. and S. Holmes (2020), 'Australia's stalled migrant boom derails golden economic run', Reuters, 31 May. Available from https://www.reuters.com/article/us-health-coronavirus-australia-immigrat-idUSKBN2370U1 (accessed on 2 December 2020).

Ponta, A. (2020), 'Human rights law in the time of the coronavirus', *ASIL Insights*, **24** (5), 20 April. Available from www.asil.org/insights/volume/24/issue/5/human-rights-law-time-coronavirus (accessed on 2 December 2020).

PricewaterhouseCoopers (PwC) (2020), 'Maintaining food resilience in a time of uncertainty', PwC'. Available from https://www.pwc.com/sg/en/publications/maintaining-food-resilience-in-a-time-of-uncertainty.html (accessed on 2 December 2020).

Privacy International (2020), 'Telco data and Covid-19: A primer', 21 April. Available from https://privacyinternational.org/explainer/3679/telco-data-and-covid-19-primer (accessed on 2 December 2020).

Radanliev, P., D.C.D. Roure, R. Walton, M. Van Kleek, R.M. Montalvo, O. Santos, L. Maddox and S. Cannady (2020), 'COVID-19 what have we learned? The rise of social machines and connected devices in pandemic management following the concepts of predictive, preventive and personalised medicine', *EPMA Journal*, **11** (3): 311–332.

Reidy, E. (2020), 'The COVID-19 excuse? How migration policies are hardening around the globe', *The New Humanitarian*, 17 April. Available from www.thenewhumanitarian.org/analysis/2020/04/17/coronavirus-global-migration-policies-exploited (accessed on 2 December 2020).

Rühmann, F., et al. (2020), 'Can blockchain technology reduce the cost of remittances?', *OECD Development Co-operation Working Papers*, No. 73, OECD Publishing, Paris. https://doi.org/10.1787/d4d6ac8f-en

Schloenhardt, A. (2006), 'From Black Death to bird flu: Infectious diseases and immigration restrictions in Asia', *New England Journal of International and Comparative Law*, **12** (2):33–63.

Skeldon, R. (2018), 'International migration, internal migration, mobility and urbanization: Towards more integrated approaches', Migration Research Series, MRS No. 53. Geneva: IOM. Available from https://publications.iom.int/system/files/pdf/mrs_53.pdf (accessed on 2 December 2020).

Smith, M.D. and D. Wesselbaum (2020), 'COVID-19, food insecurity, and migration', *The Journal of Nutrition*, **150** (11), 2855–2858.

Stringer, K. (2004), 'Visa diplomacy', *Diplomacy and Statecraft*, **15** (4):655–682.

Ting, D.S.W., L. Carin, V. Dzau and T. Y. Wong, (2020), 'Digital technology and COVID-19', *Nature Medicine*, **26**, 1–3. Available from: https://www.nature.com/articles/s41591-020-0824-5 (accessed on 2 December 2020).

Toh, A. (2020), 'Big data could undermine the COVID-19 response: Blind spots in location-tracking data can threaten both public health and human rights', WIRED, 12 April. Available from https://www.wired.com/story/big-data-could-undermine-the-covid-19-response/ (accessed on 2 December 2020).

United Nations Department of Economic and Social Affairs (UN DESA) (2019), 'International migrant stock 2019'. Available from https://www.un.org/en/development/desa/population/migration/data/estimates2/estimates19.asp (accessed on 2 December 2020).

United Nations Development Programme (UNDP) (2009), 'Human Development Report 2009', New York: UNDP. Available from http://www.hdr.undp.org/en/content/human-development-report-2009 (accessed on 2 December 2020).

United Nations High Commissioner for Refugees (UNHCR) (2020), 'Key Legal Considerations on access to territory for persons in need of international protection in the context of the COVID-19 response', Geneva, 16 March. Available from www.refworld.org/docid/5e7132834.html (accessed on 2 December 2020).

Vertovec, S. (2020), COVID-19 and enduring stigma. Max Planck Institute, 27 April. Available from www.mpg.de/14741776/covid-19-and-enduring-stigma (accessed on 2 December 2020).

Waever, O. (1995), 'Securitization and desecuritization'. In: *On Security* (R.D. Lipschultz, ed.), New York, NY: Columbia University Press.

Watson, S. (2009), *The Securitization of Humanitarian Migration*. Abingdon: Routledge.

Whitelaw, S., M.A. Mamas, E. Topol and H.G.C.V. Spall (2020), 'Applications of digital technology in COVID-19 pandemic planning and response', *The Lancet Digital Health*. Available from: https://www.thelancet.com/journals/landig/article/PIIS2589-7500(20)30142-4/fulltext (accessed on 2 December 2020).

World Bank (2020), 'World Bank predicts sharpest decline of remittances in recent history'. Available from: https://www.worldbank.org/en/news/press-release/2020/04/22/world-bank-predicts-sharpest-decline-of-remittances-in-recent-history#:~:text=WASHINGTON%2C%20April%2022%2C%202020%20%E2%80%94 (accessed on 2 December 2020).

World Health Organization (WHO) (2020), *WHO Director-General's statement on IHR Emergency Committee on Novel Coronavirus (2019-nCoV)*, 30 January. Available from www.who.int/dg/speeches/detail/who-director-general-s-statement-on-ihr-emergency-committee-on-novel-coronavirus-(2019-ncov) (accessed on 2 December 2020).

Index

Abel, G. J. 35, 78
access to justice 236, 237, 248, 346
Achilli 128
adaptive algorithms 189
adaptive learning environments 189
adaptive learning methods 190
adaptive learning systems 189, 190
Ade-Odutola 21
ad hoc publics 323
administrative data sets 30
administrative law
    frameworks 144
    principles 137
advocacy groups 228, 335
Africa(n)
    migrants and religious transnationalism 195
    migrants' economic activities 202–3
    migrants' economic and religious
        transnationalism 197–9
    virtual/online transnationalism 199–202
agency of migration 362, 364, 370, 372
agent-based models 385, 386
age of misinformation, data outputs in 51
Agresti, A. 387
agricultural sector, automation in 415–16
Ahas, R. 113
AI see artificial intelligence (AI)
Aker, J. 257
aleatory uncertainty 377
Alexander, M. 379
algorithmic community detection 76
algorithmic decision making 137, 142, 144, 229
Alinejad, D. 17
'alt-right' 317, 318
American Border Patrol 155, 159, 162
American Civil Liberties Union 19
Amoore, L. 20, 367
Analysis of Mobile Phone Networks Conference
    (NETMOB) 111
anonymization of CDR data 109
anti-austerity movement 333
anti-austerity protests 333
anti-immigrants
    activism 324
    disinformation 316–22, 324, 325
    extremism 324
    global landscape 136
    groups 336, 338
    mobilisation 319

narratives 317, 318, 417
parties 336, 340
publics 339, 340
sentiment 339
stance 320
anti-immigration
    activism 324
    agenda 153
    narrative 322
    policies 136
    sentiments 316
anti-migrant decision making 310
anti-migrant xenophobia 301
anti-racism 318
APIs see application programming interfaces
    (APIs)
AppCID 347
'appification of migration' 5, 8, 301, 368
application programming interfaces (APIs) 37,
    65, 110
archetypical visualizations 60
Arnold, F. 77
Aron, J. 255, 258, 259
Article 3 of the Protocol against Human
    Smuggling 125
Article 17 of the ICCPR 143
Article 19 of the Universal Declaration of Human
    Rights 345
Article 22 of the European Union's (EU) 141
artificial intelligence (AI) 134–7, 160, 161, 171,
    309, 311, 398, 410
    algorithms 310
    blockchain and 309–10
    divide 139
    machine learning 50, 51, 81, 83, 107, 135,
        142–3, 189, 407, 411
    proliferation of 415
artificial intelligence-based recognition systems
    305
ASEAN Trade Union Council 244
Asian Trade Union Council (ATUC) Information
    System for Migrant Workers (ATIS) 239,
    244
ASRC see Asylum Seeker Resource Centre
    (ASRC)
assisted voluntary return and reintegration
    (AVRR) programme 225
Asylum Seeker Resource Centre (ASRC) 351